Arithmetic, Geometry, Cryptography and Coding Theory

CONTEMPORARY MATHEMATICS

770

Arithmetic, Geometry, Cryptography and Coding Theory

17th International Conference
Arithmetic, Geometry, Cryptography and Coding Theory
June 10–14, 2019
Centre International de Rencontres Mathématiques,
Marseilles, France

Stéphane Ballet
Gaetan Bisson
Irene Bouw
Editors

AMS AMERICAN MATHEMATICAL SOCIETY

2020 *Mathematics Subject Classification*. Primary 11G10, 11G20, 11G25, 11G30, 11T71, 14H45, 14K15, 20C33, 51E20, 94B27.

Library of Congress Cataloging-in-Publication Data

Names: International Conference on Arithmetic, Geometry, Cryptography and Coding Theory (17th : 2019 : Marseille, France). | Ballet, Stéphane, 1971– editor.

Title: Arithmetic, geometry, cryptography and coding theory : 17th International Conference on Arithmetic, Geometry, Cryptography and Coding Theory, June 10–14, 2019, Centre International de Rencontres Mathématiques, Marseille, France / Stéphane Ballet, Gaetan Bisson, Irene Bouw, editors.

Description: Providence, Rhode Island : American Mathematical Society, [2021] | Series: Contemporary mathematics, 0271-4132 ; volume 770 | Includes bibliographical references.

Identifiers: LCCN 2020043187 | ISBN 9781470454265 (paperback) | ISBN 9781470464264 (ebook)

Subjects: LCSH: Coding theory–Congresses. | Geometry, Algebraic–Congresses. | Cryptography–Congresses. | Number theory–Congresses. | AMS: Number theory – Arithmetic algebraic geometry (Diophantine geometry) [See also 11Dxx, 14Gxx, 14Kxx] – Abelian varieties of dimension > 1 [See also 14Kxx]. | Number theory – Arithmetic algebraic geometry (Diophantine geometry) [See also 11Dxx, 14Gxx, 14Kxx] – Curves over finite and local fields [See also 14H25]. | Number theory – Arithmetic algebraic geometry (Diophantine geometry) [See also 11Dxx, 14Gxx, 14Kxx] – Varieties over finite and local fields [See also 14G15, 14G20]. | Number theory – Arithmetic algebraic geometry (Diophantine geometry) [See also 11Dxx, 14Gxx, 14Kxx] – Curves of arbitrary genus or genus $\neq 1$ over global fields [See also 14H25]. | Number theory – Finite fields and commutative rings (number-theoretic aspects) – Algebraic coding theory; cryptography. | Algebraic geometry – Curves – Special curves and curves of low genus. | Algebraic geometry – Abelian varieties and schemes – Arithmetic ground fields [See also 11Dxx, 11Fxx, 11G10, 14Gxx]. | Group theory and generalizations – Representation theory of groups [See also 19A22 (for representation rings and Burnside rings)] – Representations of finite groups of Lie type. | Geometry For algebraic geometry, see 14-XX – Finite geometry and special incidence structures – Combinatorial structures in finite projective spaces [See also 05Bxx]. | Information and communication, circuits – Theory of error-correcting codes and error-detecting codes – Geometric methods (including applications of algebraic geometry) [See also 11T71, 14G50].

Classification: LCC QA268 .I57 2019 | DDC 510–dc23

LC record available at https://lccn.loc.gov/2020043187

Contemporary Mathematics ISSN: 0271-4132 (print); ISSN: 1098-3627 (online)

DOI: https://doi.org/10.1090/conm/770

Color graphic policy. Any graphics created in color will be rendered in grayscale for the printed version unless color printing is authorized by the Publisher. In general, color graphics will appear in color in the online version.

∞ The paper used in this book is acid-free and falls within the guidelines
established to ensure permanence and durability.
Visit the AMS home page at `https://www.ams.org/`

10 9 8 7 6 5 4 3 2 1 26 25 24 23 22 21 20 21

In memory of Gilles Lachaud.

Contents

Preface

Since 1987, the international conference *Arithmetic, Geometry, Cryptography, and Coding Theory* (AGC^2T) has been held biennially at the *Centre International de Rencontres Mathématiques* in Marseille, France. It brings together the world's best experts on arithmetic and algebraic geometry to foster interactions between pure mathematics and computer science and information theory, specifically error-correcting codes, cryptography and algorithmic complexity.

This volume contains the proceedings of this event's 17th edition, held 10–14 June 2019. It is composed of original research articles which reflect recent developments on a wide range of topics. All share the common goal of connecting arithmetic and algebraic geometry, through explicit aspects, to its many fields of applications.

AGC^2T-17 welcomed a hundred participants from around the world and we would like to pay special tribute to the speakers: Yves Aubry, Alp Bassa (plenary), Elena Berardini, Frits Beukers, Nils Bruin, Alina Bucur (plenary), Xavier Caruso (plenary), Alain Couvreur, John Cremona (plenary), Iwan Duursma, Bas Edixhoven, Sudhir Ghorpade, Alejandro Giangreco, Richard Griffon, Annamaria Iezzi, Sorina Ionica, Kiran Kedlaya (plenary), Jean Kieffer, Dmitrii Koshelev, Elisa Lorenzo García, Jade Nardi, Fabien Pazuki, Ruud Pellikaan (plenary), Matthieu Rambaud, Hugues Randriambololona, Christophe Ritzenthaler, Sergey Rybakov, Jean-Pierre Serre (plenary), Ben Smith, Andrew Sutherland, Michael Tsfasman, Christelle Vincent, Bianca Viray (plenary), and Serge Vladuts; as well as the chairmen: Peter Beelen, Jean-Marc Couveignes, Marc Hindry, James Hirschfeld, David Kohel, Ruud Pellikaan, Christophe Ritzenthaler, René Schoof, and Serge Vladuts.

The topics of the talks ranged from algebraic number theory to Diophantine geometry, and from curves and abelian varieties over finite fields to applications to codes and cryptography. They highlighted the impact of the most recent advances in computational algebraic geometry as well as algorithmic number theory.

This conference was exceptional in more ways than one. First, it was dedicated to the memory of Gilles Lachaud, one of the founding fathers of the AGC^2T series, who passed away in 2018 at the age of seventy. It was an opportunity to celebrate his brilliant career as well as his latest work, with the present volume containing Gilles' last paper. We were also honored by the presence of Jean-Pierre Serre, who presented his latest book extending the notes from his acclaimed Harvard course *Algebraic curves over finite fields.* Finally, we want to pay tribute to our close friend Alexey Zykin, tragically deceased in 2017, while he was a member of the AGC^2T organizing committee.

We are grateful to a great number of colleagues for making this conference a successful event. In particular we wish to acknowledge the members of the Steering

Committee, our colleagues in the Program Committee, the many reviewers who carefully evaluated submissions, and most of all the authors for submitting high-quality papers.

We are also indebted to the staff of CIRM (Olivia Barbarroux, Muriel Milton, and Laure Stefanini) and of the Institut de Mathématiques de Marseille (Jessica Bouanane, Eric Lozingot and Corinne Roux) for their remarkable professionalism and invaluable help in organizing this conference. Special thanks are also due to Christine Thivierge from American Mathematical Society, who helped us to publish the present volume in the Contemporary Mathematics series.

Last but not least, we are grateful to the sponsors of AGC^{2}T-17, namely Aix-Marseille University (AMU), the Institute of Mathematics of Marseille (I2M), the LABEX Archimède, the GAATI Laboratory of the University of French Polynesia, and the city of Marseille.

Stéphane Ballet
Gaetan Bisson
Irene Bouw

List of Participants

Samuele Anni
Aix-Marseille Université

Yves Aubry
Aix-Marseille & Toulon Université

Christine Bachoc
Université de Bordeaux

Stéphane Ballet
Aix-Marseille Université

Alp Bassa
Bogazici University

Peter Beelen
Technical University of Denmark

Jean-Robert Belliard
Université de Franche-Comté

Elena Berardini
Aix-Marseille Université

Frits Beukers
Utrecht University

Gaetan Bisson
Université de la Polynésie

Régis Blache
Université des Antilles-Guyane

Alexis Bonnecaze
Aix-Marseille Université

Irene Bouw
Universität Ulm

Nils Bruin
Simon Fraser University

Alina Bucur
University of California, San Diego

Mireille Car
Aix-Marseille Université

Xavier Caruso
CNRS & Université de Bordeaux

Leonardo Colò
Aix-Marseille Université

Jean-Marc Couveignes
Université de Bordeaux

Alain Couvreur
INRIA & École Polytechnique

John Cremona
University of Warwick

Thanh-Hung Dang
Aix-Marseille Université

Luca De Feo
Université de Versailles Saint Quentin

Bogdan Dina
Universität Ulm

Iwan Duursma
University of Illinois

Bas Edixhoven
University of Leiden

Elie Eid
Université de Rennes

Daniel Fiorilli
CNRS & Université d'Orsay

Francesc Fité
Institute for Advanced Study

Sudhir Ghorpade
IIT Bombay

Alejandro Giangreco
Aix-Marseille Université

Heidi Goodson
Brooklyn College (CUNY)

Richard Griffon
University of Basel

Emmanuel Hallouin
Université Toulouse

Johan P. Hansen
Aarhus University

Thierry Henocq
Université Toulouse

Marc Hindry
Université Paris-Diderot

James Hirschfeld
University of Sussex

Annamaria Iezzi
University of South Florida

Sorina Ionica
Université de Picardie Jules Verne

Valentijn Karemaker
University of Pennsylvania

Kiran Kedlaya
University of California, San Diego

Jean Kieffer
INRIA & Université de Bordeaux

Pinar Kilicer
University of Groningen

David Kohel
Aix-Marseille Université

Julien Koperecz
Université de Franche-Comté

Dmitrii Koshelev
Université de Versailles Saint Quentin

Philippe Langevin
Université de Toulon

Julien Lavauzelle
Université de Rennes

Philippe Lebacque
Université de Franche-Comté

Reynald Lercier
Université de Rennes

Elisa Lorenzo García
Université de Rennes

Stéphane Louboutin
Aix-Marseille Université

David Lubicz
Université de Rennes

Jean-François Mestre
Université Paris-Diderot

Fabien Narbonne
Université de Rennes

Jade Nardi
Université Paul Sabatier

Alessandro Neri
University of Zurich

Anca Nitulescu
Aarhus University

Roger Oyono
Université de la Polynésie

Bastien Pacifico
Aix-Marseille Université

Isabella Panaccione
INRIA & École Polytechnique

Fabien Pazuki
University of Copenhagen

Ruud Pellikaan
Technical University Eindhoven

Marc Perret
Université de Toulouse

Julia Pieltant
Conservatoire National des Arts et Métiers

Ivan Pogildiakov
Université de la Polynésie

Bjorn Poonen
MIT

Matthieu Rambaud
Télécom ParisTech

Hugues Randriambololona
Télécom ParisTech

Christophe Ritzenthaler
Université de Rennes

François Rodier
CNRS & Aix-Marseille Université

Robert Rolland
Aix-Marseille Université

Xavier Roulleau
Aix-Marseille Université

Edouard Rousseau
Télécom ParisTech

Sergey Rybakov
IITP & NRU HSE, Moscow

René Schoof
University Roma Tor Vergata

Jean-Pierre Serre
Collège de France

Kaloyan Slavov
ETH Zurich

Benjamin Smith
INRIA & École Polytechnique

Patrick Solé
CNRS & Télécom ParisTech

Katherine Stange
University of Colorado, Boulder

Peter Stevenhagen
University of Leiden

Andrew Sutherland
MIT

Jaap Top
University of Groningen

Michael Tsfasman
CNRS & IITP & IUM

Christelle Vincent
University of Vermont

Bianca Viray
University of Washington

Serge Vladuts
Aix-Marseille Université

Jose Felipe Voloch
University of Canterbury

Contemporary Mathematics
Volume **770**, 2021
https://doi.org/10.1090/conm/770/15427

A new upper bound for the largest complete (k,n)-arc in $\mathrm{PG}(2,q)$

Salam A. F. Alabdullah and James W. P. Hirschfeld

ABSTRACT. The non-existence of some (k,n)-arcs in $\mathrm{PG}(2,q)$ is proved for $q=19$ and $q=23$ when $n>\frac{1}{2}(q+3)$. Also, a new largest bound is proved and applied to $\mathrm{PG}(2,47)$.

1. Introduction

In $\mathrm{PG}(2,q)$, the projective plane over the field $\mathbf{F}_q$ of q elements, a (k,n)-*arc* is a set $\mathcal{K}$ of k points with at most n points on any line of the plane but containing n points on some line.

An (s,t)-blocking set $\mathcal{B}$ in $\Pi=\mathrm{PG}(2,q)$ is a set of s points meeting every line in at least t points; $\mathcal{B}$ is also called a *t-fold* blocking set. A blocking set is *minimal* or *irreducible* if $\mathcal{B}\backslash\{P\}$ is not a blocking set for every $P\in\mathcal{B}$. The smallest blocking sets are just the lines and any blocking set containing a line is *trivial.* For more details see [**11**, Chapter 12].

An (s,t)-blocking set $\mathcal{B}$ is the complement of a (k,n)-arc $\mathcal{K}$ in with $t=q+1-n$ and $s=q^2+q+1-k$.

Blocking sets have been first studied in 1969 by Di Paola [**8**], where the author has calculated the minimum size of a non-trivial blocking set in $\mathrm{PG}(2,q)$, for $q=4,5,7,8,9$. The major challenge was finding the minimum size of a blocking set. In 1970, Bruen [**5, 6**] proved that $|\mathcal{B}|\geq q+\sqrt{q}+1$ for any non-trivial blocking set $\mathcal{B}$ in $\mathrm{PG}(2,q)$. In Lemma 13.6 of [**11**], it is shown that, when q is odd, there is a blocking set, the *projective triangle*, with size $3(q+1)/2$, while for even q there is a blocking set, the *projective triad*, with size $(3q+2)/2$. Hill and Mason studied multiple blocking sets in $\mathrm{PG}(2,q)$; in [**9**], it is shown that, for even q, there are 2-blocking sets of size $3q$ and 3-blocking sets of size $4q$.

From Ball [**1**], the $(78,8)$-arcs and $(90,9)$-arcs are the largest complete arcs in $\mathrm{PG}(2,11)$. For $\mathrm{PG}(2,13)$, there exists no $(106,9)$-arc, $(110,10)$-arc, $(134,11)$-arc. For a triple blocking set in $\mathrm{PG}(2,q)$, a lower bound has been found for $q<11$ by Ball [**3**]. For a double blocking set in $\mathrm{PG}(2,q)$, Ball and Blokhuis [**4**] established a new lower bound for $q\geq 11$. Daskalov [**7**] investigated $\mathrm{PG}(2,17)$, and found the largest complete (k,n)-arc for $n=11,\dots,16$.

The main purpose of this paper is to find the upper bound $m_n(2,q)$ for k of a (k,n)-arc in $\mathrm{PG}(2,q)$ in the case that q is prime and $\frac{1}{2}(q+3)<n<q-1$.

2. Some basic equations

THEOREM 2.1 (Ball [**1**]). *Let $\mathcal{K}$ be (k,r)-arc in* PG$(2,p)$*, where p is prime.*

(1) *If $r \leq (p+1)/2$, then $k \leq (r-1)p+1$.*
(2) *If $r \geq (p+3)/2$, then $k \leq (r-1)p+r-(p+1)/2$.*

THEOREM 2.2 (Ball [**2**]). *Let $\mathcal{B}$ be t-fold blocking set in* PG$(2,p)$*, p prime and $p > 3$.*

(1) *If $t < p/2$, then $|\mathcal{B}| \geq (t+\frac{1}{2})(p+1)$.*
(2) *If $t > p/2$, then $|\mathcal{B}| \geq (t+1)p$.*

THEOREM 2.3 (Ball [**2**]). *Let $\mathcal{B}$ be a t-fold blocking set in* PG$(2,q)$ *that contains a line.*

(1) *If $(t-1,q) = 1$, then $|\mathcal{B}| \geq q(t+1)$.*
(2) *If $(t-1,q) > 1$ and $t \leq q/2+1$, then $|\mathcal{B}| \geq tq+q-t+2$.*
(3) *If $(t-1,q) > 1$ and $t \geq q/2+1$, then $|\mathcal{B}| \geq t(q+1)$.*

DEFINITION 2.4. [**1**] A polynomial in $\mathbf{F}_q[x]$ is *fully reducible* if it factors completely into linear factors over $\mathbf{F}_q$. If in the sequence of coefficients of a polynomial a long run of zeros occurs, this polynomial is *lacunary*.

THEOREM 2.5 (Ball [**2**]). *Let $f \in \mathbf{F}_q[x]$ be fully reducible, and suppose that f has the form $f(x) = x^q v(x) + w(x)$, where v and w have no common factor. Let $m < q$ be the maximum of the degrees of v and w. Let e be maximal such that f and hence also v and w are p^e-th powers. Then one of the following holds:*

(1) *$e = h$ and $m = 0$;*
(2) *$e \geq h/2$ and $m \geq p^e$;*
(3) *$e < h/2$ and $m \geq p^e[(p^{h-e}+1)/(p^e+1)]$;*
(4) *$e = 0$, $m = 1$ and $f(x) = a(x^q - x)$.*

Theorem 2.6 gives a slight improvement of Theorem 2.5.

THEOREM 2.6 (Daskalov [**7**]). *Let $\mathcal{B}$ be an (l,t)-blocking set in* PG$(2,p)$*, p prime.*

(1) *If $t < p/2$, and $p > 3$, then $l \geq n(p+1)+(p+1)/2$.*
(2) *If $l = t(p+1)+(p+1)/2$, then*
   (a) *through each point of $\mathcal{B}$ there are exactly $(p+3)/2$ lines that are not t-secants;*
   (b) *through each point of $\mathcal{B}$ there are exactly $(p-1)/2$ lines that are t-secants;*
   (c) *the total number of t-secants is $\mu = l(p-1)/(2t)$.*

LEMMA 2.7 (Chaper 12 [**11**]). *For any set $\mathcal{K}$ of k points in* PG$(2,q)$*, with τ_i the number of i-secants, the following hold:*

$$\sum_{i=0}^{q+1} \tau_i = q^2+q+1; \tag{2.1}$$

$$\sum_{i=1}^{q+1} i\tau_i = k(q+1); \tag{2.2}$$

$$\sum_{i=2}^{q+1} i(i-1)\tau_i = k(k-1). \tag{2.3}$$

3. Non-existence of some (k, n)-arcs in PG$(2, q)$

In this section, the non-existence of (k, n)-arcs in PG$(2, q)$ is proved for q=19, 23 and $n > \frac{1}{2}(q + 3)$.

3.1. Non-existence of some arcs in PG(2, 19).

THEOREM 3.1.
(1) *There exists no* $(211, 12)$-*arc in* PG$(2, 19)$; *so* $m_{12}(2, 19) \leq 210$.
(2) *There exists no* $(231, 13)$-*arc in* PG$(2, 19)$; *so* $m_{13}(2, 19) \leq 230$.
(3) *There exists no* $(291, 16)$-*arc in* PG$(2, 19)$; *so* $m_{16}(2, 19) \leq 290$.

PROOF. (1) Finding a maximum $(k, 12)$-arc in PG$(2, 19)$ is equivalent to finding a minimum 8-fold blocking set. Theorem 2.2 implies that $\mathcal{B}$ must have at least 170 points. Theorem 2.6 gives that the total number of 8-secants is $(170*18)/16$, which is not an integer. Therefore a $(211, 12)$-arc does not exist and

$$m_{12}(2, 19) \leq 210.$$

The other bounds are proved in the same way. □

THEOREM 3.2.
(1) *There exists no* $(251, 14)$-*arc in* PG$(2, 19)$; *so* $m_{14}(2, 19) \leq 250$.
(2) *There exists no* $(271, 15)$-*arc in* PG$(2, 19)$; *so* $m_{15}(2, 19) \leq 270$.
(3) *There exists no* $(311, 17)$-*arc in* PG$(2, 19)$; *so* $m_{17}(2, 19) \leq 310$.
(4) *There exists no* $(331, 18)$-*arc in* PG$(2, 19)$; *so* $m_{18}(2, 19) \leq 330$.

PROOF. (1) Finding a maximum $(251, 14)$-arc is equivalent to finding a $(130, 6)$-blocking set $\mathcal{B}$. Theorem 2.6 implies that the total number of 6-secants is 195. Let r be the length of the longest secant. If $r = 20$, then $\mathcal{B}$ contains a line and Theorem 2.3 implies that $|\mathcal{B}| \geq 133$, a contradiction. If $16 < r \leq 19$, then considering lines through a point on the longest secant but not in $\mathcal{B}$, so $\mathcal{B}$ must have at least $6*19+r$ points. This contradicts that $|\mathcal{B}| = 130$.

Consider the intersection of the 6-secants through $P \notin \mathcal{B}$ with the longest secant. So,

$$\tau_6 \geq 9r + (20 - r)(19 - i). \tag{3.1}$$

The values of τ_6 are calculated from (3.1) for $i \leq 8$, and give Table 1.

TABLE 1. The values of τ_6 for $i \leq 8$

r	16	15	14	13	12	11	10	9	8
i	0	1	2	3	4	5	6	7	8
$\tau_6 \geq$	220	225	228	229	228	225	220	213	204

This shows that all values of τ_6 for $r = 8, \ldots, 16$ give contradictions. This is because the total number of 6-secants is 195. For $r = 6, 7$, Lemma 2.7 gives the following:

$$\begin{aligned} \tau_6 + \tau_7 &= 381, \\ 6\tau_6 + 7\tau_7 &= 2600, \\ 3\tau_6 + 42\tau_7 &= 16770. \end{aligned}$$

There is no solution for this system. Therefore, no 130-point 6-blocking set exists and hence no $(251, 14)$-arc exists.

The remaining cases are proved similarly. □

3.2. Non-existence of some arcs in PG(2, 23).

THEOREM 3.3.

(1) *There exists no* $(301, 14)$*-arc in* $\mathrm{PG}(2, 23)$*; so* $m_{14}(2, 23) \leq 300$.
(2) *There exists no* $(325, 15)$*-arc in* $\mathrm{PG}(2, 23)$*; so* $m_{15}(2, 23) \leq 324$.
(3) *There exists no* $(349, 16)$*-arc in* $\mathrm{PG}(2, 23)$*; so* $m_{16}(2, 23) \leq 348$.
(4) *There exists no* $(373, 17)$*-arc in* $\mathrm{PG}(2, 23)$*; so* $m_{17}(2, 23) \leq 372$.
(5) *There exists no* $(421, 19)$*-arc in* $\mathrm{PG}(2, 23)$*; so* $m_{19}(2, 23) \leq 420$.

PROOF. (1) Finding a maximum $(k, 14)$-arc in $\mathrm{PG}(2, 23)$ is equivalent to finding a minimum 10-fold blocking set. Theorem 2.2 implies, since 23 is prime, that such a set must have at least 252 points. Theorem 2.6 shows that the total number of 10-secants is $(252 * 22)/20$, which is not an integer. Therefore there exists no $(301, 14)$-arc in $\mathrm{PG}(2, 23)$ and $m_{14}(2, 23) \leq 300$. The other bounds are shown in the same way. □

THEOREM 3.4.

(1) *There exists no* $(397, 18)$*-arc in* $\mathrm{PG}(2, 23)$*; so* $m_{18}(2, 23) \leq 396$.
(2) *There exists no* $(445, 20)$*-arc in* $\mathrm{PG}(2, 23)$*; so* $m_{20}(2, 23) \leq 444$.
(3) *There exists no* $(469, 21)$*-arc in* $\mathrm{PG}(2, 23)$*; so* $m_{21}(2, 23) \leq 468$.
(4) *There exists no* $(493, 22)$*-arc in* $\mathrm{PG}(2, 23)$*; so* $m_{22}(2, 23) \leq 492$.

PROOF. (1) Finding a maximum $(397, 18)$-arc is equivalent to finding a $(156, 6)$-blocking set $\mathcal{B}$. Theorem 2.6 implies that the total number of 6-secants is 286. Let r be the length of the longest secant. If $r = 24$, then $\mathcal{B}$ contains a line and Theorem 2.3 can be applied. It follows from Theorem 2.3 that $|\mathcal{B}| \geq 161$, a contradiction. If $18 < r \leq 23$ then consider lines through a point on the longest secant but not in $\mathcal{B}$. Since $\mathcal{B}$ must have at least $6 * 23 + r$ points, then the values $18 < r \leq 23$ do not give $|\mathcal{B}|$.

Consider the intersection of the 6-secants through $P \notin \mathcal{B}$ with the longest secant. So,

$$\tau_6 \geq 11r + (24 - r)(23 - i). \tag{3.2}$$

The values of τ_6 are calculated from (3.2) for $i = 0, \dots, 10$, and give Table 2.

This shows that all values of τ_6 for $r = 8, \dots, 18$ give contradictions; this is because the total number of 6-secants is 286. For $r = 6, 7$, Lemma 2.7 gives

$$\begin{aligned} \tau_6 + \tau_7 &= 553, \\ 6\tau_6 + 7\tau_7 &= 3744, \\ 3\tau_6 + 42\tau_7 &= 24180. \end{aligned}$$

There is no solution for this system. So, no 156-point 6-blocking set exists and hence no $(397, 18)$-arc exists.

TABLE 2. The values of τ_6 for $i \leq 10$

r	18	17	16	15	14	13	12	11	10	9	8
i	0	1	2	3	4	5	6	7	8	9	10
$\tau_6 \geq$	336	341	344	345	344	341	336	329	320	309	296

The proof of the remaining cases is similar. □

4. New largest bound

THEOREM 4.1. *For $\frac{1}{2}(q+3) < n < q$, with q prime,*

$$m_n(2,q) \leq \frac{(q+1)(2n-3)}{2}.$$

PROOF. From Theorem 2.1, a (k,n)-arc satisfies

$$k \leq (q+1)(n - \tfrac{3}{2}) + 1.$$

Suppose that there exists a $((q+1)(n-\frac{3}{2})+1, n)$-arc $\mathcal{K}$.

Let $\mathcal{B}$ be an (l,t)-blocking set that is the complement of $\mathcal{K}$. Since $l = q^2+q+1-k$ and $t = q+1-n$, so

$$l \geq (q+1)(q-n+\tfrac{3}{2}).$$

This implies that $\mathcal{B}$ is a $((q+1)(q-n+\frac{3}{2}), q+1-n)$-blocking set, and

$$\begin{aligned} |\mathcal{B}| &= t(q+1) + \tfrac{1}{2}(q+1) \\ &= (q+1)(q-n+\tfrac{3}{2}). \end{aligned} \tag{4.1}$$

Let T be the total number of t-secants of $\mathcal{B}$. From Theorem 2.5, $f(x) = x^q v(x) + w(x)$. Since $|\mathcal{B}| = (q+1)(\frac{1}{2}+t)$, then the lacunary polynomial from a point of $\mathcal{B}$ is $xv(x) * qw(x)$ which satisfies $f(x) = (vx+w)(v_1 w - w_1 v)$. This implies that the number of different factors in $f(x)$ is precisely $(q+3)/2$.

So, the number of t-secants through each point of $\mathcal{B}$ is precisely $(q-1)/2$. Then, counting $\{(x,L)\}$, where x is in $\mathcal{B}$ and L is a t-secant, shows that $Tt = |\mathcal{B}|(q-1)/2$; so $T = |\mathcal{B}|(q-1)/(2t)$. Hence

$$T = (q+1)(t+\tfrac{1}{2})(q-1)/(2t) \tag{4.2}$$

$$= (q^2-1)(2t+1)/(4t). \tag{4.3}$$

If $\mu = (q^2-1)(2t+1)/(4t)$ is not an integer, this implies that there exists no

$$((q+1)(n-\tfrac{3}{2})+1, n) - \text{arc}.$$

So

$$m_n(2,q) \leq (q+1)(n-\tfrac{3}{2}).$$

Suppose that μ is an integer. Let L be an r-secant and $P \in L \backslash \mathcal{B}$, where r is the largest number of points of $\mathcal{B}$ on any line through P. If there are s lines that are

t-secants to $\mathcal{B}$, and s_i lines through P that are $(t+i)$-secants, for $1 \le i \le m$, then

$$\begin{aligned} |\mathcal{B}| &= st + s_1(t+1) + s_2(t+2) + \cdots + s_m(t+m) + r, \\ &= st + s_1(t+1) + s_2(t+1) + s_2 + \cdots + s_m(t+1) + (m-1)s_m + r \\ &= st + (s_1 + s_2 + \cdots + s_m)(t+1) + s_2 + 2s_3 + \cdots + (m-1)s_m + r \\ &= st + s'(t+1) + s'' + r, \end{aligned}$$

where $s + s' = q$, $s' = s_1 + s_2 + \cdots + s_m$, $s'' = s_2 + 2s_3 + \cdots + (m-1)s_m$. So,

$$st + (q-s)(t+1) + r \le |\mathcal{B}|.$$

This implies that

$$s \ge q(t+1) + r - |\mathcal{B}|.$$

Since $|\mathcal{B}| = t(q+1) + (q+1)/2 = (t + \frac{1}{2})(q+1)$, so

$$\begin{aligned} s &\ge r + q(t+1) - (t+\tfrac{1}{2})(q+1) \\ &\ge (q-1)/2 + r - t. \end{aligned} \tag{4.4}$$

Now, the number of t-secants is at least $s(q+1-r) + r(q-1)/2$. So

$$l(q-1)/(2t) \ge r(q-1)/2 + (\tfrac{1}{2}(q-1) + r - t)(q+1-r) \tag{4.5}$$

$$\ge (q^2-1)/2 + (r-t)(q+1-r). \tag{4.6}$$

So, the inequality (4.6) becomes

$$r^2 - r(q+1+t) + t(q+1) - (q^2-1)/(4t) \ge 0. \tag{4.7}$$

To solve (4.7) for $r = t, t+1, \ldots, q+1$, the values of r can be divided into the following cases.

(1) When $r = q+1$, then $\mathcal{B}$ contains a line and Theorem 2.3 implies that $|\mathcal{B}| \ge q(t+1)$, a contradiction.

(2) When $\frac{1}{2}(q+1) + t < r \le q$, then consider lines through a point on the longest secant but not on $\mathcal{B}$. So $\mathcal{B}$ must have at least $qt + r$ points. This contradicts that $|\mathcal{B}| = (t+\frac{1}{2})(q+1)$.

(3) When $t + 2 \le r \le \frac{1}{2}(q+1) + t$, then, since $t = q+1-n$, so $n > \frac{1}{2}(q+3)$ and $q > 2t+1$.

Let $f(r) = r^2 - (q+1+t)r + t(q+1) + (q^2-1)/(4t)$; so

$$\begin{aligned} f(r) &> r^2 - (3t+2)r + (2t+1)(t+1), \\ &> r^2 - (3t+2)r + ((\tfrac{1}{2}(3t+2))^2 - ((\tfrac{1}{2}(3t+2))^2 + (2t+1)(t+1), \\ &> (r - \tfrac{1}{2}(3t+2))^2 - t^2/4. \end{aligned}$$

Here, $f(r)$ is positive for some values of t and negative for others.

Therefore (4.7) is not true for $t+2 \le r \le \frac{1}{2}(q+1) + t$.

(4) When $t \le r \le t+1$, according to Lemma 2.7,

$$\tau_t + \tau_{t+1} = q^2 + q + 1; \tag{4.8}$$

$$t\tau_t + (t+1)\tau_{t+1} = |\mathcal{B}|(q+1); \tag{4.9}$$

$$t(t-1)\tau_t + t(t+1)\tau_{t+1} = |\mathcal{B}|(|\mathcal{B}|-1). \tag{4.10}$$

Multiplying equation (4.8) by t and subtracting equation (4.9) gives

$$\begin{aligned} \tau_{t+1} &= |\mathcal{B}|(q+1) - t(q^2+q+1) \\ &= \tfrac{1}{2}(q^2+2q+2qt+1); \qquad (4.11) \\ \tau_t &= q^2+q+1+t(q^2+q+1) - |\mathcal{B}|(q+1) \\ &= \tfrac{1}{2}(q^2-2qt+1). \qquad (4.12) \end{aligned}$$

Substituting the values of τ_t and τ_{t+1} in (4.10) implies that

$$\begin{aligned} t(t-1)\tau_t + t(t+1)\tau + t + 1 &= tq + t^2(q^2+3q+1) \\ &\neq |\mathcal{B}|(|\mathcal{B}|-1). \end{aligned}$$

Therefore, there exists no $((q+1)(n-\frac{3}{2})+1, n)$-arc in $\mathrm{PG}(2,q)$ for $n > (q+3)/2$. Hence

$$m_n(2,q) \leq (q+1)(n-\tfrac{3}{2}) \quad \text{for } n > \tfrac{1}{2}(q+3).$$

□

5. Application of Theorem 4.1

Case I : Bounds for complete (k,n)-arcs when μ is a non-integer.

THEOREM 5.1. *In* $\mathrm{PG}(2,47)$, *there exists no* (k,n)*-arc for the following values of* k, *giving corresponding upper bounds for* $m_n(2,47)$.

TABLE 3. The values of n when μ is not an integer

k	1177	1225	1273	1321	1369	1417	1465	1513
n	26	27	28	29	30	31	32	33
$m_n(2,47) \leq$	1176	1224	1272	1320	1368	1416	1464	1512
k	1561	1609	1705	1753	1801	1897	1993	
n	34	35	37	38	39	41	43	
$m_n(2,47) \leq$	1560	1608	1704	1752	1800	1896	1992	

PROOF. Since $k = 1177$ and $n = 26$, then l=1080, t=22;

$$\begin{aligned} |\mathcal{B}| &= t(q+1) + \tfrac{1}{2}(q+1) \\ &= 22 * 48 + 44 \\ &= 1080. \end{aligned}$$

This implies that $|\mathcal{B}| = l$.

Assume that the total number of t-secants is T. Then, from (4.3),

$$\begin{aligned} T &= |\mathcal{B}|(q-1)/2t \\ &= (1080 * 46)/44. \end{aligned}$$

As $\mu = (1080 * 46)/24$ is not an integer, then there exists no $(1177, 26)$-arc. So

$$m_{26}(2, 47) \leq 1176.$$

The remaining cases are proved similarly. □

Case II: Bounds for complete (k, n)-arcs when T is integer.

THEOREM 5.2. *In* PG$(2, 47)$, *there exists no (k, n)-arc for the following values of k. Hence the upper bound for $m_n(2, 47)$ is established in the corresponding cases.*

TABLE 4. The values of n when μ is integer

k	1657	1849	1945	2041	2089	2137
n	36	40	42	44	45	46
$m_n(2, 47) \leq$	1656	1848	1944	2040	2088	2136

PROOF. Finding a $(1657, 36)$-arc is equivalent to finding a $(600, 12)$-blocking set $\mathcal{B}$. The total number of 12-secants is 1150. Let r be the length of the longest secant. If $r = 48$, then $\mathcal{B}$ contains a line and $|\mathcal{B}| \geq 611$, a contradiction. If $35 \leq r \leq 47$, considering lines through a point on the longest secant but not in $\mathcal{B}$, then $\mathcal{B}$ must have at least $12 * 47 + r$ points. This contradicts that $|\mathcal{B}| = 600$.

Now, consider the intersection of the 12-secants through $P \notin \mathcal{B}$ with the longest secant. Then (4.7) becomes

$$\tau_{12} \geq (r - 12)(48 - r). \tag{5.1}$$

The lower bounds for τ_{12} are calculated according to (5.1) as shown in Table 5.

TABLE 5. The values of $14 \leq r \leq 36$

r	36	35	34	33	32	31	30	29	28	27	26	25
$\tau_{12} \geq$	288	299	308	315	320	323	324	323	320	315	308	299
r	24	23	22	21	20	19	18	17	16	15	14	
$\tau_{12} \geq$	288	275	260	243	224	203	180	155	128	99	68	

This shows that all values of r for $r = 14, \ldots, 36$ give a contradiction. This is because the total number of 12-secants is 46.

However, for $r = 12$ and $r = 13$, Equations (2.1), (2.2), (2.3) of Lemma 2.7 become the following:

$$\begin{aligned} \tau_{12} + \tau_{13} &= 2257, \\ 12\tau_{12} + 13\tau_{13} &= 28800, \\ 132\tau_{12} + 156\tau_{13} &= 359400. \end{aligned}$$

As there is no solution of this system, so no $(600, 12)$-blocking set exists and hence no $(1657, 36)$-arc exists.

The other bounds are established similarly. □

Acknowledgment

Salam Alabdullah's PhD studentship is funded by the Ministry of Higher Education and Scientific Research of the Government of Iraq via the University of Basra. He thanks Professor S. Ball for his support.

References

[1] S. Ball, *On sets of points in finite planes.* PhD thesis, University of Sussex, 1994.
S. Ball,
[2] Simeon Ball, *Multiple blocking sets and arcs in finite planes*, J. London Math. Soc. (2) **54** (1996), no. 3, 581–593, DOI 10.1112/jlms/54.3.581. MR1413900
S. Ball,
[3] Simeon Ball, *On the size of a triple blocking set in* $\mathrm{PG}(2, q)$, European J. Combin. **17** (1996), no. 5, 427–435, DOI 10.1006/eujc.1996.0036. MR1397150
S. Ball, A. Blokhuis,
[4] Simeon Ball and Aart Blokhuis, *On the size of a double blocking set in* $\mathrm{PG}(2, q)$, Finite Fields Appl. **2** (1996), no. 2, 125–137, DOI 10.1006/ffta.1996.9999. MR1384155
A.A. Bruen,
[5] A. Bruen, *Baer subplanes and blocking sets*, Bull. Amer. Math. Soc. **76** (1970), 342–344, DOI 10.1090/S0002-9904-1970-12470-3. MR251629
A.A. Bruen,
[6] A. Bruen, *Blocking sets in finite projective planes*, SIAM J. Appl. Math. **21** (1971), 380–392, DOI 10.1137/0121041. MR303406
[7] R. Daskalov, On the existence and the non-existence of some (k, r)-arcs in $\mathrm{PG}(2, 17)$, *Ninth International Workshop on Algebraic Combinatorial Coding Theory* (19-25 June, 2004), 95–100.
J.W. Di Paola,
[8] Jane W. Di Paola, *On minimum blocking coalitions in small projective plane games*, SIAM J. Appl. Math. **17** (1969), 378–392, DOI 10.1137/0117036. MR247879
R. Hill and J.R.M. Mason, On (k, n)-arcs and the falsity
[9] R. Hill and J. R. M. Mason, *On (k, n)-arcs and the falsity of the Lunelli-Sce conjecture*, Finite geometries and designs (Proc. Conf., Chelwood Gate, 1980), London Math. Soc. Lecture Note Ser., vol. 49, Cambridge Univ. Press, Cambridge-New York, 1981, pp. 153–168. MR627497
J.W.P. Hirschfeld,
[10] J. W. P. Hirschfeld, *Projective geometries over finite fields*, The Clarendon Press, Oxford University Press, New York, 1979. Oxford Mathematical Monographs. MR554919
J.W.P. Hirschfeld,
[11] J. W. P. Hirschfeld, *Projective geometries over finite fields*, 2nd ed., Oxford Mathematical Monographs, The Clarendon Press, Oxford University Press, New York, 1998. MR1612570

Salam A. F. Alabdullah, College of Engineering, University of Basra, Basra, Iraq
Email address: salam.abdulqader@uobasrah.edu.iq

James W. P. Hirschfeld, Department of Mathematics, University of Sussex, Brighton BN1 9QH, United Kingdom
Email address: JWPH@sussex.ac.uk

Contemporary Mathematics
Volume **770**, 2021
https://doi.org/10.1090/conm/770/15428

Bounds on the minimum distance of algebraic geometry codes defined over some families of surfaces

Yves Aubry, Elena Berardini, Fabien Herbaut, and Marc Perret

To the memory of Gilles Lachaud.

Abstract. We prove lower bounds for the minimum distance of algebraic geometry codes over surfaces whose canonical divisor is either nef or anti-strictly nef and over surfaces without irreducible curves of small genus. We sharpen these lower bounds for surfaces whose arithmetic Picard number equals one, surfaces without curves with small self-intersection and fibered surfaces. Finally we specify our bounds to the case of surfaces of degree $d \geq 3$ embedded in $\mathbb{P}^3$.

1. Introduction

The construction of Goppa codes over algebraic curves ([**8**]) has enabled Tsfasman, Vlăduţ and Zink to beat the Gilbert-Varshamov bound ([**19**]). Since then, algebraic geometry codes over curves have been largely studied. Even though the same construction holds on varieties of higher dimension, the literature is less abundant in this context. However one can consult [**15**] for a survey of Little and [**12**] for an extensive use of intersection theory involving the Seshadri constant proposed by S. H. Hansen. Some work has also been undertaken in the direction of surfaces. Rational surfaces yielding to good codes were constructed by Couvreur in [**7**] from some blow-ups of the plane and by Blache *et al.* in [**5**] from Del Pezzo surfaces. Codes from cubic surfaces where studied by Voloch and Zarzar in [**20**], from toric surfaces by J. P. Hansen in [**11**], from Hirzebruch surfaces by Nardi in [**16**], from ruled surfaces by one of the authors in [**1**] and from abelian surfaces by Haloui in [**10**] in the specific case of simple Jacobians of genus 2 curves, and by the authors in [**2**] for general abelian surfaces. Furthermore Voloch and Zarzar ([**20**], [**21**]) and Little and Schenck ([**14**]) have studied surfaces whose arithmetic Picard number is one.

The aim of this paper is to provide a study of the minimum distance $d(X, rH, S)$ of the algebraic geometry code $\mathcal{C}(X, rH, S)$ constructed from an algebraic surface X, a set S of rational points on X, a rational effective ample divisor H on X avoiding S and an integer $r > 0$.

2020 *Mathematics Subject Classification.* 14J99, 14G15, 14G50.
Funded by ANR grant ANR-15-CE39-0013-01 "Manta".

We prove in Section 3 lower bounds for the minimum distance $d(X, rH, S)$ under some specific assumptions on the *geometry of the surface* itself. Two quite wide families of surfaces are studied. The first one is that of surfaces whose canonical divisor is either nef or anti-strictly nef. The second one consists of surfaces which do not contain irreducible curves of low genus. We obtain the following theorem, where we denote, as in the whole paper, the finite field with q elements by $\mathbb{F}_q$ and the virtual arithmetic genus of a divisor D by π_D, and where we set $m := \lfloor 2\sqrt{q} \rfloor$.

THEOREM. *(Theorem 3.2 and Theorem 3.4) Let X be an absolutely irreducible smooth projective algebraic surface defined over $\mathbb{F}_q$ whose canonical divisor is denoted by K_X. Consider a set S of rational points on X, a rational effective ample divisor H avoiding S, and a positive integer r. In order to compare the following bounds, we set*

$$d^*(X, rH, S) := \sharp S - rH^2(q + 1 + m) - m(\pi_{rH} - 1).$$

1) (i) *If K_X is nef, then*

$$d(X, rH, S) \geq d^*(X, rH, S).$$

(ii) *If $-K_X$ is strictly nef, then*

$$d(X, rH, S) \geq d^*(X, rH, S) + mr(\pi_H - 1).$$

2) *If there exists an integer $\ell > 0$ such that any $\mathbb{F}_q$-irreducible curve lying on X and defined over $\mathbb{F}_q$ has arithmetic genus strictly greater than ℓ, then*

$$d(X, rH, S) \geq d^*(X, rH, S) + \left(rH^2 - \frac{\pi_{rH} - 1}{\ell} \right) (q + 1 + m).$$

Inside both families, adding some extra geometric assumptions on the surface yields in Section 4 to some improvements for these lower bounds. This is the case for surfaces whose arithmetic Picard number is one, for surfaces without irreducible curves defined over $\mathbb{F}_q$ with small self-intersection, so as for fibered surfaces. Theorems 4.8 and 4.9 (that hold for fibered surfaces) improve the bounds of Theorems 3.2 and 3.4 (that hold for the whole wide families). Indeed the bound on the minimum distance $d(X, rH, S)$ is increased by the non-negative *defect* $\delta(B) = q + 1 + mg_B - \sharp B(\mathbb{F}_q)$ of the base curve B. Finally in Section 5 we specify our bounds to the case of surfaces of degree $d \geq 3$ embedded in $\mathbb{P}^3$.

Characterizing surfaces that yield good codes seems to be a complex question. It is not the goal of our paper to produce good codes: we aim to give theoretical bounds on the minimum distance of algebraic geometry codes on general surfaces. However one can derive from our work one or two heuristics. Indeed, Theorem 3.4 suggests to look for surfaces with no curves of small genus and fibered surfaces provide natural examples of such surfaces (see Theorem 4.9).

2. Background

Codes from algebraic surfaces are defined in the same way as on algebraic curves: we evaluate some functions with prescribed poles on some sets of rational points. Whereas the key tool for the study of the minimum distance in the 1-dimensional case is the mere fact that a function has as many zeroes as poles, in the 2-dimensional case most of the proofs rest on intersection theory.

We sum up in this section the few results on intersection theory we need. Following the authors cited in the Introduction we recall the definition of the algebraic geometry codes. We also recall quickly how the dimension of the code can be bounded from below under the assumption of the injectivity of the evaluation map. Then we prove a lemma that will be used in the course of the paper to bound from below the minimum distance of the code for several families of surfaces. Finally, we recall some results on the number of rational points on curves over finite fields.

2.1. Intersection theory. Intersection theory has almost become a mainstream tool to study codes over surfaces (see [**1**], [**12**], [**20**], [**21**], [**14**], [**2**]) and it is also central in our proofs. We do not recall here the classical definitions of the different equivalent classes of divisors and we refer the reader to [**13**, §V] for a presentation. We denote by $\mathrm{NS}(X)$ the *arithmetic* Néron-Severi group of a smooth surface X defined over $\mathbb{F}_q$ whose rank is called the arithmetic Picard number of X, or Picard number for short. Recall that a divisor D on X is said to be *nef* (respectively *strictly nef*) if $D.C \geq 0$ (respectively $D.C > 0$) for any irreducible curve C on X. A divisor D is said to be *anti-ample* if $-D$ is ample, *anti-nef* if $-D$ is nef and *anti-strictly nef* if $-D$ is strictly nef. Let us emphasize three classical results we will use in this paper.

The first one is (a generalisation of) the adjunction formula (see [**13**, §V, Exercise 1.3]). For any $\mathbb{F}_q$-irreducible curve D on X of arithmetic genus π_D, we have

$$D.(D+K_X) = 2\pi_D - 2 \tag{1}$$

where K_X is the canonical divisor on X. This formula allows us to define the virtual arithmetic genus of any divisor D on X.

The second one is the corollary of the Hodge index theorem stating that if H and D are two divisors on X with H ample, then

$$H^2D^2 \leq (H.D)^2, \tag{2}$$

where equality holds if and only if H and D are numerically proportional.

The last one is a simple outcome of Bézout's theorem in projective spaces (and the trivial part of the Nakai-Moishezon criterion). It ensures that for any ample divisor H on X and for any irreducible curve C on X, we have $H^2 > 0$ and $H.C > 0$.

2.2. Algebraic geometry codes.

2.2.1. *Definition of AG codes.* We study, as in the non-exhaustive list of papers [**1**], [**20**], [**7**], [**12**], [**21**], [**10**], [**14**] and [**2**], the generalisation of Goppa algebraic geometry codes from curves to surfaces. In the whole paper we consider an absolutely irreducible smooth projective algebraic surface X defined over $\mathbb{F}_q$ and a set S of rational points on X. Given a rational effective ample divisor G on X avoiding S, the algebraic geometry code, or AG code for short, is defined by evaluating the elements of the Riemann-Roch space $L(G)$ at the points of S. Precisely we define the linear code $\mathcal{C}(X,G,S)$ as the image of the evaluation map $\mathrm{ev} : L(G) \longrightarrow \mathbb{F}_q^{\sharp S}$.

2.2.2. *Length and dimension of AG codes.* From the very definition, the length of the code is $\sharp S$. As soon as the morphism ev is injective - see (7) for a sufficient condition - the dimension of the code equals $\ell(G) = \dim_{\mathbb{F}_q} L(G)$ which can be easily bounded from below using standard algebraic geometry tools as follows. By Riemann-Roch theorem (see [**13**, V, §1]), we have

$$\ell(G) - s(G) + \ell(K_X - G) = \frac{1}{2}G.(G-K_X) + 1 + p_a(X)$$

where $p_a(X)$ is the arithmetic genus of X, and where the so-called *superabundance* $s(G)$ of G in X is non-negative (as it is the dimension of some vector space). Now, under the assumption that

$$K_X.A < G.A, \tag{3}$$

for some ample divisor A, we have from [**13**, V, Lemma 1.7] that $\ell(K_X - G) = 0$. Thus, if the evaluation map ev is injective and under assumption (3), we get the lower bound

$$\dim \mathcal{C}(X,G,S) = \ell(G) \geq \frac{1}{2}G.(G-K_X) + 1 + p_a(X) \tag{4}$$

for the dimension of the code $\mathcal{C}(X,G,S)$.

2.2.3. *Toward the minimum distance of AG codes.* It follows that the difficulty lies in the estimation of the minimum distance $d(X,G,S)$ of the code. For any non-zero $f \in L(G)$, we introduce the number $N(f)$ of rational points of the divisor of zeroes of f. The Hamming weight $w(\mathrm{ev}(f))$ of the codeword $\mathrm{ev}(f)$ satisfies

$$w(\mathrm{ev}(f)) \geq \sharp S - N(f), \tag{5}$$

from which it follows that

$$d(X,G,S) \geq \sharp S - \max_{f\in L(G)\setminus\{0\}} N(f). \tag{6}$$

We also deduce from (5) that

$$\mathrm{ev} \text{ is injective if } \max_{f\in L(G)\setminus\{0\}} N(f) < \sharp S. \tag{7}$$

We now broadly follow the way of [**10**]. We associate to any non-zero function $f \in L(G)$ the rational effective divisor

$$D_f := G + (f) = \sum_{i=1}^{k} n_i D_i \geq 0, \tag{8}$$

where (f) is the principal divisor defined by f, the n_i are positive integers and each D_i is a reduced $\mathbb{F}_q$-irreducible curve.

Note that in this setting, the integer k and the curves D_i's depend on $f \in L(G)$. Several lower bounds for the minimum distance $d(X,G,S)$ in this paper will follow from the key lemma below.

LEMMA 2.1. *Let X be a smooth projective surface defined over $\mathbb{F}_q$, let S be a set of rational points on X and let G be a rational effective divisor on X avoiding S. Set $m = \lfloor 2\sqrt{q} \rfloor$ and keep the notations introduced in (8). If there exist non-negative real numbers a, b_1, b_2, c, such that for any non-zero $f \in L(G)$ the three following assumptions are satisfied*

(1) $k \leq a$,
(2) $\sum_{i=1}^{k} \pi_{D_i} \leq b_1 + kb_2$ *and*
(3) *for any* $1 \leq i \leq k$ *we have* $\sharp D_i(\mathbb{F}_q) \leq c + m\pi_{D_i}$

then the minimum distance $d(X,G,S)$ of $\mathcal{C}(X,G,S)$ satisfies

$$d(X,G,S) \geq \sharp S - a(c + mb_2) - mb_1.$$

PROOF. Let us write the principal divisor $(f) = (f)_0 - (f)_\infty$ as the difference of its effective divisor of zeroes minus its effective divisor of poles. Since G is effective and f belongs to $L(G)$, we have $(f)_\infty \leq G$. Hence, formula (8) reads $G + (f)_0 - (f)_\infty = \sum_{i=1}^k n_i D_i$, that is

$$(f)_0 = \sum_{i=1}^k n_i D_i + (f)_\infty - G \leq \sum_{i=1}^k n_i D_i.$$

This means that any $\mathbb{F}_q$-rational point of $(f)_0$ lies in some D_i so

$$N(f) \leq \sum_{i=1}^k \sharp D_i(\mathbb{F}_q). \tag{9}$$

Then it follows successively from the assumptions of the lemma that

$$N(f) \leq \sum_{i=1}^k (c + m\pi_{D_i}) \leq kc + m(b_1 + kb_2) \leq mb_1 + a(c + mb_2).$$

Finally Lemma 2.1 follows from (6). □

REMARK 2.2. In several papers, the point of departure to estimate the minimum distance is a bound on the number of components k, which corresponds to the condition (1) of Lemma 2.1 above. In the special case where $\mathrm{NS}(S) = \mathbb{Z}H$ and $G = rH$, for H an ample divisor on X, Voloch and Zarzar have proven in [**20**] that $k \leq r$. In the present paper we obtain a bound on k in a more general context, that is when the Néron-Severi group has rank greater than one (see for example Lemma 3.1, point (1) of Lemma 3.3 and point (1) of Lemma 4.5).

2.3. Two upper bounds for the number of rational points on curves. We manage to fulfill assumption (3) in Lemma 2.1 using the bounds on the number of rational points given in Theorem 2.3 and Proposition 2.4 below. Point (2) of Theorem 2.3 appears in the proof of Theorem 3.3 of Little and Schenck in [**14**] within a more restrictive context, whereas point (1) follows from [**3**]. We state a general theorem and give here the full proof for the sake of completeness following [**14**].

THEOREM 2.3 (Aubry-Perret [**3**] and Little-Schenck [**14**]). *Let D be an $\mathbb{F}_q$-irreducible curve of arithmetic genus π_D lying on a smooth projective algebraic surface. Then,*

(1) *we have $\sharp D(\mathbb{F}_q) \leq q + 1 + m\pi_D$.*
(2) *(Little-Schenck) If moreover D is* not *absolutely irreducible, we have*

$$\sharp D(\mathbb{F}_q) \leq \pi_D + 1.$$

PROOF. We first prove the second item, following the proof of [**14**, Th. 3.3]. Since D is $\mathbb{F}_q$-irreducible, the Galois group $\mathrm{Gal}(\overline{\mathbb{F}}_q/\mathbb{F}_q)$ acts transitively on the set of its $\bar{r} \geq 1$ absolutely irreducible components $D_1, \ldots, D_{\bar{r}}$. Since a $\mathbb{F}_q$-rational point on D is stable under the action of $\mathrm{Gal}(\overline{\mathbb{F}}_q/\mathbb{F}_q)$, it lies in the intersection $\cap_{1\leq i\leq \bar{r}} D_i$. Under the assumption that D is not absolutely irreducible, that is $\bar{r} \geq 2$, it follows that $\sharp D(\mathbb{F}_q) \leq \sharp(D_i \cap D_j)(\overline{\mathbb{F}}_q) \leq D_i.D_j$ for every couple (i, j) with $i \neq j$.

As a divisor, D can be written over $\overline{\mathbb{F}}_q$ as $D = \sum_{i=1}^{\bar{r}} a_i D_i$. By transitivity of the Galois action, we have $a_1 = \cdots = a_{\bar{r}} = a$. Now since D can be assumed to be reduced, we have $a = 1$, so that finally $D = \sum_{i=1}^{\bar{r}} D_i$. Using the adjonction formula (1) for D and each D_i, and taking into account that $\pi_{D_i} \geq 0$ for any i, we get

$$\begin{aligned} 2\pi_D - 2 &= (K_X + D).D \\ &= \sum_{i=1}^{\bar{r}} (K_X + D_i).D_i + \sum_{i \neq j} D_i.D_j \\ &= \sum_{i=1}^{\bar{r}} (2\pi_{D_i} - 2) + \sum_{i \neq j} D_i.D_j \\ &\geq -2\bar{r} + \sum_{i \neq j} D_i.D_j. \end{aligned}$$

Since there are $\bar{r}(\bar{r}-1)$ pairs (i,j) with $i \neq j$, we deduce that for at least one such pair (i_0, j_0), we have

$$D_{i_0}.D_{j_0} \leq \frac{2(\pi_D - 1 + \bar{r})}{\bar{r}(\bar{r}-1)}.$$

It is then easily checked that the left hand side of the former inequality is a decreasing function of $\bar{r} \geq 2$, so that we obtain

$$\sharp D(\mathbb{F}_q) \leq D_{i_0}.D_{j_0} \leq \frac{2(\pi_D - 1 + 2)}{2(2-1)} = \pi_D + 1$$

and the second item is proved.

The first item follows from Aubry-Perret's bound in [**3**] in case D is absolutely irreducible, that is in case $\bar{r} = 1$, and from the second item in case D is not absolutely irreducible since $\pi_D + 1 \leq q + 1 + m\pi_D$. $\square$

The following bound will be useful in Subsection 4.3 for the study of codes from fibered surfaces.

PROPOSITION 2.4 (Aubry-Perret [**4**]). *Let C be a smooth projective absolutely irreducible curve of genus g_C over $\mathbb{F}_q$ and D be an $\mathbb{F}_q$-irreducible curve having $\bar{r}$ absolutely irreducible components $\overline{D}_1, \ldots, \overline{D}_{\bar{r}}$. Suppose there exists a regular map $D \to C$ in which none absolutely irreducible component maps onto a point. Then*

$$|\sharp D(\mathbb{F}_q) - \sharp C(\mathbb{F}_q)| \leq (\bar{r}-1)q + m(\pi_D - g_C).$$

PROOF. Since C is smooth and none geometric component of D maps onto a point, the map $D \to C$ is flat. Hence by [**4**, Th.14] we have

$$|\sharp D(\mathbb{F}_q) - \sharp C(\mathbb{F}_q)| \leq (\bar{r}-1)(q-1) + m\left(\sum_{i=1}^{\bar{r}} g_{\overline{D}_i} - g_C\right) + \Delta_D$$

where $\Delta_D = \sharp\tilde{D}(\overline{\mathbb{F}}_q) - \sharp D(\overline{\mathbb{F}}_q)$ with $\tilde{D}$ the normalization of D. The result follows from [**4**, Lemma 2] where it is proved that $m\sum_{i=1}^{\bar{r}} g_{\overline{D}_i} + \Delta_D - \bar{r} + 1 \leq m\pi_D$. $\square$

3. The minimum distance of codes over some families of algebraic surfaces

We are unfortunately unable to fulfill simultaneously assumptions (1) and (2) of Lemma 2.1 for general surfaces. So we focus on two families of algebraic surfaces where we do succeed. To begin with, let us fix some common notations.

We consider a rational effective ample divisor H on the surface X avoiding a set S of rational points on X and for a positive integer r we consider $G = rH$. We study, in accordance to Section 2.2, the evaluation code $\mathcal{C}(X, rH, S)$ and we denote by $d(X, rH, S)$ its minimum distance.

3.1. Surfaces whose canonical divisor is either nef or anti-strictly nef. We study in this section codes defined over surfaces such that either the canonical divisor K_X is nef, or its opposite $-K_X$ is strictly nef. This family is quite large. It contains, for instance:

- surfaces whose canonical divisor K_X is anti-ample.
- Minimal surfaces of Kodaira dimension 0, for which the canonical divisor is numerically zero, hence nef. These are abelian surfaces, $K3$ surfaces, Enriques surfaces and hyperelliptic or quasi-hyperelliptic surfaces (see [**6**]).
- Minimal surfaces of Kodaira dimension 2. These are the so called minimal surfaces of general type. For instance, surfaces in $\mathbb{P}^3$ of degree $d \geq 4$, without curves C with $C^2 = -1$, are minimal of general type.
- Surfaces whose arithmetic Picard number is one.
- Surfaces of degree 3 embedded in $\mathbb{P}^3$.

The main theorem of this section (Theorem 3.2) rests mainly on the next lemma designed to fulfill assumptions (1) and (2) of Lemma 2.1.

LEMMA 3.1. *Let $D = \sum_{i=1}^{k} n_i D_i$ be the decomposition as a sum of $\mathbb{F}_q$-irreducible and reduced curves of an effective divisor D linearly equivalent to rH. Then we have:*

(1) $k \leq rH^2$;
(2) (i) *if K_X is nef, then* $\sum_{i=1}^{k} \pi_{D_i} \leq \pi_{rH} - 1 + k$;
(ii) *if $-K_X$ is strictly nef, then* $\sum_{i=1}^{k} \pi_{D_i} \leq \pi_{rH} - 1 - \frac{1}{2} rH.K_X + \frac{1}{2}k$.

PROOF. Using that D is numerically equivalent to rH, that $n_i > 0$ and $D_i.H > 0$ for every $i = 1, \ldots, k$ since H is ample, we prove item (1):

$$rH.H = D.H = \sum_{i=1}^{k} n_i D_i.H \geq \sum_{i=1}^{k} D_i.H \geq k.$$

Now we apply inequality (2) to H and D_i for every i, to get $D_i^2 H^2 \leq (D_i.H)^2$. We thus have, together with adjunction formula (1) and inequality $H^2 > 0$,

$$\pi_{D_i} - 1 \leq (D_i.H)^2 / 2H^2 + D_i.K_X/2. \tag{10}$$

To prove point (i) of item (2) we sum from $i = 1$ to k and thus obtain

$$\begin{aligned}
\sum_{i=1}^{k} \pi_{D_i} - k &\leq \frac{1}{2H^2} \sum_{i=1}^{k} (D_i.H)^2 + \frac{1}{2} \sum_{i=1}^{k} D_i.K_X \\
&\leq \frac{1}{2H^2} \left(\sum_{i=1}^{k} n_i D_i.H \right)^2 + \frac{1}{2} \sum_{i=1}^{k} n_i D_i.K_X \qquad (11) \\
&\leq \frac{(rH.H)^2}{2H^2} + \frac{rH.K_X}{2} \\
&= \pi_{rH} - 1,
\end{aligned}$$

where we use the positivity of the coefficients n_i, the numeric equivalence between D and $\sum_{i=1}^k n_i D_i$, the fact that H is ample and the hypothesis taken on K_X.

Under the hypothesis of point (ii) we have $D_i.K_X \leq -1$. Replacing in the first line of (11) gives $\sum_{i=1}^k \pi_{D_i} - k \leq \frac{1}{2H^2}\sum_{i=1}^k (D_i.H)^2 - \frac{k}{2}$. We conclude in the same way. $\square$

THEOREM 3.2. *Let H be a rational effective ample divisor on a surface X avoiding a set S of rational points and let r be a positive integer. We set*

$$d^*(X, rH, S) := \sharp S - rH^2(q + 1 + m) - m(\pi_{rH} - 1). \tag{12}$$

(i) *If K_X is nef, then*

$$d(X, rH, S) \geq d^*(X, rH, S).$$

(ii) *If $-K_X$ is strictly nef, then*

$$d(X, rH, S) \geq d^*(X, rH, S) + mr(\pi_H - 1).$$

PROOF. The theorem follows from Lemma 2.1 for which assumptions (1) and (2) hold from Lemma 3.1 and assumption (3) holds from Theorem 2.3. $\square$

3.2. Surfaces without irreducible curves of small genus. We consider in this section surfaces X with the property that there exists an integer $\ell \geq 1$ such that any $\mathbb{F}_q$-irreducible curve D lying on X and defined over $\mathbb{F}_q$ has arithmetic genus $\pi_D \geq \ell + 1$. It turns out that under this hypothesis, we can fulfill assumptions (1) and (2) of Lemma 2.1 without any hypothesis on K_X contrary to the setting of Section 3.1.

Examples of surfaces with this property do exist. For instance:

- simple abelian surfaces satisfy this property for $\ell = 1$ (see [**2**] for abelian surfaces with this property for $\ell = 2$).
- Fibered surfaces on a smooth base curve B of genus $g_B \geq 1$ and generic fiber of arithmetic genus $\pi_0 \geq 1$, and whose singular fibers are $\mathbb{F}_q$-irreducible, do satisfy this property for $\ell = \min(g_B, \pi_0) - 1$.
- Smooth surfaces in $\mathbb{P}^3$ of degree d whose arithmetic Picard group is generated by the class of an hyperplane section do satisfy this property for $\ell = \frac{(d-1)(d-2)}{2} - 1$ (see Lemma 5.2).

LEMMA 3.3. *Let X be a surface without $\mathbb{F}_q$-irreducible curves of arithmetic genus less than or equal to ℓ for ℓ a positive integer. Consider a rational effective ample divisor H on X and a positive integer r. Let $D = \sum_{i=1}^k n_i D_i$ be the decomposition as a sum of $\mathbb{F}_q$-irreducible and reduced curves of an effective divisor D linearly equivalent to rH. Then we have*

(1) $k \leq \frac{\pi_{rH} - 1}{\ell}$;

(2) $\sum_{i=1}^k \pi_{D_i} \leq \pi_{rH} - 1 + k$.

In case X falls in both families of Section 3.1 and of this Section 3.2, the present new bound of the first item for k is better than the one of Lemma 3.1 if and only if $\pi_{rH} - 1 < \ell r H^2$, that is if and only if $\ell > \frac{H.K_X}{2H^2} + \frac{r}{2}$. In the general setting, this inequality sometimes holds true, sometimes not. As a matter of example, supposed K_X to be ample and let us consider $H = K_X$. In this setting the inequality holds if and only if $r < 2\ell - 1$.

PROOF. By assumption, we have $0 \leq \ell \leq \pi_{D_i} - 1$ and $n_i \geq 1$ for any $1 \leq i \leq k$, hence using adjunction formula (1), we have

$$2\ell k \leq 2\sum_{i=1}^{k}(\pi_{D_i} - 1) \leq 2\sum_{i=1}^{k} n_i(\pi_{D_i} - 1) = \sum_{i=1}^{k} n_i D_i^2 + \sum_{i=1}^{k} n_i D_i.K_X.$$

Moreover using (2) and (8), we get

$$2\ell k \leq \sum_{i=1}^{k} n_i \frac{(D_i.H)^2}{H^2} + \left(\sum_{i=1}^{k} n_i D_i\right).K_X \leq \sum_{i=1}^{k} n_i^2 \frac{(D_i.H)^2}{H^2} + rH.K_X.$$

Since H is ample, we obtain

$$2\ell k \leq \sum_{i,j=1}^{k} n_i n_j \frac{(D_i.H)(D_j.H)}{H^2} + rH.K_X = \frac{(\sum_{i,=1}^{k} n_i D_i.H)^2}{H^2} + rH.K_X.$$

By (8), we conclude that

$$2\ell k \leq \frac{(rH.H)^2}{H^2} + rH.K_X = 2(\pi_{rH} - 1),$$

and both items of Lemma 3.3 follow. □

THEOREM 3.4. *Let X be a surface without $\mathbb{F}_q$-irreducible curves of arithmetic genus less than or equal to ℓ for ℓ a positive integer. Consider a rational effective ample divisor H on X avoiding a finite set S of rational points and let r be a positive integer. Then we have*

$$d(X, rH, S) \geq d^*(X, rH, S) + \left(rH^2 - \frac{\pi_{rH} - 1}{\ell}\right)(q + 1 + m).$$

PROOF. The theorem follows from Lemma 2.1, for which items (1) and (2) hold from Lemma 3.3 and item (3) holds from Theorem 2.3. □

4. Four improvements

In this section we manage to obtain better parameters for conditions (1), (2) or (3) of Lemma 2.1 in four cases: for surfaces of arithmetic Picard number one, for surfaces which do not contain $\mathbb{F}_q$-irreducible curves of small self-intersection and whose canonical divisor is either nef or anti-nef, for fibered surfaces with nef canonical divisor, and for fibered surfaces whose singular fibers are $\mathbb{F}_q$-irreducible curves.

4.1. Surfaces with Picard number one. As mentioned in the Introduction, the case of surfaces X whose arithmetic Picard number equals one has already attracted some interest (see [**21**], [**20**], [**14**] and [**5**]). We prove in this subsection Lemma 4.1 and Theorem 4.3 which improve, under this rank one assumption, the bounds of Lemma 3.1 and Theorem 3.2. These new bounds depend on the sign of $3H^2 + H.K_X$, where H is the ample generator of $\mathrm{NS}(X)$.

LEMMA 4.1. *Let X be a smooth projective surface of arithmetic Picard number one. Let H be the ample generator of* NS(X) *and let r be a positive integer. For any non-zero function $f \in L(rH)$ consider the decomposition $D_f = \sum_{i=1}^{k} n_i D_i$ into $\mathbb{F}_q$-irreducible and reduced curves D_i with positive integer coefficients n_i as in*

(8). *Then the sum of the arithmetic genera of the curves* D_i *satisfies:*

(i) $\sum_{i=1}^k \pi_{D_i} \leq (k-1)\pi_H + \pi_{(r-k+1)H}$ *if* $3H^2 + H.K_X \geq 0$;

(ii) $\sum_{i=1}^k \pi_{D_i} \leq H^2(r-k)^2/2 + H^2(r-2k) + k$ *if* $3H^2 + H.K_X < 0$.

REMARK 4.2. Note that the condition $3H^2 + H.K_X \geq 0$ is satisfied as soon as $H.K_X \geq 0$. It is also satisfied in the special case where $K_X = -H$ which corresponds to Del Pezzo surfaces.

PROOF. In order to prove the first item, we consider a non-zero function $f \in L(rH)$ and we keep the notation already introduced in (8), namely $D_f = \sum_{i=1}^k n_i D_i$. As $\mathrm{NS}(X) = \mathbb{Z}H$, for all i we have $D_i = a_i H$ and we know by Lemma 2.2 in [**21**] that $k \leq r$. Intersecting with the ample divisor H enables to prove that for all i we have $a_i \geq 1$ and that $\sum_{i=1}^k n_i a_i = r$. Thus to get an upper bound for $\sum_{i=1}^k \pi_{D_i} = \sum_{i=1}^k \pi_{a_i H}$, we are reduced to bounding $\left(\sum_{i=1}^k a_i^2\right) H^2/2 + \left(\sum_{i=1}^k a_i\right) H.K_X/2 + k$ under the constraint $\sum_{i=1}^k a_i n_i = r$. Our strategy is based on the two following arguments.

First, the condition $3H^2 + H.K_X \geq 0$ guarantees that $a \mapsto \pi_{aH}$ is an increasing sequence. Indeed, for integers $a' > a \geq 1$ we have $\pi_{a'H} \geq \pi_{aH}$ if and only if $(a+a')H^2 \geq -H.K_X$, which is true under the condition above because $a + a' \geq 3$. As a consequence, if we fix an index i between 1 and k and if we consider that the product $n_i a_i$ is constant, then the value of $\pi_{a_i H}$ is maximum when a_i is, that is when $a_i = n_i a_i$ and $n_i = 1$.

Secondly, assume that all the n_i equal 1 and that $\sum_{i=1}^k a_i = r$. We are now reduced to bounding $\sum_{i=1}^k a_i^2$. We can prove that the maximum is reached when all the a_i equal 1 except one which equals $r-k+1$. Otherwise, suppose for example that $2 \leq a_1 \leq a_2$. Then $a_1^2 + a_2^2 < (a_1-1)^2 + (a_2+1)^2$ and $\sum_{i=1}^k a_i^2$ is not maximum, and the first item is thus proved.

For the second item, using the adjonction formula we get

$$\sum_{i=1}^k \pi_{D_i} - k \leq \frac{1}{2H^2} \sum_{i=1}^k (D_i.H)^2 + \frac{1}{2} \sum_{i=1}^k D_i.K_X.$$

Again as $\mathrm{NS}(X) = \mathbb{Z}H$, for all i we have $D_i = a_i H$. Thus we get

$$\sum_{i=1}^k \pi_{D_i} - k \leq \frac{1}{2H^2} \sum_{i=1}^k a_i^2 (H^2)^2 + \frac{1}{2} \sum_{i=1}^k a_i H.K_X.$$

Now using that $H.K_X \leq -3H^2$ by hypothesis, that $\sum_{i=1}^k a_i \geq k$ since every a_i is positive and that since $\sum_{i=1}^k a_i \leq r$ we can prove again that $\sum_{i=1}^k a_i^2 \leq (r-k+1)^2 + (k-1)$, we get

$$\sum_{i=1}^k \pi_{D_i} - k \leq \frac{H^2}{2}((r-k+1)^2 + (k-1)) - \frac{3H^2}{2}k.$$

Some easy calculation shows that this is equivalent to our second statement. □

THEOREM 4.3. *Let* X *be a smooth projective surface of arithmetic Picard number one. Let* H *be the ample generator of* $\mathrm{NS}(X)$ *and* S *a finite set of rational points avoiding* H. *For any positive integer* r, *the minimum distance* $d(X, rH, S)$ *of the code* $\mathcal{C}(X, rH, S)$ *satisfies:*

(i) *if* $3H^2 + H.K_X \geq 0$, *then*

$$d(X, rH, S) \geq \begin{cases} \sharp S - (q + 1 + m\pi_{rH}) \text{ if } r > 2(q+1+m)/mH^2, \\ \sharp S - r(q + 1 + m\pi_H) \text{ otherwise.} \end{cases}$$

(ii) *If* $3H^2 + H.K_X < 0$, *then*

$$d(X, rH, S) \geq \begin{cases} \sharp S - (q + 1 + m) - mH^2(r^2 - 3)/2 \text{ if } r > 2(q+1+m)/mH^2 - 3, \\ \sharp S - r(q + 1 + m - mH^2) \text{ otherwise.} \end{cases}$$

PROOF. For any non-zero $f \in L(rH)$, we have by (9) and by point (1) of Theorem 2.3 the following inequality

$$N(f) \leq k(q+1) + m \sum_{i=1}^{k} \pi_{D_i}.$$

We apply Lemma 4.1 to bound $\sum_{i=1}^{k} \pi_{D_i}$. We get in the first case $N(f) \leq \phi(k)$ where $\phi(k) := m\pi_{(r-k+1)H} + k(q + 1 + m\pi_H) - m\pi_H$. Remark that $\pi_{(r-k+1)H}$ is quadratic in k and so $\phi(k)$ is a quadratic function with positive leading coefficient. In [**20**, Lemma 2.2] Voloch and Zarzar proved that if X has arithmetic Picard number one then $k \leq r$. Thus $\phi(k)$ attends its maximum for $k = 1$ or for $k = r$ and $N(f) \leq \max\{\phi(1), \phi(r)\}$. A simple calculus shows that $\phi(1) - \phi(r) > 0$ if and only if $r > 2(q+1+m)/mH^2$. Since we have $d(X, rH, S) \geq \sharp S - \max_{f \in L(rH)\setminus\{0\}} N(f)$, part (i) of the theorem is proved.

The treatment of part (ii) is the same, except that we use Lemma 4.1 to bound $\sum_{i=1}^{k} \pi_{D_i}$.

□

REMARK 4.4. Little and Schenck have given bounds in [**14**, §3] for the minimum distance of codes defined over algebraic surfaces of Picard number one. In particular, they obtain (if we keep the notations of Theorem 4.3): $d(X, rH, S) \geq \sharp S - (q+1+m\pi_H)$ for $r = 1$ ([**14**, Th. 3.3]) and $d(X, rH, S) \geq \sharp S - r(q+1+m\pi_H)$ for $r > 1$ and q large ([**14**, Th. 3.5]). Comparing their bounds with Theorem 4.3, one can see that when $3H^2 + H.K_X \geq 0$ we get the same bound for $r = 1$ and also for $r > 1$ without any hypothesis on q. Moreover, when $3H^2 + H.K_X < 0$, our bounds improve the ones given by Little and Schenck, again without assuming large enough q when $r > 1$.

4.2. Surfaces without irreducible curves over $\mathbb{F}_q$ with small self-intersection and whose canonical divisor is either nef or anti-nef. We consider in this section surfaces X such that there exists some integer $\beta \geq 0$ for which any $\mathbb{F}_q$-irreducible curve D lying on X and defined over $\mathbb{F}_q$ has self-intersection $D^2 \geq \beta$. We prove in this case Lemma 4.5 below, from which we can tackle assumption (1) in Lemma 2.1 in case $\beta > 0$. Unfortunately, Lemma 4.5 enables to fulfill assumption (2) of Lemma 2.1 only in case the intersection of the canonical divisor with $\mathbb{F}_q$-irreducible curves has constant sign, that is for surfaces of Section 3.1. The lower bound for the minimum distance we get is better than the one given in Theorem 3.2.

Let us propose some examples of surfaces with this property:

- simple abelian surfaces satisfy this property for $\beta = 2$.

- Surfaces whose arithmetic Picard number is one. Indeed consider a curve D defined over $\mathbb{F}_q$ on X, and assume that $\mathrm{NS}(X) = \mathbb{Z}H$ with H ample. Then we have that $D = aH$ for some integer a. Since H is ample we get $1 \leq D.H = aH^2$ hence $a \geq 1$ and $D^2 = a^2H^2 \geq H^2$.
- Surfaces whose canonical divisor is anti-nef and without irreducible curves of arithmetic genus less or equal to $\ell > 0$. Indeed the adjunction formula gives $D^2 = 2\pi_D - 2 - D.K_X \geq 2\pi_D - 2 \geq 2\ell$.

LEMMA 4.5. *Let X be a surface on which any $\mathbb{F}_q$-irreducible curve has self-intersection at least $\beta \geq 0$. Assume that H is a rational effective ample divisor on X and let r be a positive integer. Let $D = \sum_{i=1}^k n_i D_i$ be the decomposition as a sum of $\mathbb{F}_q$-irreducible and reduced curves of an effective divisor D linearly equivalent to rH. Then we have*

(1) *if $\beta > 0$ then $k \leq r\sqrt{\frac{H^2}{\beta}}$;*
(2) *$\sum_{i=1}^k (2\pi_{D_i} - 2 - D_i.K_X) \leq \varphi(k)$, with*

$$\varphi(k) := (k-1)\beta + \left(r\sqrt{H^2} - (k-1)\sqrt{\beta}\right)^2. \tag{13}$$

PROOF. Since by hypothesis we have $\sqrt{\beta} \leq \sqrt{D_i^2}$, we deduce that $k\sqrt{\beta} \leq \sum_{i=1}^k n_i \sqrt{D_i^2}$. By (2), we get $k\sqrt{\beta} \leq \sum_{i=1}^k n_i \frac{D_i.H}{\sqrt{H^2}} = \frac{rH.H}{\sqrt{H^2}} = r\sqrt{H^2}$, from which the first item follows.

Setting $x_i := \sqrt{2\pi_{D_i} - 2 - D_i.K_X}$, we have by adjunction formula $x_i = \sqrt{D_i^2} \geq \sqrt{\beta}$. Moreover the previous inequalities ensure that $\sum_{i=1}^k x_i \leq \sum_{i=1}^k n_i\sqrt{D_i^2} \leq r\sqrt{H^2}$. Then, the maximum of $\sum_{i=1}^k (2\pi_{D_i} - 2 - D_i.K_X) = \sum_{i=1}^k x_i^2$ under the conditions $x_i \geq \sqrt{\beta}$ and $\sum_{i=1}^k x_i \leq r\sqrt{H^2}$ is reached when each but one x_i equals the minimum $\sqrt{\beta}$, and only one is equal to $r\sqrt{H^2} - (k-1)\sqrt{\beta}$, and this concludes the proof. □

THEOREM 4.6. *Let X be a surface on which any $\mathbb{F}_q$-irreducible curve has self-intersection at least $\beta > 0$. Consider a rational effective ample divisor H on X avoiding a set S of rational points and let r be a positive integer. Then*

$$d(X, rH, S) \geq \begin{cases} \sharp S - \max\left\{\psi(1), \psi\left(r\sqrt{\frac{H^2}{\beta}}\right)\right\} - \frac{m}{2} r\sqrt{\frac{H^2}{2\beta}} & \text{if } K_X \text{ is nef,} \\ \sharp S - \max\left\{\psi(1), \psi\left(r\sqrt{\frac{H^2}{\beta}}\right)\right\} & \text{if } -K_X \text{ is nef} \end{cases}$$

with

$$\psi(k) := \frac{m}{2}\varphi(k) + k(q+1+m),$$

where $\varphi(k)$ is given by equation (13).

PROOF. For any non-zero $f \in L(rH)$, we have by (9) and by point (1) of Theorem 2.3 that $N(f) \leq k(q+1) + m\sum_{i=1}^k \pi_{D_i}$. Lemma 4.5 implies that $N(f) \leq k(q+1) + \frac{m}{2}\left(2k + \varphi(k) + \sum_{i=1}^k D_i.K_X\right)$. In case K_X is nef, we have $\sum_{i=1}^k D_i.K_X \leq \sum_{i=1}^k n_i D_i.K_X = rH.K_X$, and in case $-K_X$ is nef, we get $\sum_{i=1}^k D_i.K_X \leq 0$, and the theorem follows. □

4.3. Fibered surfaces with nef canonical divisor. We consider in this subsection AG codes from fibered surfaces whose canonical divisor is nef. We adopt the vocabulary of [**18**, III, §8] and we refer the reader to this text for the basic notions we recall here. A fibered surface is a surjective morphism $\pi : X \to B$ from a smooth projective surface X to a smooth absolutely irreducible curve B. We denote by π_0 the common arithmetic genus of the fibers and by g_B the genus of the base curve B. Elliptic surfaces are among the first non-trivial examples of fibered surfaces. For such surfaces we have $\pi_0 = 1$ and the canonical divisor is always nef (see [**6**]).

We recall that on a fibered surface every divisor can be uniquely written as a sum of *horizontal* curves (that is mapped onto B by π) and *fibral* curves (that is mapped onto a point by π).

LEMMA 4.7. *Let $\pi : X \to B$ be a fibered surface. Let H be a rational effective ample divisor on X and let r be a positive integer. For any effective divisor D linearly equivalent to rH, consider its decomposition $D = \sum_{i=1}^k n_i D_i$ as a sum of reduced $\mathbb{F}_q$-irreducible curves as in* (8)*. Denote by $\overline{r}_i$ the number of absolutely irreducible components of D_i. Then, we have*

$$\sum_{i=1}^{k} \overline{r}_i \leq rH^2.$$

PROOF. Write $D = \sum_{i=1}^k n_i D_i = \sum_{i=1}^k n_i \sum_{j=1}^{\overline{r}_i} D_{i,j}$ where the $D_{i,j}$ are the absolutely irreducible components of D_i.

We use that $n_i > 0$, that D is numerically equivalent to rH and that $D_{i,j}.H > 0$ to get

$$\sum_{i=1}^{k} \overline{r}_i \leq \sum_{i=1}^{k} \sum_{j=1}^{\overline{r}_i} D_{i,j}.H \leq \sum_{i=1}^{k} n_i \sum_{j=1}^{\overline{r}_i} D_{i,j}.H = \sum_{i=1}^{k} n_i D_i.H = rH.H,$$

which proves the lemma. □

The next theorem involves the *defect* $\delta(B)$ of a smooth absolutely irreducible curve B defined over $\mathbb{F}_q$ of genus g_B, which is defined by

$$\delta(B) := q + 1 + mg_B - \sharp B(\mathbb{F}_q).$$

By the Serre-Weil theorem this defect is a non-negative number. The so-called maximal curves have defect 0, and the smaller the number of rational points on B is, the greater the defect is.

THEOREM 4.8. *Let $\pi : X \to B$ be a fibered surface whose canonical divisor K_X is nef. Assume that H is a rational effective ample divisor on X having at least one horizontal component and avoiding a set S of rational points. For any positive integer r the minimum distance of $\mathcal{C}(X, rH, S)$ satisfies*

$$d(X, rH, S) \geq d^*(X, rH, S) + \delta(B)$$

where $d^*(X, rH, S)$ is given by formula (12).

Recall that the general bound we obtain in Theorem 3.2 in Section 3 for surfaces with nef canonical divisor is $d(X, rH, S) \geq d^*(X, rH, S)$, thus the bound from Theorem 4.8 is always equal or better. Actually Theorem 4.8 is surprising, since it says that the lower bound for the minimum distance is all the more large because the defect $\delta(B)$ is. Consequently it looks like considering fibered surfaces on curves with few rational points and large genus could lead to potentially good codes.

PROOF. Recall that for any non-zero $f \in L(rH)$, we have $d(X, rH, S) \geq \sharp S - N(f)$, and that $N(f) \leq \sum_{i=1}^{k} \sharp D_i(\mathbb{F}_q)$ if we use the notation $D_f := rH + (f) = \sum_{i=1}^{k} n_i D_i$ introduced in (8). We again denote by $\overline{r}_i$ the number of absolutely irreducible components of D_i. In order to introduce the $\mathbb{F}_q$-irreducible components of D_f, write $k = h + v$, where h (respectively v) is the number of horizontal curves denoted by $H_1, \ldots, H_h$, (respectively fibral curves denoted by $F_1, \ldots, F_v$). Then we get $N(f) \leq \sum_{i=1}^{h} \sharp H_i(\mathbb{F}_q) + \sum_{i=1}^{v} \sharp F_i(\mathbb{F}_q)$. Since B is a smooth curve, the morphisms $H_i \to B$ are flat. Now applying Proposition 2.4 to horizontal curves and Theorem 2.3 to fibral curves gives

$$\begin{aligned} N(f) &\leq h(\sharp B(\mathbb{F}_q) - mg_B) + m\sum_{i=1}^{h} \pi_{H_i} + q\sum_{i=1}^{h}(\overline{r}_i - 1) + qv + v + m\sum_{i=1}^{v} \pi_{F_i} \\ &= h(\sharp B(\mathbb{F}_q) - mg_B - q) + m\sum_{i=1}^{k} \pi_{D_i} + q\sum_{i=1}^{k} \overline{r}_i + v, \end{aligned} \tag{14}$$

where we used the fact that $v \leq \sum_{i=h+1}^{k} \overline{r}_i$.

Since the canonical divisor of the fibered surface is assumed to be nef, Lemma 3.1 gives a bound for $\sum_{i=1}^{k} \pi_{D_i}$. We set $v = k - h$ and we use Lemma 4.7 with (14) to obtain

$$\begin{aligned} N(f) &\leq h(\sharp B(\mathbb{F}_q) - mg_B - q) + m(\pi_{rH} - 1) + mk + qrH^2 + v \\ &= h(\sharp B(\mathbb{F}_q) - mg_B - q - 1) + m(\pi_{rH} - 1) + mk + qrH^2 + k \\ &= -h\delta(B) + m(\pi_{rH} - 1) + mk + qrH^2 + k. \end{aligned}$$

Now, $D_f.F \equiv rH.F > 0$ since F is a generic fiber and rH is assumed to have at least one horizontal component. Thus, D_f has also at least one horizontal component, that is $h \geq 1$. Moreover, again from Lemma 3.1 we have $k \leq rH^2$. As the defect $\delta(B)$ is non-negative it follows that

$$N(f) \leq -\delta(B) + rH^2(q + 1 + m) + m(\pi_{rH} - 1)$$

and the theorem is proved. □

4.4. Fibered surfaces whose singular fibers are irreducible. In this subsection we drop off the condition on the canonical divisor. Instead, we assume that every singular fiber on X is $\mathbb{F}_q$-irreducible. To construct examples of such surfaces, fix any $d \geq 1$ and recall that the dimension of the space of degree d homogeneous polynomials in three variables is $\binom{d+2}{2}$. Hence the space $\mathcal{P}_d$ of plane curves of degree d is $\mathcal{P}_d = \mathbb{P}^{\binom{d+2}{2}-1}$. Any curve B drawn in $\mathcal{P}_d$ gives rise to a fibered surface, whose fibers are plane curves of degree d, that is with $\pi_0 = \frac{(d-1)(d-2)}{2}$. The locus of singular curves being a subvariety of $\mathcal{P}_d$, choosing B not contained in this singular locus yields to a fibered surface with smooth generic fiber. As the locus of reducible curves has high codimension in $\mathcal{P}_d$, choosing B avoiding this locus yields to fibered surfaces without reducible fibers.

We consider the case where π_0 and g_B are both at least 2 and we set $\ell = \min(\pi_0, g_B) - 1 \geq 1$. We recall again that every divisor on X can be uniquely written as a sum of horizontal and fibral curves. If we denote by H an horizontal curve and by V a fibral curve defined over $\mathbb{F}_q$, we have that $\pi_H \geq g_B$ and $\pi_V = \pi_0$. Therefore, in this setting, X contains no $\mathbb{F}_q$-irreducible curves defined over $\mathbb{F}_q$ of arithmetic genus smaller than or equal to ℓ. Thus Lemma 3.3 applies and gives the same bound for $\sum_{i=1}^k \pi_i$ as when K_X is nef and the bound $k \leq (\pi_{rH} - 1)/\ell$ for the number k of $\mathbb{F}_q$-irreducible components of D_f. We consider this new bound for k in the proof of Theorem 4.8 and we get instead the following result.

THEOREM 4.9. *Let $\pi : X \to B$ be a fibered surface. We consider a rational effective ample divisor H on X having at least one horizontal component and avoiding a set S of rational points. Let r be a positive integer. We denote by g_B the genus of B and by π_0 the arithmetic genus of the fibers and we set $\ell = \min(\pi_0, g_B) - 1$. Suppose that every singular fiber is $\mathbb{F}_q$-irreducible and that $\ell \geq 1$. Then the minimum distance of $\mathcal{C}(X, rH, S)$ satisfies*

$$d(X, rH, S) \geq d^*(X, rH, S) + \left(rH^2 - \frac{\pi_{rH} - 1}{\ell}\right)(q + 1 + m) + \delta(B),$$

where $d^*(X, rH, S)$ is given by formula (12).

Naturally this bound is better than the one in Theorem 4.8 if and only if $\pi_{rH} - 1 < \ell r H^2$. Furthermore it improves the bound of Theorem 3.4 by the addition of the non-negative term $\delta(B)$.

5. An example: surfaces in $\mathbb{P}^3$

This section is devoted to the study of the minimum distance of AG codes over a surface X of degree $d \geq 3$ embedded in $\mathbb{P}^3$.

We consider the class L of an hyperplane section of X. So L is ample, $L^2 = d$ and the canonical divisor on X is $K_X = (d-4)L$ (see [**17**, p.212]). In this setting, we fix a rational effective ample divisor H and r a positive integer. We apply our former theorems to this context to give bounds on the minimum distance of the code $\mathcal{C}(X, rH, S)$.

We recall that cubic surfaces are considered by Voloch and Zarzar in [**20**] and [**21**] to provide computationally good codes. In Section 4 of [**14**] Little and Schenck propose theoretical and experimental results for surfaces in $\mathbb{P}^3$ always in the prospect of finding good codes. We also contribute to this study with a view to bounding the minimum distance according to the geometry of the surface.

PROPOSITION 5.1. *Let X be a surface of degree $d \geq 3$ embedded in $\mathbb{P}^3$. Consider a rational effective ample divisor H avoiding a set S of rational points on X and let r be a positive integer. Then the minimum distance of the code $\mathcal{C}(X, rH, S)$ satisfies*

(1) *if X is a cubic surface, then*

$$d(X, rH, S) \geq d^*(X, rH, S) + mr(\pi_H - 1).$$

(2) *If X has degree $d \geq 4$ then*

$$d(X, rH, S) \geq d^*(X, rH, S),$$

where

$$d^*(X, rH, S) = \sharp S - rH^2(q + 1 + m) - m(\pi_{rH} - 1)$$

is the function defined in (12).

PROOF. Since $K_X = (d-4)L$ we have for cubic surfaces that $K_X = -L$ and thus the canonical divisor is anti-ample, while for surfaces of degree $d \geq 4$ the canonical divisor is ample or the zero divisor, thus is nef. Hence we can apply Theorem 3.2 from which the proposition follows. □

5.1. Surfaces in $\mathbb{P}^3$ without irreducible curves of low genus. In the complex setting, the Noether-Lefschetz theorem asserts that a general surface X of degree $d \geq 4$ in $\mathbb{P}^3$ is such that $\mathrm{Pic}(X) = \mathbb{Z}L$, where L is the class of an hyperplane section (see [**9**]). Here, general means outside a countable union of proper subvarieties of the projective space parametrizing the surfaces of degree d in $\mathbb{P}^3$. Even if we do not know an analog of this statement in our context, it suggests us the strong assumptions we take in this subsection, namely in Lemma 5.2 and Proposition 5.3.

LEMMA 5.2. *Let X be a surface of degree $d \geq 4$ in $\mathbb{P}^3$ of arithmetic Picard number one. Suppose that* $\mathrm{NS}(X)$ *is generated by the class of an hyperplane section L. Consider an $\mathbb{F}_q$-irreducible curve D on X of arithmetic genus π_D. Then*

$$\pi_D \geq (d-1)(d-2)/2.$$

PROOF. Let a be the integer such that $D = aL$ in $\mathrm{NS}(X)$. Since D is an $\mathbb{F}_q$-irreducible curve and L is ample, we must have $a > 0$. Then, using the adjonction formula, we get

$$\begin{aligned} 2\pi_D - 2 = D^2 + D.K &= a^2L^2 + aL.(d-4)L \\ &= a^2d + ad(d-4) \geq d + d(d-4), \end{aligned}$$

and thus $\pi_D \geq (d-1)(d-2)/2$. □

By the previous lemma it is straightforward that in our context X does not contain any $\mathbb{F}_q$-irreducible curves of arithmetic genus smaller than or equal to ℓ for $\ell = (d-1)(d-2)/2 - 1 = d(d-3)/2$. This allows us to apply Theorem 3.4, and get the following proposition.

PROPOSITION 5.3. *Let X be a degree $d \geq 4$ surface in $\mathbb{P}^3$ of arithmetic Picard number one whose Néron-Severi group* $\mathrm{NS}(X)$ *is generated by the class of an hyperplane section L. Assume that S is a set of rational points avoiding L. For any positive integer r the minimum distance of the code $\mathcal{C}(X, rL, S)$ satisfies*

$$d(X, rL, S) \geq d^*(X, rL, S, L) + rd\left(1 - \frac{r+d-4}{d(d-3)}\right)(q+1+m)$$

where

$$d^*(X, rL, S, L) = \sharp S - rd(q+1+m) - mrd(r+d-4)/2.$$

5.2. Surfaces in $\mathbb{P}^3$ of arithmetic Picard number one. In this subsection we suppose that the arithmetic Picard number of X is one, but we do not take the assumption that the Néron-Severi group is generated by an hyperplane section. Also in this case we can apply Theorem 4.3 which brings us to the following proposition.

PROPOSITION 5.4. *Let X be a surface of degree $d \geq 4$ in $\mathbb{P}^3$. Assume that* $\mathrm{NS}(X) = \mathbb{Z}H$ *for an ample divisor H. Consider $L = hH$, the class of an hyperplane section of X, for h a positive integer. Let S be a set of rational points on X*

avoiding H and let r be a positive integer. Then the minimum distance of the code $\mathcal{C}(X, rH, S)$ satisfies

$$d(X, rH, S) \geq \begin{cases} \sharp S - (q+1+m) - rH^2(r + h(d-4))/2 \text{ if } r > 2(q+1+m)/mH^2, \\ \sharp S - r(q+1+m) - rH^2(1+h(d-4))/2 \text{ otherwise.} \end{cases}$$

PROOF. Since we have $3H^2 + H.K_X = 3H^2 + H.(d-4)L = 3H^2 + h(d-4)H^2 = H^2(3 + h(d-4)) \geq 0$, we can apply point (i) of Theorem 4.3 from which the proposition follows. □

Acknowledgments.

The authors would like to thank the anonymous referee for relevant observations.

References

[1] Y. Aubry, *Algebraic geometric codes on surfaces*, talk at Eurocode'92 - International symposium on coding theory and applications (1992, Udine, Italie), in Ph.D. thesis of the University of Aix-Marseille II, France (1993), hal-00979000.

[2] Y. Aubry, E. Berardini, F. Herbaut, and M. Perret, *Algebraic geometry codes over abelian surfaces containing no absolutely irreducible curves of low genus*, Finite Fields Appl. **70** (2021), 101791, 20, DOI 10.1016/j.ffa.2020.101791. MR4192804

[3] Y. Aubry and M. Perret, *A Weil theorem for singular curves*, Arithmetic, geometry and coding theory (Luminy, 1993), de Gruyter, Berlin, 1996, pp. 1–7. MR1394921

[4] Y. Aubry and M. Perret, *On the characteristic polynomials of the Frobenius endomorphism for projective curves over finite fields*, Finite Fields Appl. **10** (2004), no. 3, 412–431, DOI 10.1016/j.ffa.2003.09.005. MR2067606

[5] R. Blache, A. Couvreur, E. Hallouin, D. Madore, J. Nardi, M. Rambaud, and H. Randriam, *Anticanonical codes from del Pezzo surfaces with Picard rank one*, Trans. Amer. Math. Soc. **373** (2020), no. 8, 5371–5393, DOI 10.1090/tran/8119. MR4127880

[6] E. Bombieri and D. Mumford, *Enriques' classification of surfaces in char. p. III*, Invent. Math. **35** (1976), 197–232, DOI 10.1007/BF01390138. MR491720

[7] A. Couvreur, *Construction of rational surfaces yielding good codes*, Finite Fields Appl. **17** (2011), no. 5, 424–441, DOI 10.1016/j.ffa.2011.02.007. MR2831703

[8] V. D. Goppa, *Codes on algebraic curves* (Russian), Dokl. Akad. Nauk SSSR **259** (1981), no. 6, 1289–1290. MR628795

[9] P. Griffiths and J. Harris, *On the Noether-Lefschetz theorem and some remarks on codimension-two cycles*, Math. Ann. **271** (1985), no. 1, 31–51, DOI 10.1007/BF01455794. MR779603

[10] S. Haloui, *Codes from Jacobian surfaces*, Arithmetic, geometry, cryptography and coding theory, Contemp. Math., vol. 686, Amer. Math. Soc., Providence, RI, 2017, pp. 123–135, DOI 10.1090/conm/686. MR3630612

[11] J. P. Hansen, *Toric surfaces and error-correcting codes*, Coding theory, cryptography and related areas (Guanajuato, 1998), Springer, Berlin, 2000, pp. 132–142. MR1749454

[12] S. H. Hansen, *Error-correcting codes from higher-dimensional varieties*, Finite Fields Appl. **7** (2001), no. 4, 531–552, DOI 10.1006/ffta.2001.0313. MR1866342

[13] R. Hartshorne, *Algebraic geometry*, Springer-Verlag, New York-Heidelberg, 1977. Graduate Texts in Mathematics, No. 52. MR0463157

[14] J. Little and H. Schenck, *Codes from surfaces with small Picard number*, SIAM J. Appl. Algebra Geom. **2** (2018), no. 2, 242–258, DOI 10.1137/17M1128277. MR3797729

[15] J. B. Little, *Algebraic geometry codes from higher dimensional varieties*, Advances in algebraic geometry codes, Ser. Coding Theory Cryptol., vol. 5, World Sci. Publ., Hackensack, NJ, 2008, pp. 257–293, DOI 10.1142/9789812794017_0007. MR2509126

[16] J. Nardi, *Algebraic geometric codes on minimal Hirzebruch surfaces*, J. Algebra **535** (2019), 556–597, DOI 10.1016/j.jalgebra.2019.06.022. MR3986402

[17] I. R. Shafarevich, *Basic algebraic geometry. 1*, second ed., Springer-Verlag, Berlin, 1994, Varieties in projective space, Translated from the 1988 Russian edition and with notes by Miles Reid. MR1328833

[18] J. H. Silverman, *Advanced topics in the arithmetic of elliptic curves*, Graduate Texts in Mathematics, vol. 151, Springer-Verlag, New York, 1994, DOI 10.1007/978-1-4612-0851-8. MR1312368

[19] M. A. Tsfasman, S. G. Vlăduţ, and Th. Zink, *Modular curves, Shimura curves, and Goppa codes, better than Varshamov-Gilbert bound*, Math. Nachr. **109** (1982), 21–28, DOI 10.1002/mana.19821090103. MR705893

[20] J. F. Voloch and M. Zarzar, *Algebraic geometric codes on surfaces* (English, with English and French summaries), Arithmetics, geometry, and coding theory (AGCT 2005), Sémin. Congr., vol. 21, Soc. Math. France, Paris, 2010, pp. 211–216. MR2856569

[21] M. Zarzar, *Error-correcting codes on low rank surfaces*, Finite Fields Appl. **13** (2007), no. 4, 727–737, DOI 10.1016/j.ffa.2007.05.001. MR2359313

Yves Aubry, Institut de Mathématiques de Toulon - IMATH, Université de Toulon and Institut de Mathématiques de Marseille - I2M, Aix Marseille Université, CNRS, Centrale Marseille, UMR 7373, France

Email address: `yves.aubry@univ-tln.fr`

Elena Berardini, Institut de Mathématiques de Marseille - I2M, Aix Marseille Université, CNRS, Centrale Marseille, UMR 7373, France

Email address: `elena_berardini@hotmail.it`

Fabien Herbaut, INSPE Nice-Toulon, Université Côte d'Azur, Institut de Mathématiques de Toulon - IMATH, Université de Toulon, France

Email address: `fabien.herbaut@univ-cotedazur.fr`

Marc Perret, Institut de Mathématiques de Toulouse, UMR 5219, Université de Toulouse, CNRS, UT2J, F-31058 Toulouse, France

Email address: `perret@univ-tlse2.fr`

Contemporary Mathematics
Volume **770**, 2021
https://doi.org/10.1090/conm/770/15429

On the number of effective divisors in algebraic function fields defined over a finite field

Stéphane Ballet, Gilles Lachaud, and Robert Rolland

ABSTRACT. We study the number of effective divisors of a given degree of an algebraic function field defined over a finite field. We first give somme lower bounds and upper bounds when the function field, the degree and the underlying finite field are fixed. Then we study the behavior of the number of effective divisors when some of the parameters, namely the underlying finite field, the degree of the effective divisors, the algebraic function field can be variable.

1. Introduction

The algebraic properties of algebraic function fields defined over a finite field is somehow reflected by their numerical properties, namely their numerical invariants such as the number of places of degree one over a given ground field extension, the number of classes of its Picard group, the number of effective divisors of a given degree and so on. In this paper, we are interested in the study of the number of effective divisors of a given degree and in the asymptotic behavior of this number under various assumptions.

The context of our study is as follows. We consider a function field $F/\mathbb{F}_q$ of genus g over the finite field $\mathbb{F}_q$ with q elements. Sometimes we will use the dual language of curves. We will denote by X an irreducible smooth curve defined over $\mathbb{F}_q$, having $F/\mathbb{F}_q$ for algebraic function field over $\mathbb{F}_q$ and by $X(\mathbb{F}_q)$ the set of $\mathbb{F}_q$-rational points of X, corresponding to the set of places of degree one of $F/\mathbb{F}_q$. Throughout this article, curve will mean irreducible smooth curve.

For any integer $n \geq 0$, let $A_n(F/\mathbb{F}_q)$ be the number of effective divisors of degree n of $F/\mathbb{F}_q$, $h(F/\mathbb{F}_q)$ its class number and $B_n = B_n(F/\mathbb{F}_q)$ its number of places of degree n. If there is no ambiguity we will set $A_n = A_n(F/\mathbb{F}_q)$, $h = h(F/\mathbb{F}_q)$ and $B_n = B_n(F/\mathbb{F}_q)$.

In the study of the quantity A_n, we need distinguish the two following cases:

(a) $n \leq g - 1$;
(b) arbitrary n.

2020 *Mathematics Subject Classification.* Primary 11G30; Secondary 14H05.

Our friend Gilles Lachaud passed away on February 21, 2018 while this work was almost in its final version.

Indeed, in the first case the quantity A_n is linked to the functional equation (1) involving several fundamental invariants, in particular the class number of the algebraic function field and the following $S(F/\mathbb{F}_q)$ and $R(F/\mathbb{F}_q)$ quantities:

$$S(F/\mathbb{F}_q) = \sum_{n=0}^{g-1} A_n + \sum_{n=0}^{g-2} q^{g-1-n} A_n \quad \text{and} \quad R(F/\mathbb{F}_q) = \sum_{i=1}^{g} \frac{1}{\mid 1 - \alpha_i \mid^2},$$

where $(\alpha_i, \overline{\alpha_i})_{1 \leq i \leq g}$ are the reciprocal roots of the numerator of the zeta-function $Z(F/\mathbb{F}_q, T)$ of $F/\mathbb{F}_q$. By a result due to G. Lachaud and M. Martin-Deschamps [**8**], we know that

$$S(F/\mathbb{F}_q) = hR(F/\mathbb{F}_q). \tag{1}$$

Let us also recall that the zeta-funtion of $F/\mathbb{F}_q$ is given by:

$$Z(t) = \sum_{m=0}^{+\infty} A_m t^m = \frac{L(t)}{(1-t)(1-qt)}$$

where $L(t) = \prod_{i=1}^{2g}(1 - \alpha_i t) = \sum_{i=0}^{2g} a_i t^i$ is in $\mathbb{Z}[t]$.

2. Contents

The paper is organized in the following way. In Section 3 the study is done when the finite field $\mathbb{F}_q$ and the algebraic function field $F/\mathbb{F}_q$ are fixed. We first give in Subsection 3.1 general results and general *formulae* on the numbers A_n for any positive n. Then in Subsection 3.2 we present some lower bounds on A_n for any positive n and finally in Subsection 3.3 we give when $1 \leq n \leq g-1$ some upper bounds on $\frac{A_n}{h}$.

Next in Section 4 we study the asymptotic behavior of $A_n(F/\mathbb{F}_q)$ when some of the parameters n, F, q are variable. More precisely, in Subsection 4.1 we study the case of a fixed curve curve X, a fixed degree n for the effective divisors and q growing to infinity, namely, starting from a finite field $\mathbb{F}_{q_1}$ we consider a sequence of extensions $\mathbb{F}_{q_i}$ of $\mathbb{F}_{q_1}$ where q_i is growing to infinity. In Subsection 4.2 we suppose that the field $\mathbb{F}_q$ is fixed and we consider a sequence of curves $(X_k)_k$ of genus $g_k = g(X_k)$ growing to infinity. For each curve X_k we fix a degree d_k. In this case we study the behavior of the sequence $\left(A_{d_k}(F_k/\mathbb{F}_q)\right)_k$ where $F_k/\mathbb{F}_q$ is the algebraic function field associated with the curve X_k and where d_k is linked to g_k in some way.

3. Non-asymptotical case

3.1. General results. In this section, we consider the case where the degree n is an arbitrary integer. Let us set

$$\Delta = \{i \in \mathbb{N} \mid 1 \leq i \leq g-1 \text{ and } B_i \geq 1\}.$$

$$U_n = \left\{ b = (b_i)_{i \in \Delta} \mid b_i \geq 0 \text{ and } \sum_{i \in \Delta} i b_i = n \right\}.$$

Note first that if $B_i \geq 1$ and $b_i \geq 0$, the number of solutions of the equation $n_1 + n_2 + \cdots + n_{B_i} = b_i$ with integers ≥ 0 is:

$$\binom{B_i + b_i - 1}{b_i} = \binom{B_i + b_i - 1}{B_i - 1}. \tag{2}$$

Then the number of effective divisors of degree n is given by the following result, introduced by S. Vlăduţ in [**13**] and already mentioned in [**12**], [**3**] and [**4**]:

PROPOSITION 3.1. *The number of effective divisors of degree n of an algebraic function field $F/\mathbb{F}_q$ is:*

$$A_n = \sum_{b \in U_n} \left[\prod_{i \in \Delta} \binom{B_i + b_i - 1}{b_i} \right].$$

PROOF. It is sufficient to consider that in the formula, b_i is the sum of coefficients that are applied to the places of degree i. So, the sum of the terms ib_i is the degree n of the divisor. The number of ways to get a divisor of degree ib_i with some places of degree i is given by the binomial coefficient (2). For a given b , the product of the second member is the number of effective divisors for which the weight corresponding to the places of degree i is ib_i. Then it remains to compute the sum over all possible b to get the number of effective divisors. □

PROPOSITION 3.2. *Let $\mathbf{F}/\mathbb{F}_q$ be a function field of genus g and let $L(t) = \sum_{i=0}^{2g} a_i t^i$ be the numerator of its zeta-function. Moreover, let us set $a_i = 0$ for any integer $i > 2g$. Then, for any integer $n \geq 0$, we have:*

$$A_n = \sum_{i=0}^{n} \frac{q^{n-i+1} - 1}{q - 1} a_i \tag{3}$$

and

$$\delta_n = A_{n+1} - A_n = \sum_{i=0}^{n+1} q^{n-i+1} a_i. \tag{4}$$

In particular, if $B_1 > 0$, we have $\delta_n \geq 0$.

PROOF. The zeta-function can be written as

$$Z(t) = \sum_{n=0}^{+\infty} A_n t^n = \frac{L(t)}{(1-t)(1-qt)} = \frac{\sum_{i=0}^{2g} a_i t^i}{(1-t)(1-qt)}.$$

From the equality $\dfrac{1}{(1-t)(1-qt)} = \dfrac{H}{1-t} + \dfrac{G}{1-qt}$ where $H = \dfrac{-1}{q-1}$ and $G = \dfrac{q}{q-1}$, and the power series expansions

$$\frac{1}{1-t} = 1 + t + t^2 + \ldots + t^k + \ldots$$

and

$$\frac{1}{1-qt} = 1 + qt + q^2t^2 + \ldots + q^k t^k + \ldots,$$

we obtain:

$$\sum_{n=0}^{\infty} A_n t^n = \left(\sum_{n=0}^{\infty} (H + q^n G) t^n \right) \times \sum_{n=0}^{\infty} a_n t^n.$$

Hence,

$$A_n = \sum_{i+j=n} (H + q^j G)a_i = \sum_{i+j=n} \left(\frac{-1}{q-1} + \frac{q^{j+1}}{q-1} \right) a_i.$$

Then, we have:

$$A_n = \sum_{i=0}^{n} \left(\frac{q^{n-i+1} - 1}{q-1} \right) a_i.$$

The value of δ_n follows.

If $B_1 > 0$, let P be a place of degree one. The map ϕ_P from the set of effective divisors of degree n to the set of effective divisors of degree $n+1$ defined by $\phi_P(\mathcal{D}) = \mathcal{D} + P$ is injective. Hence $A_{n+1} \geq A_n$. □

3.2. Lower bounds on A_n. From Proposition 3.1 we obtain in the next proposition a lower bound on the number of effective divisors of degree n containing in their support only places of some fixed distinct degrees $r_1, r_2, ..., r_k \geq 1$.

PROPOSITION 3.3. *Let $(r_\mu)_{\mu=1,\dots,p}$ be a family of distinct integers ≥ 1 such that $B_{r_\mu} > 0$ and n be an integer > 0. Suppose that $B_1 > 0$. Then*

$$A_n \geq \sum_{\mu=1}^{p} \binom{B_{r_\mu} + m_{r_\mu}(n) - 1}{B_{r_\mu} - 1} \binom{B_1 + s_{r_\mu}(n) - 1}{B_1 - 1} \tag{5}$$

where $m_{r_\mu}(n)$ and $s_{r_\mu}(n)$ are respectively the quotient and the remainder of the Euclidean division of n by r_μ, namely $m_{r_\mu}(n)$ is the integer part of n/r_μ.

PROOF. For any integer r_μ, let $a_\mu = (a_{\mu,i})_{i\in\Delta} \in U_n$ such that $a_{\mu,i} = 0$ for $i \in \Delta \setminus \{1, r_\mu\}$, $a_{\mu,1} = s_{r_\mu}(n)$ and $a_{\mu,r_\mu} = m_{r_\mu}(n)$. Then by Proposition (3.1), we have:

$$A_n = \sum_{b\in U_n} \prod_{i\in\Delta} \binom{B_i + b_i - 1}{b_i} \geq \sum_{\mu=1}^{p} \prod_{i\in\Delta} \binom{B_i + a_{\mu,i} - 1}{a_{\mu,i}} =$$

$$\sum_{\mu=1}^{p} \binom{B_{r_\mu} + m_{r_\mu}(n) - 1}{B_{r_\mu} - 1} \binom{B_1 + s_{r_\mu}(n) - 1}{B_1 - 1}.$$

□

In particular, if $p = 1$ and $r_1 = 1$ we obtain:

$$A_n \geq \binom{B_1 + n - 1}{n} = \binom{B_1 + n - 1}{B_1 - 1}.$$

Moreover, if the degrees r_μ are > 1 and divide n, we do not need the assumption of the existence of places of degree one.

PROPOSITION 3.4. *Let n be an integer > 0. Let $(r_\mu)_{\mu=1,\dots,p}$ be a family of distinct integers ≥ 1 dividing n. Suppose that $B_{r_\mu} > 0$ for any $\mu = 1, ..., p$. Then*

$$A_n \geq \sum_{\mu=1}^{p} \binom{B_{r_\mu} + m_{r_\mu}(n) - 1}{B_{r_\mu} - 1} \tag{6}$$

where $m_r(n)$ is the quotient of the Euclidean division of n by r_μ.

PROOF. For any integer r_μ, let $a_\mu = (a_{\mu_i})_{i\in\Delta} \in U_n$ such that $a_{\mu,i} = 0$ for $i \in \Delta \setminus \{r_\mu\}$ and $a_{\mu,r_\mu} = m_{r_\mu}(n)$. Then by Proposition (3.1), we have:

$$A_n = \sum_{b\in U_n}\prod_{i\in\Delta}\begin{pmatrix} B_i + b_i - 1 \\ b_i \end{pmatrix} \geq \sum_{\mu=1}^{p}\prod_{i\in\Delta}\begin{pmatrix} B_i + a_{\mu_i} - 1 \\ a_{\mu_i} \end{pmatrix} =$$

$$\sum_{\mu=1}^{p}\begin{pmatrix} B_{r_\mu} + m_{r_\mu}(n) - 1 \\ B_{r_\mu} - 1 \end{pmatrix}.$$

□

PROPOSITION 3.5. *Let $(r_\mu)_{\mu=1,\dots,p}$ be a family of distinct integers > 1 such that $B_{r_\mu} > 0$ and n be an integer > 0. Suppose that $B_1 \geq 1$. Let $m = (m_{r_\mu})_{\mu=1,\dots,p}$ be a family of integers ≥ 0 such that*

$$\sum_{\mu=1}^{p} m_{r_\mu} r_\mu \leq n.$$

Then

$$A_n \geq \begin{pmatrix} B_1 + n - 1 \\ B_1 \end{pmatrix} + \prod_{\mu=1}^{p}\begin{pmatrix} B_{r_\mu} + m_{r_\mu} \\ B_{r_\mu} \end{pmatrix}. \tag{7}$$

PROOF. Let

$$V_n = \left\{ b = (b_{r_\mu})_{\mu=1,\dots,p} \mid b_{r_\mu} \geq 0, \sum_{\mu=1}^{p} r_\mu b_{r_\mu} \leq n \right\}.$$

As $B_1 \geq 1$ the following holds:

$$A_n = \sum_{b\in U_n}\left[\prod_{i\in\Delta}\begin{pmatrix} B_i + b_i - 1 \\ b_i \end{pmatrix}\right] \geq \sum_{b\in V_n}\left[\prod_{\mu=1}^{p}\begin{pmatrix} B_{r_\mu} + b_{r_\mu} - 1 \\ b_{r_\mu} \end{pmatrix}\right].$$

Let $C_n = \prod_{\mu=1}^{p}\{0,..,m_{r_\mu}\}$. Then $C_n \subset V_n$, hence

$$A_n \geq \sum_{b\in C_n}\left[\prod_{\mu=1}^{p}\begin{pmatrix} B_{r_\mu} + b_{r_\mu} - 1 \\ b_{r_\mu} \end{pmatrix}\right].$$

But

$$\sum_{b\in C_n}\left[\prod_{\mu=1}^{p}\begin{pmatrix} B_{r_\mu} + b_{r_\mu} - 1 \\ b_{r_\mu} \end{pmatrix}\right] = \prod_{\mu=1}^{p}\sum_{b_{r_\mu}=0}^{m_{r_\mu}}\begin{pmatrix} B_{r_\mu} + b_{r_\mu} - 1 \\ b_{r_\mu} \end{pmatrix}$$

$$= \prod_{\mu=1}^{p}\begin{pmatrix} B_{r_\mu} + m_{r_\mu} \\ m_{r_\mu} \end{pmatrix}.$$

In the previous estimate, we did not take into account the effective divisors built only with places of degree one. Hence, the result is obtained by adding the number of such divisors. □

EXAMPLE 3.6. Let us suppose that $r_\mu = \mu + 1$ for any $1 \leq \mu \leq p$ and that we only know the value of B_1 and that $B_{\mu+1} \geq 1$. Then the m_i are such that $\sum_{i=2}^{p+1} im_i \leq n$ and

$$A_n \geq \binom{B_1 + n - 1}{B_1} + \prod_{i=2}^{p+1}(1 + m_i).$$

Let us set $x_i = 1 + m_i$. Then

$$\sum_{i=2}^{p+1} ix_i \leq \frac{(p+1)(p+2)}{2} - 1 + n.$$

To optimize the choice of the m_i, we will optimize the product $\prod_{i=2}^{p+1} x_i$ under the constraint $\sum_{i=1}^{p+1} ix_i = K$ where $K = \frac{(p+1)(p+2)}{2} - 1 + n$. This is done by the method of Lagrange's multipliers. Let us introduce the following function:

$$L(x_2, \cdots, x_{p+1}, \lambda) = \prod_{i=2}^{p+1} x_i - \lambda \left(\sum_{i=2}^{p+1} ix_i - K \right).$$

Let us denote by π_j the incomplete product:

$$x_2 x_3 \cdots x_{j-1} x_{j+1} \cdots x_{p+1}.$$

We have to solve the system:

$$\begin{cases} \frac{\partial L(x_2, \cdots, x_{p+1}, \lambda)}{\partial x_2} & = & \pi_2 - 2\lambda & = & 0, \\ \vdots & \vdots & \vdots & \vdots & \vdots \\ \frac{\partial L(x_2, \cdots, x_{p+1}, \lambda)}{\partial x_{p+1}} & = & \pi_{p+1} - (p+1)\lambda & = & 0, \\ \frac{\partial L(x_2, \cdots, x_{p+1}, \lambda)}{\partial \lambda} & = & \sum_{i=2}^{p+1} ix_i - K & = & 0. \end{cases}$$

Hence $x_i = \frac{2}{i} x_2$. Hence

$$\sum_{i=2}^{p+1} ix_i = 2px_2 = K.$$

This gives a value for x_2 and then for the x_i. These values are not always integers. Then we have to choose the best way to give to each x_i a integer value near the computed value, in order to obtain an optimal solution for the $m_i = x_i - 1$. For example if $p = 3$ and $n = 9$, then $K = 18$. We conclude that $x_2 = 3$, $x_3 = 2$, $x_4 = 3/2$. Then we can try $m_2 = 3$, $m_3 = 1$ and $m_4 = 0$. or $m_2 = 1$, $m_3 = 1$ and $m_4 = 1$ or $m_2 = 2$, $m_3 = 0$ and $m_4 = 1$. The two first solutions give the maximum 8 for the product $(1 + m_2)(1 + m_3)(1 + m_4)$ (it is impossible to do better).

3.3. Upper bounds in the case $n \leq g - 1$.

PROPOSITION 3.7. *Let $F/\mathbb{F}_q$ be a function field of genus g and let $L(t) = \sum_{i=0}^{2g} a_i t^i$ be the numerator of its zeta-function. Then*

$$A_{g-k} = \frac{1}{q-1} \left[q^{-k+1} \left(h - \sum_{i=0}^{g+k-1} a_i \right) - \sum_{i=0}^{g-k} a_i \right].$$

PROOF. From

$$Z(t) = \sum_{m=0}^{+\infty} A_m t^m = \frac{L(t)}{(1-t)(1-qt)} = \frac{\sum_{i=0}^{2g} a_i t^i}{(1-t)(1-qt)}$$

we deduce that for all $0 \leq m \leq 2g$,

$$A_m = \sum_{i=0}^{m} \frac{q^{m-i+1} - 1}{q - 1} a_i.$$

In particular,

$$(q-1)A_{g-k} = \sum_{i=0}^{g-k} (q^{g-k-i+1} - 1)a_i.$$

Since $a_i = q^{i-g}a_{2g-i}$, for all $i = 0, \ldots g$, we get

$$(q-1)A_{g-k} = q^{g-k+1} \sum_{i=0}^{g-k} q^{-i} a_i - \sum_{i=0}^{g-k} a_i = q^{g-k+1} \sum_{i=0}^{g-k} q^{-i} q^{i-g} a_{2g-i} - \sum_{i=0}^{g-k} a_i.$$

Hence

$$(q-1)A_{g-k} = q^{-k+1} \sum_{i=0}^{g-k} (a_{2g-i} - a_i) - \sum_{i=0}^{g-k} a_i + q^{-k+1} \sum_{i=0}^{g-k} a_i.$$

Furthermore, we know that $h = L(1) = \sum_{i=0}^{2g} a_i$, therefore

$$\sum_{i=0}^{g-k} (a_{2g-i} - a_i) = h - \sum_{i=0}^{g+k-1} a_i - \sum_{i=0}^{g-k} a_i,$$

which completes the proof. □

REMARK 3.8. For $k = 1$ one obtains the two following equalities in the interesting particular case of divisors of degree $g - 1$:

(8) $$A_{g-1} = \frac{1}{q-1} \left[h - \left(a_g + 2 \sum_{i=0}^{g-1} a_i \right) \right].$$

(9) $$A_{g-1} = \sum_{i=0}^{g-1} (a_{2g-i} - a_i).$$

Now we can give general bounds about the quantity $\frac{A_d}{h}$ which can be of interest. Let us give different useful bounds for $R(F/\mathbb{F}_q)$.

PROPOSITION 3.9.

(10) $$R(F/\mathbb{F}_q) \leq \frac{g}{(\sqrt{q} - 1)^2}.$$

(11) $$R(F/\mathbb{F}_q) \geq \frac{g}{(\sqrt{q} + 1)^2}.$$

(12) $$R(F/\mathbb{F}_q) \leq \frac{1}{(q-1)^2} \left((g+1)(q+1) - B_1(F/\mathbb{F}_q) \right).$$

(13) $$R(F/\mathbb{F}_q) \leq \frac{1}{(q-1)^2} \left((g+1)(q+1) \right).$$

(14) $$R(F/\mathbb{F}_q) \geq \frac{1}{(q+1)^2} \left((g+1)(q+1) - B_1(F/\mathbb{F}_q) \right).$$

PROOF. It is known by [8] that the quantity $R(F/\mathbb{F}_q)$ is bounded by the following upper bound:

$$R(F/\mathbb{F}_q) \leq \frac{1}{(q-1)^2}\left((g+1)(q+1) - B_1(F/\mathbb{F}_q)\right). \tag{15}$$

The inequality (12) is obtained as follows:

$$R(F/\mathbb{F}_q) = \sum_{i=1}^{g} \frac{1}{(1-\alpha_i)(1-\overline{\alpha_i})} = \sum_{i=1}^{g} \frac{1}{1+q-(\alpha_i+\overline{\alpha_i})}.$$

Multiplying the denominators by the corresponding conjugated quantities, we get:

$$R(F/\mathbb{F}_q) \leq \frac{1}{(q-1)^2}\sum_{i=1}^{g}(1+q+\alpha_i+\overline{\alpha_i}).$$

This last inequality associated to the following formula deduced from the Weil's formulas:

$$\sum_{i=1}^{g}(\alpha_i+\overline{\alpha_i}) = 1+q-B_1(F/\mathbb{F}_q),$$

gives the inequality (12). The inequality (12) cannot be improved in the general case. Remark that in the same way we can prove that

$$R(F/\mathbb{F}_q) \geq \frac{1}{(q+1)^2}\left((g+1)(q+1) - B_1(F/\mathbb{F}_q)\right). \tag{16}$$

□

REMARK 3.10. Note that Bound (12) is better than Bound (10) because of the lower Weil bound. Indeed,

$$B_1(F/\mathbb{F}_q) \geq q+1-2g\sqrt{q},$$

then

$$\frac{1}{(q-1)^2}\left((g+1)(q+1) - B_1(F/\mathbb{F}_q)\right)$$
$$\leq \frac{1}{(q-1)^2}\left((g+1)(q+1) - (q+1) + 2g\sqrt{q}\right).$$

and we can conclude thanks to the following equality:

$$\frac{1}{(q-1)^2}\left((g+1)(q+1) - (q+1) + 2g\sqrt{q}\right) = \frac{g}{(\sqrt{q}-1)^2}.$$

Moreover (13) is interesting when the number of places of degree one is unknown and $q+1-2g\sqrt{q} < 0$. Indeed in this case (13) is better than (10). On the contrary, if $q+1-2g\sqrt{q} > 0$ (10) is better than (13).

The following theorem gives upper bounds on $\frac{A_d}{h}$.

THEOREM 3.11. *For any function field $F/\mathbb{F}_q$ of genus g defined over $\mathbb{F}_q$ and any degree n such that $1 \leq n \leq g-1$, the following holds:*

$$\begin{cases} \frac{A_n}{h} < \frac{1}{2q^{\frac{g-n-1}{2}}(\sqrt{q}-1)^2} & \text{if} \quad n < g-1, \\ \frac{A_n}{h} < \frac{1}{(\sqrt{q}-1)^2} & \text{for} \quad n = g-1. \end{cases} \tag{17}$$

(18)
$$\frac{A_n}{h} \leq \frac{g}{q^{g-n-1}\left(\sqrt{q}-1\right)^2}.$$

(19)
$$\frac{A_n}{h} \leq \frac{(g+1)(q+1)}{q^{g-n-1}(q-1)^2}.$$

PROOF. Inequality (17) was obtained in the proof of [**2**, Theorem 3.3]. We prove here inequalities (18) and (19). Note that Inequality (18) was also proved by I. Cascudo, R. Cramer, C. Xing in [**5**, Proposition 3.4]. Let us denote by

$$L(t) = \prod_{j=1}^{g}(1-\alpha_j t)(1-\overline{\alpha_j}t)$$

the numerator of the zeta-function of $F/\mathbb{F}_q$. Then

$$h = L(1) = \prod_{j=1}^{g}|1-\alpha_j|^2.$$

The Hecke formula (see [**8**]) implies

(20)
$$A_{g-1} + \sum_{d=0}^{g-2} A_d\left(1+q^{g-1-d}\right) = h\sum_{j=1}^{g}\frac{1}{|1-\alpha_j|^2}.$$

But by Formula (10)

$$\sum_{j=1}^{g}\frac{1}{|1-\alpha_j|^2} \leq \frac{g}{(\sqrt{q}-1)^2},$$

then

$$A_{g-1} + \sum_{d=0}^{g-2} A_d\left(1+q^{g-1-d}\right) \leq \frac{gh}{(\sqrt{q}-1)^2},$$

hence for any n such that $1 \leq n \leq g-1$ the following holds:

$$A_n\, q^{g-1-n} \leq \frac{gh}{(\sqrt{q}-1)^2}.$$

From this last inequality we get (18). Replacing Formula (10) by Formula (13) we obtain (19). □

REMARK 3.12. If the algebraic function field $F/\mathbb{F}_q$ of genus $g \geq 1$ is ordinary then by [**2**, Proposition 4.3] there exists a non-special divisor of degree $g-1$. Hence $A_{g-1} < h$ which improves for $q = 2$ or 3 the second part of Formula 17.

If $B_1 \geq 1$ inequalities (18) and (19) can be improved. by the following proposition:

PROPOSITION 3.13. *Let $F/\mathbb{F}_q$ be a function field of genus g defined over $\mathbb{F}_q$ and n an integer such that $1 \leq n < g-1$. Suppose that $B_1 \geq 1$. Then the following holds:*

(21)
$$\frac{A_n}{h} \leq \frac{1}{2(\sqrt{q}-1)^2\left(\frac{1}{2}+\sqrt{q}\,\frac{\sqrt{q}^{g-n-1}-1}{\sqrt{q}-1}\right)}.$$

(22)
$$\frac{A_n}{h} \leq \frac{(g+1)(q+1)-B_1}{(q-1)\left[(q-1)(g-n)+q\left(q^{g-n-1}-1\right)\right]}.$$

PROOF. In order to prove Formula (21) we use the following inequality established in [**2**, Formula (6)]:

$$2\sum_{d=0}^{g-2} q^{(g-1-d)/2} A_d + A_{g-1} \leq \frac{h}{(\sqrt{q}-1)^2}.$$

Then, as $B_1 \geq 1$, we know by Proposition 3.2 that $A_{n+1} \geq A_n$. Hence

$$2A_n \sum_{d=n}^{g-2} q^{(g-1-d)/2} + A_n \leq \frac{h}{(\sqrt{q}-1)^2},$$

from which we deduce (21).

To prove Formula (22), first we can replace inequalities (10) and (13) by the better inequality (12). Next we know by Proposition 3.2 that $A_{n+1} \geq A_n$. Hence we can deduce from formula (20) the following ones:

$$A_n \left((g-n) + \sum_{d=n}^{g-2} q^{g-d-1} \right) \leq \frac{h}{(q-1)^2} \left((g+1)(q+1) - B_1 \right),$$

$$A_n \left((g-n) + \frac{q\left(q^{g-n-1}-1\right)}{(q-1)} \right) \leq \frac{h}{(q-1)^2} \left((g+1)(q+1) - B_1 \right).$$

This last inequality leads to the result. □

If we compare Inequalities (17), (18) and (19) we can see that each of them can be better than the other depending on the parameters. A complete study is done in Annexe A.

We can also obtain bounds concerning directly the quantity A_n from bounds on $\dfrac{A_n}{h}$ by using the Weil bounds [**14**] [**15**]:

$$(\sqrt{q}-1)^{2g} \leq h \leq (\sqrt{q}+1)^{2g}. \tag{23}$$

A better upper bound for h than the Weil bound, due to P. Lebacque and A. Zykin [**9**] can be used if we know upper bounds for the number of rational points of the curve $X(\mathbb{F}_{q^k})$ over the fields $\mathbb{F}_{q^k}$ for $1 \leq k \leq N$ where N is an integer ≥ 1 :

$$h \leq q^g \exp\left(\sum_{k=1}^{N} \frac{1}{kq^k} \mid X(\mathbb{F}_{q^k}) \mid - \sum_{k=1}^{N} \frac{1+q^{-k}}{k} + \frac{2g}{(\sqrt{q}-1)(N+1)q^{\frac{N}{2}}} \right). \tag{24}$$

Moreover, in the special case $n = g-1$, the estimates for $\frac{A_{g-1}}{h}$ can be improved. Let us introduce

$$Q_{r,s} = \sum_{k=0}^{s} \frac{1}{q^{\frac{rk}{2}}} \begin{pmatrix} B_r + k - 1 \\ k \end{pmatrix}.$$

LEMMA 3.14.

$$Q_{r,s} = \left(\frac{q^{\frac{r}{2}}}{q^{\frac{r}{2}}-1} \right)^{B_r} - B_r \begin{pmatrix} B_r + s \\ B_r \end{pmatrix} \int_0^{\frac{1}{q^{\frac{r}{2}}}} \frac{(\frac{1}{q^{\frac{r}{2}}} - t)^s}{(1-t)^{B_r+s+1}} \mathrm{d}t.$$

PROOF. Let us set

$$S_r(X) = \sum_{k=0}^{\infty} X^k \binom{B_r + k - 1}{k}, \quad T_r(X,s) = \sum_{k=0}^{s} X^k \binom{B_r + k - 1}{k},$$

$$\text{and } R_r(X,s) = \sum_{k=s+1}^{\infty} X^k \binom{B_r + k - 1}{k}.$$

Let us remark that

$$Q_{r,s} = T_r\left(\frac{1}{q^{\frac{r}{2}}}, s\right) \quad \text{and} \quad S_r(X) = \frac{1}{(1-X)^{B_r}}$$

which converges for $|X| < 1$ and moreover

$$S_r(X) = T_r(X,s) + R_r(X,s).$$

By the Taylor Formula, we get

$$R_r(X,s) = B_r \binom{B_r + s}{B_r} \int_0^X \frac{(X-t)^s}{(1-t)^{B_r+s+1}} \mathrm{d}t.$$

Then

$$Q_{r,s} = \left(\frac{q^{\frac{r}{2}}}{q^{\frac{r}{2}} - 1}\right)^{B_r} - B_r \binom{B_r + s}{B_r} \int_0^{\frac{1}{q^{\frac{r}{2}}}} \frac{(\frac{1}{q^{\frac{r}{2}}} - t)^s}{(1-t)^{B_r+s+1}} \mathrm{d}t.$$

□

LEMMA 3.15. *Let $F/\mathbb{F}_q$ be a function field of genus g defined over $\mathbb{F}_q$. Let let $m = (m_r)_{r\in\Delta}$ be a finite sequence of integers such that $m_r \geq 0$ and $\sum_{r\in\Delta} rm_r \leq g-2$. Then the following inequality holds:*

$$\sum_{k=0}^{g-2} \frac{A_k}{q^{\frac{k}{2}}} \geq \prod_{r\in\Delta} Q_{r,m_r}.$$

PROOF. By Proposition 3.1 we know that

$$A_k = \sum_{b\in U_k} \left[\prod_{i\in\Delta} \binom{B_i + b_i - 1}{b_i}\right].$$

If one set

$$V = \bigcup_{k=0}^{g-2} U_k = \left\{ b = (b_r)_{r\in\Delta} \mid b_r \geq 0 \text{ and } \sum_{r\in\Delta} rb_r \leq g-2 \right\}$$

the following holds:

$$\sum_{k=0}^{g-2} \frac{A_k}{q^{\frac{k}{2}}} = \sum_{b\in V} \prod_{r\in\Delta} \frac{1}{q^{\frac{rb_r}{2}}} \binom{B_r + b_r - 1}{b_r}.$$

Let C be the subset of V defined by

$$C = \prod_{r\in\Delta} \{0, \cdots, m_r\}.$$

Then

$$\sum_{k=0}^{g-2} \frac{A_k}{q^{\frac{k}{2}}} \geq \sum_{b\in C} \prod_{r\in\Delta} \frac{1}{q^{\frac{rb_r}{2}}} \binom{B_r + b_r - 1}{b_r} =$$

$$\prod_{r\in\Delta}\sum_{k=0}^{m_r}\frac{1}{q^{\frac{rk}{2}}}\left(\begin{array}{c}B_r+k-1\\k\end{array}\right)=\prod_{r\in\Delta}Q_{r,m_r}.$$

□

PROPOSITION 3.16. *Let $F/\mathbb{F}_q$ be a function field of genus g defined over $\mathbb{F}_q$. Let $m=(m_r)_{r\in\Delta}$ be a finite sequence of integers such that $m_r\geq 0$ and $\sum_{r\in\Delta}rm_r\leq g-2$.*

Then the following inequalities holds:

$$A_{g-1}\leq\frac{h}{(\sqrt{q}-1)^2}-2q^{\frac{g-1}{2}}\prod_{r\in\Delta}Q_{r,m_r},\tag{25}$$

$$A_{g-1}\leq\frac{(\sqrt{q}+1)^{2g}}{(\sqrt{q}-1)^2}-2q^{\frac{g-1}{2}}\prod_{r\in\Delta}Q_{r,m_r},\tag{26}$$

where for any $r\geq 1$ and $s\geq 0$ the following holds:

$$Q_{r,s}=\left(\frac{q^{\frac{r}{2}}}{q^{\frac{r}{2}}-1}\right)^{B_r}-B_r\left(\begin{array}{c}B_r+s\\B_r\end{array}\right)\int_0^{\frac{1}{q^{\frac{r}{2}}}}\frac{(\frac{1}{q^{\frac{r}{2}}}-t)^s}{(1-t)^{B_r+s+1}}\mathrm{d}t.$$

PROOF. Let us recall the following inequality established in [**2**, Formula (6)]:

$$2q^{(g-1)/2}\sum_{d=0}^{g-2}\frac{A_d}{q^{d/2}}+A_{g-1}\leq\frac{h}{(\sqrt{q}-1)^2}.$$

We know by Proposition 3.15 that

$$\sum_{k=0}^{g-2}\frac{A_k}{q^{\frac{k}{2}}}\geq\prod_{r\in\Delta}Q_{r,m_r}.$$

Then the inequality (25) holds.

The inequality (26) directly follows from the inequality (25) and from the upper Weil bound.

Finally, Lemma 3.14 gives the last equality. □

REMARK 3.17. The inequality (26) can be improved by using, if possible, the upper bound (24).

THEOREM 3.18. *For any curve X of genus g defined over $\mathbb{F}_q$. If $1\leq d\leq g-1$ the following holds:*

$$A_d\leq\frac{g\left(\sqrt{q}+1\right)^{2g}}{q^{g-d-1}\left(\sqrt{q}-1\right)^2}.\tag{27}$$

PROOF. From (18) and (23) we obtain (27). As previously remarked, if possible, we can use (24) instead of (23) and get the bound:

$$A_d\leq\frac{gq^g\exp\left(\sum_{k=1}^N\frac{1}{kq^k}\mid X(\mathbb{F}_{q^k})\mid-\sum_{k=1}^N\frac{1+q^{-k}}{k}+\frac{2g}{(\sqrt{q}-1)(N+1)q^{\frac{N}{2}}}\right)}{q^{g-d-1}\left(\sqrt{q}-1\right)^2}.\tag{28}$$

□

4. Asymptotical case

The study of the asymptotic behavior of certain quantities associated to curves or function fields can be done from many point of view depending on the parameter tending to infinity. The simpler cases are described by the two following situations:

- **Increasing the size of the definition function field of one fixed curve.** The first case corresponds to the situation of a unique fixed curve X_0 of genus g and a fixed degree d_0. Let us remark that if the curve X_0 is defined on the finite field $\mathbb{F}_{q_1}$, it is also defined on any extension $\mathbb{F}_q$ of $\mathbb{F}_{q_1}$, and then we can study the asymptotic behavior of quantities related to the curve X_0 when q is growing to infinity. In particular in section 4.1 we will study the asymptotic behaviour of
$$\frac{A_{d_0}(F_0/\mathbb{F}_q)}{h(F_0/\mathbb{F}_q)}$$
when q is growing to infinity, where d_0 is a fixed degree and $F_0/\mathbb{F}_q$, the function field over $\mathbb{F}_q$ associated to the curve X_0.
- **Case of a family of curves defined over the same finite field $\mathbb{F}_q$.** Let $(X_k)_{k\geq 1}$ be a family of curves defined over $\mathbb{F}_q$. We study the sequence of function fields $(F_k/\mathbb{F}_q)_k$. Let us denote by $g_k = g(X_k)$ the genus of the curve X_k. We will suppose in the following that the genus sequence $(g_k)_k$ is growing to infinity. In section 4.2 we will study the asymptotic behaviour of many interesting quantities when k (and then g_k) is growing to infinity, and when the degree d is linked to g_k by a relation. We will study in particular the case where d is a linear function of g_k.

Some asymptotic behaviours are deduced from absolute *formulae*, namely true for any value of the variables $(g(X), q, d)$. Many such *formulae* exist, each of them being mainly adapted to a particular asymptotic study. We can consider this point of view by using the results obtained in the section 3.

4.1. Case of a fixed curve and a fixed degree. In this section the curve X of genus g is fixed and d is a fixed integer such that $d \leq g-1$. Let us recall that $A_d(F/\mathbb{F}_q)$ is the number of degree d effective divisors of $F/\mathbb{F}_q$ and that $h(F/\mathbb{F}_q)$ is its class number. We give here the asymptotic behaviour of the quotient
$$\frac{A_d(F/\mathbb{F}_q)}{h(F/\mathbb{F}_q)}.$$

THEOREM 4.1. *Let us suppose that $g \geq 1$ and $d \leq g-1$. Then when q is growing to infinity, the following holds:*
$$\frac{A_d(F/\mathbb{F}_q)}{h(F/\mathbb{F}_q)} = \frac{1}{q^{g-d}}\left(1 + O\left(\frac{1}{q}\right)\right),$$
where the O Landau function depends on X and d.

This theorem is a consequence of the two following *lemmata*. Let us denote by W_d^0 the following set:
$$W_d^0 = \{[D] \in \mathcal{P}\mathrm{ic}_d(X) \mid \dim(D) > 0\}.$$
If $d \leq g-1$, the elements of W_d^0 are special divisor classes.
$$|X(\mathbb{F}_q)| = q + 1 - \mathrm{trace}\,(\pi).$$

LEMMA 4.2. *If $g \geq 1$ and $d \leq g$ then when q tends to infinity*

$$\left|W_d^0(\mathbb{F}_q)\right| = q^d - q^{d-1}\text{trace}\,(\pi) + O(q^{d-1}).$$

PROOF. The restriction of the projection

$$\Pi : \mathcal{D}\mathrm{iv}_{\,d}(X) \to \mathcal{P}\mathrm{ic}_d(X)$$

gives a surjective morphism

$$\Pi : \mathcal{D}\mathrm{iv}_{\,d}^{\,+}(X) \to W_d^0.$$

The scheme W_d^0 is a dimension d irreducible normal sub-variety of $\mathcal{P}\mathrm{ic}_d(X)$ defined over $\mathbb{F}_q$ (see [**1**, p. 190], [**10**, Prop. 5.1, p. 182]). The Albanese variety of W_d^0 is $\mathcal{J}\mathrm{ac}(X)$ (see [**10**, Prop. 5.3, p. 183]). We conclude by [**6**, Cor. 11.4]. □

LEMMA 4.3. *If $g \geq 1$ and $d \leq g-1$ then when q tends to infinity*

$$A_d(F/\mathbb{F}_q) = \left|W_d^0(\mathbb{F}_q)\right| + O(q^{d-1}).$$

PROOF. Recall that

$$A_d(F/\mathbb{F}_q) = \sum_{[D]\in\mathcal{P}\mathrm{ic}_d(\mathbb{F}_q)} \frac{q^{\dim(D)}-1}{q-1}.$$

The dimension of included varieties

$$W_d^r = \{[D] \in \mathcal{P}\mathrm{ic}_d(\mathbb{F}_q) \mid \dim(D) \geq r+1\}$$

is the Brill-Noether number (see [**1**, p. 180])

$$\rho(r) = g - (r+1)(g-d+r),$$

and $W_d^r = \emptyset$ if $r > d/2$ by the Clifford's theorem. When q is growing to infinity the following holds:

$$A_d(F/\mathbb{F}_q) = \sum_{l=1}^{+\infty} \frac{q^l-1}{q-1}\left(\left|W_d^{l-1}(\mathbb{F}_q)\right| - \left|W_d^l(\mathbb{F}_q)\right|\right),$$

$$A_d(F/\mathbb{F}_q) = \sum_{l=1}^{+\infty}\sum_{i=0}^{l-1} q^i\left(\left|W_d^{l-1}(\mathbb{F}_q)\right| - \left|W_d^l(\mathbb{F}_q)\right|\right),$$

$$A_d(F/\mathbb{F}_q) = \sum_{l=0}^{+\infty}\sum_{i=0}^{l} q^i\left|W_d^l(\mathbb{F}_q)\right| - \sum_{l=1}^{+\infty}\sum_{i=0}^{l-1} q^i\left|W_d^l(\mathbb{F}_q)\right|,$$

$$A_d(F/\mathbb{F}_q) = \left|W_d^0(\mathbb{F}_q)\right| + \sum_{l=1}^{+\infty} q^l\left|W_d^l(\mathbb{F}_q)\right|,$$

$$A_d(F/\mathbb{F}_q) \leq \left|W_d^0(\mathbb{F}_q)\right| + \sum_{l=1}^{+\infty} q^{l+\rho(l)}\left(1+O(q^{-1/2})\right),$$

$$A_d(F/\mathbb{F}_q) \leq \left|W_d^0(\mathbb{F}_q)\right| + q^d\sum_{l=1}^{+\infty} q^{-l^2}q^{-(g-d)l}\left(1+O(q^{-1/2})\right).$$

But when q tends to infinity

$$\sum_{l=1}^{+\infty} q^{-l^2} \sim q^{-1}$$

hence

$$A_d(F/\mathbb{F}_q) \le \left|W_d^0(\mathbb{F}_q)\right| + O(q^{d-1}).$$

On the other hand $A_d(F/\mathbb{F}_q) \ge \left|W_d^0(\mathbb{F}_q)\right|$, then

$$A_d(F/\mathbb{F}_q) = \left|W_d^0(\mathbb{F}_q)\right| + O(q^{d-1}).$$

□

Proof of Theorem 4.1. Let us remark that $h(F/\mathbb{F}_q) = |\mathcal{P}\mathrm{ic}_d(\mathbb{F}_q)| = q^g - q^{g-1}\mathrm{trace}\,(\pi) + O(q^{g-1})$. Then we deduce the result from Lemma 4.2 and Lemma 4.3. □

4.2. Case of a fixed finite field.

4.2.1. *Introduction.* When, for a given finite ground field, the sequence of the genus of a sequence of algebraic function fields tends to infinity, there exist asymptotic *formulae* for different numerical invariants. In this section, we are interested by the asymptotic study with respect to the genus g of the number of effective divisors of certain degrees.

Let $(X_k)_k$ be a sequence of smooth irreducible curves defined over the finite field $\mathbb{F}_q$. We denote by $\mathcal{F}/\mathbb{F}_q = (F_k/\mathbb{F}_q)_k$ the corresponding sequence of algebraic function fields defined over $\mathbb{F}_q$. We denote by g_k the genus of X_k and we suppose that the sequence $(g_k)_k$ is growing to infinity. For any integer k, let d_k be an integer. We denote by $A_{d_k,k}$ the number $A_{d_k}(F_k/\mathbb{F}_q)$ of degree d_k effective divisors of $F_k/\mathbb{F}_q$ and by h_k its class number.

In this case, the problem was first studied by M. Tsfasman in the article [**11**, pp 184–185]. Next, M. Tsfasman and S. Vladut mainly give in [**12**], for asymptotically exact families two kind of estimates:

- the first one is the "exponentiel estimate" which computes the asymptotic value of $A_{d_k,k}^{1/g_k}$ (see Theorem 4.1 in [**12**]);
- the second one is the "linear estimate" which computes for $A_{d_k,k}/h_k$ a more precise estimate (see Theorem 5.1 in [**12**]). More specifically, under some assumption on the behaviour of $\frac{d_k}{g_k}$ when k is growing to infinity, the following asymptotic estimate holds when the family $\mathcal{F}/\mathbb{F}_q = (F_k/\mathbb{F}_q)_k$ is an asymptotically exact family:

$$\frac{A_{d_k,k}}{h_k} = \frac{1}{q^{g_k-d_k}}\frac{q}{q-1}\left(1+o(1)\right). \tag{29}$$

Remark 4.4. The assumption done in [**12**, Teorem 5.1] is made precise in [**12**, Lemma 5.1]. It turns out that it is verified if there is an $\epsilon > 0$ and an integer $k_0 \ge 1$ such that for all $k \ge k_0$ the inequality $\dfrac{d_k}{g_k} \ge 2\lambda + \epsilon$ holds, where λ is the unique root of the equation $H_{1+\frac{1}{\sqrt{q}}}(x) = 0$ on $[0,1]$, $H_y(x)$ being the entropy function defined by

$$H_y(x) = x\log_y(y-1) - x\log_y(x) - (1-x)\log_y(1-x).$$

Remark 4.5. If the family is a tower, by [**7**] it is asymptotically exact. If moreover this tower is ordinary then for any k, $A_{g_k-1,k} < h_k$ by Remark 3.12. Then for $q = 2$ one can deduce from Formula (29) that $\lim_k \frac{A_{g_k-1,k}}{h_k} = 1$ and that this limit is reached by lower values.

4.2.2. *General sequences.* In this section, we consider general sequences of algebraic function fields namely which are not necessarily asymptotically exact. Then, we obtain a result in two parts which respectively follows from *formulae* (17) and (18) or (19) of Section 3.

THEOREM 4.6. *Let $(d_k)_k$ be a sequence of integers such that $1 \leq d_k \leq g_k - 1$. Let us set $d_k = g_k - \phi(g_k)$ where ϕ is a function with integer values.*

- *If there is an integer k_0 such that for any $k \geq k_0$ the inequality $d_k < g_k - 1$ holds then*

$$0 \leq \limsup_{k \to +\infty} \frac{A_{d_k,k}}{h_k} \leq \limsup_{k \to +\infty} \frac{\sqrt{q}}{2(\sqrt{q}-1)^2} q^{-\frac{\phi(g_k)}{2}}, \tag{30}$$

- *else*

$$0 \leq \limsup_{k \to +\infty} \frac{A_{d_k,k}}{h_k} \leq \frac{1}{(\sqrt{q}-1)^2}. \tag{31}$$

PROOF. These inequalities follow from *Formulae* (17). □

THEOREM 4.7. *Let us suppose that the following limit exists $\delta = \lim_{k \to +\infty} \frac{d_k}{g_k}$ and satisfies $0 \leq \delta \leq 1$, then*

$$\limsup_{k \to +\infty} \left(\frac{A_{d_k,k}}{h_k} \right)^{\frac{1}{g_k}} \leq \frac{1}{q^{1-\delta}}. \tag{32}$$

PROOF. This inequality follows from Formula (18) or from Formula (19). □

COROLLARY 4.8. *Let us set $d_k = g_k - \phi(g_k)$ where ϕ is a function with integer values. If $\lim_{k \to +\infty} \phi(g_k) = +\infty$, then we have*

$$\lim_{k \to +\infty} \frac{A_{d_k,k}}{h_k} = 0. \tag{33}$$

PROOF. The result is a straightforward consequence of formula (30). For asymptotically exact sequences it is also a direct consequence of Formula (29). □

REMARK 4.9. Formula (29) for asymptotically exact sequences gives a more accurate asymptotic estimation than Formula (30). But this last formula is valid for any sequence, not only for asymptotically exact sequences and without condition on d_k/g_k.

Appendix A

A.1. Comparison of bounds. Let us denote by K_{17}, K_{18} and K_{19} the respective second members of the inequalities (17), (18) and (19). In Remark 3.10 we proved that $K_{18} \leq K_{19}$ if and only if $g \leq \frac{1}{2}\left(\sqrt{q} + \frac{1}{\sqrt{q}}\right)$. In the following we compare K_{17} to K_{18} and to K_{19}.

(1) Case $d = g - 1$.

$$K_{17} = \frac{1}{(\sqrt{q}-1)^2}, K_{18} = \frac{g}{(\sqrt{q}-1)^2}, K_{19} = \frac{(g+1)(q+1)}{(q-1)^2}.$$

We remark that in this case $K_{17} \leq K_{18}$.

(a) For $g \leq \frac{1}{2}\left(\sqrt{q}+\frac{1}{\sqrt{q}}\right)$: from the previous results the following inequalities hold:

$$K_{17} \leq K_{18} \leq K_{19}.$$

(b) For $g \geq \frac{1}{2}\left(\sqrt{q}+\frac{1}{\sqrt{q}}\right)$:

$$\frac{K_{17}}{K_{19}} = \frac{(\sqrt{q}+1)^2}{(g+1)(q+1)}.$$

But

$$(g+1)(q+1) - (\sqrt{q}+1)^2 = gq - 2\sqrt{q} + g.$$

The last right member is a quadratic polynomial in $\sqrt{q}$ which has a discriminant ≤ 0. The sign is constant and ≥ 0. We conclude that:

$$K_{17} \leq K_{19} \leq K_{18}.$$

(2) Case $d = g-2$. In this case

$$K_{17} = \frac{1}{2\sqrt{q}(\sqrt{q}-1)^2}, K_{18} = \frac{g}{q(\sqrt{q}-1)^2},$$

$$K_{19} = \frac{(g+1)(q+1)}{q(q-1)^2}.$$

(a) For $g \leq \frac{1}{2}\left(\sqrt{q}+\frac{1}{\sqrt{q}}\right)$:

$$\frac{K_{17}}{K_{18}} = \frac{\sqrt{q}}{2g}.$$

(i) If $\frac{1}{2}\sqrt{q} \leq g \leq \frac{1}{2}\left(\sqrt{q}+\frac{1}{\sqrt{q}}\right)$, then the following inequalities hold:

$$K_{17} \leq K_{18} \leq K_{19}.$$

(ii) If $g \leq \frac{1}{2}\sqrt{q}$ then

$$\frac{K_{17}}{K_{19}} = \frac{\sqrt{q}(\sqrt{q}+1)^2}{2(g+1)(q+1)}.$$

But as $g \leq \frac{1}{2}\sqrt{q}$, the following holds:

$$\sqrt{q}(\sqrt{q}+1)^2 - 2(g+1)(q+1) \geq$$
$$2gq + 2q + 2g - (2gq + 2g + 2q + 2)$$

and

$$\sqrt{q}(\sqrt{q}+1)^2 - 2(g+1)(q+1) \geq 2.$$

Hence $K_{19} \leq K_{17}$ and

$$K_{18} \leq K_{19} \leq K_{17}.$$

(b) For $g \geq \frac{1}{2}\left(\sqrt{q}+\frac{1}{\sqrt{q}}\right)$:

$$\begin{aligned} 2(g+1)(q+1) - \sqrt{q}(\sqrt{q}+1)^2 \geq \\ \left(\sqrt{q}+\frac{1}{\sqrt{q}}\right)(q+1) + 2(q+1) - \sqrt{q}(\sqrt{q}+1)^2 = \\ (\sqrt{q}+1)^2. \end{aligned}$$

Then $K_{17} \leq K_{19}$ and

$$K_{17} \leq K_{19} \leq K_{18}.$$

(3) Case $d \leq g-3$.

(a) The case $q=2$ and $d=g-3$ ($g \geq 3$) is the only case such that

$$\left(\sqrt{q}+\frac{1}{\sqrt{q}}\right) > q^{\frac{g-d-1}{2}}.$$

By a simple computation we obtain:

$$K_{17} \leq K_{19} \leq K_{18}.$$

(b) For $q \neq 2$ or $d < g-3$ we have

$$\frac{K_{17}}{K_{19}} = \frac{q^{\frac{g-d-1}{2}}}{2g}.$$

(i) If we have the following inequalities

$$\frac{1}{2}\left(\sqrt{q}+\frac{1}{\sqrt{q}}\right) \leq g \leq \frac{1}{2}q^{\frac{g-d-1}{2}}$$

then

$$K_{19} \leq K_{18} \leq K_{17}.$$

(ii) If

$$g \leq \frac{1}{2}\left(\sqrt{q}+\frac{1}{\sqrt{q}}\right),$$

then $K_{18} \leq K_{17}$, $K_{18} \leq K_{19}$ and

$$\frac{K_{17}}{K_{19}} = \frac{q^{\frac{g-d-1}{2}}(\sqrt{q}+1)^2}{2(g+1)(q+1)}.$$

But

$$\begin{aligned} q^{\frac{g-d-1}{2}}(\sqrt{q}+1)^2 - 2(g+1)(q+1) \geq \\ q^{\frac{g-d}{2}}(q+2\sqrt{q}+1) - (q+1+2\sqrt{q})(q+1) = \\ (q+2\sqrt{q}+1)\left(q^{\frac{g-d}{2}} - (q+1)\right). \end{aligned}$$

(A) If $q=2$ and $g-d \geq 4$ then

$$q^{\frac{g-d}{2}} - (q+1) \geq q^2 - q - 1 > 0.$$

(B) If $q \geq 3$ (and $g - d \geq 3$) then
$$q^{\frac{g-d}{2}} - (q+1) \geq q^{\frac{3}{2}} - (q+1) > 0.$$
In any cases $K_{19} \leq K_{17}$. Hence
$$K_{18} \leq K_{19} \leq K_{17}.$$

(iii) If
$$g \geq \frac{1}{2} q^{\frac{g-d-1}{2}},$$
then $K_{17} \leq K_{18}$, $K_{19} \leq K_{18}$ and
$$\frac{K_{17}}{K_{19}} = \frac{q^{\frac{g-d-1}{2}} (\sqrt{q}+1)^2}{2(g+1)(q+1)}.$$
Let us set
$$g = \frac{1}{2} q^{\frac{g-d-1}{2}} + a$$
where $a \geq 0$. Then the sign of $K_{17} - K_{19}$ is the sign of
$$2(g-a)(\sqrt{q}+1)^2 - 2(g+1)(q+1),$$
namely the sign of
$$2g\sqrt{q} - (q+1) - a(\sqrt{q}+1)^2.$$
Remark that $2g\sqrt{q} - (q+1)$ is ≥ 0. Hence:

(A) if
$$a \leq \frac{2g\sqrt{q} - (q+1)}{(\sqrt{q}+1)^2}$$
then $K_{19} \leq K_{17}$ and
$$K_{19} \leq K_{17} \leq K_{18}.$$

(B) else if
$$a \geq \frac{2g\sqrt{q} - (q+1)}{(\sqrt{q}+1)^2}$$
then $K_{19} \geq K_{17}$ and
$$K_{17} \leq K_{19} \leq K_{18}.$$

A.2. Examples. Let us give some examples where we compare bounds K_{17}, K_{18}, and K_{19}:

(1) g and $d < g - 1$ are fixed, q large enough, then Bound K_{18} is the best of the three bounds: we are in cases (2) (a) (ii) or (3) (b) (ii);

(2) g is fixed and $d = g - 1$, for any q, Bound K_{17} is the best one: we are in case (1);

(3) q and d are fixed, g large enough, then Bound K_{19} is the best one: we are in the case (3) (b) (i);

(4) q fixed, g and d large:

(a) $d = g - c$ where c is a constant, then Bound K_{17} is the best one; for $c = 1$ we are in the case (1), for $c = 2$ we are in the case (2) (b) and for $c < 2$, we are in the case (3) (a) or in the case (3) (b) (iii) (B);

(b) $d = g(1-\epsilon)$ where ϵ is a constant, then Bound K_{19} is the best one; we are in the case (3) (b) (i);

(c) $d = g - \alpha \log(g) - 1$, then

$$\frac{g-d-1}{2} = \frac{\alpha}{2}\log(g) \quad \text{and} \quad q^{\frac{g-d-1}{2}} = g^{\frac{\alpha}{2}\log(q)}.$$

(i) if $\alpha > \frac{2}{\log(q)}$ then we are in the case (3) (b) (i), hence the best bound is (19).

(ii) if $\alpha \leq \frac{2}{\log(q)}$ then we are in the case (3) (b) (iii) (B), hence the best bound is (17).

The following example is an example of the case (3) (b) (iii) (A). Set $q = 4$, $g = 520$ and $d = 509$. Hence $(g-d-1)/2) = 5$. Then

$$\frac{1}{2}q^{\frac{g-d-1}{2}} = 512 < g.$$

We can compute

$$a = g - \frac{1}{2}q^{\frac{g-d-1}{2}} = 8.$$

Now

$$\frac{2g\sqrt{q}-(q+1)}{(\sqrt{q}+1)^2} = \frac{2075}{9}.$$

Hence

$$a \leq \frac{2g\sqrt{q}-(q+1)}{(\sqrt{q}+1)^2}.$$

Here the best bound is K_{19}.

References

[1] E. Arbarello, M. Cornalba, P. A. Griffiths, and J. Harris, *Geometry of algebraic curves. Vol. I*, Grundlehren der Mathematischen Wissenschaften [Fundamental Principles of Mathematical Sciences], vol. 267, Springer-Verlag, New York, 1985, DOI 10.1007/978-1-4757-5323-3. MR770932

[2] S. Ballet, C. Ritzenthaler, and R. Rolland, *On the existence of dimension zero divisors in algebraic function fields defined over* $\mathbb{F}_q$, Acta Arith. **143** (2010), no. 4, 377–392, DOI 10.4064/aa143-4-4. MR2652586

[3] Stéphane Ballet and Robert Rolland, *Lower bounds on the class number of algebraic function fields defined over any finite field* (English, with English and French summaries), J. Théor. Nombres Bordeaux **24** (2012), no. 3, 505–540. MR3010627

[4] S. Ballet, R. Rolland, and S. Tutdere, *Lower bounds on the number of rational points of Jacobians over finite fields and application to algebraic function fields in towers* (English, with English and Russian summaries), Mosc. Math. J. **15** (2015), no. 3, 425–433, 604, DOI 10.17323/1609-4514-2015-15-3-425-433. MR3427433

[5] Ignacio Cascudo, Ronald Cramer, and Chaoping Xing, *Torsion limits and Riemann-Roch systems for function fields and applications*, IEEE Trans. Inform. Theory **60** (2014), no. 7, 3871–3888, DOI 10.1109/TIT.2014.2314099. MR3225937

[6] Sudhir R. Ghorpade and Gilles Lachaud, *Étale cohomology, Lefschetz theorems and number of points of singular varieties over finite fields*, Mosc. Math. J. **2** (2002), no. 3, 589–631, DOI 10.17323/1609-4514-2002-2-3-589-631. Dedicated to Yuri I. Manin on the occasion of his 65th birthday. MR1988974

[7] Takehiro Hasegawa, *A note on optimal towers over finite fields*, Tokyo J. Math. **30** (2007), no. 2, 477–487, DOI 10.3836/tjm/1202136690. MR2376523

[8] Gilles Lachaud and Mireille Martin-Deschamps, *Nombre de points des jacobiennes sur un corps fini* (French), Acta Arith. **56** (1990), no. 4, 329–340, DOI 10.4064/aa-56-4-329-340. MR1096346

[9] Philippe Lebacque and Alexey Zykin, *On the number of rational points of Jacobians over finite fields*, Acta Arith. **169** (2015), no. 4, 373–384, DOI 10.4064/aa169-4-5. MR3371766

[10] J. S. Milne, *Jacobian varieties*, Arithmetic geometry (Storrs, Conn., 1984), Springer, New York, 1986, pp. 167–212. MR861976

[11] Michael Tsfasman, *Some remarks on the asymptotic number of points*, Lecture Notes in Mathematics **1518** (2006), pp. 178–192.

[12] M. A. Tsfasman and S. G. Vlăduţ, *Asymptotic properties of zeta-functions*, J. Math. Sci. (New York) **84** (1997), no. 5, 1445–1467, DOI 10.1007/BF02399198. Algebraic geometry, 7. MR1465522

[13] Serguei Vlăduţ, *An exhaustion bound for algebraic-geometric modular codes*, Problems of Information Transmission **23** (1987), no. 1, 22–34.

[14] André Weil, *Sur les courbes algébriques et les variétés qui s'en déduisent* (French), Actualités Sci. Ind., no. 1041 = Publ. Inst. Math. Univ. Strasbourg **7** (1945), Hermann et Cie., Paris, 1948. MR0027151

[15] André Weil, *Basic number theory*, Die Grundlehren der mathematischen Wissenschaften, Band 144, Springer-Verlag New York, Inc., New York, 1967. MR0234930

Stéphane Ballet, Aix Marseille Univ, CNRS, Centrale Marseille, I2M, Marseille, France, Case 907, 13288 MARSEILLE Cedex 9, France

Email address: `ballet@univ-amu.fr`

Gilles Lachaud, Aix Marseille Univ, CNRS, Centrale Marseille, I2M, Marseille, France, Case 907, 13288 MARSEILLE Cedex 9, France.

Robert Rolland, Aix Marseille Univ, CNRS, Centrale Marseille, I2M, Marseille, France, Case 907, 13288 MARSEILLE Cedex 9, France

Email address: `robert.rolland@acrypta.fr`

Contemporary Mathematics
Volume **770**, 2021
https://doi.org/10.1090/conm/770/15430

The absolute discriminant of the endomorphism ring of most reductions of a non-CM elliptic curve is close to maximal

Alina Carmen Cojocaru and Matthew Fitzpatrick

ABSTRACT. Let $E/\mathbb{Q}$ be a non-CM elliptic curve. Assuming GRH, we prove that, for a set of primes p of density 1, the absolute discriminant of the $\mathbb{F}_p$-endomorphism ring of the reduction of E modulo p is close to maximal.

1. Introduction

Let $E/\mathbb{Q}$ be an elliptic curve defined over the field of rational numbers, of conductor N_E, and let $p \nmid N_E$ be a prime of good reduction for E. We denote by $E_p/\mathbb{F}_p$ the reduction of E modulo p and we recall that it is an elliptic curve defined over the finite field $\mathbb{F}_p$ with p elements, with the property that $|E_p(\mathbb{F}_p)| = p + 1 - a_p$ for some integer a_p satisfying $|a_p| < 2\sqrt{p}$. Consequently, the polynomial $X^2 - a_pX + p$ has two complex conjugate roots, π_p and $\overline{\pi}_p$, satisfying $|\pi_p| = \sqrt{p}$. Upon identifying any one of these roots, say π_p, with the p-th power Frobenius endomorphism of $E_p/\mathbb{F}_p$, we obtain the embeddings of imaginary quadratic orders $\mathbb{Z}[\pi_p] \leq \mathrm{End}_{\mathbb{F}_p}(E_p) \leq \mathcal{O}_{\mathbb{Q}(\pi_p)}$ in the field $\mathbb{Q}(\pi_p)$, with $\mathcal{O}_{\mathbb{Q}(\pi_p)}$ denoting the maximal order of $\mathbb{Q}(\pi_p)$. Focusing on the discriminants of these orders, we obtain the relation

$$a_p^2 - 4p = b_p^2 \Delta_p, \tag{1}$$

where Δ_p denotes the discriminant of $\mathrm{End}_{\mathbb{F}_p}(E_p)$ and b_p denotes the unique positive integer satisfying the $\mathbb{Z}$-module isomorphism $\mathrm{End}_{\mathbb{F}_p}(E_p)/\mathbb{Z}[\pi_p] \simeq \mathbb{Z}/b_p\mathbb{Z}$. If $p \geq 5$ is supersingular, then $\Delta_p \in \{-p, -4p\}$, while if p is ordinary and $\mathrm{End}_{\overline{\mathbb{Q}}}(E) \not\simeq \mathbb{Z}$, then Δ_p equals the discriminant of the imaginary quadratic order $\mathrm{End}_{\overline{\mathbb{Q}}}(E)$.

The goal of this article is to focus on the setting p ordinary and $\mathrm{End}_{\overline{\mathbb{Q}}}(E) \simeq \mathbb{Z}$ and to investigate the growth of the absolute discriminant $|\Delta_p|$ as a function of p, in particular in relation to the upper bound $4p - a_p^2$ arising from (1). In this setting, it was shown in [**14**] that $|\Delta_p|$ does indeed grow with p: there exists a positive constant $c(E)$ such that, for any prime $p \nmid N_E$,

$$|\Delta_p| \geq c(E)\frac{(\log p)^2}{(\log\log p)^4}.$$

2020 *Mathematics Subject Classification.* Primary 11G05, 11G20,11N05; Secondary 11N36, 11N37, 11N56.

Key words and phrases. Elliptic curves, endomorphism rings, distribution of primes, sieve methods.

The first author was partially supported by a Collaboration Grant for Mathematicians from the Simons Foundation under Award No. 318454.

Moreover, it was shown in [**14**] under the assumption of the Generalized Riemann Hypothesis (GRH for short) that there exists a positive constant $c'(E)$ and there exist infinitely many primes p such that

$$|\Delta_p| \leq c'(E) p^{\frac{2}{3}} \log p.$$

Under similar hypotheses, we will prove that, in fact, the growth of $|\Delta_p|$ is very close to the growth of $4p - a_p^2$ for most primes:

THEOREM 1. *Let $E/\mathbb{Q}$ be an elliptic curve of conductor N_E with* $\operatorname{End}_{\overline{\mathbb{Q}}}(E) \simeq \mathbb{Z}$*. Assume that GRH holds for the division fields of E. Then, for any function $f : (0, \infty) \longrightarrow (0, \infty)$ with $\lim_{x\to\infty} f(x) = \infty$,*

$$\#\left\{p \leq x : p \nmid N_E, |\Delta_p| \geq \frac{4p - a_p^2}{f(p)}\right\} \sim \pi(x), \tag{2}$$

where $\pi(x)$ denotes the number of primes up to x.

The growth of $|\Delta_p|$ has also been investigated in other settings, including that of arbitrary elliptic curves over finite fields – see [**13**], [**16**], [**17**] – and that of finite Drinfeld modules – see [**5**].

Regarding Theorem 1, the proximity of $|a_p|$ to $\left[2\sqrt{p}\right]$ was studied in several papers by K. James and his co-authors, such as [**11**] and [**8**] (see also the recent follow-up [**6**]). In [**11**], it is conjectured that, when $\operatorname{End}_{\overline{\mathbb{Q}}}(E) \simeq \mathbb{Z}$, the number of primes $p \leq x$ with $|a_p| = \left[2\sqrt{p}\right]$, called extremal primes, is asymptotically equal to $C(E)\frac{x^{\frac{1}{4}}}{\log x}$ for some constant $C(E)$; in [**8**], it is proved that this conjecture holds on average over two-parameter families of elliptic curves $E/\mathbb{Q}$ (the majority of which have a trivial endomorphism ring $\operatorname{End}_{\overline{\mathbb{Q}}}(E)$). Thus extremal primes are not expected to contribute to the left hand side of (2).

The proof of Theorem 1 relies on the intimate connection between the integer b_p and the discriminant Δ_p provided by equation (1), as well as on a characterization criterion of the divisors of b_p through splitting conditions on p in certain subfields of the division fields of E. Thanks to these connections, we approach the study of the growth of $|\Delta_p|$ as a potential application of the Chebotarev Density Theorem in an infinite family of number fields. As such, the assumption of GRH facilitates best possible error terms. Even under this assumption, the accumulation of all occurring error terms is overbearing. This we circumvent by resorting to an application of the Square Sieve, which, itself, incorporates another application of the Chebotarev Density Theorem.

Notation. In what follows, we use the standard o, O, $\ll$, $\asymp$, and $\sim$ notation: given suitably defined real functions h_1, h_2, we say that $h_1 = \mathrm{o}(h_2)$ if $\lim_{x\to\infty} \frac{h_1(x)}{h_2(x)} = 0$; we say that $h_1 = \mathrm{O}(h_2)$ or $h_1 \ll h_2$ if h_2 is positive valued and there exists a positive constant C such that $|h_1(x)| \leq C h_2(x)$ for all x in the domain of h_1 and h_2; we say that $h_1 \asymp h_2$ if h_1, h_2 are positive valued and $h_1 \ll h_2 \ll h_1$; we say that $h_1 = \mathrm{O}_D(h_2)$ or $h_1 \ll_D h_2$ if $h_1 = \mathrm{O}(h_2)$ and the implied O-constant C depends on priorly given data D; similarly, we say that $h_1 \asymp_D h_2$ if the implied constant in either one of the $\ll$-bounds $h_1 \ll h_2 \ll h_1$ depends on priorly given data D; we say that $h_1 \sim h_2$ if $\lim_{x\to\infty} \frac{h_1(x)}{h_2(x)} = 1$.

2. Proof of the main theorem

Let $E/\mathbb{Q}$ be an elliptic curve of conductor N_E, with $\mathrm{End}_{\overline{\mathbb{Q}}}(E) \simeq \mathbb{Z}$. Let $f : (0, \infty) \longrightarrow (0, \infty)$ be a function satisfying $\lim_{x \to \infty} f(x) = \infty$. Without loss of generality, we may assume that $f(x)$ is increasing, for we may replace $f(x)$ with $\sup\{f(z) : z \leq x\}$. With notation as in Section 1, we observe that, thanks to (1), in order to prove (2) it is enough to prove

$$\#\left\{p \leq x : p \nmid N_E, b_p > \sqrt{f(p)}\right\} = \mathrm{o}(\pi(x)). \tag{3}$$

This we will do by exploring the divisibility properties of b_p.

As usual, for a positive integer n, we denote by $E[n]$ the group of $\overline{\mathbb{Q}}$-rational points of E of order dividing n and by $\mathbb{Q}(E[n])$ the finite, Galois extension of $\mathbb{Q}$ generated by the x and y coordinates of the points of $E[n]$. We view the Galois group $\mathrm{Gal}(\mathbb{Q}(E[n])/\mathbb{Q})$ as a subgroup of $\mathrm{GL}_2(\mathbb{Z}/n\mathbb{Z})$ under the residual modulo n Galois representation of E. With this notation, the main result of [**7**] specialized to elliptic curves over $\mathbb{Q}$ states that, for any prime $p \nmid nN_E$, the reduction modulo n of the integral matrix

$$\begin{pmatrix} \frac{a_p + b_p\delta_p}{2} & b_p \\ \frac{b_p(\Delta_p - \delta_p)}{4} & \frac{a_p - b_p\delta_p}{2} \end{pmatrix},$$

with $\delta_p := 0, 1$ according to whether $\Delta_p \equiv 0, 1 (\mathrm{mod}\, 4)$, gives a representative of the conjugacy class defined by the Artin symbol $\left(\frac{\mathbb{Q}(E[n])/\mathbb{Q}}{p}\right)$ in $\mathrm{Gal}(\mathbb{Q}(E[n])/\mathbb{Q})$. Consequently, upon defining

$$J_n := \{z \in \mathbb{Q}(E[n]) : \sigma(z) = z \quad \forall \sigma \in \mathrm{Gal}(\mathbb{Q}(E[n])/\mathbb{Q}) \text{ a scalar element}\},$$

we obtain the criterion

$$n \mid b_p \;\Leftrightarrow\; p \text{ splits completely in } J_n/\mathbb{Q}. \tag{4}$$

For each prime p, there are unique positive integers r_p and m_p, with m_p square-free, such that

$$b_p^2 |\Delta_p| = r_p^2 m_p.$$

Observe that we must have $b_p \mid r_p$, which gives $b_p \leq r_p$. Recalling (1), observe that

$$4p - a_p^2 = r_p^2 m_p, \tag{5}$$

which gives $r_p < 2\sqrt{p}$ and

$$m_p = \frac{4p - a_p^2}{r_p^2} \leq \frac{4p}{b_p^2}. \tag{6}$$

Furthermore, observe that the divisibility $n \mid b_p$ implies that $n \mid r_p$ and, in particular, that $n \leq r_p$.

Now let us proceed to bounding from above the left hand side of (3). We fix an arbitrary parameter $z = z(x)$ satisfying $0 < z < x$ and define

$$g(z) := \inf\{f(p) : z < p \leq x\}.$$

Note that $f(z) \leq g(z)$. We have the bounds

$$
\begin{aligned}
\#\left\{p \leq x : p \nmid N_E, b_p > \sqrt{f(p)}\right\} &\leq \pi(z) + \#\left\{z < p \leq x : p \nmid N_E, b_p > \sqrt{f(p)}\right\} \\
&\leq \pi(z) + \#\left\{z < p \leq x : p \nmid N_E, b_p > \sqrt{g(z)}\right\} \\
&\leq \pi(z) + \sum_{\sqrt{g(z)} < n \leq 2\sqrt{x}} \#\left\{p \leq x : p \nmid N_E, b_p = n\right\} \\
&= \pi(z) + S\left(\sqrt{g(z)}, 2\sqrt{x}\right), \qquad (7)
\end{aligned}
$$

where, for any $u < v \leq 2\sqrt{x}$,

$$S(u,v) := \sum_{u<n\leq v} \#\left\{p \leq x : p \nmid N_E, b_p \equiv 0 (\operatorname{mod} n)\right\}.$$

To estimate $S\left(\sqrt{g(z)}, 2\sqrt{x}\right)$, it is natural to use criterion (4) and employ an effective version of the Chebotarev Density Theorem, such as the one in [**12**]. However, this strategy leads to limitations on the range of n – indeed, looking at (8) below, we see that an application of the effective version of the Chebotarev Density Theorem to the entire sum $S\left(\sqrt{g(z)}, 2\sqrt{x}\right)$ gives rise to an error term of $\mathrm{O}_E(x \log x)$. As such, we adjust our strategy as follows: we fix an arbitrary parameter $y = y(x)$ with $\sqrt{g(z)} \leq y \leq 2\sqrt{x}$ and analyze each of the sums $S\left(\sqrt{g(z)}, y\right)$ and $S\left(y, 2\sqrt{x}\right)$ separately, using criterion (4) for the first sum and using observation (5) for the second sum.

Specifically, rewriting $S\left(\sqrt{g(z)}, y\right)$ via (4) and using the effective Chebotarev Density Theorem, conditional upon GRH, as in [**2**, Prop. 4.2, p. 523], we obtain that

$$
\begin{aligned}
S\left(\sqrt{g(z)}, y\right) &= \sum_{\sqrt{g(z)}<n\leq y} \frac{1}{[J_n : \mathbb{Q}]}\pi(x) + \mathrm{O}\left(\sum_{\sqrt{g(z)}<n\leq y} x^{\frac{1}{2}} \log(N_E n x)\right) \\
&= \sum_{\sqrt{g(z)}<n\leq y} \frac{1}{[J_n : \mathbb{Q}]}\pi(x) + \mathrm{O}_E\left(y x^{\frac{1}{2}} \log(yx)\right) \\
&= \sum_{\sqrt{g(z)}<n\leq y} \frac{1}{[J_n : \mathbb{Q}]}\pi(x) + \mathrm{O}_E\left(y x^{\frac{1}{2}} \log x\right), \qquad (8)
\end{aligned}
$$

where in the last line we also used that $y < 2\sqrt{x}$.

We estimate the first term on the right hand side of the above equation by using that, thanks to Serre's Open Image Theorem [**15**], the assumption $\operatorname{End}_{\overline{\mathbb{Q}}}(E) \simeq \mathbb{Z}$ gives

$$n^3 \ll_E [J_n : \mathbb{Q}] \ll n^3. \qquad (9)$$

The upper bound is an immediate consequence of the inequality

$$[J_n : \mathbb{Q}] \leq |\mathrm{PGL}_2(\mathbb{Z}/n\mathbb{Z})|,$$

while the lower bound requires a more detailed analysis, which is presented in [**4**].

Together with Chebyshev's Theorem, the lower bound in (9) gives that

$$\begin{aligned} S\left(\sqrt{g(z)}, y\right) &\ll_E \sum_{\sqrt{g(z)}<n\leq y} \frac{\pi(x)}{n^3} + yx^{\frac{1}{2}} \log x \\ &\ll_E \pi(x) \int_{\sqrt{g(z)}-1}^{y} \frac{1}{t^3}\, dt + yx^{\frac{1}{2}} \log x \\ &\ll_E \frac{x}{g(z) \log x} + yx^{\frac{1}{2}} \log x. \end{aligned} \tag{10}$$

To estimate $S\left(y, 2\sqrt{x}\right)$, we use (5) - (6) and obtain that

$$\begin{aligned} S\left(y, 2\sqrt{x}\right) &= \sum_{y<n\leq 2\sqrt{x}} \#\left\{p \leq x : p \nmid N_E, b_p \equiv 0 (\operatorname{mod} n)\right\} \\ &\leq \sum_{y<n\leq 2\sqrt{x}} \#\left\{p \leq x : p \nmid N_E, r_p \equiv 0 (\operatorname{mod} n)\right\} \\ &= \sum_{y<n\leq 2\sqrt{x}} \sum_{\substack{n<r\leq 2\sqrt{x} \\ n|r}} \#\left\{p \leq x : p \nmid N_E, r_p = r\right\} \\ &\leq \sum_{y<r\leq 2\sqrt{x}} \tau(r) \#\left\{p \leq x : p \nmid N_E, r_p = r\right\} \\ &= \sum_{y<r\leq 2\sqrt{x}} \tau(r) \#\left\{p \leq x : p \nmid N_E, 4p - a_p^2 = r^2 m_p\right\} \\ &\leq \sum_{\substack{m\leq \frac{4x}{y^2} \\ m \text{ squarefree}}} \sum_{y<r\leq 2\sqrt{x}} \tau(r) \#\left\{p \leq x : p \nmid N_E, 4p - a_p^2 = r^2 m\right\} \\ &= \sum_{\substack{m\leq \frac{4x}{y^2} \\ m \text{ squarefree}}} \sum_{y<r\leq 2\sqrt{x}} \tau(r) \#\left\{p \leq x : p \nmid N_E, m(4p - a_p^2) = (mr)^2\right\}, \end{aligned}$$

where $\tau(\cdot)$ is the divisor function. Using the standard bound $\tau(r) \ll_\varepsilon r^\varepsilon$ for any $\varepsilon > 0$ and observing that, for each squarefree m and each prime p, there is at most one $y < r \leq 2\sqrt{x}$ such that $4p - a_p^2 = r^2 m$, we deduce that

$$S\left(y, 2\sqrt{x}\right) \ll_\varepsilon x^\varepsilon \sum_{\substack{m\leq \frac{4x}{y^2} \\ m \text{ squarefree}}} S_m(x), \tag{11}$$

where

$$S_m(x) := \#\left\{p \leq x : p \nmid N_E, m\left(4p - a_p^2\right) \text{ is a square}\right\}.$$

The sum of the counting functions $S_m(x)$ can be estimated as an indirect application of the conditional effective Chebotarev Density Theorem of [**12**] via the Square Sieve of [**9**] (see also the version of this sieve in [**3**]), by relying once again on GRH and on the assumption $\operatorname{End}_{\overline{\mathbb{Q}}}(E) \simeq \mathbb{Z}$. Such an application was pursued in [**2**], [**3**], [**1**], and [**16**]. While any of the estimates proved in these papers suffices to conclude the proof of our theorem, we use the best among all of them, which is the one of [**16**, Lem. 3.2, p. 260]; its proof relies on [**1**, Thm. 3, p. 4] and [**10**, Thm.

1, p. 237]. We deduce that

$$\sum_{\substack{m \le \frac{4x}{y^2} \\ m \text{ squarefree}}} S_m(x) \ll_\varepsilon \frac{x^{\frac{13}{7}+\varepsilon}}{y^{\frac{13}{7}}}. \tag{12}$$

Next, by putting together (7), (10), (11), and (12), we obtain that

$$\#\left\{p \le x : p \nmid N_E, b_p > \sqrt{f(p)}\right\} \ll_{E,\varepsilon} \frac{z}{\log z} + \frac{x}{g(z)\log x} + yx^{\frac{1}{2}}\log x + \frac{x^{\frac{13}{7}+\varepsilon}}{y^{\frac{13}{7}}}.$$

The trivial endomorphism assumption is reflected in each of the second, third and fourth terms on the right hand side of the above estimate, while the GRH assumption is reflected in each of the third and fourth terms.

Finally, we choose $y \asymp x^{\frac{19}{40}}$ and $z \asymp x^{\frac{1}{2}}$, we recall that $f(z) \le g(z)$, and we use the assumption that f is increasing. We deduce that

$$\#\left\{p \le x : p \nmid N_E, b_p > \sqrt{f(p)}\right\} \ll_{E,\varepsilon} \frac{x^{\frac{1}{2}}}{\log x} + \frac{x}{f\left(x^{\frac{1}{2}}\right)\log x} + x^{\frac{39}{40}+\varepsilon}.$$

Note that, since $\lim_{x\to\infty} f(x) = \infty$, the right hand side of the above estimate is $o(\pi(x))$. This completes the proof.

Acknowledgments

We thank Nathan Jones for helpful comments on a previous draft of this manuscript. We thank the anonymous referee for suggestions that added to the clarity of the paper.

References

[1] A. C. Cojocaru and C. David, *Frobenius fields for elliptic curves*, Amer. J. Math. **130** (2008), no. 6, 1535–1560, DOI 10.1353/ajm.0.0028. MR2464027

[2] A. C. Cojocaru and W. Duke, *Reductions of an elliptic curve and their Tate-Shafarevich groups*, Math. Ann. **329** (2004), no. 3, 513–534, DOI 10.1007/s00208-004-0517-2. MR2127988

[3] A. C. Cojocaru, E. Fouvry, and M. R. Murty, *The square sieve and the Lang-Trotter conjecture*, Canad. J. Math. **57** (2005), no. 6, 1155–1177, DOI 10.4153/CJM-2005-045-7. MR2178556

[4] A. C. Cojocaru and N. Jones, *Degree bounds for projective division fields associated to elliptic modules with a trivial endomorphism ring*, to appear in *Journal de Théorie des Nombres de Bordeaux*; `https://arxiv.org/abs/2002.08411`

[5] A. C. Cojocaru and M. Papikian, *The growth of the discriminant of the endomorphism ring of the reduction of a rank 2 generic Drinfeld module*, to appear in *Journal of Number Theory*; `https://arxiv.org/abs/2002.09582`

[6] C. David, A. Gafni, A. Malik, N. Prabhu, and C. L. Turnage-Butterbaugh, *Extremal primes for elliptic curves without complex multiplication*, Proc. Amer. Math. Soc. **148** (2020), no. 3, 929–943, DOI 10.1090/proc/14748. MR4055924

[7] W. Duke and Á. Tóth, *The splitting of primes in division fields of elliptic curves*, Experiment. Math. **11** (2002), no. 4, 555–565 (2003). MR1969646

[8] L. Giberson and K. James, *An average asymptotic for the number of extremal primes of elliptic curves*, Acta Arith. **183** (2018), no. 2, 145–165, DOI 10.4064/aa170406-30-1. MR3798785

[9] D. R. Heath-Brown, *The square sieve and consecutive square-free numbers*, Math. Ann. **266** (1984), no. 3, 251–259, DOI 10.1007/BF01475576. MR730168

[10] D. R. Heath-Brown, *A mean value estimate for real character sums*, Acta Arith. **72** (1995), no. 3, 235–275, DOI 10.4064/aa-72-3-235-275. MR1347489

[11] K. James, B. Tran, M.-T. Trinh, P. Wertheimer, and D. Zantout, *Extremal primes for elliptic curves*, J. Number Theory **164** (2016), 282–298, DOI 10.1016/j.jnt.2016.01.009. MR3474389

[12] J. C. Lagarias and A. M. Odlyzko, *Effective versions of the Chebotarev density theorem*, Algebraic number fields: L-functions and Galois properties (Proc. Sympos., Univ. Durham, Durham, 1975), Academic Press, London, 1977, pp. 409–464. MR0447191

[13] F. Luca and I. E. Shparlinski, *Discriminants of complex multiplication fields of elliptic curves over finite fields*, Canad. Math. Bull. **50** (2007), no. 3, 409–417, DOI 10.4153/CMB-2007-039-2. MR2344175

[14] R. Schoof, *The exponents of the groups of points on the reductions of an elliptic curve*, Arithmetic algebraic geometry (Texel, 1989), Progr. Math., vol. 89, Birkhäuser Boston, Boston, MA, 1991, pp. 325–335. MR1085266

[15] J.-P. Serre, *Propriétés galoisiennes des points d'ordre fini des courbes elliptiques* (French), Invent. Math. **15** (1972), no. 4, 259–331, DOI 10.1007/BF01405086. MR387283

[16] I. E. Shparlinski, *Tate-Shafarevich groups and Frobenius fields of reductions of elliptic curves*, Q. J. Math. **61** (2010), no. 2, 255–263, DOI 10.1093/qmath/hap001. MR2646088

[17] I. E. Shparlinski, *Small discriminants of complex multiplication fields of elliptic curves over finite fields*, Czechoslovak Math. J. **65(140)** (2015), no. 2, 381–388, DOI 10.1007/s10587-015-0183-4. MR3360434

Alina Carmen Cojocaru, Department of Mathematics, Statistics and Computer Science, University of Illinois at Chicago, 851 S Morgan St, 322 SEO, Chicago, 60607, IL, USA, Alina Carmen Cojocaru, Institute of Mathematics "Simion Stoilow" of the Romanian Academy, 21 Calea Grivitei St, Bucharest, 010702, Sector 1, Romania

Email address: `cojocaru@uic.edu`

Matthew Fitzpatrick, Department of Mathematics, Statistics and Computer Science, University of Illinois at Chicago, 851 S Morgan St, 322 SEO, Chicago, 60607, IL, USA

Email address: `mfitzp6@uic.edu`

Contemporary Mathematics
Volume **770**, 2021
https://doi.org/10.1090/conm/770/15431

Toward good families of codes from towers of surfaces

Alain Couvreur, Philippe Lebacque, and Marc Perret,
with an appendix by Alexander Schmidt

To our late teacher, colleague and friend,
Gilles Lachaud.

ABSTRACT. We introduce in this article a new method to estimate the minimum distance of codes from algebraic surfaces. This lower bound is generic, i.e. can be applied to any surface, and turns out to be "liftable" under finite morphisms, paving the way toward the construction of good codes from towers of surfaces. In the same direction, we establish a criterion for a surface with a fixed finite set of closed points $\mathcal{P}$ to have an infinite tower of ℓ–étale covers in which $\mathcal{P}$ splits totally. We conclude by stating several open problems. In particular, we relate the existence of asymptotically good codes from general type surfaces with a very ample canonical class to the behaviour of their number of rational points with respect to their K^2 and coherent Euler characteristic.

Introduction

In the early 80's, V. D. Goppa proposed a construction of codes from algebraic curves [**21**]. A pleasant feature of these codes from curves is that they benefit from an elementary but rather sharp lower bound for the minimum distance, the so–called *Goppa designed distance*. In addition, an easy lower bound for the dimension can be derived from Riemann-Roch theorem. The very simple nature of these lower bounds for the parameters permits to formulate a concise criterion for a sequence of curves to provide asymptotically good codes : roughly speaking, *the curves should have a large number of rational points compared to their genus.* Ihara [**32**] and independently Tsfasman, Vlăduţ and Zink [**49**] proved that this criterion was satisfied by some modular and Shimura curves. This led to an impressive breakthrough in coding theory : the unexpected existence of sequences of codes beating the asymptotic Gilbert-Varshamov bound.

As suggested by Manin and Vlăduţ [**50**], Goppa's original construction extends to higher dimensional varieties. However, in comparison to the rich literature involving codes from curves, codes from higher dimensional varieties have been subject to very few developments. One of the reasons for this gap is that for varieties of dimension two and higher, divisors and points are objects of different nature, and this difference makes the estimate of the minimum distance much more difficult.

2020 *Mathematics Subject Classification.* Primary 11G25, 11G45, 14G50, 19F05, 94B27.
Funded by ANR grant ANR-15-CE39-0013 "Manta", ANR grant ANR-17-CE40-0012 "Flair" and BFC grant GA CROCOCO.

For this reason, most of the works on codes from surfaces in the literature concern specific subfamilies of surfaces, and estimate the parameters by ad hoc techniques relying on specific arithmetic and geometric properties of the involved surfaces. For instance, codes from quadric and Hermitian surfaces are considered in [**13**, **16**, **17**], toric surface codes are studied in [**27**] and codes from Hirzebruch surfaces in [**42**], codes from rational surfaces obtained by specific blowups of the projective planes are analysed in [**5**, **6**, **12**, **14**] and codes from abelian surfaces are investigated in [**4**, **26**]. Toward an analog of *having a large number of points compared to their genus*, Voloch and Zarzar suggested to seek surfaces with a low Picard number [**51**, **53**], leading to further discoveries of good codes [**9**, **38**]. A significant part of the previously described code constructions appear in the excellent survey [**37**].

On the other hand, a few references propose "generic" lower bounds for the minimum distance, i.e. bounds that could be applied to codes from any surface. The first one is proposed by Aubry in [**3**], resting on a counting argument on the maximum number of points of a curve in a very ample linear system. Next, S. H. Hansen proposed two other approaches to bound the minimum distance from below [**28**]. The first one involves an auxiliary set of curves, while the second one involves the Seshardi constant at the evaluation points. Note that the latter approach is very close to that of Bouganis [**10**], while Bouganis' code construction is not *stricto sensu* an algebraic geometry code from a surface. Unfortunately, it should be emphasized that, even with the knowledge of these bounds, no result is known on the possible asymptotic performance of codes from algebraic surfaces, and the principle of getting good sequences of codes from sequences of curves *whose number of rational points grows quickly compared to their genus* has no known counterpart in the surfaces' setting.

It is worth noting that besides the highly difficult objective of getting new families of codes with excellent asymptotic parameters, the geometry of algebraic surfaces could provide codes with many other interesting features. Indeed, in the last decade, new questions brought by distributed storage problems arose and motivated a focus on codes with *good local properties*. Concerning the use of algebraic geometry codes for constructing locally recoverable codes, one can cite for instance (and the list is very far from being exhaustive) [**7**, **8**, **30**, **36**, **41**, **44**, **48**]. It should be pointed out that algebraic geometry codes from surfaces are of deep interest from this point of view. Indeed, given a code from a surface X, to access to some digit of the code, which corresponds to some point of the surface, one can consider a curve contained in the surface and containing this point and restrict to the code on the curve. This point of view, which is in particular investigated in [**44**], provides another motivation to study codes from surfaces or higher dimensional varieties. Note finally that these interesting local properties of codes on surfaces have been remarked more than 15 years ago by Bouganis [**10**].

Our contribution. This article expects to be a first step toward the construction of asymptotically good families codes from sequences of algebraic surfaces. For this sake, we provide several tools and results that would be helpful to progress in this direction.

In § 1, we propose a new "generic" lower bound for the minimum distance of an algebraic geometry code from a surface. This bound involves a linear system of divisor with a property we called $\mathcal{P}$*–interpolating* (see Definition 1.4). The existence and the construction of $\mathcal{P}$–interpolating divisors could appear as a limitation to handle this result. However, we also prove that for a surface X over the finite field $\mathbf{F}_q$, the linear system associated to $\mathcal{O}_X(q+1)$ is always $\mathcal{P}$–covering. Further we compare this new bound with previous ones in the literature. Finally, we discuss the behaviour of these generic bounds, including ours, under finite morphisms of projective surfaces and show that our bound such as those involving Seshadri constants are easy to "lift" under such a finite map, and hence could be used to study codes from towers of surfaces.

In § 2.1, we propose a class field–like criterion for a surface with a fixed set of rational points $\mathcal{P}$ to have an infinite tower of ℓ–étale covers in which $\mathcal{P}$ splits totally. Next, in § 2.2 we sieve the Kodaiara classification of algebraic surfaces to determine which types of surfaces may have infinite totally split towers of étale covers. Despite we acknowledge that we have not been able to apply the criterion established in § 2.1 in order to provide new families of codes, we however show in § 2.3 that this criterion can be applied on a product of two hyperelliptic curves.

We conclude the article by presenting some open problems in § 3. In particular, we show that the existence of asymptotically good families of codes can be deduced from that of families of general type surfaces with very ample canonical class K whose number N of rational points goes to infinity, whose K^2 and coherent Euler characteristic $\chi(\mathcal{O}_X)$ are proportional to N and whose asymptotic ratio $\chi(\mathcal{O}_X)/K_X$ is the largest possible.

1. Codes from surfaces

1.1. Context and notation.

1.1.1. *Context.* In what follows, X denotes a smooth projective geometrically connected surface over $\mathbf{F}_q$ and $\mathcal{P}$ a non empty set of rational points on X of cardinality n. The surface $\overline{X}$ is defined as $\overline{X} := X \times_{\mathrm{Spec}(\mathbf{F}_q)} \mathrm{Spec}(\overline{\mathbf{F}}_q)$. In addition, G is a divisor on X whose support avoids the elements of $\mathcal{P}$. Our point is to study the parameters of the linear code $C(X, \mathcal{P}, G)$ introduced in [**21, 50**] defined as the image of the map:

$$\mathrm{ev}_{\mathcal{P}} : \left\{ \begin{array}{ccc} \mathrm{H}^0(X,G) & \longrightarrow & \mathbf{F}_q^n \\ f & \longmapsto & (f(P))_{P\in\mathcal{P}} \end{array} \right. ,$$

where elements of $\mathrm{H}^0(X,G)$ are regarded as rational fractions on X.

REMARK 1.1. Actually, the condition "the support of G avoids the elements of $\mathcal{P}$" can be removed. In such a case, one needs to choose at each point P a generator s_P of the stalk of the sheaf $\mathcal{O}(G)$ at P. Then, any global section f can be locally written $f_P s_P$ for some $f_P \in \mathcal{O}_{X,P}$, and one can evaluate f_P at P.

1.1.2. *Codes.* Recall that a *code* is a vector subspace C of $\mathbf{F}_q^n$ for some finite field $\mathbf{F}_q$ and some positive integer n, called the *length* of C. The *dimension* of the code is the dimension of C as a $\mathbf{F}_q$-vector space. The *Hamming weight* $\mathrm{w}_H(\mathbf{x})$ of a vector $\mathbf{x} \in \mathbf{F}_q^n$ is the number of nonzero entries of $\mathbf{x}$, and that the *minimum distance* of a code $C \subseteq \mathbf{F}_q^n$ is the minimum weight of a nonzero vector of C. The objective of this first section is to investigate various manners to estimate the parameters of codes from algebraic surfaces, namely the dimension and the minimum distance.

1.1.3. *Equivalence of divisors and cycles.* In the sequel, we frequently deal with linear and numerical equivalence of divisors. One is denoted by $\sim$ and the second one by $\equiv$. We refer to [**29**, Chapters II.6 & V.1] for the definitions. Recall that linear equivalence entails numerical equivalence. Further we also deal with rational equivalence of cycles which is a generalisation of linear equivalence of divisors and hence will also be denoted by $\sim$. We refer to [**19**] for the definition of rational equivalence.

1.2. The dimension of codes from surfaces. To bound from below the dimension of a code $C(X,\mathcal{P},G)$, the natural tool is Riemann-Roch theorem which asserts that

$$h^0(X,G) - h^1(X,G) + h^2(X,G) = \frac{1}{2} G \cdot (G - K_X) + \chi(\mathcal{O}_X),$$

and hence

$$h^0(X,G) + h^2(X,G) \geqslant \frac{1}{2} G \cdot (G - K_X) + \chi(\mathcal{O}_X).$$

The following lemma is useful in what follows. This criterion can be found in [**29**] and has been previously suggested for applications to codes on surfaces by Bouganis in [**10**].

LEMMA 1.2. *Let H be an ample divisor on X. If $G \cdot H > K_X \cdot H$ then*

$$h^0(X,G) \geqslant \frac{1}{2} G \cdot (G - K_X) + \chi(\mathcal{O}_X).$$

PROOF. From [**29**, Lemma V.1.7], if $G \cdot H > K_X \cdot H$ then $h^2(X,G) = 0$ which yields the proof. □

1.3. A new lower bound for the minimum distance. In this subsection, we present a general manner to bound from below the minimum distance of a code from a surface. This bound can be regarded as the counterpart for codes from surfaces of Goppa's designed distance for codes from curves. Indeed, in the case of curves, an algebraic geometry code of length n associated to a divisor G has minimum distance bounded from below by $n - \deg G$. In the sequel, we prove that on a surface, a code of length n associated to a divisor G has minimum distance bounded from below by $n - \Gamma \cdot G$ for some divisor Γ with a special property. Here, such as in many situations in algebraic geometry, the degree for divisors on a curve is replaced by the intersection product for divisors on surfaces. For this sake, the introduction of an auxiliary divisor Γ is necessary. Our work to follow consists in providing de relevant definition for Γ. Therefore, the statement of this new bound requires first to recall some basic notions on linear systems of divisors.

1.3.1. *Linear systems on a surface.* Recall that a *linear system* of divisors Γ on $\overline{X}$ is a family of positive divisors which are all linearly equivalent, and which is parametrised by a projective space. Given a divisor D on $\overline{X}$, the *complete linear system* denoted as $|D|$ is the set of all positive divisors on $\overline{X}$ which are linearly equivalent to D. This set is parametrised by $\mathbf{P}(\mathrm{H}^0(\overline{X},D))$ and a general linear system is a subset of some complete linear system $|D|$ parametrised by some proper linear subspace of $\mathbf{P}(\mathrm{H}^0(\overline{X},D))$.

1.3.2. *$\mathcal{P}$–interpolating linear systems.*

NOTATION 1.3. Let Γ be a linear system of curves on $\overline{X}$ and let $\overline{Y}$ be a proper closed sub-scheme of $\overline{X}$. Then, we denote by $\Gamma - \overline{Y}$ the maximal linear subsystem of Γ of elements whose base locus contains $\overline{Y}$.

DEFINITION 1.4. Given a surface X and a set of rational points $\mathcal{P}$, a linear system Γ of divisors on $\overline{X}$ is said to be *$\mathcal{P}$–interpolating* if

(i) $\Gamma - \mathcal{P}$ is non empty;
(ii) the base locus of $\Gamma - \mathcal{P}$ has dimension 0.

REMARK 1.5. About condition (ii), obviously, the base locus of $\Gamma - \mathcal{P}$ contains $\mathcal{P}$. In addition it may contain a finite number of other points. In particular, this base locus cannot be empty.

REMARK 1.6. Note that the notion of $\mathcal{P}$–interpolating system such as lower bound for the minimum distance we get in Theorem 1.10 to follow are of **geometric** and not arithmetic nature: the $\mathcal{P}$-interpolating linear system Γ is **not** assumed to be defined over $\mathbf{F}_q$.

The following statement provides an equivalent definition of $\mathcal{P}$–interpolating systems that will be useful in the sequel.

PROPOSITION 1.7. *Conditions* (i) *and* (ii) *of Definition* 1.4 *hold if and only if the following condition holds*

(i') *There exist two sections $\overline{A}, \overline{B}$ of $\Gamma - \mathcal{P}$ such that the supports of $\overline{A}$ and $\overline{B}$ have no common component.*

PROOF. Suppose (i') holds. Then, the base locus of $\Gamma - \mathcal{P}$ is contained in the intersection of the supports of $\overline{A}$ and $\overline{B}$. From Remark 1.5, this intersection contains $\mathcal{P}$, then it is non empty and by (i') has dimension 0.

Conversely, suppose that (i) and (ii) hold. From (i), there exists at least one section $\overline{A}$ of $\Gamma - \mathcal{P}$. Let $\overline{A}_1, \ldots, \overline{A}_s$ be the geometrically irreducible components of its support. Then, by (ii), for any $i \in \{1, \ldots, s\}$, the linear system $\Gamma - \mathcal{P} - \overline{A}_i$ is strictly contained in $\Gamma - \mathcal{P}$. Consequently, consider the projective space $\mathbf{P}^r$ parametrizing $\Gamma - \mathcal{P}$, then the sublinear systems $\Gamma - \mathcal{P} - \overline{A}_i$ correspond to finitely many of proper linear subvarieties of $\mathbf{P}^r(\overline{\mathbf{F}}_q)$ and since $\overline{\mathbf{F}}_q$ is infinite, there exists a point in $\mathbf{P}^r(\overline{\mathbf{F}}_q)$ avoiding them all. Therefore, there exists a section $\overline{B}$ of $\Gamma - \mathcal{P}$ whose support has no common component with that of $\overline{A}$. □

REMARK 1.8. If Γ is a sub–linear system of a linear system Δ, then, Δ is $\mathcal{P}$–interpolating too. On the other hand, for any $\mathcal{P}' \subseteq \mathcal{P}$, any $\mathcal{P}$–interpolating linear system is $\mathcal{P}'$–interpolating.

The following statement is useful in what follows.

PROPOSITION 1.9. *Let Γ be a $\mathcal{P}$–interpolating system, then*

$$\Gamma^2 \geqslant |\mathcal{P}|.$$

PROOF. From Proposition 1.7, there exist $\overline{A}, \overline{B} \in \Gamma - \mathcal{P}$ with no common irreducible component. The points of $\mathcal{P}$ lying at the intersection of $\overline{A}$ and $\overline{B}$, we have thus $\Gamma^2 = \overline{A} \cdot \overline{B} \geqslant |\mathcal{P}|$. □

1.3.3. *A lower bound for the minimum distance.*

THEOREM 1.10. *Let X be a smooth geometrically connected surface over $\mathbf{F}_q$ with a set of rational points $\mathcal{P}$ and G be a divisor on X whose support avoids $\mathcal{P}$. Let Γ be a $\mathcal{P}$–interpolating linear system on X. Then the minimum distance d of $C_L(X, \mathcal{P}, G)$ satisfies*

$$d \geqslant n - \Gamma \cdot G,$$

where $n = |\mathcal{P}|$.

PROOF. Let f be an element of $\mathrm{H}^0(X,G)$ providing a nonzero codeword $\mathbf{c}$. Let D be the positive divisor $D=(f)+G$. Then,

$$n - \mathrm{w}_H(\mathrm{ev}(f)) \leqslant |\mathbf{Supp}(D)\cap \mathcal{P}|,$$

where $\mathbf{Supp}(D)$ denotes the support of the divisor D.

We claim that there exists $E \in \Gamma - \mathcal{P}$ whose support has no common irreducible component with that of D. Indeed, by Definition 1.4(i), $\Gamma - \mathcal{P}$ is nonempty. Therefore it is parametrised by some projective space $\mathbf{P}^\ell$ for $\ell \geqslant 0$. Next, by Definition 1.4(ii), for any $\overline{\mathbf{F}}_q$–irreducible component $\overline{Y}$ of the support of D, the linear system $\Gamma - (\mathcal{P}\cup \overline{Y})$ is distinct from $\Gamma - \mathcal{P}$. Thus, there is a proper linear subvariety $V_{\overline{Y}}$ of $\mathbf{P}^\ell(\overline{\mathbf{F}}_q)$ parametrising $\Gamma - (\mathcal{P}\cup\overline{Y})$. Let $\overline{W}\subseteq \mathbf{P}^\ell(\overline{\mathbf{F}}_q)$ be defined as

$$\overline{W} := \bigcup_{\overline{Y}} V_{\overline{Y}},$$

where $\overline{Y}$ runs over all the $\overline{\mathbf{F}}_q$–irreducible components of the support of D. Then, any element of $\mathbf{P}^\ell(\overline{\mathbf{F}}_q)\smallsetminus \overline{W}$ provides such a divisor E. Consequently,

$$|\mathbf{Supp}(D)\cap\mathcal{P}| \leqslant |\mathbf{Supp}(D)\cap \mathbf{Supp}(E)| \leqslant D\cdot E = G\cdot\Gamma.$$

□

Several examples of applications of this lower bound are given further in § 1.5 and 1.6. Before, let us state a brief summary of our estimates.

1.4. Summary on the parameters of codes for a surface. Our previous results lead to the following statement.

THEOREM 1.11. *Let Γ be a $\mathcal{P}$–interpolating linear system such that $n > \Gamma{\cdot}G$, and assume moreover that there exists an ample divisor H on X such that $G{\cdot}H > K_X{\cdot}H$. Then the code $C(X,\mathcal{P},G)$ has parameters*

$$k \geqslant \frac{1}{2}G\cdot(G-K_X)+\chi(\mathcal{O}_X) \qquad \text{and} \qquad d\geqslant n-\Gamma\cdot G.$$

PROOF. The condition $n>\Gamma{\cdot}G$ entails that the evaluation map is injective and hence asserts that the minimum distance of the code is that of the Riemann-Roch space. □

1.5. A "universal" example of application of our bound. A natural question is : *how to find a $\mathcal{P}$–interpolating system?* The following lemma provides an example that can be applied to any surface with a very ample divisor.

THEOREM 1.12. *Let $\mathcal{L}$ be a very ample sheaf on X, then for any $\mathcal{P}$ contained in $X(\mathbf{F}_q)$, the sheaf $\mathcal{L}^{\otimes(q+1)}$ is $\mathcal{P}$–interpolating and, in the conditions of Theorem* 1.10, *we get*

$$d\geqslant n-(q+1)G\cdot\mathcal{L}.$$

*If in addition there exists $L\in|\mathcal{L}|$ such that $\mathcal{P}$ is contained in the affine chart $X\smallsetminus$ **Supp**(L), then $\mathcal{L}^{\otimes q}$ is $\mathcal{P}$–interpolating and*

$$d\geqslant n-qG\cdot\mathcal{L}.$$

REMARK 1.13. In particular, if X is embedded in $\mathbf{P}^\ell$, then $\mathcal{O}_X(q+1)$ is $\mathcal{P}$–interpolating.

The proof of Theorem 1.12 rests on the two following well–known statements.

LEMMA 1.14. *Let $x_1, \dots, x_\ell$ be a system of coordinates in the affine space $\mathbf{A}^\ell$, then the intersection of the hypersurfaces of equation $x_i^q - x_i$ for $i = 1, \dots, \ell$ equals $\mathbf{A}^\ell(\mathbf{F}_q)$.*

LEMMA 1.15. *Let ℓ be a positive integer and $x_0, \dots, x_\ell$ be a system of homogeneous coordinates in $\mathbf{P}^\ell_{\mathbf{F}_q}$. For any pair (i,j) with $0 \leqslant i < j \leqslant \ell$, let $\mathcal{H}_{ij}$ be the hypersurface of equation $x_i^q x_j - x_i x_j^q$. Then, the variety $\bigcap_{0\leqslant i<j\leqslant \ell} \mathcal{H}_{ij}$ has dimension 0 and is the union of the rational points of $\mathbf{P}^\ell$.*

PROOF OF THEOREM 1.12. Let $\phi : X \hookrightarrow \mathbf{P}^\ell$ with $\ell = h^0(X, \mathcal{L}) - 1$ be the embedding associated to $\mathcal{L}$. Hence $\mathcal{L}$ is isomorphic to $\phi^* \mathcal{O}_{\mathbf{P}^n}(1)$ and we denote by $x_0, \dots, x_\ell$ a system of homogeneous coordinates of $\mathbf{P}^\ell$.

The reduced base locus of $|\mathcal{L}^{\otimes(q+1)}| - \mathcal{P}$ is contained in the intersection of the zero loci of the sections $(x_i^q - x_i x_j^{q-1})x_j$ of $\mathcal{L}^{\otimes(q+1)}$ for $0 \leqslant i < j \leqslant \ell$. From Lemma 1.15, this zero locus equals $X(\mathbf{F}_q)$. Therefore, the sheaf $\mathcal{L}^{\otimes(q+1)}$ is $\mathcal{P}$–interpolating.

For the second situation, we can apply Lemma 1.14 to get the result. □

1.6. Further examples. In § 1.5 we proposed a generic linear system that is $\mathcal{P}$–interpolating for any set $\mathcal{P}$ of rational points. When we have further information on the geometry of the surface it is possible to construct more specific $\mathcal{P}$–interpolating linear systems as suggested in the examples to follow.

1.6.1. *Product of curves.* Let $X = C \times D$ be a product of curves and denote by π_C, π_D the corresponding projections. Assume moreover that $\mathcal{P}$ is a "grid" of rational points, i.e. there is a set $\mathcal{P}_C$ of rational points of C and a set $\mathcal{P}_D$ of rational points of D such that $\mathcal{P} = \mathcal{P}_C \times \mathcal{P}_D$. Note that $\mathcal{P}_C$, $\mathcal{P}_D$ can be regarded as reduced positive divisors respectively of C and D. Assume moreover that these divisors are base point free. For instance, assume that $\mathcal{P}_C, \mathcal{P}_D$ have respective degrees t_C, t_D larger than or equal to $2g_C$ and $2g_D$ where g_C, g_D denote the respective genera of C and D. Then, the linear system

$$\Gamma := |\pi_C^*(\mathcal{P}_C) + \pi_D^*(\mathcal{P}_D)|$$

is $\mathcal{P}$–interpolating. Indeed, by hypothesis, $\mathcal{P}_C$ is base point free and hence, there exists a positive divisor E_C on C (resp. E_D on D) which is linearly equivalent to $\mathcal{P}_C$ (resp $\mathcal{P}_D$) and with disjoint support. Hence, the linear system Γ contains $\pi_C^* E_C + \pi_D^* \mathcal{P}_D$ and $\pi_C^* \mathcal{P}_C + \pi_D^* E_D$. The supports of these divisors have no common component. We deduce from Proposition 1.7 that Γ is $\mathcal{P}$–interpolating.

As a consequence, Γ is numerically equivalent to

$$\Gamma \equiv t_C F_C + t_D F_D,$$

where F_C, F_D denote the respective numerical equivalences of a fibre by π_C and π_D and t_C, t_D denote the respective degrees of the divisors $\mathcal{P}_C, \mathcal{P}_D$ respectively on C and D. Hence, given a divisor G on X, the code $C(X, \mathcal{P}, G)$ has minimum distance

$$d \geqslant n - t_C F_C \cdot G - t_D F_D \cdot G$$

by Theorem 1.10.

EXAMPLE 1.16. On the product of two projective lines, the divisor class group is generated by two classes H, V. Let $G = aH + bV$. For the choice of Γ, one can easily see that $H + V$ is very ample and hence, from Theorem 1.12, the system $|(q+1)(H+V)|$ is $\mathcal{P}$–interpolating. Therefore, we get

$$d \geqslant n - (q+1)(a+b).$$

Actually the exact minimum distance is known to be $n - (q+1)(a+b) + ab$ (see [**13**, Theorem 2.1 & Remark 2.2]).

1.6.2. *Fibred surfaces.* Let $\pi : X \to C$ be a fibred surface over a curve C of genus g. Let $F_1, \dots, F_r$ be the fibres under π of r distinct rational points of C, let $C_1, \dots, C_s$ be s distinct sections of π and consider the set of rational points $\mathcal{P}$ given by

$$\mathcal{P} = (F_1 \cup \dots \cup F_r) \cap (C_1 \cup \dots \cup C_s).$$

Suppose in addition that $r > 2g_C$ where g_C denotes the genus of C and that there exists $C \in |C_1 + \dots + C_s|$ whose support avoids any element of $\mathcal{P}$. Then, the complete linear system $|F_1 + \dots + F_r + C_1 + \dots + C_s|$ is $\mathcal{P}$–interpolating.

Indeed, similarly to the previous case, $F_1 + \dots + F_r$ is the pullback by π of a base point free divisor on C and there exists a positive divisor $F \sim F_1 + \dots + F_r$ whose support does not contain any of the F_i's. Next, the divisors

$$F_1 + \dots + F_r + C \qquad \text{and} \qquad F + C_1 + \dots + C_s$$

have no common component and are both in $|F_1 + \dots + F_r + C_1 + \dots + C_s|$. Thus, according to Proposition 1.7, the linear system Γ is $\mathcal{P}$–interpolating.

1.6.3. *Hirzebruch surfaces.* To get a more explicit example, consider the case of a rational ruled surface, i.e. a Hirzebruch surface Σ_e. Such a surface is ruled, i.e. there is a morphism $\pi_e : \Sigma_e \to \mathbf{P}^1$ with a section whose image in Σ_e is denoted by S_e. This surface has a discrete Picard group generated by S_e and a fibre F_e and

$$F_e^2 = 0, \qquad F_e \cdot S_e = 1 \qquad \text{and} \qquad S_e^2 = -e.$$

The surface can be obtained from $\mathbf{P}^2$ by a sequence of blow up and blow down:

- Σ_1 is the blow up of $\mathbf{P}^2$ at a point P. Fix a line $L \subseteq \mathbf{P}^2$ containing P. We denote by S_1 the exceptional divisor and by F_1 the strict transform of L. We have $S_1^2 = -1$ and $F_1^2 = 0$.
- Σ_{e+1} is obtained from Σ_e as follows. Let P be the point at the intersection of F_e and S_e. Blow up Σ_e at P and denote by E the exceptional divisor and by $\widetilde{S}_e, \widetilde{F}_e$ the respective strict transforms of S, F by the blowup map. Blow down $\widetilde{F}_e$ and set F_{e+1} and S_{e+1} the respective images of E and $\widetilde{S}_e$ by the blow down map. See Figure 1 for an illustration.

Such surfaces are rational, in particular there exists a birational map

$$\psi : \mathbf{P}^2 \overset{\sim}{\dashrightarrow} \Sigma_e \tag{1.1}$$

which induces an isomorphism between the affine chart $\mathbf{P}^2 \smallsetminus L$ and the affine chart $\Sigma_e \smallsetminus (S_e \cup F_e)$.

Consider in $\mathbf{P}^2$ a line L' distinct from L and not containing P. Let Q be the point at the intersection of L and L'. The Zariski closure of the image of $L' \smallsetminus \{Q\}$ by the map ψ of (1.1) is a genus 0 curve C on Σ_e which is another section of π_e which does not meet S_e. Therefore, it satisfies

$$C \cdot S_e = 0 \qquad \text{and} \qquad C \cdot F_e = 1.$$

Thus,

$$C \sim eF_e + S_e. \tag{1.2}$$

Let $(X : Y : Z)$ be a system of homogeneous coordinates in $\mathbf{P}^2$ such that $L = \{X = 0\}$, $P = (0:1:0)$ and $Q = (0:0:1)$. Let $\mathcal{P}_0$ be the set of rational points $\{(1:x:y) \mid x \in A,\ y \in B\}$ for some sets $A \subseteq \mathbf{F}_q$ and $B \subseteq \mathbf{F}_q$ of respective

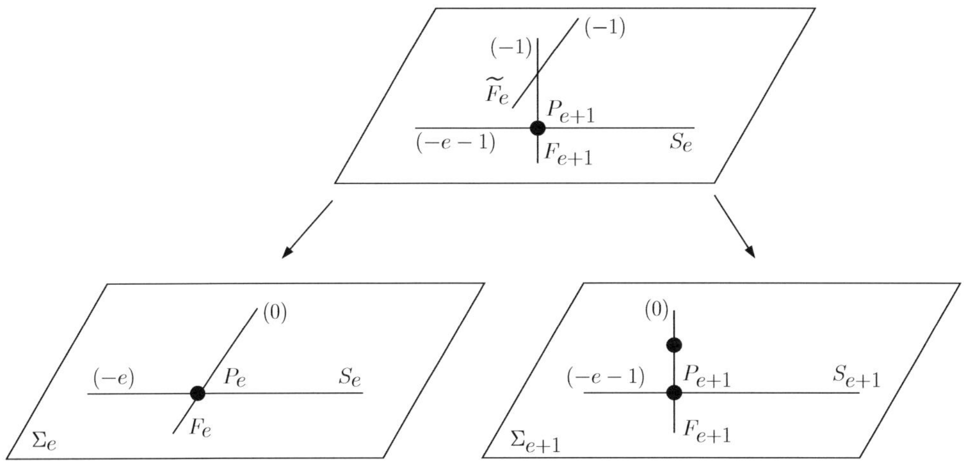

FIGURE 1. From Σ_e to Σ_{e+1}

cardinalities a, b. Finally, let $\mathcal{P}$ be the image of $\mathcal{P}_0$ by ψ. The points in $\mathcal{P}$ and $\mathcal{P}_0$ are represented by filled black dots ● in Figure 2.

Note first that in $\mathbf{P}^2 \setminus L$ the set of points $\mathcal{P}_0$ is contained in the following two unions of parallel lines:

$$\bigcup_{t \in A} \{Y = tX\} \qquad \text{and} \qquad \bigcup_{u \in B} \{X = uZ\}.$$

On the left–hand part of Figure 2, the first family of lines is represented by - - - - - and the second one by -.-.-.-. The Zariski closures of the respective images of these lines by ψ are denoted as V and H and, from (1.2), we have

$$V \sim aC \sim a(eF_e + S_e) \qquad \text{and} \qquad H \sim bF_e.$$

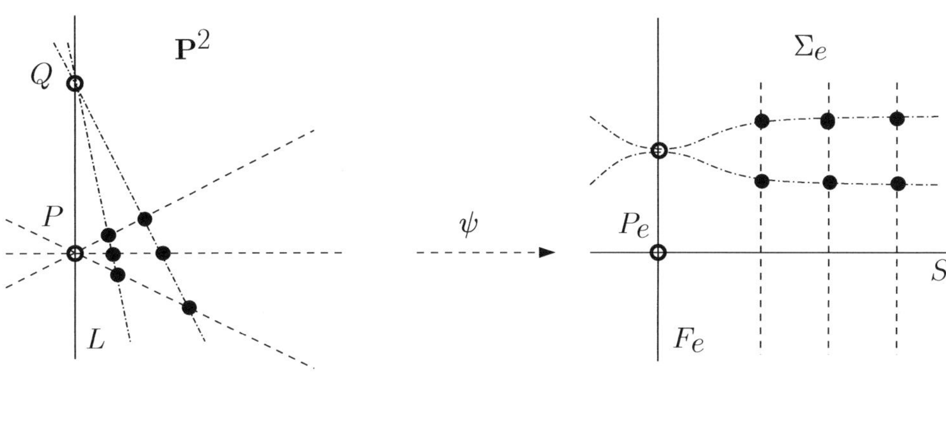

FIGURE 2. Behaviour of the family of lines under ψ

We claim that the linear system $\Gamma := |(a+be)F_e + bS_e|$ is $\mathcal{P}$–interpolating. Indeed, since $|F_e|$ is base point free, then so is $|bF_e|$ and there exists H' in $|bF_e|$ such that $\mathcal{P} \cap \mathbf{Supp}(V') = \varnothing$. Similarly, the divisor $eF_e + S_e$ is very ample [**29**, Corollary 2.18] and hence the corresponding linear system is base point free. Thus there exists $V' \sim V$ whose support avoids any point of $\mathcal{P}$. Finally, the divisors

$$V + F' \qquad \text{and} \qquad V' + F$$

are both in Γ and the intersection of their supports is $\mathcal{P}$. Therefore, Γ is $\mathcal{P}$–interpolating.

Now, consider a divisor $G \sim uF_e + vS_e$ on Σ_e. Using the previous context, we get a code $C(\Sigma_e, \mathcal{P}, G)$ with length $n = ab$ and minimum distance

$$d \geqslant n - (a + be)v - ub + ebv.$$

On the other hand if one wants to take $\mathcal{P} = \Sigma_e(\mathbf{F}_q)$, that is to say evaluating global sections at all the rational points, one can use § 1.5. For this sake, one needs a very ample divisor. According to [**29**, Corollary V.2.18(a)], the divisor $H = (e+1)F_e + S_e$ is very ample. Therefore, one ca take $\Gamma = |(q+1)(e+1)F_e + (q+1)S_e|$. For a divisor $G \sim uF_e + vS_e$ leads to the following lower bound for the minimum distance.

$$\begin{aligned} d &\geqslant \overbrace{(q+1)^2}^{=n} - (q+1)\left((e+1)v + u - ev\right) \\ &\geqslant (q+1)^2 - (q+1)(u+v). \end{aligned}$$

REMARK 1.17. The exact minimum distance and dimension of codes from Hirzebruch surfaces when the evaluation set is the full set of rational points have been computed by Nardi in [**42**]. According to this reference, when $v \geqslant 1$, the genuine minimum distance is $d = q(q - u + 1)$. In particular, it does not depend on v. Here we can estimate the defect of our bound compare to the actual minimum distance :

$$q(q - u + 1) - (q + 1)^2 + (q + 1)(u + v) = u + v - 1 + q(v - 1).$$

In particular, the larger the v, the worst our estimate.

1.7. Previous estimates for the minimum distance in the literature. Let us first list some previous estimates given in the literature, which can roughly speaking be divided into three categories.

1.7.1. *Using the maximum number of points of curves in a linear system: Aubry's bound.* The first category of bounds consists in observing that the minimum distance is related to the maximum number of rational points of an element of the linear system $|G|$, that is to say,

$$d \geqslant n - \max_{C \in |G|} |C(\mathbf{F}_q)|.$$

The problem is then translated into that of getting an upper bound on such a maximum. This is the approach used for instance by Aubry in [**3**] who obtained the following result.

PROPOSITION 1.18 ([**2**, Proposition 3.1(ii)]). *Let D be a very ample divisor on a nonsingular projective surface X. Then the minimum distance of $C(X, \mathcal{P}, D)$ satisfies*

$$d \geqslant n - D^2(q + 1).$$

When $G \sim \mathcal{O}_X(1)$, and taking as $\mathcal{P}$–interpolating system $\Gamma \sim \mathcal{O}_X(q + 1)$ as suggested in § 1.5, our bound yields the very same result as Aubry's bound. The exemples to follow show that our bound can actually be much better than Aubry's one in other situations. Another advantage of Theorem 1.10 is that it does not require the very ampleness of the divisor G.

EXAMPLE 1.19. Consider the case of a surface X together with a very ample sheaf $\mathcal{O}_X(1)$ and G be a divisor such that $G \sim \mathcal{O}_X(d)$ for some positive integer d. Thus, Aubry's result yields

$$d \geqslant n - (q+1)G^2 = n - (q+1)d^2 \deg(X), \tag{1.3}$$

while our bound yields

$$d \geqslant n - G \cdot \mathcal{O}(q+1) = n - d(q+1)\deg(X). \tag{1.4}$$

Thus a d^2 is replaced by a d.

EXAMPLE 1.20. Back to Example 1.16, for any positive integers a, b, the divisor $G = aH + bV$ is very ample and Aubry's bound yields

$$d \geqslant n - 2ab(q+1).$$

Here our "$a+b$" term replaces an "ab" term in the above lower bound.

Caution. In the sequel, we discuss two lower bounds for the minimum distance, that are due to S. H. Hansen. One is presented in § 1.7.2 and rests on the use of a set of auxiliary curves. The second one is introduced in § 1.7.3 and involves the Seshardi constant. To avoid confusion, the first kind of bound will be referred to *Hansen bound (A)*, where (A) stands for *Auxiliary curves* and the second one to *Hansen bound (S)*, where the (S) stands for *Seshadri constant.*

1.7.2. *Using an auxiliary set of curves : Hansen's bound (A).* The second category of estimates consists in using an auxiliary family of irreducible curves $C_1, \ldots, C_s$ such that the curve $C := C_1 \cup \cdots \cup C_s$ contains $\mathcal{P}$. Bounds of this kind appear in [**10**, Theorem 4] and [**28**, §3.2]. For instance, we have the following statement.

PROPOSITION 1.21 (Hansen bound (A), [**28**, Proposition 3.2]). *Let X be a normal projective variety defined over $\mathbf{F}_q$ of dimension at least two. Let $C_1, \ldots, C_a$ be (irreducible) curves on X and $\mathcal{P} = \{P_1, \ldots, P_n\}$ be a set of $\mathbf{F}_q$-rational points of X. Assume that for all $1 \leqslant i \leqslant a$, the number of $\mathbf{F}_q$-rational points on C_i is less than an integer N. Let $\mathcal{L}$ be a line bundle on X over $\mathbf{F}_q$ such that $\mathcal{L} \cdot C_i \geqslant 0$ for all i. Let*

$$\ell := \sup_{s \in \mathrm{H}^0(X, \mathcal{L})} |\{i \mid Z(s) \text{ contains } C_i\}|,$$

where $Z(s)$ denotes the vanishing locus of s. Then the code $C(X, \mathcal{P}, \mathcal{L})$ has length n and minimum distance

$$d \geqslant n - \ell N - \sum_{i=1}^{a} \mathcal{L} \cdot C_i.$$

Moreover, if $\mathcal{L} \cdot C_i = \eta \leqslant N$ for all i, then

$$d \geqslant n - \ell N - (a - \ell)\eta.$$

1.7.3. *Using Seshadri constants, Hansen bound (S).* The third kind of estimate is based on Seshadri constants whose definition is recalled below.

DEFINITION 1.22 (Seshadri constant). Let $\mathcal{L}$ be a line bundle on X and $\mathcal{P}$ be a union of closed points of X. Let $\pi : \mathrm{Bl}_{\mathcal{P}} X \to X$ be the blowup of X at $\mathcal{P}$. Then the local Seshadri constant is defined as

$$\varepsilon(X, \mathcal{P}, \mathcal{L}) := \sup\{\varepsilon \in \mathbf{Q} \mid \pi^*\mathcal{L} - \varepsilon E \text{ is nef}\},$$

where E denotes the exceptional divisor of $\mathrm{Bl}_{\mathcal{P}}(X)$.

PROPOSITION 1.23 (Hansen bounds (S), [**28**, Proposition 3.1]). *Let X be a smooth projective surface defined over $\mathbf{F}_q$. Let $\mathcal{L}$ be a line bundle on X.*

(S1) *Suppose $\mathcal{L}$ is ample with Seshadri constant $\varepsilon(X, \mathcal{P}, \mathcal{L}) \geqslant \varepsilon$. Then the corresponding code has minimum distance*

$$d \geqslant n - \frac{\mathcal{L}^2}{\varepsilon}\cdot$$

(S2) *Let $\mathcal{I}$ be the ideal sheaf of the $\mathbf{F}_q$–rational points $\mathcal{P} = \{P_1, \ldots, P_n\}$. Suppose $\mathcal{L}^{\otimes\xi} \otimes \mathcal{I}$ is generated by global sections (such $\xi \in \mathbf{N}$ exists for instance if $\mathcal{L}$ is ample). Then,*

$$d \geqslant n - \xi\mathcal{L}^2.$$

REMARK 1.24. Actually, in [**28**] the author states the result for codes on an arbitrary variety.

Note that using Proposition 1.21, Hansen [**28**] obtains a better lower bound, namely

$$d \geqslant n - (q+1)(a+b) + ab$$

which turns out to be the actual minimum distance as proved in [**13**, Theorem 2.1 & Remark 2.2].

REMARK 1.25. Actually, Hansen bound (S2) is very close to Aubry's one. In particular, when $\mathcal{L}$ is a very ample line bundle on X, then $\mathcal{L}^{\otimes(q+1)} \otimes \mathcal{I}$ is always generated by its global sections. Indeed, consider the embedding $X \hookrightarrow \mathbf{P}^\ell$ associated to $\mathcal{L}$, using the notation of Lemma 1.15 one can check that the global sections of the form $x_i^q x_j - x_i x_j^q$ generate locally the sheaf $\mathcal{L}^{\otimes(q+1)} \otimes \mathcal{I}$ at any point. Consequently, for the choice $\xi = q+1$, Hansen's bound of Proposition 1.23(2) is nothing but Aubry's bound.

1.8. Further discussion about Seshadri constants. Our Theorem 1.10, which provides a lower for the minimum distance can be related with the Seshadri constant as follows.

THEOREM 1.26. *Let $d^* = n - \Gamma \cdot G$ be the lower bound for the minimum distance given by Theorem* 1.10 *of $C(X, \mathcal{P}, G)$. We have*

$$d^* \leqslant (1 - \varepsilon(X, \mathcal{P}, G))n.$$

PROOF. Let $a \in \mathbf{Q}$ such that $\pi^*G - aE$ is nef. Then, consider the strict transform $\widetilde{\Gamma}$ of Γ by the blowup map $\pi : \mathrm{Bl}_{\mathcal{P}}(X) \to X$. We have

$$(\pi^*G - aE) \cdot \widetilde{\Gamma} \geqslant 0,$$

that is to say,

$$(\pi^*G - aE) \cdot (\pi^*\Gamma - \sum_{p \in \mathcal{P}} \mathrm{mult}_P(\Gamma) E_P) \geqslant 0,$$

where $\mathrm{mult}_P(\Gamma) = \min\{\mathrm{mult}_P(C) \mid C \in \Gamma\}$. This leads to

$$\Gamma \cdot G - a \sum_{P \in \mathcal{P}} \mathrm{mult}_P(\Gamma) \geqslant 0.$$

By definition of a $\mathcal{P}$–interpolating system, $\mathrm{mult}_P(\Gamma) \geqslant 1$ for all $P \in \mathcal{P}$. This entails $\Gamma \cdot G \geqslant an$. This holds for any $a \in \mathbf{Q}$ such that $\pi^*G - aE$ is nef, and hence, we get

$$\Gamma \cdot G \geqslant \varepsilon(X, \mathcal{P}, G) n. \tag{1.5}$$

□

REMARK 1.27. Inequality (1.5) suggest another interesting application of $\mathcal{P}$–interpolating linear systems. They permit to get upper bounds for Seshadri constants.

1.9. Behaviour in towers. Another interest of our approach is that our criterion can be lifted by a finite morphism. Hence it can be used to estimate the asymptotic parameters of codes from towers of surfaces.

PROPOSITION 1.28. *Let $\varphi : Y \to X$ be a finite morphism of smooth projective surfaces. Let $\mathcal{P}$ be a set of rational points of X, and $\mathcal{P}_0$ be a set of rational points of Y such that $\varphi(\mathcal{P}_0) \subseteq \mathcal{P}$. If Γ is a $\mathcal{P}$–interpolating linear system on X, then $\varphi^*\Gamma$ is $\mathcal{P}_0$–interpolating.*

PROOF. Let $H \in \Gamma - \mathcal{P}$, then φ^*H is in $\varphi^*\Gamma - \mathcal{P}_0$ since $\varphi(\mathcal{P}_0) \subseteq \mathcal{P}$. Assume now that $\varphi^*\Gamma - \mathcal{P}_0$ had a base curve and let Y_0 be an irreducible component of this base curve. This entails that $\varphi(Y_0)$ is in the base locus of $\Gamma - \mathcal{P}$. Since by assumption this linear system has no base curve, φ maps Y into a single point. This contradicts the finiteness of φ. □

Note that no previous work in the literature investigated the behaviour of a lower bound for the minimum distance in such a relative case. For this reason, and in order to push further the comparison with other known bounds, we conclude this section by investigating the behaviour of these bounds under morphisms.

1.9.1. *Aubry's bound.* This bound is simple to apply and hence could be used for estimates in towers. However, it requires to have a very ample divisor at each level of the tower. Note that even for a finite morphism, $\pi : Y \to X$, given a very ample divisor G on X, then, from [**23**, Prop 5.1.12], π^*G is ample on Y but not necessarily very ample.

1.9.2. *Hansen bound (A).* Contrarily to our bound, Hansen's bound (A) does not seem to be usable for asymptotics. Even in the case of a single morphism $\pi : Y \to X$. Let $\mathcal{P}_X$ be a set of rational points of X, G be a divisor and set of curves $C_1, \dots, C_a$ as in Proposition 1.21. In addition, suppose there exists a set of rational points $\mathcal{P}_Y$ of Y such that $\pi(\mathcal{P}_Y) = \mathcal{P}_X$. To study the code $C(P, \mathcal{P}_Y, \pi^*G)$ one could consider the set of curves $\pi^*C_1, \dots, \pi^*C_a$. In the lower bound from

Proposition 1.21,

$$\begin{aligned} d &\geqslant |\mathcal{P}_Y| - \ell_Y N_Y - \sum_{i=1}^{a} \pi^* G \cdot \pi^* C_i \\ &\geqslant |\mathcal{P}_Y| - \ell_Y N_Y - (\deg \pi) \sum_{i=1}^{a} G \cdot C_i. \end{aligned}$$

Thus, a part of the lower bound can be deduced from the lower bound for the minimum distance of $C(X, \mathcal{P}, G)$ and $\deg \pi$. Even N_Y can roughly be bounded above by $(\deg \pi)N$. However, there does not seem to be a manner to deduce ℓ_Y from ℓ.

1.9.3. *Hansen bounds (S).* Contrarily to the two previous ones, the bounds given in Proposition 1.23 behave well under finite morphisms, and can be applied to towers of finite morphisms. To prove it, first notice that the criterion of Proposition 1.23(1) rests on an assumption of ampleness. Fortunately, as we already noticed, the pullback of an ample divisor by a finite map between smooth surfaces is ample [**23**, Prop 5.1.12]. Next, to prove the good behaviour of Hansen criteria in the relative case, we use the following statements.

PROPOSITION 1.29. *Let Y be a smooth projective geometrically connected surface over $\mathbf{F}_q$ and $\pi : Y \to X$ be a finite morphism. Then,*

$$\varepsilon(X, \mathcal{P}, G) = \varepsilon(Y, \mathcal{Q}, \pi^* G),$$

where $\mathcal{Q}$ is the set of points of $\overline{Y}$ defined as $\mathcal{Q} := \pi^{-1}(\mathcal{P})$.

PROOF. Consider the following diagram:

$$\begin{array}{ccc} Y & \xleftarrow{p_Y} & \mathrm{Bl}_{\mathcal{Q}}(Y) \\ \big\downarrow{\scriptstyle \pi} & & \big\downarrow{\scriptstyle \widetilde{\pi}} \\ X & \xleftarrow[p_X]{} & \mathrm{Bl}_{\mathcal{P}}(X) \end{array},$$

where $\mathrm{Bl}_{\mathcal{P}}(X)$ (resp. $\mathrm{Bl}_{\mathcal{Q}}(Y)$) denotes the blowup of X at $\mathcal{P}$ (resp. the blowup of Y at $\mathcal{Q}$). In addition, we denote respectively by E_X and E_Y the exceptional divisor of $\mathrm{Bl}_{\mathcal{P}}(X)$ and $\mathrm{Bl}_{\mathcal{Q}}(Y)$. Let $s \in \mathbf{Q}$ be such that $p_X^* G - sE_X$ is nef. That is $s \leqslant \varepsilon(X, \mathcal{P}, G)$. Let C be a curve on $\mathrm{Bl}_{\mathcal{Q}}(Y)$ and suppose that

$$C \cdot (p_Y^* \pi^* G - sE_Y) < 0.$$

By the commutativity of the diagram, we get

$$C \cdot (\widetilde{\pi}^* p_X^* G - s\widetilde{\pi}^* E_X) < 0.$$

By the projection formula,

$$\widetilde{\pi}_* C \cdot (p_X^* G - sE_X) < 0,$$

which contradicts the original assumption on s. As a consequence, we get

$$\varepsilon(X, \mathcal{P}, G) \leqslant \varepsilon(Y, \mathcal{Q}, \pi^* G).$$

Conversely, consider a rational number $s \leqslant \varepsilon(Y, \mathcal{Q}, \pi^* G)$ and a curve D on $\mathrm{Bl}_{\mathcal{P}}(X)$ such that $D \cdot (p_X^* G - sE_X) < 0$. Then,

$$\widetilde{\pi}^* D \cdot (\widetilde{\pi}^* p_X^* G - s\widetilde{\pi}^* E_X) < 0,$$

$$\text{that is } \widetilde{\pi}^* D \cdot (p_Y^* \pi^* G - sE_Y) < 0,$$

which contradicts the original assumption on s and concludes the proof. $\square$

PROPOSITION 1.30. *Let Y be a smooth projective geometrically connected surface over $\mathbf{F}_q$ and $\pi : Y \to X$ be a finite morphism. Let $\mathcal{Q}$ be the set of points of $\overline{Y}$ defined as $\mathcal{Q} := \pi^{-1}(\mathcal{P})$. Let $\mathcal{I}_{\mathcal{P}}$ (resp. $\mathcal{I}_{\mathcal{Q}}$) be the ideal sheaf of $\mathcal{O}_X$ (resp. $\mathcal{O}_Y$) associated to the set of rational points $\mathcal{P}$ (resp. $\mathcal{Q}$). Suppose that the rational points in $\mathcal{P}$ do not lie in the ramification locus of π. Let $\mathcal{L}$ be a line bundle on X and ξ a positive integer such that $\mathcal{L}^{\otimes \xi} \otimes \mathcal{I}_{\mathcal{P}}$ is generated by its global sections, then $\pi^*\mathcal{L}^{\otimes \xi} \otimes \mathcal{I}_{\mathcal{Q}}$ is generated by its global sections.*

PROOF. The sheaf $\mathcal{L}^{\otimes \xi} \otimes \mathcal{I}_{\mathcal{P}}$ is generated by its global sections, which is equivalent to the existence of a surjective morphism

$$\mathcal{O}_X^m \to \mathcal{L}^{\otimes \xi} \otimes \mathcal{I}_{\mathcal{P}}$$

for some positive integer m. Since π is finite, it is flat ([**29**, Exercise 9.3(a)]) and hence the functor π^* is exact. Thus, by pulling back by π and using the exactness of π^* we get a surjective morphism

$$\mathcal{O}_Y^m \to \pi^*\mathcal{L}^{\otimes \xi} \otimes \pi^*\mathcal{I}_{\mathcal{P}}$$

and the non ramification hypothesis entails that $\pi^*\mathcal{I}_{\mathcal{P}} = \mathcal{I}_{\mathcal{Q}}$. $\square$

Thanks to the previous results, we can assert that Hansen's bounds (S) stated in Proposition 1.23 applies in the relative case. However, one needs to be careful about one fact: to get a code over the same ground field, the set of points $\mathcal{Q} = \pi^{-1}(\mathcal{P})$ should contain only rational points. In addition, for Proposition 1.23 (S2) to hold, π should not be ramified at the points of $\mathcal{P}$ and hence, $\mathcal{P}$ should split completely.

1.9.4. *Summary of the behaviours of various bounds under finite morphisms.* Given a surface X with a finite set of rational points $\mathcal{P}$ and a divisor G, an interesting question would be the following. Consider a tower of finite covers

$$\begin{array}{c} \vdots \\ \downarrow \\ X_n \\ \downarrow \\ \vdots \\ \downarrow \\ X_1 \\ \downarrow \\ X \end{array}$$

in which $\mathcal{P}$ split totally. Is it possible to deduce lower bounds for the parameters of the code $C(X_n, \pi_n^{-1}(\mathcal{P}), \pi_n^* G)$, where π_n denotes the composed morphism $\pi_n : X_n \to X$, from some information defined only on the bottom surface X?

- *Aubry's bound (Proposition* 1.18*).* Partially possible but a very ampleness condition on $\pi_n^* G$ to assert for every n;
- *Hansen bound (A) (Proposition* 1.21*).* Seems difficult;

- *Hansen bounds (S) (Proposition* 1.23*).* Yes : one only needs to know some lower bound for a Seshadri constant on the bottom surface X or the integer ξ such that $\mathcal{L}^{\otimes \xi} \otimes \mathcal{I}_{\mathcal{P}}$ is globally generated.
- *Our bound (Theorem* 1.10*).* Yes, one only needs a $\mathcal{P}$–interpolating linear system on the bottom surface X.

1.10. How to get good towers of surfaces? This is clearly a natural question since codes from towers of curves are well–known to provide the best known sequences of codes. In the case of curves, one can see three approaches in the literature:

- modular curves, see for instance [**49**];
- recursive towers of function fields, see for instance [**20**];
- class field towers, see for instance [**35**].

In § 2.1, we investigate a class field like approach towards by providing a criterion for a surface to have an infinite tower of étale ℓ–covers that splits totally above a fixed finite set of points.

2. Infinite étale towers of surfaces

In § 2.1 to follow, we investigate a class–field like approach for the existence of an infinite tower of étale covers of a given surface X. In particular, we prove Theorem 2.7, a sufficient condition for the existence of such a tower in which some finite set T of points splits totally.

This criterion raises the following question: *which explicit surfaces may have an infinite tower of étale covers in which some non-empty fixed set of rational points splits totally?* For this sake, we prove in § 2.2 that one can find such an example on a given surface X if, and only if, one can find an example on some of its relatively minimal models. Then, we consider the Kodaira classification of minimal surfaces, explaining why some cases can quickly be excluded. In particular, note that for a surface to have an infinite étale tower that splits completely at some finite set of rational points, a necessary condition is that its geometric étale fundamental group is infinite.

We conclude this Section with an example in § 2.3 of surface for which Theorem 2.7 can be entirely worked out.

2.1. Étale covers of marked surfaces.

2.1.1. *The étale site of a marked scheme.* Throughout this section, let X be a scheme and T a finite set of closed points. For a morphism $\phi : Y \to X$, we denote by T_Y the set $\phi^{-1}(T)$.

Caution. Compared to the previous section where the finite set of points on the surfaces was referred to as $\mathcal{P}$, in this section we denote it by T to be coherent with the usual notation and to emphasize that what follows holds for a finite set of closed points and not only a set of rational points.

Basic definitions and properties. In this paragraph, we briefly recall the definition of the marked étale site of a marked scheme which has been introduced by A. Schmidt [**45**] in the case of curves and later generalized to general schemes (see Appendix). A marked scheme is a pair (Y, S) where Y is a scheme and S is a set of points of Y. A morphism of marked schemes $f : (Z, R) \to (Y, S)$ is a morphism of schemes $f : Z \to Y$ such that $f(R) \subset S$.

We consider the marked scheme (X,T) and we denote by $(X,T)_{\text{ét}}$ its marked étale site. It consists in the category of morphisms $\phi : (U,S) \to (X,T)$ such that $\phi : U \to X$ is étale and $S = T_U$ (called T-marked étale morphisms) together with the following coverings: surjective families $(p_i : (U_i, S_i) \to (U,S))_i$ such that, for any $s \in S$, there is a i and a $u_i \in S_i$ such that $p_i(u_i) = s$ and the induced field homomorphism $\kappa(s) \to \kappa(u_i)$ is an isomorphism.

We can easily check that these data define a site (*i.e.* a Grothendieck topology in [**1**, Def. 1.1.1]). Thus, the general properties of Grothendieck topologies apply here.

Let $P(X,T)$ and $S(X,T)$ denote respectively the category of presheaves and sheaves of abelian groups on $(X,T)_{\text{ét}}$. An example of such a sheaf is given by $\mathcal{O}_{(X,T)}(U) = \Gamma(U, \mathcal{O}_U)$. We denote by $i : S(X,T) \to P(X,T)$ the inclusion functor and by $\# : P(X,T) \to S(X,T)$ its left adjoint (see [**1**, §2.1] for details of the construction). It is a general fact for Grothendieck topologies that i is left exact and $\#$ is exact (see [**1**, Th. 2.1.4]).

As usual, if $\pi : Y \to X$ is a morphism and $\mathcal{G}$ is a presheaf on $(Y, T_Y)_{\text{ét}}$, we define a presheaf $\pi_*\mathcal{G}$ by putting $\pi_*\mathcal{G}(U) = \mathcal{G}(U \times_X Y)$ for any T-marked étale morphism $U \to X$. It is clear that it satisfies the axiom of being a sheaf as soon as $\mathcal{G}$ is a sheaf, and that the induced functor $\pi_* : S(Y, T_Y) \to S(X,T)$ is left exact.

We can define (see [**1**, Th. 1.3.1]) a left adjoint π^* to π_*. Indeed, for a presheaf $\mathcal{G}$ on $(X,T)_{\text{ét}}$, we define first a presheaf $\pi'(\mathcal{G})$ on (Y, T_Y) by putting, for any T_Y-marked étale map $V \to Y$, $\pi'(\mathcal{G})(V) = \varinjlim \mathcal{G}(U)$, where the direct limit is taken on the T-marked étale maps $U \to X$ making the diagram

$$\begin{array}{ccc} V & \longrightarrow & U \\ \downarrow & & \downarrow \\ Y & \longrightarrow & X \end{array}$$

commute. If $\mathcal{F}$ is a sheaf, we define $\pi^*(\mathcal{F}) = \#\pi'(i(\mathcal{F}))$.

As in any Grothendieck topology, $S(X,T)$ has enough injectives (see [**1**, Misc. 2.1.1] or [**22**, Th. 1.10.1]). For a sheaf $\mathcal{F}$ of abelian groups on $(X,T)_{\text{ét}}$, the functor $\mathcal{F} \mapsto \mathcal{F}(X,T) =: \Gamma(X,T,\mathcal{F})$ is left exact and we denote by $\mathrm{H}^i(X,T,\mathcal{F})$ its associated cohomology groups. One can also define the cohomology groups with support in a closed subscheme Z, as the right derivatives $\mathrm{H}^i_Z(X,T,\mathcal{F})$ of the left exact functor $\mathcal{F} \mapsto \ker(\mathcal{F}(X,T) \to \mathcal{F}(U, T \cap U))$, where U is $X \smallsetminus Z$.

As well as for the étale site, we can prove excision (see Appendix, §A.2.–A.3.) and a limit result, which is the main tool to compute marked étale cohomology groups. For $x \in T$, put $X^h_x = \mathrm{Spec}(\mathcal{O}^h_{X,x})$, where $\mathcal{O}^h_{X,x}$ denotes the henselianization of the local ring $\mathcal{O}_{X,x}$. Let us quote here the results of the Appendix we will use.

PROPOSITION 2.1. *Let X be a scheme and T a finite set of closed points. Then, for every sheaf of abelian groups $\mathcal{F}$ on (X,T), there is a long exact sequence*

$$\cdots \to \mathrm{H}^i_T(X,T,\mathcal{F}) \to \mathrm{H}^i_{\acute{e}t}(X,T,\mathcal{F}) \to \mathrm{H}^i_{\acute{e}t}(X \smallsetminus T, \mathcal{F}) \to \dots$$

and, for all $i \geqslant 0$,

$$\mathrm{H}^i_T(X,T,\mathcal{F}) \simeq \bigoplus_{x \in T} \mathrm{H}^i_{\{x\}}(X^h_x, x, \mathcal{F}).$$

Following Schmidt (see Appendix, A.5.) one can define, for a noetherian, normal and connected scheme X a fundamental group $\pi_1(X,T)$ which is profinite and classifies étale covers of X where the points of T split completely. Be cautious that fundamental groups are defined in the Appendix for more general schemes and that what we denote here by $\pi_1(X,T)$ for convenience is denoted by $\hat{\pi}_1^{\mathrm{et}}(X,T)$ in Appendix, A.5. (in our setting, there is no possible confusion, thanks to Appendix, Proposition A.15).

We denote by $\widetilde{(X,T)}(\ell)$ the universal pro-ℓ-cover of (X,T). The projection $\widetilde{(X,T)}(\ell) \to X$ is Galois with Galois group the maximal pro-ℓ-quotient $\pi_1(X,T)(\ell)$ of $\pi_1(X,T)$. Finally, the Hochschild-Serre spectral sequence relates Galois cohomology and étale cohomology groups (see §A.4.–A.5. of the Appendix). Let us also recall it here.

PROPOSITION 2.2. *Let X be a connected noetherian scheme, T a finite set of closed points of X. Let M be a discrete ℓ-torsion $\pi_1(X,T)(\ell)$-module. Then we have the following spectral sequence:*

$$E_2^{pq} = \mathrm{H}^p(\pi_1(X,T)(\ell), \mathrm{H}^q(\widetilde{(X,T)}(\ell), T, M)) \Rightarrow \mathrm{H}^{p+q}(X,T,M).$$

It induces a five term exact sequence:

$$0 \to \mathrm{H}^1(\pi_1(X,T)(\ell), M) \to \mathrm{H}^1_{\text{ét}}(X,T,M) \to \mathrm{H}^1(\widetilde{(X,T)}(\ell), M)^{\pi_1(X,T)(\ell)} \to$$
$$\to \mathrm{H}^2(\pi_1(X,T)(\ell), M) \to \mathrm{H}^2_{\text{ét}}(X,T,M).$$

As $\mathrm{H}^1(\widetilde{(X,T)}(\ell), M) = 0$, because $\widetilde{(X,T)}(\ell)$ admits no non-trivial ℓ-cover, we see that we have the following isomorphisms.

COROLLARY 2.3. *In the setting of the proposition, the Hochschild-Serre spectral sequence induces isomorphisms*

$$\mathrm{H}^i(\pi_1(X,T)(\ell), M) \simeq \mathrm{H}^i_{\text{ét}}(X,T,M) \textit{ for } i = 0,1$$

and an injection

$$\mathrm{H}^2(\pi_1(X,T)(\ell), M) \hookrightarrow \mathrm{H}^2_{\text{ét}}(X,T,M).$$

2.1.2. *The cohomology of marked surfaces.* Let X be a 2-dimensional noetherian regular scheme defined over $\mathbf{F}_q$ and let T be a finite set of closed points. Let $\Lambda = \mathbf{F}_\ell$, where $\ell \nmid q$ is prime number. The aim of this paragraph is to compute the étale cohomology groups $\mathrm{H}^i(X,T,\Lambda)$ of the marked surface (X,T). Notice that this can be immediately generalized to higher dimensions.

Local computations. Let X_x^h denote the spectrum $\mathrm{Spec}(\mathcal{O}_{X,x}^h)$ of the henselisation of the local ring $\mathcal{O}_{X,x}$ of X at a point $x \in T$ with finite residue field k. The scheme X_x^h is henselian with closed point x, and therefore

$$\mathrm{H}^i_{\text{ét}}(X_x^h, \Lambda) = \mathrm{H}^i_{\text{ét}}(x, \Lambda) = \mathrm{H}^i(k, \Lambda) \simeq \begin{cases} \Lambda & \text{for } i = 0,1 \\ 0 & \text{for } i \geqslant 2. \end{cases}$$

Let $U_x = X_x^h - x$. We wish to understand the groups

$$\mathrm{H}^i_x(X_x^h, \Lambda) \text{ and } \mathrm{H}^i_x(X_x^h, x, \Lambda).$$

The first groups arise in the sequence

$$\cdots \to \mathrm{H}^i_x(X_x^h, \Lambda) \to \mathrm{H}^i(X_x^h, \Lambda) \to \mathrm{H}^i(U_x, \Lambda) \to \cdots.$$

They are computed thanks to Gabber purity theorem [**18**]:

$$\mathrm{H}^i_x(X_x^h, \Lambda) \simeq \mathrm{H}^{i-4}(x, \Lambda(-2))$$

as x is regular of codimension 2 in the regular affine scheme X_x^h. Then, they are 0 for $i \neq 4,5$ and $\mathrm{H}_x^4(X_x^h,\Lambda) \simeq \mathrm{Hom}(\mu_\ell(k)\otimes\mu_\ell(k),\Lambda)$, $\mathrm{H}_x^5(X_x^h,\Lambda) \simeq \mathrm{H}^1(k,\Lambda(-2))$ this latter having the same cardinality as $\mathrm{Hom}(\mu_\ell(k)\otimes\mu_\ell(k),\Lambda)$ (thus Λ-vector spaces of dimension 1 for $i=4,5$ if ℓ divides $q-1$ and trivial otherwise).

We deduce that $\mathrm{H}^1(U_x,\Lambda) \simeq \mathrm{H}^1(X_x^h,\Lambda)$, that

$$\mathrm{H}^i(U_x,\Lambda) \simeq \mathrm{H}_x^{i+1}(X_x^h,\Lambda) \text{ for } i=3,4$$

and that $\mathrm{H}^i(U_x,\Lambda)=0$ for $i=2$ or $i\geqslant 5$.

The second groups appear in

$$\cdots \to \mathrm{H}_x^i(X_x^h,x,\Lambda) \to \mathrm{H}^i(X_x^h,x,\Lambda) \to \mathrm{H}^i(U_x,\Lambda) \to \cdots.$$

Since the identity of (X_x^h,x) is cofinal among the covering families of (X_x^h,x), we have $\mathrm{H}^i(X_x^h,x,\Lambda)=0$ for $i\geqslant 1$. Indeed, if $f:Y\to X_x^h$ is a finite étale morphism marked at x, then it admits a section (see [**39**, Th. I.4.2.]) inducing an isomorphism between X_x^h and a connected component of Y (as f is separated). If Y is connected then the map f is an isomorphism.

It implies that $\mathrm{H}_x^1(X_x^h,x,\Lambda) = \ker(\mathrm{H}^1(X_x^h,x,\Lambda)\to\mathrm{H}^1(U_x,\Lambda)) = 0$ and that $\mathrm{H}_x^i(X_x^h,x,\Lambda)\simeq \mathrm{H}^{i-1}(U_x,\Lambda)$ for any $i\geqslant 2$. Therefore, we have shown the following:

PROPOSITION 2.4. *Let T be $\{x\}$ or $\varnothing$. The groups $\mathrm{H}_x^i(X_x^h,T,\Lambda)$ are trivial for $i\leqslant 1$, $i=3$ or $i\geqslant 6$. Moreover, we have:*

$$\dim \mathrm{H}_x^i(X_x^h,T,\Lambda) = \begin{cases} \sharp T & \text{if } i=2 \\ 1 & \text{if } i=4,5 \text{ and } \ell \mid q-1 \\ 0 & \text{otherwise} \end{cases}$$

Global computations. Let us now use excision and the local computations in order to get information on the cohomology groups. Put $h^i(X,\dots)$ for the dimension $\dim_\Lambda \mathrm{H}^i(X,\dots)$. As soon as it is well defined, put $\chi(X,\dots) := \sum_{i\geqslant 0}(-1)^i h^i(X,\dots)$.

PROPOSITION 2.5. *Let X be a 2-dimensional noetherian regular scheme defined over $\mathbf{F}_q$ and let T be a finite set of closed points. Let $\Lambda=\mathbf{F}_\ell$.*

$$h^2(X,T,\Lambda)-h^1(X,T,\Lambda) = \sharp T + h^2(X,\Lambda)-h^1(X,\Lambda) = \dim \mathrm{H}^2(\overline{X},\Lambda)^{G_{\mathbf{F}_q}} + \sharp T - 1,$$

and

$$h^5(X,T,\Lambda)-h^4(X,T,\Lambda)+h^3(X,T,\Lambda) = h^5(X,\Lambda)-h^4(X,\Lambda)+h^3(X,\Lambda),$$

where $G_{\mathbf{F}_q}=\mathrm{Gal}(\overline{\mathbf{F}}_q,\mathbf{F}_q)$ and $\overline{X}=X\times_{\mathbf{F}_q}\overline{\mathbf{F}}_q$.

PROOF. We use excision to compute the $\mathrm{H}^i(X,T,\Lambda)$. It provides exact sequences

$$\cdots \to \bigoplus_{x\in T}\mathrm{H}_x^i(X_x^h,\Lambda) \to \mathrm{H}^i(X,\Lambda) \to \mathrm{H}^i(X-T,\Lambda) \to \cdots \text{ and}$$

$$\cdots \to \bigoplus_{x\in T}\mathrm{H}_x^i(X_x^h,T_x,\Lambda) \to \mathrm{H}^i(X,T,\Lambda) \to \mathrm{H}^i(X-T,\Lambda) \to \cdots.$$

The first sequence leads to isomorphisms $\mathrm{H}^i(X,\Lambda)\simeq\mathrm{H}^i(X-T,\Lambda)$ for $i=1,2$, and from the second one can deduce the first equality of the proposition by the local computations.

Moreover, the Hochschild-Serre spectral sequence leads to the following, for any $i\geqslant 1$:

$$0\to \mathrm{H}^1(G_{\mathbf{F}_q},\mathrm{H}^{i-1}(\overline{X},\Lambda)) \to \mathrm{H}^i(X,\Lambda) \to \mathrm{H}^i(\overline{X},\Lambda)^{G_{\mathbf{F}_q}} \to 0.$$

Together with the fact that for any finite module M, $\mathrm{H}^0(G_{\mathbf{F}_q}, M)$ and $\mathrm{H}^1(G_{\mathbf{F}_q}, M)$ have same cardinality, we obtain

$$h^2(X,\Lambda) - h^1(X,\Lambda) = \dim \mathrm{H}^2(\overline{X},\Lambda)^{G_{\mathbf{F}_q}} - 1,$$

which leads to

$$h^2(X,T,\Lambda) - h^1(X,T,\Lambda) = \dim \mathrm{H}^2(\overline{X},\Lambda)^{G_{\mathbf{F}_q}} + \sharp T - 1.$$

□

The second part of the theorem comes from the isomorphisms $\mathrm{H}^i_x(X^h_x, x, \Lambda) \simeq \mathrm{H}^i_x(X^h_x, \Lambda)$ for $i \geqslant 3$, and their triviality for $i = 3$ and 6. Note that if $\ell \nmid q-1$, then the exact sequences imply that $\mathrm{H}^i(X,T,\Lambda) \simeq \mathrm{H}^i_{\text{ét}}(X,\Lambda)$ for $i \geqslant 3$.

COROLLARY 2.6. *Let X be a smooth projective surface defined over $\mathbf{F}_q$. Let T be a finite set of closed points of X. Then, the cohomology groups $\mathrm{H}^i(X,T,\mathbf{F}_\ell)$ vanish for $i \geqslant 6$, and we have $\chi(X,T,\mathbf{F}_\ell) = \sharp T$.*

PROOF. Indeed, we have for $i \geqslant 6$, $\mathrm{H}^i(X,T,\mathbf{F}_\ell) \simeq \mathrm{H}^i(X-T,\mathbf{F}_\ell) \simeq \mathrm{H}^i(X,\mathbf{F}_\ell) = 0$, as the local cohomology groups vanish, as X is smooth. Moreover, since X is projective, $\chi(X,\mathbf{F}_\ell)$ vanishes by the Poincaré duality, which gives the desired result by the proposition. □

Choice of the closed points.

Let X be a smooth projective (absolutely irreducible) surface defined over $\mathbf{F}_q$ and T be a finite set of closed points containing at least a point x_0. Class field theory (see [**46**, VI.§4–5] or [**47**]) for a concise exposition) defines a reciprocity map ρ from the Chow group $\mathrm{CH}_0(X)$ of zero-cycles of X to $\pi_1^{ab}(X)$, by sending closed points x to their Frobenius (the image in $\pi_1(X)^{ab}$ of the Frobenius of $Gal(\overline{\kappa(x)}/\kappa(x))^{ab}$). Its image is exactly the set $\pi_1^{ab,f}(X)$ of elements of $\pi_1^{ab}(X)$ whose image in $Gal(\overline{\mathbf{F}_q}/\mathbf{F}_q)$ by the natural map is an element of $\mathbb{Z}$ (in $\hat{\mathbb{Z}}$). By [**46**, VI.n° 16 Théorème 1]), we have the following isomorphism of short exact sequences:

$$\begin{array}{ccccccccc}
0 & \longrightarrow & \mathrm{CH}_0^0(X) & \longrightarrow & \mathrm{CH}_0(X) & \xrightarrow{deg} & \mathbb{Z} & \longrightarrow & 0 \\
 & & \rho \downarrow \simeq & \circlearrowleft & \simeq \downarrow \rho & \circlearrowleft & \downarrow = & & \\
0 & \longrightarrow & \pi_1^{ab,0}(X) & \longrightarrow & \pi_1^{ab,f}(X) & \longrightarrow & \mathbb{Z} & \longrightarrow & 0
\end{array}$$

The map ρ induces an isomorphism from $\mathrm{CH}_0(X)/(x_0)$ to $\pi_1(X)^{ab}/(\rho(x_0))$. Then the group $\pi_1(X)^{ab}/(\rho(x_0))$ is the finite quotient of $\pi_1(X)^{ab}$ corresponding to the maximal étale abelian cover of X, totally split at x_0 (see [**46**, VI.§5] : a finite étale cover is totally split at x_0 if and only if its associated Frobenius is trivial). Therefore, we have an isomorphism $\mathrm{CH}_0(X)/(\ell, x_0) \simeq \pi_1(X,\{x_0\})^{ab}/\ell$.

Similarly, adding one more point x to T lowers h^1 by one or enlarge h^2 by one, depending on the fact that x belongs to the group generated by the previous split points in $\mathrm{CH}_0(X)/\ell$ or not.

Therefore, we have $h^1(X,T,\Lambda) = h^1(X) - r_T$, where r_T is the $\mathbf{F}_\ell$-dimension of the space generated by the points of T in $\mathrm{CH}_0(X)/\ell$. If $T \neq \varnothing$, $1 \leqslant r_T \leqslant \sharp T$.

2.1.3. *Golod-Shafarevich criterion.* Our strategy to produce infinite covers of X with splitting properties is the Golod-Shafarevich criterion. For this purpose, we need the following inequality:

$$h^2(\pi_1(X,T)(\ell),\mathbb{Z}/\ell\mathbb{Z}) \leqslant \frac{h^1(\pi_1(X,T)(\ell),\mathbb{Z}/\ell\mathbb{Z})^2}{4}.$$

Because of the previous comparison maps (see Corollary 2.3), it is sufficient to check that it is satisfied for the étale cohomology groups of the marked surface (X,T).

THEOREM 2.7. *If*

$$\dim \mathrm{H}^1(\overline{X},\Lambda)^{G_{\mathbf{F}_q}} \geqslant r_T + 1 + 2\sqrt{\dim \mathrm{H}^2(\overline{X},\Lambda)^{G_{\mathbf{F}_q}} + \sharp T},$$

then $\pi_1(X,T)(\ell)$ *is infinite, i.e.* X *admits a* $T-\ell$ *infinite classfield tower.*

PROOF. The group $\pi_1(X,T)(\ell)$ is infinite if

$$h^2(\pi_1(X,T)(\ell),\mathbb{Z}/\ell\mathbb{Z}) \leqslant \frac{h^1(\pi_1(X,T)(\ell),\mathbb{Z}/\ell\mathbb{Z})^2}{4}.$$

As $h^1(\pi_1(X,T)(\ell),\mathbb{Z}/\ell\mathbb{Z}) = h^1_{\text{ét}}(X,T;\Lambda) := h^1_T$ and

$$h^2(\pi_1(X,T)(\ell),\mathbb{Z}/\ell\mathbb{Z}) \leqslant h^2(X,T,\mathbb{Z}/\ell\mathbb{Z}) =: h^2_T$$

because of Corollary 2.3. Therefore it suffices to have $4h^2_T \leqslant (h^1_T)^2$.

Let $\bar{h}^i := \dim \mathrm{H}^i(\overline{X},\Lambda)$ and $\bar{h}^{i,G} := \dim \mathrm{H}^i(\overline{X},\Lambda)^{G_{\mathbf{F}_q}}$. The previous inequality is equivalent to

$$(h^1_T - 2)^2 \geqslant 4(h^2_T - h^1_T + 1)$$

and to

$$(\bar{h}^{1,G} - r_T - 1)^2 \geqslant 4(\bar{h}^{2,G} + \sharp T)$$

and the proposition follows. □

The next proposition seems to be more useful in a computing point of view, as the Euler-Poincaré characteristic is computable. Let $\alpha = \bar{h}^1 - \bar{h}^{1,G}$ and $\bar{\chi} = \chi(\overline{X},\mathbf{F}_\ell) = \chi(\overline{X},\mathbf{Q}_\ell)$ (cf. [**39**] p166).

PROPOSITION 2.8. *The conclusion also holds if*

$$(\bar{h}^1 - \alpha + r_T - 5)^2 \geqslant 4(\bar{\chi} + 2\alpha + 2r_T + 4 + \sharp T).$$

PROOF. We introduce $\bar{\chi} = \bar{h}^2 - 2\bar{h}^1 + 2$ in what precedes and note that $\bar{h}^{2,G} \leqslant \bar{h}^2$. □

Remark that the first ℓ-adic Betti number b_1 of $\overline{X}$ satisfies $b_1 \leqslant \bar{h}^1$. We may hope to be able to compute b_1 and $\bar{\chi}$ by computing the Zeta function of X, and thus to verify whether the previous inequality holds for b_1, in particular after extending the constants so that $\alpha = 0$. When $\bar{\chi}$ is very small the condition should be satisfied.

2.2. Using the classification of surfaces. Theorem 2.7 raises the question: *which surfaces may have an infinite tower of étale covers in which some finite set* T *splits completely?* In the sequel, we sieve the Kodaiara classification of surfaces in order to study which classes should be excluded and which ones may provide good candidates. The following statement will be useful to exclude some cases.

PROPOSITION 2.9. *Let* $\pi : \overline{Y} \to \overline{X}$ *be a finite morphism of smooth connected projective surfaces over an algebraically closed field. Suppose that* $\overline{Y}$ *has a finite fundamental group, then* $\overline{X}$ *cannot have an infinite tower of étale covers.*

PROOF. Since $\overline{Y}$ has a finite fundamental group, then its universal cover $\widetilde{\overline{Y}} \to \overline{Y}$ is a finite map and hence, replacing $\overline{Y}$ by its universal cover, one can suppose that $\overline{Y}$ is simply connected.

Now, denote by s the degree of the morphism $\overline{Y} \to \overline{X}$ and consider an étale cover $\overline{X}_1 \to \overline{X}$ of degree m. Its pullback on $\overline{Y}$ provides an étale cover of $\overline{Y}$ and since $\overline{Y}$ is supposed to be simply connected, this cover is the disjoint union of m copies of $\overline{Y}$

$$\begin{array}{ccc} \overline{Y} & \xleftarrow{f} & \bigsqcup_{i=1}^{m} \overline{Y} \\ \downarrow & & \downarrow g \\ \overline{X} & \longleftarrow & \overline{X}_1 \end{array}$$

and the restriction of f to any component is an isomorphism. Consider the restriction of g to a connected component of $\bigsqcup_{i=1}^{m} \overline{Y}$, then the map $\overline{Y} \to \overline{X}$ is factorised by a dominant map $\overline{Y} \to \overline{X}_1$. As a consequence the degree s of $\overline{Y} \to \overline{X}$ divides m and hence any finite étale cover of $\overline{X}$ has a bounded degree. □

2.2.1. *Relatively minimal models.* Recall that a smooth surface X over a field k is said to be *relatively minimal* if there is no (-1)–curve of genus 0 in $\overline{X}$. Equivalently, for such a surface, there does not exist a regular surface $\overline{Y}$ over $\bar{k}$ and a $\bar{k}$–morphism $\overline{X} \to \overline{Y}$ which consists in blowing down a finite set of curves in $\overline{X}$.

The following result asserts that any étale cover of a surface comes from an étale cover of a relatively minimal model, so that it is sufficient to investigate relatively minimal surfaces. It is a refinement for relatively minimal models of surfaces of the well known birational invariance property of the fundamental group. This result is probably well–known by the experts, we give a proof because of a lack of a reference.

THEOREM 2.10. *Let $\overline{X}_0$ be a surface over an algebraically closed field $\bar{k}$ and $\pi : \overline{X} \to \overline{X}_0$ be a surface obtained from $\overline{X}_0$ after blowing up a point. Then for any étale cover $f : \overline{Y} \to \overline{X}$, there exists an étale cover $f_0 : \overline{Y}_0 \to \overline{X}_0$ from a smooth surface $\overline{Y}_0$, such that $\overline{Y} = \overline{Y}_0 \times_{\overline{X}_0} \overline{X}$.*

PROOF. From Stein Factorization Theorem [**29**, Corollary III.11.5], the map $f \circ \pi$ factorizes as

$$\begin{array}{ccc} \overline{Y} & \xrightarrow{f} & \overline{X} \\ {\scriptstyle \pi'}\downarrow & & \downarrow{\scriptstyle \pi} \\ \overline{Y}_0 & \xrightarrow{f_0} & \overline{X}_0 \end{array}$$

where π' has connected fibres and f_0 is finite.

Let us prove that $f_0 : \overline{Y}_0 \to \overline{X}_0$ is étale. For any variety V in this diagram, we denote by U_V the open subvariety of V above the open subscheme $U_{\overline{X}_0}$ of $\overline{X}_0$ obtained by puncturing the blown-up point x_0. Let $\imath : \overline{C} \hookrightarrow \overline{X}_0$ be a geometric curve on $\overline{X}_0$, and $U_{\overline{C}} = \imath^* U_{\overline{X}_0}$. Since f is étale, $\pi'^* \circ f_0^* U_{\overline{C}} = f^* \circ \pi^* U_{\overline{C}}$ is reduced, so that $f_0^* U_{\overline{C}}$ itself is reduced and f_0 is étale outside x_0. By Zarisky purity Theorem [**52**], it follows that f_0 is étale. In particular, $\overline{Y}_0$ is smooth since $\overline{X}_0$ is.

By the universal property of fiber products, there exists a unique morphism φ such that the following diagram commutes.

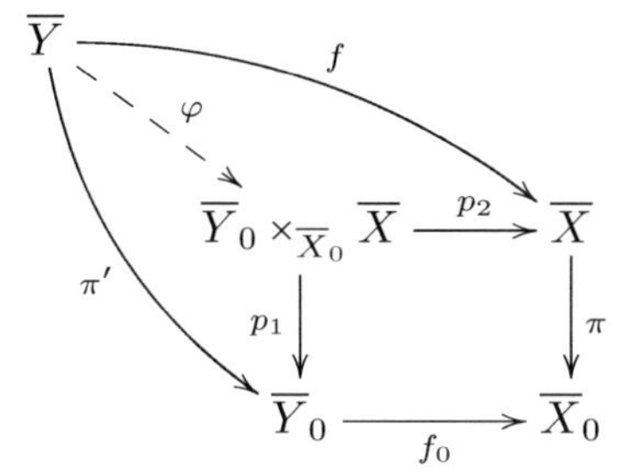

Let us prove that φ is an isomorphism, concluding the proof. Remind that $f = p_2 \circ \varphi$ is étale and p_2 is étale as a base change of the étale map f_0 [**39**, Proposition I.3.3.(b)]. Thus, from [**39**, Corollary I.3.6], the morphism φ is étale too. Therefore, to prove that φ is an isomorphism it is sufficient to prove that any geometric closed point $P = (y_0, x) \in \overline{Y}_0 \times_{\overline{X}_0} \overline{X}$ has a unique preimage. By the commutativity of the diagram, the pullback of P by φ is contained in $\pi'^*(y_0)$, which is connected by Stein Theorem. In the same way, it is contained in $f^*(x)$ which is finite since f is finite. Consequently, P has a single inverse image by φ and hence φ is an isomorphism. □

2.2.2. *Classification of surfaces.* It follows from Theorem 2.10 that a surface X admits an infinite tower in which a non-empty set T of rational points splits if and only if any relatively minimal model $\pi : X \to X_0$ of X also admits such an infinite tower, in which $\pi(T)$ splits. This leads us to investigate the Kodaira classification of minimal surfaces given in Table 1 (see for instance [**34**, §II.8.1] for the characteristic 0 case and or [**11**] for the general case), in which the numerical equivalence of divisors is denoted by $\equiv$. For each class of surface in the table, we ask whether such a tower can exists. We indicate in a last column the conclusion of the discussion that follows.

TABLE 1. Classification of algebraic surfaces

Structure	κ	canonical class K	K^2	existence of infinite splitting tower
Rational	-1	-	-	no
Ruled above a curve of genus g		-	$8(1-g)$	comes from the base curve
Abelian	0	$K \equiv 0$	$K^2 = 0$	no
$K3$				no
Enriques				no
Hyperelliptic				no
Quasi–elliptic (only in char. 2 and 3)				?
Elliptic	1	$\forall n > 0, nK \not\equiv 0$	$K^2 = 0$	?
General type	2	$K \geqslant 0$	$K^2 > 0$	?

We recall that the Kodaira dimension is invariant under étale covers [**33**, Theorem 10.9]. Thus, for an étale tower of surfaces, the Kodaira dimension is constant and hence is that of the bottom surfaces. So let us investigate surfaces of Kodaira dimension up to one.

Beginning with surfaces of Kodaira dimension -1, it is well–known that rational surfaces are simply connected (see [**24**, Corollaire XI.1.2] for instance). Regarding ruled surfaces X over a base C, since for complete varieties, the fundamental group is a birational invariant [**39**, Example 5.2(h)] and X is birational to $C \times \mathbf{P}^1$, then, from [**24**, Corollary X.1.7], $\pi_1(X) \simeq \pi_1(C) \times \pi_1(\mathbf{P}^1)$ and since $\mathbf{P}^1$ is simply connected, we conclude that $\pi_1(X) = \pi_1(C)$. More precisely, it can be checked, with a proof similar to that of Theorem 2.10 above using Stein factorisation, that any étale cover of a ruled surface is the pullback of some étale cover of the base curve. It follows that the splitting property of some infinite tower of a ruled surface would comes from some of the base, so that we are reduced to the classical problem on curves.

Next, we continue with surfaces of Kodaira dimension 0. First, K3 surfaces are simply connected [**31**, Remark I.2.3]. Second, in odd characteristic, Enriques surfaces are known (see [**15**]) to have an étale cover of degree 2 by a K3 surface. In characteristic 2 (see [**15**] again) an Enriques surface is either covered by a surface which is birational to a K3 surface or by a non normal rational surface. Therefore, in any characteristic, an Enriques surface is covered by a simply connected surface and hence Proposition 2.9 asserts that Enriques surfaces are bad candidates. Third, whereas abelian surfaces can have infinite towers of étale covers, such towers cannot satisfy a non trivial splitting property because each stage being an abelian surface (see [**40**, Chapter 4 §18]), it cannot have more than $(\sqrt{q}+1)^4$ rational points from Weil bound. This means that the marked fundamental group of an abelian surface is finite. Fourth, hyperelliptic surfaces are a finite quotient of a product of two elliptic curves, hence are dominated under a finite map by an abelian surface whose marked fundamental group is finite. Hence, from Proposition 2.9, their marked fundamental group is finite too. Last, we are not able to conclude for quasi-elliptic surfaces.

Then, we end with surfaces of Kodaira dimension 1, that is for elliptic surfaces. Here, we are only able to prove, again with a proof very close to that of Theorem 2.10 above using Stein factorisation, that any étale cover of an elliptic surface is an elliptic surface whose base is an étale cover of the original base curve.

2.3. Example: product of hyperelliptic curves. The aim of this section is to provide an illustrative example of application of Theorem 2.7 in which any computation can be made explicit. It turns out that, for most surfaces X for which the answer in the last column in Table 1 is not negative, it is a hard task to compute the Galois invariants of $\mathrm{H}^1(\overline{X}, \mathbf{F}_\ell)$ and $\mathrm{H}^2(\overline{X}, \mathbf{F}_\ell)$, and even to bound from below the first one and to bound from above the second one in an efficient way.

The authors acknowledge that they do succeed only in the case of the product of two hyperelliptic curves of genus larger than 2. Unfortunately, for a product of two curves $C \times D$ with canonical projections p_C, p_D and a prime integer ℓ, one can prove that the existence of an infinite tower of ℓ–étale covers splitting totally at a set of closed points $\mathcal{P}$ entails either the existence of an infinite tower of ℓ–étale covers of C splitting totally at $p_C(\mathcal{P})$ or the existence of an infinite tower of D splitting totally at $p_D(\mathcal{P})$.

PROPOSITION 2.11. *Let C, D be two smooth curves. Consider the product $X = C \times D$ with canonical projections p_C, p_D. Let P be a rational point of C and Q be a rational point of D and $C_0 := p_D^* Q$ and $D_0 := p_C^* P$. Let ℓ be a prime integer and Y be a connected smooth surface and $\pi : Y \to X$ be an ℓ–étale Galois cover of X, then at least one of the two curves $\pi^* C_0$ or $\pi^* D_0$ is connected. In addition, if $\mathcal{P}$ is a set of rational points of X that is totally split in Y, then $p_C(\mathcal{P})$ (resp. $p_D(\mathcal{P})$) is totally split in $\pi^* C_0 \to C_0$ (resp. $\pi^* D_0 \to D_0$).*

PROOF. **Step 1.** Let us prove that the divisor $C_0 + D_0$ is ample. Indeed, let g be a positive integer larger than the genus of C and that of D. Then we will prove that $(2g+1)(C_0 + D_0)$ is very ample.

Clearly, $(2g+1)P$ is very ample on C and hence provides a closed immersion $C \hookrightarrow \mathbf{P}^{n_C}$ for some positive integer n_C. Similarly we $(2g+1)Q$ provides a closed immersion $D \hookrightarrow \mathbf{P}^{n_D}$. This yields a closed immersion $C \times D \hookrightarrow \mathbf{P}^{n_C} \times \mathbf{P}^{n_D}$, then by Segré embedding, we get a closed immersion $\varphi : C \times D \hookrightarrow \mathbf{P}^N$ for some positive integer N and $(2g+1)(C_0 + D_0) \sim \varphi^* \mathcal{O}_{\mathbf{P}^n}(1)$.

Step 2. Since the pullback of an ample divisor is ample ([**23**, Prop 5.1.12]), the divisor $\pi^* C_0 + \pi^* D_0$ is ample and hence has a connected support ([**29**, Corollary III.7.9]).

Step 3. Suppose that both $\pi^* C_0$ and $\pi^* D_0$ are not connected. Then, using base change principle and since π is a Galois cover, one deduces that both are ℓ copies of their base. In addition, by the projection formula we have

$$\pi^* C_0 \cdot \pi^* D_0 = \ell$$

and hence each copy of C_0 in $\pi^* C_0$ meets a single copy of D_0 in $\pi^* D_0$ with multiplicity 1 and avoids the other ones. Thus, one deduces that $\pi^* C_0 + \pi^* D_0$ has a support which is isomorphic to the disjoint union of ℓ copies of the support of $C_0 + D_0$ on X. Therefore, it is non connected which contradicts its ampleness.

Step 4. Suppose that the pullback of C_0 is connected and that some rational point of $p_C(\mathcal{P})$ has a non rational inverse image R under $\pi^* C_0 \to C_0$. Then, some elements of $\pi^* \mathcal{P}$ lie in the fibre $\{R\} \times \pi^* D_0$ which does not contain any rational point. This contradicts the assumption that $\mathcal{P}$ splits completely under π. □

Therefore, the application of our infiniteness-criterion in the sequel yields an existence result that could have been proved with usual class field theory in dimension one. Moreover, it is even true that, under the usual assumptions, the marked (at a product of sets of rational points) étale fundamental group of a product is the fibre product of the corresponding marked étale fundamental groups, as shown in Appendix, Proposition A.17. It follows that if the marked étale fundamental group of a product $\pi_1(C \times D, S \times T)$ is infinite, then at least one of the two marked étale fundamental groups $\pi_1(C, S)$ and $\pi_1(D, T)$ has to be infinite and this also holds for maximal pro-p-quotients.

Thus, we emphasize that the example to follow is only illustrative. The objective in only to show that the objects involved in Theorem 2.7 can be explicitly computed.

PROPOSITION 2.12. *Let q be a power of an odd prime number. Let $f(t) \in \mathbf{F}_q[t]$ be the product of $2g_1+2$ distinct linear factors and $g(t) \in \mathbf{F}_q[t]$ be the product of g_2+1 distinct irreducible quadratic factors. Let $C : y^2 = f(t)$ and $D : y^2 = g(t)$ be the associated hyperelliptic curves. Let ρ be a positive integer such that :*

- $2g_1 + 2 + 2\rho \leqslant \sharp C(\mathbf{F}_q)$;
- $2\rho \leqslant \sharp D(\mathbf{F}_q)$;
- $2g_1 + g_2 \geqslant 3\rho + 2 + 2\sqrt{2g_1g_2 + 2 + 4\rho}$.

Then, there exists an infinite étale tower of $C \times D$ in which some set of 4ρ rational points splits totally.

Before proving this Proposition below, let us fix some notations and state two preliminary lemmas. We begin by choosing the prime number $\ell = 2$, prime to q. We denote by σ and τ the hyperelliptic involutions on C and D respectively. Those extend on $X = C \times D$ by $\sigma(p,q) = (\sigma(p), q)$ and $\tau(p,q) = (p, \tau(q))$, generating the abelian group $A = \langle \sigma, \tau \rangle \simeq (\mathbb{Z}/2\mathbb{Z})^2$. We denote by $A.P$ the orbit of a point $P = (p,q) \in C \times D$ under A. Then neither p nor q is a Weierstrass point if, and only if, this orbit have order 4.

LEMMA 2.13. *Let $P_1, \dots, P_\rho$ be ρ rational points on $X = C \times D$ whose orbits under A have 4 elements and are disjoints. Let T be the union of these orbits. Then, $\sharp T = 4\rho$ and*

$$r_T \leqslant 3\rho + 1 = \frac{3}{4}\sharp T + 1.$$

PROOF. We denote $P_i = (p_i, q_i) \in C \times D$. Since all divisors of the form $p + \sigma(p)$ on C are equivalent in $\mathrm{CH}_0(C) = \mathrm{Jac}_C$, and considering both horizontal curves $C \times \{q_1\}$ and $C \times \{\tau(q_1)\}$ on X, we have

$$(p_1, q_1) + (\sigma(p_1), q_1) \sim (p_i, q_1) + (\sigma(p_i), q_1)$$

and

$$(p_1, \tau(q_1)) + (\sigma(p_1), \tau(q_1)) \sim (p_i, \tau(q_1)) + (\sigma(p_i), \tau(q_1))$$

for any $2 \leqslant i \leqslant \rho$.

Summing these relations, we get, for any $2 \leqslant i \leqslant \rho$,

$$\begin{aligned}(2.1)\qquad (p_1, q_1) + (\sigma(p_1), q_1) + (p_1, \tau(q_1)) + (\sigma(p_1), \tau(q_1)) \sim \\ (p_i, q_1) + (\sigma(p_i), q_1) + (p_i, \tau(q_1)) + (\sigma(p_i), \tau(q_1)).\end{aligned}$$

In the same way, considering both vertical curves $\{p_i\} \times D$ and $\{\sigma(p_i)\} \times D$, we get

$$\begin{aligned}(2.2)\qquad (p_i, q_1) + (\sigma(p_i), q_1) + (p_i, \tau(q_1)) + (\sigma(p_i), \tau(q_1)) \sim \\ (p_i, q_i) + (\sigma(p_i), q_i) + (p_i, \tau(q_i)) + (\sigma(p_i), \tau(q_i))\end{aligned}$$

for any $2 \leqslant i \leqslant \rho$. Both equations (2.1) and (2.2) give, for any $2 \leqslant i \leqslant \rho$,

$$\begin{aligned}(p_1, q_1) + (\sigma(p_1), q_1) + (p_1, \tau(q_1)) + (\sigma(p_1), \tau(q_1)) \sim \\ (p_i, q_i) + (\sigma(p_i), q_i) + (p_i, \tau(q_i)) + (\sigma(p_i), \tau(q_i))\end{aligned}$$

from which we deduce that, for any $2 \leqslant i \leqslant \rho$, we can express one point in the orbit $A.P_i$ in term of the three other ones and those of $A.P_1$. It follows that

$$r_T \leqslant 4\rho - (\rho - 1),$$

and the lemma is proved. □

Now, let us consider the Galois invariants of the groups $\mathrm{H}^1(\overline{C}\times\overline{D},\mathbf{F}_2)$ and $\mathrm{H}^2(\overline{C}\times\overline{D},\mathbf{F}_2)$, with coefficients in $\Lambda=\mathbf{F}_2$, for the action of $G_{\mathbf{F}_q}=\mathrm{Gal}(\overline{\mathbf{F}}_q/\mathbf{F}_q)$. The first one is easily tackled. We have $\mathrm{H}^1(\overline{C},\mathbf{F}_2)=\mathrm{Jac}_C[2]$, and so as for $\overline{D}$. From Künneth formula

$$\mathrm{H}^1(\overline{C}\times\overline{D},\mathbf{F}_2)\simeq\mathrm{H}^1(\overline{C},\mathbf{F}_2)\oplus\mathrm{H}^1(\overline{D},\mathbf{F}_2)$$

as a sum of $G_{\mathbf{F}_q}$-modules, we deduce

(2.3)
$$\mathrm{H}^1(\overline{C}\times\overline{D},\mathbf{F}_2)^{G_{\mathbf{F}_q}}\simeq\mathrm{H}^1(\overline{C},\mathbf{F}_2)^{G_{\mathbf{F}_q}}\oplus\mathrm{H}^1(\overline{D},\mathbf{F}_2)^{G_{\mathbf{F}_q}}\simeq\mathrm{Jac}_C[2](\mathbf{F}_q)\oplus\mathrm{Jac}_D[2](\mathbf{F}_q).$$

The Galois invariants of $\mathrm{H}^2(\overline{C}\times\overline{D},\mathbf{F}_2)$ also come from Künneth formula

$$\mathrm{H}^2(\overline{C}\times\overline{D},\mathbf{F}_2)=\mathrm{H}^1(\overline{C},\mathbf{F}_2)\otimes\mathrm{H}^1(\overline{D},\mathbf{F}_2)\oplus K$$

as a sum of $G_{\mathbf{F}_q}$-modules, where K is a 2-dimensional vector space over $\mathbf{F}_2$ on which $G_{\mathbf{F}_q}$ acts trivially. It follows that

$$\dim\mathrm{H}^2(\overline{C}\times\overline{D},\mathbf{F}_2)^{G_{\mathbf{F}_q}}=\dim(\mathrm{H}^1(\overline{C},\mathbf{F}_2)\otimes\mathrm{H}^1(\overline{D},\mathbf{F}_2))^{G_{\mathbf{F}_q}}+2.\tag{2.4}$$

In the following Lemma (holding in fact for any prime ℓ, prime to q), we denote by $\mathrm{Sp}_2(C)$ (resp. $\mathrm{Sp}_2(D)$) the spectrum in $\overline{\mathbf{F}}_2$ of the Frobenius on $\mathrm{Jac}_C[2]\simeq\mathbf{F}_2^{2g_1}$ (resp. on $\mathrm{Jac}_D[2]\simeq\mathbf{F}_2^{2g_2}$), by $m_{\lambda,C}$ the dimension of the eigenspace of $\mathrm{Jac}_C[2]$ for the eigenvalue $\lambda\in\mathrm{Sp}_2(C)$, and by $\mathrm{Sp}_2(C,D)=\{\lambda\in\overline{\mathbf{F}}_2;\lambda\in\mathrm{Sp}_2(C)\text{ and }\lambda^{-1}\in\mathrm{Sp}_2(D)\}$.

LEMMA 2.14. *We have*

$$\dim\left(\mathrm{H}^1(\overline{C},\mathbf{F}_2)\otimes\mathrm{H}^1(\overline{D},\mathbf{F}_2)\right)^{G_{\mathbf{F}_q}}=\sum_{\lambda\in\mathrm{Sp}_2(C,D)}m_{\lambda,C}.m_{\lambda^{-1},D}.$$

PROOF. We consider the decomposition of $\mathrm{H}^1(\overline{C},\mathbf{F}_2)=\mathrm{Jac}_C[2]$ and that of $\mathrm{H}^1(\overline{D},\mathbf{F}_2)=\mathrm{Jac}_D[2]$ as a sum of their *characteristic* subspaces for the action of the Frobenius

$$\mathrm{H}^1(\overline{C},\mathbf{F}_2)=\bigoplus_{\lambda\in\mathrm{Sp}_2(C)}E_\lambda(C)$$

and

$$\mathrm{H}^1(\overline{D},\mathbf{F}_2)=\bigoplus_{\mu\in\mathrm{Sp}_2(D)}E_\mu(D).$$

The decomposition of $\mathrm{H}^1(\overline{C},\mathbf{F}_2)\otimes\mathrm{H}^1(\overline{D},\mathbf{F}_2)$ for the action of the Frobenius is then

$$\mathrm{H}^1(\overline{C},\mathbf{F}_2)\otimes\mathrm{H}^1(\overline{D},\mathbf{F}_2)=\bigoplus_{(\lambda,\mu)\in\mathrm{Sp}_2(C)\times\mathrm{Sp}_2(D)}E_\lambda(C)\otimes E_\mu(D).$$

The eigenvalue on $E_\lambda(C)\otimes E_\mu(D)$ is the product $\lambda\mu$, and the uniqueness of the decomposition of a vector in the decomposition above shows that the invariants are those in

$$\bigoplus_{\lambda.\mu=1}E_{0,\lambda}(C)\otimes E_{0,\mu}(D),$$

where the $E_{0,\lambda}(C)$ and $E_{0,\mu}(D)$ are the *eigenspaces*, hence the Lemma. □

We can now prove Proposition 2.12.

PROOF. By hypothesis on $\sharp C(\mathbf{F}_q)$, there exists at least 2ρ non-Weierstrass rational points on C. Since the hyperelliptic involution σ on C is defined over $\mathbf{F}_q$, one can choose ρ distinct pairs of non σ-conjugated points $p_1, \sigma(p_1), \dots, p_\rho, \sigma(p_\rho)$ in $C(\mathbf{F}_q)$. In the same way, since no Weierstrass point of D is rational by assumption on g and by hypothesis on $\sharp D(\mathbf{F}_q)$, one can choose ρ distinct pairs of non τ-conjugated points $q_1, \tau(q_1), \dots, q_\rho, \tau(q_\rho)$ in $D(\mathbf{F}_q)$. It follows that Lemma 2.13 applies to $P_1 = (p_1, q_1), \dots, P_\rho = (p_\rho, q_\rho) \in (C \times D)(\mathbf{F}_q)$, and we choose T to be the union of the A-orbits of $P_1, \dots, P_\rho$.

By hypothesis on f, the 2–torsion of C is rational and hence

$$\mathrm{Jac}_C[2](\mathbf{F}_q) \simeq \mathbf{F}_2^{2g_1}$$

on which the Frobenius acts by identity. By hypothesis on g, we have

$$\mathrm{Jac}_D[2](\mathbf{F}_{q^2}) \simeq \mathbf{F}_2^{2g_2},$$

on which the Frobenius acts by a block-diagonal matrix with $\begin{pmatrix} 0 & 1 \\ 1 & 0 \end{pmatrix}$ on the diagonal. Since $\begin{pmatrix} 0 & 1 \\ 1 & 0 \end{pmatrix}$ is conjugated to $\begin{pmatrix} 1 & 1 \\ 0 & 1 \end{pmatrix}$ over $\mathbf{F}_2$, we deduce from (2.3) that

$$\mathrm{H}^1(\overline{C} \times \overline{D}, \mathbf{F}_2)^{G_{\mathbf{F}_q}} \simeq \mathbf{F}_2^{2g_1} \oplus \mathbf{F}_2^{g_2}, \tag{2.5}$$

and from (2.4) together with Lemma 2.14 that

$$\dim \mathrm{H}^2(\overline{X}, \mathbf{F}_\ell)^{G_{\mathbf{F}_q}} = 2g_1g_2 + 2 \tag{2.6}$$

since the only eigenvalue on C is 1, with $m_{1,C} = 2g_1$ and $m_{1,D} = g_1$.

We can then deduce from Theorem 2.7, Lemma 2.13, (2.5) and (2.6) that their exists an infinite 2-tower in which T splits as soon as

$$2g_1 + g_2 \geqslant 3\rho + 2 + 2\sqrt{2g_1g_2 + 2 + 4\rho},$$

completing the proof of Proposition 2.12. □

3. Open problems

In the theory of codes from curves, the usual problem to get asymptotically good families of codes reduces to construct sequences of curves whose sequence of numbers of rational points grows "quickly", and whose sequence of genera grows "slowly". To get asymptotically good families of codes from surfaces, we first need sequences of surfaces whose number of rational points goes to infinity. Next, the natural question is: *the number of rational points should be large, but compared to what?* What will play the part of the genus in the case of surfaces? In [**43**], Papikian studies families of surfaces whose number of rational points is large compared to the sum of the Betti numbers. Despite the interest of this problem for itself, it is up to now unclear that such a feature would permit to construct good codes.

In the present section, we show that for families of surfaces of general type, both invariants K^2 and the topological (i.e. étale) Euler characteristic could be an analog of the genus for this coding theoretic problem. In particular, since the sum of Betti number is larger than the topological Euler characteristic, Papikian's investigations are relevant for the coding theoretic setting.

Caution. The open problem introduced in what follows is independent from the question of existence of infinite étale towers and can be applied to any sequence of surfaces of general type with very ample canonical class.

3.1. From surfaces of general type to the domain of codes.

3.1.1. *The asymptotic domain of surfaces of general type.* We define here a domain $\mathcal{S}_q$ in $(\mathbb{R} \cup \{+\infty\})^2$ as follow.

DEFINITION 3.1. The asymptotic domain of surfaces of general type $\mathcal{S}_q$ is the set of points $(\kappa, \chi) \in (\mathbb{R} \cup \{+\infty\})^2$ for which there exists a sequence X_n of smooth projective absolutely irreducible surfaces defined over $\mathbf{F}_q$, such that:

- for any n, the canonical divisor K_{X_n} is very ample;
- $\sharp X_n(\mathbf{F}_q)$ goes to $+\infty$ as n goes to $+\infty$;
- the ratio $\frac{K_{X_n}^2}{\sharp X_n(\mathbf{F}_q)}$ goes to a limit $\kappa \in \mathbb{R}_+ \cup \{+\infty\}$;
- the ratio $\frac{\chi(\mathcal{O}_{X_n})}{\sharp X_n(\mathbf{F}_q)}$ goes to a limit $\chi \in \mathbb{R} \cup \{+\infty\}$.

REMARK 3.2. Note that K_{X_n} being ample, we have $\kappa \geqslant 0$, and that $\chi \geqslant 0$ in case q is odd, since from [**25**] we have $\chi(\mathcal{O}_X) > 0$ for any surface X of general type in odd characteristic.

REMARK 3.3. Note also that by Noether formula:

$$12\chi(\mathcal{O}_X) = K^2 + \chi_{\text{ét}}, \tag{3.1}$$

one can choose as parameters for this domain any pair among $\lim \frac{\chi(\mathcal{O}_{X_n})}{\sharp X_n(\mathbf{F}_q)}$, $\lim \frac{K_{X_n}^2}{\sharp X_n(\mathbf{F}_q)}$ and $\lim \frac{\chi_{\text{ét}}}{\sharp X_n(\mathbf{F}_q)}$.

To bound this domain, we use the following well–known inequality (3.2) listed for instance in [**25**]:

$$5K^2 + 36 \geqslant \chi_{\text{ét}} \tag{3.2}$$

which, together with Noether formula (3.1), leads asymptotically to

$$\chi \leqslant \frac{\kappa}{2}.$$

It follows that the domain $\mathcal{S}_q$ lies below the line $\chi = \frac{\kappa}{2}$. Moreover, Theorem 1.12 together with Proposition 1.9 entail the following new asymptotic bound.

THEOREM 3.4. *For any point $(\kappa, \chi) \in \mathcal{S}_q$, we have $\kappa \geqslant \frac{1}{(q+1)^2}$.*

PROOF. Let X be a member of a family of surfaces whose parameters go to (κ, χ). From Theorem 1.12, the linear system $\Gamma = (q+1)K_X$ is $(X(\mathbf{F}_q))$-interpolating. Hence by Proposition 1.9, we have $\Gamma^2 \geqslant \sharp X(\mathbf{F}_q)$, that is

$$(q+1)^2 K_X^2 \geqslant \sharp X(\mathbf{F}_q),$$

from which the Theorem follows. □

3.1.2. *Some maps between the domain $\mathcal{S}_q$ and the domain of codes.* We recall that the domain of codes $\mathcal{D}_q$ is the set of points $(\delta, R) \in [0,1]^2$, such that there exists a family of $[N_n, k_n, d_n]_q$-codes for which:

- N_n goes to infinity;
- $\frac{k_n}{N_n}$ goes to R;
- $\frac{d_n}{N_n}$ goes to δ.

A well-known Plotkin bound asserts that this domain lies under the line from $(0,1)$ to $(1 - \frac{1}{q}, 0)$.

PROPOSITION 3.5. *For any integer* $2 \leqslant g \leqslant q$, *the affine map*

$$\varphi_g : (\kappa, \chi) \longmapsto \left(\delta = 1 - g(q+1)\kappa, R = \frac{g(g-1)}{2}\kappa + \chi \right)$$

sends the part of $\mathcal{S}_q$ *for which* $\kappa < \frac{1}{g(q+1)}$ *into the domain of codes* $\mathcal{D}_q$.

The map φ_g is illustrated in Figure 3.

REMARK 3.6. We emphasize the following.

(1) If the maps φ_g are defined for any $g \geqslant 2$, they are irrelevant for $g \geqslant q+1$ since then, the part of $\mathcal{S}_q$ for which $\kappa < \frac{1}{g(q+1)}$ is empty by Theorem 3.4.
(2) It is easily checked that the whole part $\{(\kappa, \chi); \chi \leqslant \frac{\kappa}{2}\}$ does map under φ_g in the area below Singleton bound $R + \delta \leqslant 1$, and even under Plotkin bound adding the restriction $\kappa \geqslant \frac{1}{(q+1)^2}$ given by Theorem 3.4.

PROOF. Let $g \geqslant 2$ and X be a member of a family whose parameters go to (κ, χ) in $\mathcal{S}_q$. We consider the code $C(X, \mathcal{P}, G)$ for $\mathcal{P} = X(\mathbf{F}_q)$ and $G = gK$, whose length is $N = \sharp X(\mathbf{F}_q)$. Since K is very ample, $\Gamma = |(q+1)K|$ is $\mathcal{P}$-interpolating by Theorem 1.12. Moreover, since $g \geqslant 2$, we have $G \cdot K = gK^2 > K^2$ for the ample divisor $H = K$. Consequently, from Theorem 1.11, this code has minimum distance at least

$$d \geqslant N - \Gamma \cdot G = N - g(q+1)K^2,$$

and hence its asymptotic relative distance satisfies

$$\delta \geqslant 1 - g(q+1)\kappa.$$

In addition, its dimension is at least

$$k \geqslant \frac{1}{2} G \cdot (G - K) + \chi(\mathcal{O}_X) = \frac{g(g-1)}{2} K^2 + \chi(\mathcal{O}_X),$$

hence, its asymptotic transmission rate satisfies

$$R \geqslant \frac{g(g-1)}{2}\kappa + \chi,$$

and the Proposition is proved. □

3.1.3. *A first new open problem.* Considering these maps, it follows that the study of $\mathcal{S}_q$ is of interest for coding theoretic purposes. More precisely, proving the existence of a family $(X_n)_{n \in \mathbb{N}}$ of surfaces of general type with very ample canonical divisors with

$$\left(\kappa = \lim \frac{K_{X_n}^2}{\sharp X_n(\mathbf{F}_q)}, \chi = \lim \frac{\chi(\mathcal{O}_{X_n})}{\sharp X_n(\mathbf{F}_q)} \right)$$

as close as possible to the line $\chi = \frac{\kappa}{2}$ and with $\frac{1}{(q+1)^2} \leqslant \kappa < \frac{1}{2(q+1)}$ would yield to a family of codes near the line $(C_2 D_2)$ of Figure 3. Moreover, the closer the point (κ, χ) to $D_1 = (\frac{1}{(q+1)^2}, \frac{1}{2(q+1)^2})$, the closer the point to the Plotkin bound.

REMARK 3.7. For any non-hyperelliptic curves C_1 and C_2 of genus $g_1 \geqslant 3$ and $g_2 \geqslant 3$, the surface $C_1 \times C_2$ is of general type and the canonical divisor $K_{C_1 \times C_2} = (2g_1 - 2)V + (2g_2 - 2)H$ is very ample using the Segre embedding. Moreover, $\chi(\mathcal{O}_{C_1 \times C_2}) = (g_1 - 1)(g_2 - 1)$ and $K_{C_1 \times C_2}^2 = 8(g_1 - 1)(g_2 - 1)$, so that this kind of surfaces with large genus and high number of rational points seems to provide good candidates since, asymptotically $\chi = \frac{\kappa}{8}$, whose slope is a quarter of the one of the

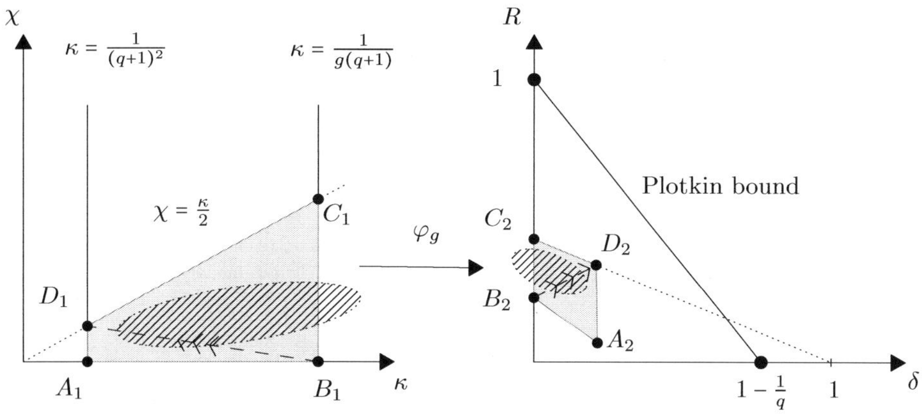

FIGURE 3. The affine map $\varphi_g(\kappa,\chi) = (1-g(q+1)\kappa, \frac{g(g-1)}{2}\kappa+\chi)$ from $\mathbb{R}^2_+$ to $\mathbb{R}^2$. The poorly known domain $\mathcal{S}_q$ is arbitrarily drawn as the hashed area in the left hand side, and its image under φ_g as the hashed area in the right hand side. The part of $\mathcal{S}_q$ lying on the left of the line $\kappa = \frac{1}{g(q+1)}$ is contained inside the grey polygon $P_1 = A_1B_1C_1D_1$. The point $A_1 = (\frac{1}{(q+1)^2}, 0)$ maps to $A_2 = (1-\frac{g}{q+1}, \frac{g^2-g}{2(q+1)^2})$, $B_1 = (\frac{1}{g(q+1)}, 0)$ maps to $B_2 = (0, \frac{g^2-g}{2g(q+1)})$, $C_1 = (\frac{1}{g(q+1)}, \frac{1}{2g(q+1)})$ maps to $C_2 = (0, \frac{g^2-g+1}{2g(q+1)})$ and $D_1 = (\frac{1}{(q+1)^2}, \frac{1}{2(q+1)^2})$ maps to $D_2 = (1-\frac{g}{q+1}, \frac{g^2-g+1}{2(q+1)^2})$. Hence the grey polygon P_1 maps onto the grey polygon $P_2 = A_2B_2C_2D_2$ inside $[0,1]^2$, below the Plotkin bound. The dotted line (C_1D_1) of equation $\chi = \frac{\kappa}{2}$ on the left maps to the dotted line (C_2D_2) of slope $-\frac{g^2-g+1}{2(q+1)^2}$ on the right; since $g \leqslant q$, this slope is less than $-\frac{1}{2}$, close to $-\frac{1}{2}$ if g is close to q. The oriented dotted path $[B_1D_1]$ on the left is mapped on the oriented doted path $[B_2D_2]$ on the right.

line $\chi = \frac{\kappa}{2}$. Unfortunately, the asymptotic Drinfeld-Vladut bounds $\frac{\sharp C_i(\mathbf{F}_q)}{g_i} \leqslant \sqrt{q}-1$ for $i = 1, 2$ yield to $\kappa \geqslant \frac{8}{(\sqrt{q}-1)^2}$, larger than $\frac{1}{2(q+1)}$ for any value of q. These product surfaces thus lie on the right of the line $\kappa = \frac{1}{2(q+1)}$, mapping under any φ_g in the area $\delta < 0$.

3.2. Asymptotic theory for surfaces : algebraic and analytic side. Kunyavskii–Tsfasman, Hindry–Pacheco, Zykin and Lebacque studied the asymptotic behavior of L-functions of elliptic curves over rational function field or of Zeta functions of varieties defined over a finite field. Zykin put it in a general context but his study is well adapted to the analytic context and his point of view does not suit the algebraic properties of families.

In the case of curves, we compare the number of their rational points with their genus. The beauty of the theory comes from the fact that everything is closely connected : good families of global fields for the algebraic point of view (unramified, with many points of small norms) are good for the analytic point of view (giving rise to good infinite Zeta functions) and they give rise to asymptotic good families of codes and sphere packings.

For surfaces, we have (at least) two reasonable choices for what plays the role of the genus : the sum of the ℓ-adic Betti-numbers or their alternate sum –their ℓ-adic Euler-Poincaré characteristic. The first appears when we consider the analytic side, and the second appears naturally in the algebraic one, as it is multiplied by the degree in étale covers. As far as we understand now, the three points of view – algebraic, analytic and applications– seem to involve different quantities and the relations between them is not so clear as in the case of curves. An asymptotically good family should be a family that is good for the algebraic and analytic point of view, and one should be able to construct from them good codes.

Questions : What should be the definition of an asymptotically good family? What should be the normalization? What should be their algebraic and analytic properties?

Acknowledgments

The authors express a deep gratitude to Olivier Dudas, David Madore, Jade Nardi, Fabrice Orgogozo, Cédric Pépin, Michel Raynaud, Alexander Schmidt and Jakob Stix, for inspiring discussions.

References

[1] Michael Artin, *Grothendieck topologies: notes on a seminar*, Harvard University, Dept. of Mathematics, 1962.

[2] Yves Aubry, *Algebraic geometric codes on surfaces*, Talk at Eurocode'92 - International symposium on coding theory and applications (1992, Udine, Italie), in Ph.D. thesis of the University of Aix-Marseille II, France (1993), hal-00979000.

[3] Yves Aubry, *Reed-Muller codes associated to projective algebraic varieties*, Coding theory and algebraic geometry (Luminy, 1991), Lecture Notes in Math., vol. 1518, Springer, Berlin, 1992, pp. 4–17, DOI 10.1007/BFb0087988. MR1186411

[4] Yves Aubry, Elena Berardini, Fabien Herbaut, and Marc Perret, *Algebraic geometry codes over abelian surfaces containing no absolutely irreducible curves of low genus*, ArXiv:1904.08227, 2019.

[5] E. Ballico, *Codes coming from a blowing up of the plane*, Afr. Mat. **24** (2013), no. 1, 93–96, DOI 10.1007/s13370-011-0042-x. MR3019809

[6] E. Ballico, *AG codes obtained from blowing-ups of the plane at non-rational points*, Int. J. Pure Appl. Math. **69** (2011), no. 1, 49–55. MR2841622

[7] Alexander Barg, Kathryn Haymaker, Everett W. Howe, Gretchen L. Matthews, and Anthony Várilly-Alvarado, *Locally recoverable codes from algebraic curves and surfaces*, Algebraic geometry for coding theory and cryptography, Assoc. Women Math. Ser., vol. 9, Springer, Cham, 2017, pp. 95–127, DOI 10.1007/978-3-319-63931-4_4. MR3775426

[8] Alexander Barg, Itzhak Tamo, and Serge Vlăduţ, *Locally recoverable codes on algebraic curves*, IEEE Trans. Inform. Theory **63** (2017), no. 8, 4928–4939, DOI 10.1109/TIT.2017.2700859. MR3683544

[9] Régis Blache, Alain Couvreur, Emmanuel Hallouin, David Madore, Jade Nardi, Matthieu Rambaud, and Hugues Randriam, *Anticanonical codes from del Pezzo surfaces with Picard rank one*, Trans. Amer. Math. Soc. **373** (2020), no. 8, 5371–5393, DOI 10.1090/tran/8119. MR4127880

[10] Thanasis Bouganis, *Error correcting codes over algebraic surfaces*, Applied algebra, algebraic algorithms and error-correcting codes (Toulouse, 2003), Lecture Notes in Comput. Sci., vol. 2643, Springer, Berlin, 2003, pp. 169–179, DOI 10.1007/3-540-44828-4_19. MR2042424

[11] Lucian Bădescu, *Algebraic surfaces*, Universitext, Springer-Verlag, New York, 2001. Translated from the 1981 Romanian original by Vladimir Maşek and revised by the author, DOI 10.1007/978-1-4757-3512-3. MR1805816

[12] Alain Couvreur, *Construction of rational surfaces yielding good codes*, Finite Fields Appl. **17** (2011), no. 5, 424–441, DOI 10.1016/j.ffa.2011.02.007. MR2831703

[13] Alain Couvreur and Iwan Duursma, *Evaluation codes from smooth quadric surfaces and twisted Segre varieties*, Des. Codes Cryptogr. **66** (2013), no. 1-3, 291–303, DOI 10.1007/s10623-012-9692-4. MR3016570

[14] Jennifer A. Davis, *Algebraic geometric codes on anticanonical surfaces*, J. Pure Appl. Algebra **215** (2011), no. 4, 496–510, DOI 10.1016/j.jpaa.2010.06.002. MR2738367

[15] Igor V. Dolgachev, *A brief introduction to Enriques surfaces*, Development of moduli theory—Kyoto 2013, Adv. Stud. Pure Math., vol. 69, Math. Soc. Japan, [Tokyo], 2016, pp. 1–32, DOI 10.2969/aspm/06910001. MR3586505

[16] Frédéric A. B. Edoukou, *Codes defined by forms of degree 2 on Hermitian surfaces and Sørensen's conjecture*, Finite Fields Appl. **13** (2007), no. 3, 616–627, DOI 10.1016/j.ffa.2006.07.001. MR2332489

[17] Frédéric A. B. Edoukou, *Codes defined by forms of degree 2 on quadric surfaces*, IEEE Trans. Inform. Theory **54** (2008), no. 2, 860–864, DOI 10.1109/TIT.2007.913450. MR2444563

[18] Kazuhiro Fujiwara, *A proof of the absolute purity conjecture (after Gabber)*, Algebraic geometry 2000, Azumino (Hotaka), Adv. Stud. Pure Math., vol. 36, Math. Soc. Japan, Tokyo, 2002, pp. 153–183, DOI 10.2969/aspm/03610153. MR1971516

[19] William Fulton, *Intersection theory*, 2nd ed., Ergebnisse der Mathematik und ihrer Grenzgebiete. 3. Folge. A Series of Modern Surveys in Mathematics [Results in Mathematics and Related Areas. 3rd Series. A Series of Modern Surveys in Mathematics], vol. 2, Springer-Verlag, Berlin, 1998, DOI 10.1007/978-1-4612-1700-8. MR1644323

[20] Arnaldo García and Henning Stichtenoth, *A tower of Artin-Schreier extensions of function fields attaining the Drinfel′d-Vlăduţ bound*, Invent. Math. **121** (1995), no. 1, 211–222, DOI 10.1007/BF01884295. MR1345289

[21] V. D. Goppa, *Codes on algebraic curves* (Russian), Dokl. Akad. Nauk SSSR **259** (1981), no. 6, 1289–1290. MR628795

[22] Alexander Grothendieck, *Sur quelques points d'algèbre homologique* (French), Tohoku Math. J. (2) **9** (1957), 119–221, DOI 10.2748/tmj/1178244839. MR102537

[23] A. Grothendieck, *Éléments de géométrie algébrique. II. Étude globale élémentaire de quelques classes de morphismes* (French), Inst. Hautes Études Sci. Publ. Math. **8** (1961), 222. MR217084

[24] Alexander Grothendieck, *Revêtements étales et groupe fondamental. Fasc. I: Exposés 1 à 5*, Séminaire de Géométrie Algébrique, vol. 1960/61, Institut des Hautes Études Scientifiques, Paris, 1963. MR0217087

[25] Yi Gu, *On algebraic surfaces of general type with negative c_2*, Compos. Math. **152** (2016), no. 9, 1966–1998, DOI 10.1112/S0010437X16007491. MR3568945

[26] Safia Haloui, *Codes from Jacobian surfaces*, Arithmetic, geometry, cryptography and coding theory, Contemp. Math., vol. 686, Amer. Math. Soc., Providence, RI, 2017, pp. 123–135, DOI 10.1090/conm/686. MR3630612

[27] Johan P. Hansen, *Toric surfaces and error-correcting codes*, Coding theory, cryptography and related areas (Guanajuato, 1998), Springer, Berlin, 2000, pp. 132–142. MR1749454

[28] Søren Have Hansen, *Error-correcting codes from higher-dimensional varieties*, Finite Fields Appl. **7** (2001), no. 4, 531–552, DOI 10.1006/ffta.2001.0313. MR1866342

[29] Robin Hartshorne, *Algebraic geometry*, Springer-Verlag, New York-Heidelberg, 1977. Graduate Texts in Mathematics, No. 52. MR0463157

[30] Kathryn Haymaker, Beth Malmskog, and Gretchen L. Matthews, *Locally recoverable codes with availability $t \geqslant 2$ from fiber products of curves*, Adv. Math. Commun. **12** (2018), no. 2, 317–336, DOI 10.3934/amc.2018020. MR3808231

[31] Daniel Huybrechts, *Lectures on K3 surfaces*, Cambridge Studies in Advanced Mathematics, vol. 158, Cambridge University Press, Cambridge, 2016, DOI 10.1017/CBO9781316594193. MR3586372

[32] Yasutaka Ihara, *Some remarks on the number of rational points of algebraic curves over finite fields*, J. Fac. Sci. Univ. Tokyo Sect. IA Math. **28** (1981), no. 3, 721–724 (1982). MR656048

[33] Shigeru Iitaka, *Algebraic geometry*, Graduate Texts in Mathematics, vol. 76, Springer-Verlag, New York-Berlin, 1982. An introduction to birational geometry of algebraic varieties; North-Holland Mathematical Library, 24. MR637060

[34] V. A. Iskovskikh and I. R. Shafarevich, *Algebraic surfaces [MR1060325 (91f:14029)]*, Algebraic geometry, II, Encyclopaedia Math. Sci., vol. 35, Springer, Berlin, 1996, pp. 127–262, DOI 10.1007/978-3-642-60925-1_2. MR1392959

[35] Philippe Lebacque and Alexey Zykin, *Asymptotic methods in number theory and algebraic geometry* (English, with English and French summaries), Actes de la Conférence "Théorie des Nombres et Applications", Publ. Math. Besançon Algèbre Théorie Nr., vol. 2011, Presses Univ. Franche-Comté, Besançon, 2011, pp. 47–73. MR2894268

[36] Xudong Li, Liming Ma, and Chaoping Xing, *Optimal locally repairable codes via elliptic curves*, IEEE Trans. Inform. Theory **65** (2019), no. 1, 108–117, DOI 10.1109/TIT.2018.2844216. MR3900980

[37] John B. Little, *Algebraic geometry codes from higher dimensional varieties*, Advances in algebraic geometry codes, Ser. Coding Theory Cryptol., vol. 5, World Sci. Publ., Hackensack, NJ, 2008, pp. 257–293, DOI 10.1142/9789812794017_0007. MR2509126

[38] John Little and Hal Schenck, *Codes from surfaces with small Picard number*, SIAM J. Appl. Algebra Geom. **2** (2018), no. 2, 242–258, DOI 10.1137/17M1128277. MR3797729

[39] James S. Milne, *Étale cohomology*, Princeton Mathematical Series, vol. 33, Princeton University Press, Princeton, N.J., 1980. MR559531

[40] David Mumford, *Abelian varieties*, Tata Institute of Fundamental Research Studies in Mathematics, No. 5, Published for the Tata Institute of Fundamental Research, Bombay; Oxford University Press, London, 1970. MR0282985

[41] Carlos Munuera, Wanderson Tenório, and Fernando Torres, *Locally recoverable codes from algebraic curves with separated variables*, Adv. Math. Commun. **14** (2020), no. 2, 265–278, DOI 10.3934/amc.2020019. MR4097440

[42] Jade Nardi, *Algebraic geometric codes on minimal Hirzebruch surfaces*, J. Algebra **535** (2019), 556–597, DOI 10.1016/j.jalgebra.2019.06.022. MR3986402

[43] Mihran Papikian, *Modular varieties of $\mathcal{D}$-elliptic sheaves and the Weil-Deligne bound*, J. Reine Angew. Math. **626** (2009), 115–134, DOI 10.1515/CRELLE.2009.004. MR2492991

[44] Cecília Salgado, Anthony Várilly–Alvarado, and José Felipe Voloch, *Locally recoverable codes on surfaces*, ArXiv:1910.13472, 2019.

[45] Alexander Schmidt, *Über Pro-p-Fundamentalgruppen markierter arithmetischer Kurven* (German, with English summary), J. Reine Angew. Math. **640** (2010), 203–235, DOI 10.1515/CRELLE.2010.025. MR2629694

[46] Jean-Pierre Serre, *Groupes algébriques et corps de classes* (French), Publications de l'institut de mathématique de l'université de Nancago, VII. Hermann, Paris, 1959. MR0103191

[47] Tamás Szamuely, *Corps de classes des schémas arithmétiques* (French, with French summary), Astérisque **332** (2010), Exp. No. 1006, viii–ix, 257–286. Séminaire Bourbaki. Volume 2008/2009. Exposés 997–1011. MR2648681

[48] Itzhak Tamo and Alexander Barg, *Bounds on locally recoverable codes with multiple recovering sets*, 2014 IEEE International Symposium on Information Theory, June 2014, pp. 691–695.

[49] M. A. Tsfasman, S. G. Vlăduţ, and Th. Zink, *Modular curves, Shimura curves, and Goppa codes, better than Varshamov-Gilbert bound*, Math. Nachr. **109** (1982), 21–28, DOI 10.1002/mana.19821090103. MR705893

[50] S. G. Vlèduts and Yu. I. Manin, *Linear codes and modular curves* (Russian), Current problems in mathematics, Vol. 25, Itogi Nauki i Tekhniki, Akad. Nauk SSSR, Vsesoyuz. Inst. Nauchn. i Tekhn. Inform., Moscow, 1984, pp. 209–257. MR770943

[51] José Felipe Voloch and Marcos Zarzar, *Algebraic geometric codes on surfaces* (English, with English and French summaries), Arithmetics, geometry, and coding theory (AGCT 2005), Sémin. Congr., vol. 21, Soc. Math. France, Paris, 2010, pp. 211–216. MR2856569

[52] Oscar Zariski, *On the purity of the branch locus of algebraic functions*, Proc. Nat. Acad. Sci. U.S.A. **44** (1958), 791–796, DOI 10.1073/pnas.44.8.791. MR95846

[53] Marcos Zarzar, *Error-correcting codes on low rank surfaces*, Finite Fields Appl. **13** (2007), no. 4, 727–737, DOI 10.1016/j.ffa.2007.05.001. MR2359313

ALAIN COUVREUR, INRIA & LABORATOIRE LIX, CNRS UMR 7161,ÉCOLE POLYTECHNIQUE, 91120 PALAISEAU CEDEX
Email address: alain.couvreur@inria.fr

PHILIPPE LEBACQUE, LABORATOIRE DE MATHÉMATIQUES DE BESANÇON, UFR SCIENCES ET TECHNIQUES, 16 ROUTE DE GRAY, 25030 BESANÇON, FRANCE
Email address: philippe.lebacque@univ-fcomte.fr

MARC PERRET, INSTITUT DE MATHÉMATIQUES DE TOULOUSE, UMR 5219, UNIVERSITÉ DE TOULOUSE, CNRS, UT2J, F-31058 TOULOUSE, FRANCE
Email address: perret@univ-tlse2.fr

Appendix A. On the étale site of marked schemes,
by Alexander Schmidt

When considering the étale site of a scheme it is often of interest to consider a variant which forces a given set of points to split in at least one member of a covering. Examples are the étale site of a marked curve used in [**11**], where a finite set of closed points is considered and the Nisnevich site [**10**], where all points are required to split. In this note we develop this approach in greater generality. Furthermore, we close a small gap in the literature by showing that any Nisnevich covering of a quasi-compact scheme has a finite subcovering.

A.1. Definition of the marked site. Let X be a scheme and let T be a set of points of X. We will loosely write $T \subset X$ and call the pair (X,T) a *marked scheme*. A morphism $f:(Y,S) \to (X,T)$ of marked schemes is a scheme morphism $f:Y \to X$ with $f(S) \subset T$.

DEFINITION A.1. Let (X,T) be a marked scheme. The *marked étale site* $(X,T)_{\text{et}}$ consists of the following data: The category $\text{Cat}(X,T)_{\text{et}}$ is the category of morphisms $f:(U,S) \to (X,T)$ such that

a) $(f:U \to X)$ is étale, and
b) $S = p^{-1}(T)$.

A family $(p_i : (U_i,S_i) \to (U,S))_{i\in I}$ of morphisms in $\text{Cat}(X,T)_{\text{et}}$ is a covering if it is surjective and any point $s \in S$ *splits*, i.e., there exists an index i and a point $u_i \in S_i$ mapping to s such that the induced field homomorphism $k(s) \to k(u_i)$ is an isomorphism.

EXAMPLE A.2. For $T = \varnothing$, we obtain the small étale site of X, for $T = X$ the Nisnevich site [**10**].

A morphism of marked schemes induces a morphism of the associated marked étale sites in the obvious way.

We consider the following "geometric points" of $(X,T)_{\text{et}}$: we fix a separable closure $k(x)^s$ of $k(x)$ for every scheme-theoretic point $x \in X$ and consider the following morphisms of marked schemes

1.) for $x \notin T$, the natural morphism $(\mathit{Spec}\, k(x)^s, \varnothing) \to (X,T)$.
2.) for $x \in T$, the natural morphisms $(\mathit{Spec}\, \kappa, \mathit{Spec}\, \kappa) \to (X,T)$ for every subextension $\kappa/k(x)$ of $k(x)^s/k(x)$.

If $f : P \to X$ is any of the morphisms described in 1.) and 2.), the assignment $F \mapsto \Gamma(P, f^*F)$ is a topos-theoretic point of $(X,T)_{\text{et}}$ and one easily verifies that this family of points is conservative. In particular, exactness of sequences of abelian sheaves can be checked stalkwise.

We denote the cohomology of a sheaf $F \in \mathit{Sh}_{\text{et}}(X,T)$ of abelian groups on $(X,T)_{\text{et}}$ by $H^*_{\text{et}}(X,T,F)$.

A.2. Excision. Let (X,T) be a marked scheme, $i:Z \hookrightarrow X$ a closed immersion and $U = X \smallsetminus Z$ the open complement. The right derivatives of the left exact functor "sections with support in Z"

$$F \mapsto \ker(F(X,T) \to F(U,T \cap U))$$

are called the *cohomology groups with support in* Z. Notation: $H^*_Z(X,T,F)$.

The proof of the next proposition is standard (cf. [**1**, III, (2.11)] for the étale case without marking).

PROPOSITION A.3. *There is a long exact sequence*

$$0 \to H^0_Z(X,T,F) \to H^0_{\mathrm{et}}(X,T,F) \to H^0_{\mathrm{et}}(U,T\cap U,F) \to$$
$$H^1_Z(X,T,F) \to H^1_{\mathrm{et}}(X,T,F) \to H^1_{\mathrm{et}}(U,T\cap U,F) \to \cdots$$

PROPOSITION A.4 (Excision). *Let $\pi : (X',T') \to (X,T)$ be a morphism of marked schemes, $Z \hookrightarrow X$, $Z' \hookrightarrow X'$ closed immersions and $U = X \smallsetminus Z$, $U' = X' \smallsetminus Z'$ the open complements. Assume that*

- *$\pi : X' \to X$ is étale,*
- *$T' = \pi^{-1}(T)$,*
- *π induces an isomorphism $Z'_{\mathrm{red}} \xrightarrow{\sim} Z_{\mathrm{red}}$,*
- *$\pi(U') \subset U$.*

Then the induced homomorphism

$$H^p_Z(X,T,F) \xrightarrow{\sim} H^p_{Z'}(X',T',\pi^*F)$$

is an isomorphism for every sheaf $F \in Sh_{\mathrm{et}}(X,T)$ and all $p \geqslant 0$.

PROOF. The standard proof for étale topology applies: By the general theory, π^* is exact. Since π belongs to $\mathrm{Cat}(X,T)_{\mathrm{et}}$, π^* has the exact left adjoint "extension by zero", hence π^* sends injectives to injectives. Therefore it suffices to deal with the case $p = 0$. Without changing the statement, we can replace all occurring schemes by their reductions. By assumption,

$$(X',T') \coprod (U,T\cap U) \longrightarrow (X,T)$$

is a covering. For $\alpha \in H^0_Z(X,T,F)$ mapping to zero in $H^0_{Z'}(X',T',\pi^*F)$ we therefore obtain $\alpha = 0$.

Now let $\alpha' \in H^0_{Z'}(X',T',\pi^*F)$ be given. We show that α' and $0 \in H^0(U,T\cap U,F)$ glue to an element in $H^0_Z(X,T,F)$. The only nontrivial compatibility on intersections is $p_1^*(\alpha') = p_2^*(\alpha')$ for $p_1,p_2 : (X' \times_X X', T' \times_T T') \to (X',T')$. This can be checked on stalks noting that $Z' \xrightarrow{\sim} Z$ implies that the two projections $Z' \times_Z Z' \to Z'$ are the same. □

A.3. Continuity.

PROPOSITION A.5. *Let X be a quasi-compact scheme and let $T \subset X$ be a closed subscheme. Then every étale covering of (X,T) admits a finite subcovering.*

PROOF. Since X has a finite affine Zariski-open covering, we may assume that X is affine, in particular X is quasi-separated. Then also T is quasi-compact and quasi-separated. Let

$$\coprod_{i\in I}(U_i,S_i) \to (X,T)$$

be an étale covering. Let, for $i \in I$, $K_i \subset T$ be the set of points in T which split in $U_i \to X$. By [**12**, Lemma 13.3], K_i is ind-constructible, i.e., open in the constructible topology of T, which is compact by [**5**, 1.9.15 (iii)]. Since $T = \bigcup_i K_i$ by assumption, we find a finite subset $J \subset I$ with $T = \bigcup_{i\in J} K_i$. Furthermore, since X is quasi-compact and étale morphisms are open, we find a finite subset $J' \subset I$ such that $\coprod_{i\in J'} U_i \to X$ is an étale covering. We conclude that $\coprod_{i\in J\cup J'}(U_i,S_i) \to (X,T)$ is a finite subcovering of $\coprod_{i\in I}(U_i,S_i) \to (X,T)$. □

As in [**2**, VII, 3.2] for the unmarked étale site, we define the *restricted marked étale site*

$$(X,T)_{\mathrm{et}}^{\mathrm{res}}$$

as the restriction of $(X,T)_{\mathrm{et}}$ to the subcategory of all $(U,S) \in (X,T)_{\mathrm{et}}$ where $U \to X$ is of finite presentation. Assume that X is quasi-compact and quasi-separated. Then the same is true for any such U and Proposition A.5 shows that the restricted site is noetherian. Moreover, the categories of sheaves on $(X,T)_{\mathrm{et}}$ and $(X,T)_{\mathrm{et}}^{\mathrm{res}}$ are naturally equivalent. Hence the same argument as in the unmarked étale case [**2**, VII, Prop. 3.3] shows

THEOREM A.6. *Let X be a quasi-compact and quasi-separated scheme and let $T \subset X$ be a closed subscheme. Let (F_i) be a filtered direct system of abelian sheaves on $(X,T)_{\mathrm{et}}$. Then*

$$\operatorname*{colim}_i H^p_{\mathrm{et}}(X,T,F_i) \cong H^p_{\mathrm{et}}(X,T,\operatorname*{colim}_i F_i)$$

for all $p \geq 0$.

Next we consider inverse limits of marked schemes.

THEOREM A.7. *Let (X,T) be a marked scheme with T closed in X and let $X_i \to X$, $i \in I$, be an inverse system of X-schemes. Assume that all X_i are quasi-separated and quasi-compact and that all transition morphisms are affine. Let T_i be the preimage of T in X_i and put $X_\infty = \varprojlim X_i$, $T_\infty = \varprojlim T_i$.*

Then the restricted site $(X_\infty, T_\infty)_{\mathrm{et}}^{\mathrm{res}}$ is the limit site of the sites $(X_i,T_i)_{\mathrm{et}}^{\mathrm{res}}$.

COROLLARY A.8. *With the notation and assumptions of Theorem A.7, let F be a sheaf of abelian groups on $(X,T)_{\mathrm{et}}$. We denote its inverse image on $(X_i,T_i)_{\mathrm{et}}$ and $(X_\infty,T_\infty)_{\mathrm{et}}$ by F_i and F_∞. Then the natural map*

$$\operatorname*{colim}_i H^p_{\mathrm{et}}(X_i,T_i,F_i) \longrightarrow H^p_{\mathrm{et}}(X_\infty,T_\infty,F_\infty)$$

is an isomorphism for all $p \geq 0$.

PROOF OF THEOREM A.7. By [**1**, III, Theorem 3.8], the site $(X_\infty)_{\mathrm{et}}^{\mathrm{res}}$ is naturally equivalent to the limit site of the $(X_i)_{\mathrm{et}}^{\mathrm{res}}$. In view of Proposition A.5, it therefore suffices to show that for every quasi-compact étale surjection $U_i \to X_i$ with the property that every point of T_∞ splits in $U_\infty = U_i \times_{X_i} X_\infty \to X_\infty$ there exist $j \geq i$ such that every point of T_j splits in $U_j = U_i \times_{X_i} X_j \to X_j$. We follow the proof of [**12**, Lemma 13.2] for Nisnevich coverings. By [**12**, Lemma 13.3], the subset $S_j \subset T_j$ of points that split in $U_j \to X_j$ is ind-constructible for all $j \geq i$. Denoting the projection by $u_j : T_\infty \to T_j$, the assumption on $U_\infty \to X_\infty$ implies $T_\infty = \bigcup_j u_j^{-1}(S_j)$. Considering the $T_j \subset X_j$ as reduced, closed subschemes, we may apply [**5**, Cor. 8.3.4] to obtain $S_j = T_j$ for some j. □

REMARK A.9. Let A be a ring and let $(A \to B_i)_{i \in I}$ be an affine Nisnevich covering. We write A as the union of its finitely generated subrings. Then, by Proposition A.5 and Theorem A.7, there exists a finite subset $J \subset I$, a subring $A' \subset A$ which is finitely generated over $\mathbb{Z}$ and a finite Nisnevich covering $(A' \to B'_j)_{j \in J}$ such that $B_j \cong A \otimes_{A'} B'_j$ for all $j \in J$.

Hence the refined definition of Nisnevich coverings for general rings introduced by Lurie in [**7**, XI, Definition 1.1 and Remark 1.15] coincides with the naive definition.

COROLLARY A.10. *Let (X,T) be a marked scheme with T closed in X and $Z = \{z_1,\dots,z_n\}$ a finite set of closed points of X. Put $X^h_{z_i} = Spec(\mathcal{O}^h_{X,z_i})$. Then, for every sheaf F of abelian groups on $(X,T)_{\text{et}}$ and all $p \geq 0$*

$$H^p_Z(X,T,F) \cong \bigoplus_{i=1}^n H^p_{\{z_i\}}(X^h_{z_i}, T \cap X^h_{z_i}, F).$$

PROOF. Since $H^p_Z(X,T,F) \cong \bigoplus_{i=1}^n H^p_{\{z_i\}}(X,T,F)$, we may assume that $Z = \{z\}$ consists of a single closed point. Excision shows that

$$H^p_{\{z\}}(X,T,F) = H^p_{\{z\}}(U, T\cap U, F)$$

for every affine étale open neighbourhood U of z. Since X^h_z is the limit over all these U, the long exact sequences of Proposition A.3 together with Corollary A.8 show the result. □

Using Corollary A.8, it is easy to calculate the stalks of the higher direct images of the site morphism $(X,T)_{\text{et}} \to (X,X)_{\text{et}} = X_{\text{Nis}}$. The Leray spectral sequence together with the fact that the Nisnevich cohomological dimension of noetherian schemes is bounded by the Krull dimension [**10**, Theorem 1.32] yields:

COROLLARY A.11. *Let X be a noetherian scheme of finite Krull dimension d, $T \subset X$ closed and assume that there exists a nonnegative integer N such that*

$$cd(k(x)) \leqslant N$$

for all points $x \in X \smallsetminus T$. Then for every abelian torsion sheaf F on $(X,T)_{\text{et}}$ we have

$$H^p_{\text{et}}(X,T,F) = 0 \qquad \textit{for } p > N + d.$$

A.4. Galois covers.

DEFINITION A.12. A *Galois cover* of X with finite Galois group G in the site $(X,T)_{\text{et}}$ is a morphism $(Y,S) \to (X,T)$ in $(X,T)_{\text{et}}$ together with a right action of G on Y over X such that the following holds:

(1) $(Y,S) \to (X,T)$ is a covering for the site $(X,T)_{\text{et}}$.
(2) $Y \to X$ is an étale Galois cover, i.e.,

$$Y \times G \to Y \times_X Y, \quad (y,g) \mapsto (y,yg)$$

is an isomorphism.

Since G acts transitively on the set of points in Y over a given point $x \in X$, we see that every $t \in T$ splits *completely* in Y/X.

PROPOSITION A.13 (Hochschild-Serre spectral sequence). *Let $(Y,S) \to (X,T)$ be a Galois cover with finite group G and $F \in Sh_{\text{et}}(X,T)$. Then there is a natural spectral sequence*

$$E_2^{pq} = H^p(G, H^q_{\text{et}}(Y,S,F)) \Longrightarrow H^{p+q}_{\text{et}}(X,T,F).$$

PROOF. The proof is word-by-word the same as for the étale cohomology, see [**8**, Theorem 2.20]. □

REMARK A.14. Assume that X is quasi-compact and quasi-separated and $T \subset X$ closed. Let

$$(Y_i, S_i) \to (X,T)$$

be a directed inverse system of Galois covers with finite Galois groups G_i and $(Y,S) = \lim(Y_i, S_i)$. Then $(Y,S) \to (X,T)$ is a pro-Galois cover with profinite Galois group $G = \lim G_i$. By Theorem A.7, for $F \in Sh_{\rm et}(X,T)$, the groups $H^q_{\rm et}(Y,S,F) = \operatorname{colim} H^q_{\rm et}(Y_i,S_i,F)$ are discrete G-modules and we obtain the profinite Hochschild-Serre sequence

$$E_2^{pq} = H^p(G, H^q_{\rm et}(Y,S,F)) \Longrightarrow H^{p+q}_{\rm et}(X,T,F),$$

where $H^*(G,-)$ is the continuous cohomology of the profinite group G with values in a discrete G-module (see [**9**, I, §2]).

A.5. Fundamental group. We recall some facts from Artin-Mazur [**3**]. Let $\mathscr{C}$ be a pointed site and $\mathrm{HR}(\mathscr{C})$ the category of pointed hypercovers of $\mathscr{C}$ [**3**, §8]. If $\mathscr{C}$ is locally connected, then the "connected component functor" π defines an object

$$\Pi\mathscr{C} = \{\pi(K_\bullet)\}_{K_\bullet \in \mathrm{HR}(\mathscr{C})}$$

in the pro-category of the homotopy category of pointed simplicial sets. By definition, the fundamental group of $\mathscr{C}$ is the pro-group $\pi_1(\Pi(\mathscr{C}))$.

Let X be a locally noetherian scheme. Then (cf. [**3**, §9]) the site $X_{\rm et}$, and hence also $(X,T)_{\rm et}$ is locally connected. Pointing $(X,T)_{\rm et}$ by choosing any "geometric" point $\bar{x}$ described at the end of Appendix A.1, we obtain the étale fundamental group $\pi_1^{\rm et}(X,T,\bar{x})$. It is independent of the choice of $\bar{x}$ up to isomorphism, which is canonical up to inner automorphisms. By [**3**, Cor. 10.7], for any group G, the set $\mathrm{Hom}(\pi_1^{\rm et}(X,T,\bar{x}),G)$ is in bijection with the set of isomorphism classes of pointed (over $\bar{x}$) G-torsors in $(X,T)_{\rm et}$. In particular, $\pi_1^{\rm et}(X,\varnothing,\bar{x})$ is the enlarged étale fundamental group of [**4**, X, §6] and its profinite completion is the usual étale fundamental group of X defined in [**6**]. If $\bar{x}$ is a geometric point of X, then $\pi_1^{\rm et}(X,T,\bar{x})$ is a factor group of $\pi_1^{\rm et}(X,\varnothing,\bar{x})$, which is profinite for normal X by [**3**, Thm. 11.1]. Hence we obtain the following result.

PROPOSITION A.15. *Let X be a noetherian, normal, connected scheme and $T \subset X$. Then (for any choice of base point) $\pi_1^{\rm et}(X,T)$ is a profinite group. Its finite quotients are in bijection with the isomorphism classes of finite connected pointed étale Galois covers of X in which every point $t \in T$ splits completely.*

EXAMPLE A.16. For general (X,T), the fundamental group need not be profinite. For example, let k be a field and $N = \mathbf{P}^1_k/(0 \sim 1)$ the node over k. Then

$$\pi_1^{\rm et}(N,T) \cong \begin{cases} \mathbb{Z} \times \mathrm{Gal}_k, & T = \varnothing \\ \mathbb{Z}, & T = X. \end{cases}$$

We will use the notation $\hat{\pi}_1^{\rm et}(X,T)$ for the profinite completion of $\pi_1^{\rm et}(X,T)$, hence we have a completion map $\pi_1^{\rm et}(X,T) \to \hat{\pi}_1^{\rm et}(X,T)$ which is an isomorphism by Proposition A.15 if X is a noetherian, normal and connected scheme.

We end this section with the following observation concerning products.

PROPOSITION A.17. *Let k be a field, X and Y geometrically connected schemes of finite type over k and $S \subset X(k)$, $T \subset Y(k)$ nonempty sets of k-rational points. Let a and b be geometric points of $X \smallsetminus S$ and $Y \smallsetminus T$ with values in a common separably closed extension field of k. Assume that at least one of the schemes X and Y is proper over k. Then the natural map*

$$\hat{\pi}_1^{\rm et}(X \times_k Y, S \times T, (a,b)) \longrightarrow \hat{\pi}_1^{\rm et}(X,S,a) \times \hat{\pi}_1^{\rm et}(Y,T,b)$$

is an isomorphism of profinite groups.

PROOF. We omit the base points from notation. For a connected scheme X, let $\widetilde{X}$ denote the profinite universal cover. For a subset $S \subset X$, the kernel of $\hat{\pi}_1^{\text{et}}(X) \to \hat{\pi}_1^{\text{et}}(X,S)$ is the (closed) normal subgroup of $\hat{\pi}_1^{\text{et}}(X) = \text{Gal}(\widetilde{X}|X)$ generated by the decomposition groups of the points in S, i.e., it is the (closed) subgroup of $\text{Gal}(\widetilde{X}|X)$ generated by all automorphisms which fix a point $\tilde{s} \in \widetilde{X}$ lying over some $s \in S$. We denote this group by $K(X,S)$.

Now assume we are in the situation of the proposition. By the topological invariance of the étale topology we may assume that k is perfect. Let $\bar{k}$ be an algebraic closure of k. We denote the base changes to $\bar{k}$ by $(\bar{X},\bar{S})$ and $(\bar{Y},\bar{T})$. By [**6**, X, 1.7], we have a natural isomorphism

$$\hat{\pi}_1^{\text{et}}(\bar{X} \times_{\bar{k}} \bar{Y}) \xrightarrow{\sim} \hat{\pi}_1^{\text{et}}(\bar{X}) \times \hat{\pi}_1^{\text{et}}(\bar{Y}).$$

Moreover, by [**6**, IX, 6.1], we have a natural exact sequence

$$1 \longrightarrow \hat{\pi}_1^{\text{et}}(\bar{X}) \longrightarrow \hat{\pi}_1^{\text{et}}(X) \longrightarrow \text{Gal}(\bar{k}|k) \longrightarrow 1.$$

This and the similar sequence for Y shows the isomorphism

$$\hat{\pi}_1^{\text{et}}(X \times_k Y) \xrightarrow{\sim} \hat{\pi}_1^{\text{et}}(X) \times_{\text{Gal}(\bar{k}|k)} \hat{\pi}_1^{\text{et}}(Y), \tag{$*$}$$

where the term on the right hand side is a fibre product in the category of profinite groups. We consider the corresponding diagram of étale Galois covers.

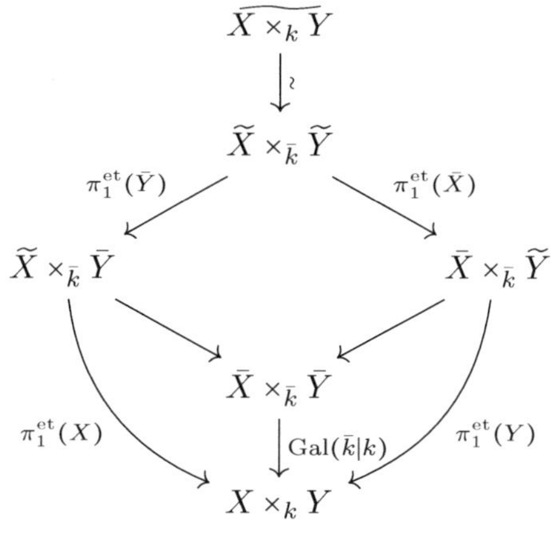

Let $(s,t) \in S \times T \subset X \times_k Y$ and let $(\tilde{s},\tilde{t}) \in \widetilde{X \times_k Y} = \widetilde{X} \times_{\bar{k}} \widetilde{Y}$ be a point lying above (s,t). An element $\sigma \in \hat{\pi}_1^{\text{et}}(X \times_k Y) = \text{Gal}(\widetilde{X \times_k Y}|X \times_k Y)$ fixes $(\tilde{s},\tilde{t})$ if and only if its image in $\hat{\pi}_1^{\text{et}}(X) = \text{Gal}(\widetilde{X} \times_{\bar{k}} \bar{Y}|X \times_k Y)$ fixes $\tilde{s} \in \widetilde{X}$ and its image in $\hat{\pi}_1^{\text{et}}(Y) = \text{Gal}(\bar{X} \times_{\bar{k}} \widetilde{Y}|X \times_k Y)$ fixes $\tilde{t} \in \widetilde{Y}$. Hence, the isomorphism $(*)$ induces an isomorphism of subgroups

$$K(X \times_k Y, S \times T) \xrightarrow{\sim} K(X,S) \times_{\text{Gal}(\bar{k}|k)} K(Y,T). \tag{$**$}$$

The isomorphisms $(*)$ and $(**)$ together induce an isomorphism

$$\hat{\pi}_1^{\text{et}}(X \times_k Y, S \times T) \xrightarrow{\sim} C$$

with

$$C = \text{coker}\big(K(X,S) \times_{\text{Gal}(\bar{k}|k)} K(Y,T) \longrightarrow \hat{\pi}_1^{\text{et}}(X) \times_{\text{Gal}(\bar{k}|k)} \hat{\pi}_1^{\text{et}}(Y)\big).$$

The natural homomorphism $C \to \hat{\pi}_1^{\text{et}}(X,S) \times \hat{\pi}_1^{\text{et}}(Y,T)$ is injective. To conclude the proof of the proposition, it remains to show surjectivity, i.e., we have to show that every element in $\hat{\pi}_1^{\text{et}}(X,S) \times \hat{\pi}_1^{\text{et}}(Y,T)$ has a preimage in $\hat{\pi}_1^{\text{et}}(X) \times_{\text{Gal}(\bar{k}|k)} \hat{\pi}_1^{\text{et}}(Y) \subset \hat{\pi}_1^{\text{et}}(X) \times \hat{\pi}_1^{\text{et}}(Y)$. For this it suffices to show that the composite map $K(X,S) \hookrightarrow \pi_1^{\text{et}}(X) \to \text{Gal}(\bar{k}|k)$ is surjective. This is true since $K(X,S)$ contains the decomposition group of a k-rational point. □

A.6. A modification. We consider a modification of the marked étale site which was used in [**11**] for one-dimensional, noetherian regular schemes.

DEFINITION A.18. The *strict marked étale site* $(X,T)_{\text{et-s}}$ consists of the following data: $\text{Cat}(X,T)_{\text{et-s}}$ is the category of morphisms $f : (U,S) \to (X,T)$ such that

a) $(f : U \to X)$ is étale,
b) $S = p^{-1}(T)$, and
c) for every $u \in S$ mapping to $t \in T$ the induced field homomorphism $k(s) \to k(u)$ is an isomorphism.

Coverings are surjective families.

PROPOSITION A.19. (i) *If $T \subset X$ consists of a finite set of closed points, then the natural morphism of sites $\varphi : (X,T)_{\text{et}} \to (X,T)_{\text{et-s}}$ induces isomorphisms*

$$H^p_{\text{et-s}}(X,T,F) \xrightarrow{\sim} H^p_{\text{et}}(X,T,\varphi^*F),\ H^p_{\text{et-s}}(X,T,\varphi_*G) \xrightarrow{\sim} H^p_{\text{et}}(X,T,G)$$

for any $F \in Sh_{\text{et-s}}(X,T)$, $G \in Sh_{\text{et-s}}(X,T)$ and $p \geq 0$.

(ii) *For locally noetherian X (and any chosen base point), the natural map*

$$\pi_1^{\text{et}}(X,T) \longrightarrow \pi_1^{\text{et-s}}(X,T)$$

is an isomorphism.

PROOF. Let $(U,S) \in \text{Cat}(X,T)_{\text{et-s}}$ and assume that $(f_i : (U_i,S_i) \to (U,S))$ is a covering in $(X,T)_{\text{et}}$. Removing for all i the finitely many points $s \in S_i$ such that $k(f(s_i)) \to k(s_i)$ is not an isomorphism from U_i, we obtain a strict covering $(f_i : (U_i',S_i') \to (U,S))$ which is a refinement of the original one. Hence $\varphi_*\varphi^*F = F$ and $R^q\varphi_*G = 0$ for $q \geq 1$. In view of the Leray spectral sequence, this shows (i). Assertion (ii) follows since both pro-groups represent the same functor: for any group G, a G-torsor in $(X,T)_{\text{et}}$ is the same as a G-torsor in $(X,T)_{\text{et-s}}$. □

Acknowledgment

The author thanks Philippe Lebacque for helpful comments and motivational discussions.

References

[1] Michael Artin, *Grothendieck topologies: notes on a seminar*, Harvard University, Dept. of Mathematics, 1962.

[2] Michael Artin, Alexander Grothendieck, and Jean-Louis Verdier, *Théorie des topos et cohomologie étale des schémas (SGA 4)*, Lecture Notes in Math., Springer-Verlag, Berlin-New York, 1972, Séminaire de Géométrie Algébrique du Bois-Marie 1963–1964 (SGA 4), Dirigé par M. Artin, A. Grothendieck, et J. L. Verdier. Avec la collaboration de N. Bourbaki, P. Deligne et B. Saint-Donat.

[3] M. Artin and B. Mazur, *Etale homotopy*, Lecture Notes in Mathematics, vol. 100, Springer-Verlag, Berlin, 1986. Reprint of the 1969 original. MR883959

[4] Michel Demazure and Alexander Grothendieck, *Schémas en groupes. II: Groupes de type multiplicatif, et structure des schémas en groupes généraux*, Séminaire de Géométrie Algébrique du Bois Marie 1962/64 (SGA 3). Lecture Notes in Math., vol. 152, Springer-Verlag, Berlin-New York, 1970.

[5] A. Grothendieck, *Éléments de géométrie algébrique. IV. Étude locale des schémas et des morphismes de schémas IV* (French), Inst. Hautes Études Sci. Publ. Math. **32** (1967), 361. MR238860

[6] ———, *Revêtements étales et groupe fondamental (SGA 1)*, Documents Mathématiques (Paris), vol. 3, Société Mathématique de France, Paris, 2003, Séminaire de Géométrie Algébrique du Bois Marie 1960–1961 (SGA 1), Dirigé par Alexandre Grothendieck. Augmenté de deux exposés de M. Raynaud, Lecture Notes in Math., Vol. 224.

[7] Jacob Lurie, *Derived algebraic geometry*, ProQuest LLC, Ann Arbor, MI, 2004. Thesis (Ph.D.)–Massachusetts Institute of Technology. MR2717174

[8] James S. Milne, *Étale cohomology*, Princeton Mathematical Series, vol. 33, Princeton University Press, Princeton, N.J., 1980. MR559531

[9] Jürgen Neukirch, Alexander Schmidt, and Kay Wingberg, *Cohomology of number fields*, 2nd ed., Grundlehren der Mathematischen Wissenschaften [Fundamental Principles of Mathematical Sciences], vol. 323, Springer-Verlag, Berlin, 2008, DOI 10.1007/978-3-540-37889-1. MR2392026

[10] Ye. A. Nisnevich, *The completely decomposed topology on schemes and associated descent spectral sequences in algebraic K-theory*, Algebraic K-theory: connections with geometry and topology (Lake Louise, AB, 1987), NATO Adv. Sci. Inst. Ser. C Math. Phys. Sci., vol. 279, Kluwer Acad. Publ., Dordrecht, 1989, pp. 241–342. MR1045853

[11] Alexander Schmidt, *Über Pro-p-Fundamentalgruppen markierter arithmetischer Kurven* (German, with English summary), J. Reine Angew. Math. **640** (2010), 203–235, DOI 10.1515/CRELLE.2010.025. MR2629694

[12] Stefan Schröer, *Geometry on totally separably closed schemes*, Algebra Number Theory **11** (2017), no. 3, 537–582, DOI 10.2140/ant.2017.11.537. MR3649361

Alexander Schmidt, Mathematisches Institut, Universität Heidelberg, Im Neuenheimer Feld 205, 69120 Heidelberg, Deutschland

Email address: schmidt@mathi.uni-heidelberg.de

Contemporary Mathematics
Volume **770**, 2021
https://doi.org/10.1090/conm/770/15432

Sato–Tate groups of abelian threefolds: a preview of the classification

Francesc Fité, Kiran S. Kedlaya, and Andrew V. Sutherland

ABSTRACT. We announce the classification of Sato–Tate groups of abelian threefolds over number fields; there are 410 possible conjugacy classes of closed subgroups of USp(6) that occur. We summarize the key points of the "upper bound" aspect of the classification, and give a more rigorous treatment of the "lower bound" by realizing 33 groups that appear in the classification as maximal cases with respect to inclusions of finite index. Further details will be provided in a subsequent paper.

1. Introduction

The *Sato–Tate group* of an abelian variety A over a number field k is a certain compact Lie group which conjecturally governs the distribution of normalized Euler factors of the L-function of A. For example, if A is an elliptic curve, then the Sato–Tate group is determined by whether or not A has complex multiplication (and if so, whether the field of complex multiplication is contained in k). In the generic case where A has no complex multiplication, its Sato–Tate group is $\mathrm{SU}(2)$ and the usual Sato–Tate conjecture predicts that the distribution of normalized Frobenius traces of A in the interval $[-2, 2]$ is the same as the distributions of traces of random matrices in $\mathrm{SU}(2)$ determined by the Haar measure on $\mathrm{SU}(2)$; this is known when k is a totally real field [**5**] or a CM field [**1**].

The present paper constitutes an announcement of the classification of Sato–Tate groups of abelian threefolds over number fields, in the style of the classification of Sato–Tate groups of abelian surfaces made in [**15**] (building upon [**22,23**]). These groups are closed subgroups of $\mathrm{USp}(6)$ that are defined up to conjugacy; we find that there are 410 Sato–Tate groups that occur for abelian threefolds (compared to 3 for elliptic curves and 52 for abelian surfaces).

2020 *Mathematics Subject Classification.* Primary 11M50; Secondary 11G10, 11G20, 14G10, 14K15.

Fité was supported by the Institute for Advanced Study during 2018–2019 via NSF grant DMS-1638352, and by the Simons Collaboration in Arithmetic Geometry, Number Theory, and Computation (Simons Foundation grant 550033).

Kedlaya was financially supported by the NSF (grant DMS-1802161), UC San Diego (Stefan E. Warschawski professorship), and IAS (visiting professorship 2018–2019).

Sutherland was financially supported by the NSF (grant DMS-1522526) and by the Simons Collaboration in Arithmetic Geometry, Number Theory, and Computation (Simons Foundation grant 550033).

Given the dramatically larger size of the classification in dimension 3, it is not feasible to give a complete account here; we have thus structured this article as a preview of our subsequent paper [**17**] in which we give a complete treatment. Here, we focus on the following steps.

- In §2, we recall some theoretical background needed to make a rigorous definition of the Sato–Tate group; recall a twisting construction that will be needed to realize certain candidate Sato–Tate groups; and formulate some necessary conditions on Sato–Tate groups that reduce the classification to a finite problem (the *Sato–Tate axioms*).
- In §3, we state the classification of closed subgroups of USp(6) satisfying the Sato–Tate axioms to be obtained in [**17**] (Theorem 3.1); there are 433 such groups. We identify the 14 connected groups that can occur, corresponding to the possible geometric endomorphism $\mathbb{R}$-algebras of an abelian threefold over an algebraically closed field of characteristic 0. For each such group G^0, we describe the maximal finite-index overgroups of G^0 that occur; the derivation of this description is given completely in all cases except when G^0 is a one-dimensional torus. In this case, we state a uniform presentation of the maximal overgroups using complex reflection groups, and summarize how the description of finite subgroups of PSU(3) by Blichfeldt–Dickson–Miller [**29**] is used to confirm this list in [**17**].
- In §4, we give a complete proof (modulo enumeration of groups; see below) that within the list of 433 groups from Theorem 3.1, exactly 410 occur as Sato–Tate groups (Theorem 4.1). More precisely, we show that exactly 23 groups must be omitted; these groups have identity components containing a factor which is a two-dimensional or three-dimensional torus, and the exclusion of these groups requires a careful study of Shimura's theory of CM types analogous to [**15**, §4.3]. Revising the list of maximal subgroups enumerated in §3 based on this exclusion yields a list of 33 candidate groups which are maximal with respect to inclusions of finite index. Using products and twisting constructions, we show that each maximal group arises from some abelian threefold over $\mathbb{Q}$ or $\mathbb{Q}(\sqrt{3})$; by base extension, we may obtain examples of abelian threefolds (over various number fields) realizing all candidate groups.

We defer to [**17**] the discussion of the following aspects, which for abelian surfaces are treated in [**15**].

- Giving a detailed proof of Theorem 3.1. This amounts to confirming that every group that can occur is contained in one of the announced maximal groups.
- Enumerating the finite-index subgroups of the announced maximal groups in order to verify that they represent 433 distinct conjugacy classes within USp(6). This is needed to confirm that the lower bound in Theorem 3.1 is indeed equal to 410.
- Computation of moment statistics and point densities. For abelian surfaces, all 52 Sato–Tate groups can be distinguished by their moments, but this fails for threefolds.
- Rigorous determination of the Sato–Tate group of an explicitly specified abelian threefold. One approach to this is to compute endomorphism algebras using the method of [**10**].

- Large-scale surveys of Sato–Tate groups. With the computational resources and software tools now available [**36**, **37**], this can now be done at much greater scale than in [**15**].
- Distinguishing between Jacobians of curves, principally polarized abelian threefolds, and arbitrary abelian threefolds. For abelian surfaces this does not change the classification of Sato–Tate groups, as was shown in [**15**] using a large scale survey of genus 2 curves over $\mathbb{Q}$. For threefolds the situation is not yet clear.
- Analysis of fields of definition. For abelian surfaces, only 34 of the 52 possible groups occur over $\mathbb{Q}$; the situation for abelian threefolds is unclear.

2. Background on Sato–Tate groups

In this section, we recall some of the theoretical results from [**15**] that form the basis of the classification of Sato–Tate groups of abelian surfaces, and extend these as required to handle the case of abelian threefolds. We then use these results to classify the possible identity components of Sato–Tate groups of abelian threefolds. See [**35**] for an overview of this circle of ideas.

2.1. The Mumford–Tate group. We begin by recalling the definition of the Mumford–Tate group of a (polarized) abelian variety A over a number field k. This construction carries enough information to determine the identity component of the Sato–Tate group.

DEFINITION 2.1. Fix an embedding of k into $\mathbb{C}$ and set $V := H^1(A^{\mathrm{top}}_{\mathbb{C}}, \mathbb{Q})$; it carries an alternating pairing ψ given by the cup product. Via the description of $A^{\mathrm{top}}_{\mathbb{C}}$ as a complex torus, we obtain an $\mathbb{R}$-linear identification of $V \otimes_{\mathbb{Q}} \mathbb{R}$ with the tangent space of $A_{\mathbb{C}}$ at the origin (or equivalently, the dual of the space of holomorphic differentials on $A_{\mathbb{C}}$). The *Mumford–Tate group* of A, denoted $\mathrm{MT}(A)$, is the smallest $\mathbb{Q}$-algebraic subgroup of $\mathrm{GL}(V)$ whose base extension to $\mathbb{R}$ contains the scalar action of $\mathbb{C}^\times$ on $V \otimes_{\mathbb{Q}} \mathbb{R}$; this group is contained in the symplectic group $\mathrm{GSp}(V, \psi)$. It follows from Deligne's theorem on absolute Hodge cycles (see [**11**]) that the definition of $\mathrm{MT}(A)$ does not depend on the choice of the embedding of k into $\mathbb{C}$; it is also clearly invariant under enlargement of k.

DEFINITION 2.2. The *Hodge group* $\mathrm{Hg}(A)$ is the intersection $\mathrm{MT}(A) \cap \mathrm{Sp}(V, \psi)$. The *Lefschetz group* $\mathrm{L}(A)$ is the connected component of the identity in the centralizer of $\mathrm{End}(A_{\overline{k}})_{\mathbb{Q}}$ in $\mathrm{Sp}(V, \psi)$. There is an obvious inclusion $\mathrm{Hg}(A) \subseteq \mathrm{L}(A)$.

PROPOSITION 2.3. For $g \leq 3$, we have $\mathrm{Hg}(A) = \mathrm{L}(A)$. (That is, the Mumford–Tate group is determined by endomorphisms.)

PROOF. See [**15**, Theorem 2.16]. □

REMARK 2.4. As shown by Mumford [**30**], Proposition 2.3 fails for $g = 4$. The case $g = 3$ considered here is the last case where the endomorphism-based methods developed in [**15**] and extended here and in [**17**] will suffice.

2.2. Definition of the Sato–Tate group. We continue with the definition of the algebraic Sato–Tate group and Sato–Tate group of an abelian variety A over a number field k in terms of ℓ-adic monodromy, and the statement of the refined Sato–Tate conjecture. This material is taken primarily from [**15**, §2.1], but includes some further results from [**3**, §2] and [**4**] that we note here.

We denote by $\mathrm{End}(A)$ the ring of endomorphisms of A (defined over k). By the endomorphism field of A, we refer to the minimal extension K of k such that $\mathrm{End}(A_K)$ coincides with the geometric endomorphism ring $\mathrm{End}(A_{\overline{k}})$ of A. Note that K/k is a finite and Galois extension.

DEFINITION 2.5. For each $\tau \in \mathrm{Gal}(K/k)$, define

$$\mathrm{L}(A,\tau) := \{\gamma \in \mathrm{Sp}(V,\psi) : \gamma^{-1}\alpha\gamma = \tau(\alpha) \text{ for all } \alpha \in \mathrm{End}(A_K)\}.$$

The *twisted Lefschetz group* $\mathrm{TL}(A)$ is the union of $\mathrm{L}(A,\tau)$ over all τ.

PROPOSITION 2.6. For $g \leq 3$, for any prime ℓ, $\mathrm{TL}(A) \otimes_{\mathbb{Q}} \mathbb{Q}_\ell$ is the kernel of the symplectic character on the Zariski closure of the image of $\rho_\ell : G_k \to H^1_{\mathrm{et}}(A_{\overline{k}}, \mathbb{Q}_\ell)$.

PROOF. This amounts to the statement that the Mumford–Tate conjecture holds for A whenever $g \leq 3$. See for example [**3**, Theorem 6.11]. □

DEFINITION 2.7. In light of Proposition 2.3 and Proposition 2.6, for $g \leq 3$, we define the *Sato–Tate group* $\mathrm{ST}(A)$ to be a maximal compact subgroup of $\mathrm{TL}(A) \otimes_{\mathbb{Q}} \mathbb{C}$.

LEMMA 2.8. *For $g \leq 3$, there is a canonical isomorphism* $\mathrm{ST}(A)/\mathrm{ST}(A)^0 \to \mathrm{Gal}(K/k)$. *In particular, this isomorphism is compatible with base change: for any finite extension k' of k, $\mathrm{ST}(A_{k'})$ is the inverse image of* $\mathrm{Gal}(k'K/k') \subseteq \mathrm{Gal}(K/k)$ *in* $\mathrm{ST}(A)$.

PROOF. This is again a consequence of [**3**, Theorem 6.11]. □

REMARK 2.9. In the definition of the Sato–Tate group, we are implicitly using the fact that for $g \leq 3$, the Mumford–Tate group of A is determined by endomorphisms. For general g, it is expected that the role of $\mathrm{TL}(A)$ in Proposition 2.6 can be filled by a certain algebraic group over $\mathbb{Q}$, the *motivic Sato–Tate group*, whose definition involves algebraic cycles on $A \times A$ of all codimensions, not just endomorphisms [**4**]. There is still a canonical surjection $\mathrm{ST}(A)/\mathrm{ST}(A)^0 \to \mathrm{Gal}(K/k)$, but it is not in general an isomorphism.

2.3. Twisting and the Sato–Tate group.

DEFINITION 2.10. For L/k a finite Galois extension and ξ a 1-cocycle of $\mathrm{Gal}(L/k)$ valued in $\mathrm{Aut}(A_L)$, there exists a unique (up to unique isomorphism) abelian variety A_ξ over k and an isomorphism $\theta : A_{\xi,L} \to A_L$ such that

$$\xi(\sigma) = \theta^\sigma \circ \theta^{-1} \qquad (\text{for all } \sigma \in \mathrm{Gal}(L/k)).$$

Moreover, A_ξ depends only on the class of ξ in the pointed set $H^1(\mathrm{Gal}(L/k), \mathrm{Aut}(A_L))$.

EXAMPLE 2.11. Let E be an elliptic curve without complex multiplication defined over $\mathbb{Q}$ and let $L/\mathbb{Q}$ be a Galois extension. Let $A = E^d$, and fix an isomorphism $\mathrm{Aut}(A_L) \simeq \mathrm{GL}_d(\mathbb{Z})$ for some integer $d \geq 1$. For every rational degree d Artin representation ρ, there is a choice of a basis for which the image of ρ lies in $\mathrm{GL}_d(\mathbb{Z})$. We may thus regard a rational degree d representation of $\mathrm{Gal}(L/\mathbb{Q})$ as a 1-cocycle

$$\xi \colon \mathrm{Gal}(L/\mathbb{Q}) \to \mathrm{Aut}(A_L)\,.$$

The endomorphism field of the abelian d-fold A_ξ defined above is the field cut out by the representation $\rho \otimes \rho^\vee$. The Galois group of the endomorphism field is thus isomorphic to the projective image of the original degree d rational representation ρ of $\mathrm{Gal}(L/\mathbb{Q})$. We will use this construction in §4.2.4.

EXAMPLE 2.12. Similarly, let E be an elliptic curve over $\mathbb{Q}$ with complex multiplication by the maximal order of the imaginary quadratic field M (which must have class number one, since E is defined over $\mathbb{Q}$). Let $L/\mathbb{Q}$ be a Galois extension containing M for which the extension

$$1 \to \mathrm{Gal}(L/M) \to \mathrm{Gal}(L/\mathbb{Q}) \to \mathrm{Gal}(M/\mathbb{Q}) \to 1$$

is split. Let $A = E^d$, and fix an isomorphism $\mathrm{Aut}(A_L) \simeq \mathrm{GL}_d(\mathcal{O}_M)$ for some integer $d \geq 1$. For the action of C_2 on $\mathrm{GL}_d(M)$ via complex conjugation on M, suppose that $\tilde{\rho} : \mathrm{Gal}(L/\mathbb{Q}) \to \mathrm{GL}_d(M) \rtimes \mathrm{C}_2$ is a representation whose restriction ρ to $\mathrm{Gal}(L/M)$ factors through $\mathrm{GL}_d(M)$. Since $\mathcal{O}_M$ has class number 1, there is a choice of basis of M^d for which ρ factors through a representation $\rho_0 : \mathrm{Gal}(L/M) \to \mathrm{GL}_d(\mathcal{O}_M)$.

Suppose further that the basis of M^d can be chosen so that $\tilde{\rho}$ factors through $\mathrm{GL}_d(\mathcal{O}_M) \rtimes \mathrm{C}_2$ (see Remark 2.13). We then obtain from $\tilde{\rho}$ a cocycle

$$\xi \colon \mathrm{Gal}(L/\mathbb{Q}) \to \mathrm{Aut}(A_L)$$

and again the Galois group of the endomorphism field of A_ξ is isomorphic to the projective image of $\mathrm{Gal}(L/\mathbb{Q})$. We will use this construction in §4.2.5.

REMARK 2.13. In Example 2.11, the condition that $\tilde{\rho}$ can be factored through $\mathrm{GL}_d(\mathcal{O}_M) \rtimes \mathrm{C}_2$ is equivalent to requiring the existence of a choice of ρ_0 for which

$$\rho_0 \cong c_M \circ \rho_0 \circ c, \tag{2.1}$$

where c denotes the action of $\mathrm{Gal}(M/\mathbb{Q})$ on $\mathrm{Gal}(L/M)$ while c_M denotes complex conjugation on $\mathcal{O}_M^d$. The following considerations will be useful.

(i) We automatically have (2.1) if ρ_0 is uniquely determined by ρ. (This can only occur if ρ is absolutely irreducible.)

(ii) We also have (2.1) if $\tilde{\rho}$ descends to $\mathrm{GL}_d(\mathbb{Z}) \times \mathrm{C}_2$ and $\mathrm{Gal}(M/\mathbb{Q})$ acts on $\mathrm{Gal}(L/M)$ as an inner automorphism.

(iii) When ρ_0 is not uniquely determined by ρ, the choices for ρ_0 will be distinguished by their reductions modulo some $\alpha \in \mathcal{O}_M$. Using this, we can sometimes track the effects of c and c_M and thus detect whether or not (2.1) holds.

REMARK 2.14. We will encounter a small number of cases where (2.1) fails. In these cases, we have an isomorphism $\rho \cong c_M \circ \rho \circ c$ which does not descend to ρ_0; that is, the image of ρ_0 is normalized by some matrix of $\mathrm{GL}_d(M) \setminus \mathrm{GL}_d(\mathcal{O}_M)$.

In general, such cases correspond to twists not of E^d but of an isogenous abelian variety (which by [**21**, Theorem 2] is itself a product of elliptic curves isogenous to E). We work out two specific constructions of this type. (Compare [**34**, §4] or [**27**, Example 5.15].)

- Let E be the elliptic curve $y^2 = x^3 - 1$; E has CM by the maximal order $\mathbb{Z}[\zeta_3]$ and contains the rational 2-torsion point $Q = (1, 0)$, with the other two 2-torsion points Q_1, Q_2 being defined over M. Let A be the quotient of the Weil restriction of E_M from M to $\mathbb{Q}$ by the rational subgroup of order 2 corresponding to $\langle Q_1 \rangle$; then

$$\begin{aligned} \mathrm{End}(A_M) &\cong \mathrm{M}_2(\mathcal{O}_M) \cap \begin{pmatrix} 2^{-1} & 0 \\ 0 & 1 \end{pmatrix} \mathrm{M}_2(\mathcal{O}_M) \begin{pmatrix} 2 & 0 \\ 0 & 1 \end{pmatrix} \\ &= \left\{ \begin{pmatrix} a & b \\ c & d \end{pmatrix} \in M_2(\mathcal{O}_M) : b \equiv 0 \pmod 2 \right\} \end{aligned}$$

with the action of $\mathrm{Gal}(M/\mathbb{Q})$ given by

$$g \mapsto \begin{pmatrix} 2^{-1} & 0 \\ 0 & 1 \end{pmatrix} \overline{g} \begin{pmatrix} 2 & 0 \\ 0 & 1 \end{pmatrix}.$$

(Note that $A_M \cong E_M \times E'_M$ where $E'_M = E_M/\langle Q_1 \rangle$ has CM by the nonmaximal order $\mathbb{Z}[2\zeta_3]$.)

- Put $k = \mathbb{Q}(\sqrt{3})$. Let E be the elliptic curve over $\mathbb{Q}(\sqrt{3})$ with LMFDB label `2.2.12.1-9.1-a3`; it has CM by $\mathbb{Z}[i]$ and has a rational point Q of order 3. Let A be the product $E \times E/\langle Q \rangle$; then

$$\begin{aligned} \mathrm{End}(A_{kM} &\cong \mathrm{M}_2(\mathcal{O}_M) \cap \begin{pmatrix} 3^{-1} & 0 \\ 0 & 1 \end{pmatrix} \mathrm{M}_2(\mathcal{O}_M) \begin{pmatrix} 3 & 0 \\ 0 & 1 \end{pmatrix} \\ &= \left\{ \begin{pmatrix} a & b \\ c & d \end{pmatrix} \in M_2(\mathcal{O}_M) : b \equiv 0 \pmod 3 \right\} \end{aligned}$$

with the action of $\mathrm{Gal}(M/\mathbb{Q})$ given by

$$g \mapsto \begin{pmatrix} 3^{-1} & 0 \\ 0 & 1 \end{pmatrix} \overline{g} \begin{pmatrix} 3 & 0 \\ 0 & 1 \end{pmatrix}.$$

(Note that $E/\langle Q \rangle$ has CM by the nonmaximal order $\mathbb{Z}[3i]$.)

2.4. Axioms for Sato–Tate groups. A key tool used to classify Sato–Tate groups in [**15**] is a list of necessary conditions called the *Sato–Tate axioms.* The formulation in [**15**, Definition 3.1] is applicable to arbitrary motives; we state here a restricted form of the three original Sato–Tate axioms applicable to the 1-motives associated to abelian threefolds. We also add a fourth axiom coming from the fact that the Sato–Tate group of an abelian threefold is determined by endomorphisms (Proposition 2.3).

DEFINITION 2.15. For a subgroup G of $\mathrm{GL}_6(\mathbb{C})$ with identity connected component G^0, the *Sato–Tate axioms (for abelian threefolds)* are as follows.

(ST1) The group G is a closed subgroup of $\mathrm{USp}(6)$. For definiteness, we take the latter to be defined with respect to the symplectic form given by the block matrix

$$J = \begin{pmatrix} 0 & I_3 \\ -I_3 & 0 \end{pmatrix},$$

where I_3 denotes a 3×3 identity matrix, unless otherwise specified.

(ST2) (Hodge condition) There exists a homomorphism $\theta\colon \mathrm{U}(1) \to G^0$ such that $\theta(u)$ has eigenvalues u, u^{-1} of multiplicity 3. The image of such a θ is a *Hodge circle*, and the set of all Hodge circles must generate a dense subgroup of G^0.

(ST3) (Rationality condition) For each component H of G and each irreducible character χ of $\mathrm{GL}_6(\mathbb{C})$, the expected value (under the Haar measure) of $\chi(\gamma)$ over $\gamma \in H$ is an integer. In particular, for any positive integers m and n, the expectation $\mathrm{E}[\mathrm{Trace}(\gamma, \wedge^m \mathbb{C}^6)^n : \gamma \in H]$ is an integer.

(ST4) (Lefschetz condition) The subgroup of $\mathrm{USp}(6)$ fixing $\mathrm{End}(\mathbb{C}^6)^{G^0}$ is G^0.

PROPOSITION 2.16. Let A be an abelian threefold over k. Then $G = \mathrm{ST}_A$ satisfies the Sato–Tate axioms.

PROOF. For (ST1), (ST2), (ST3), this is [**15**, Proposition 3.2] except that the density condition in (ST2) is not stated therein; that statement is a consequence of the definition of the Mumford–Tate group [**15**, Definition 2.11]. For (ST4), apply Proposition 2.3. □

3. Classification: an overview

This section is mainly expository. We report on several aspects of the following theorem, whose proof will appear in the upcoming work [**17**].

THEOREM 3.1 ([**17**]). *Up to conjugacy, at most* 433 *subgroups of* $\mathrm{GL}_6(\mathbb{C})$ *satisfy the Sato–Tate axioms for an abelian threefold. Among these,* 30 *are maximal with respect to finite index inclusions.*

REMARK 3.2. In §4, we will show that all 433 groups of the above theorem satisfy the Sato–Tate axioms (see Remark 4.7).

We will describe the possible connected components of these groups (§3.1), explain their connection to the possible geometric endomorphism rings of abelian threefolds (§3.2), and for each identity component present the extensions that are maximal with respect to the relation of inclusion of finite index up to conjugacy (§3.3). We should nevertheless emphasize that §3.1 and §3.2 contain complete proofs; only §3.3 depends on results in [**17**].

3.1. Connected component of the identity. We first identify the connected groups that satisfy the Sato–Tate axioms (ST1), (ST2), (ST4). Note that condition (ST3) is vacuous for a connected group, as the expected value in question is simply the dimension of the fixed subspace of the representation with character χ.

Let G^0 be a connected group satisfying the Sato–Tate axioms and let T be a maximal torus of G^0. Let $\mathfrak{h}$ denote the Lie algebra of G^0 and let $\mathfrak{h}_{\mathbb{C}}$ be the complexification of $\mathfrak{h}$. Since T is contained in a maximal torus of $\mathrm{USp}(6)$, which is 3-dimensional, $\mathfrak{h}_{\mathbb{C}}$ has rank at most 3.

The set of Lie algebras of rank at most 3 can easily be determined from the classification of Dynkin diagrams:

r	Lie algebras of rank r
1	$\mathfrak{t}_1,\ \mathfrak{sl}_2 = \mathfrak{so}_3$
2	$\mathfrak{t}_2,\ \mathfrak{t}_1{\times}\mathfrak{sl}_2,\ \mathfrak{sl}_2{\times}\mathfrak{sl}_2 = \mathfrak{so}_4,\ \mathfrak{sl}_3,\ \mathfrak{sp}_4 = \mathfrak{so}_5,\ \mathfrak{g}_2$
3	$\mathfrak{t}_3, \mathfrak{t}_2{\times}\mathfrak{sl}_2,\ \mathfrak{t}_1{\times}\mathfrak{sl}_2{\times}\mathfrak{sl}_2,\ \mathfrak{t}_1{\times}\mathfrak{sl}_3, \mathfrak{t}_1{\times}\mathfrak{sp}_4, \mathfrak{t}_1{\times}\mathfrak{g}_2,\ \mathfrak{sl}_2{\times}\mathfrak{sl}_2{\times}\mathfrak{sl}_2,$ $\mathfrak{sl}_2{\times}\mathfrak{sl}_3,\ \mathfrak{sl}_2{\times}\mathfrak{sp}_4, \mathfrak{sl}_2{\times}\mathfrak{g}_2,\ \mathfrak{sl}_4 = \mathfrak{so}_6,\ \mathfrak{so}_7,\ \mathfrak{sp}_6$

The standard representation of $\mathrm{USp}(6)$ gives rise to a faithful 6-dimensional unitary symplectic self-dual representation of $\mathfrak{h}$. This eliminates some Lie algebras from the above list.

- There are no 6-dimensional symplectic representations of $\mathfrak{sp}_4$ except if we allow some trivial summands, but these violate (ST2).
- The smallest dimension of a nontrivial representation of $\mathfrak{g}_2$ is $7 > 6$; this also rules out $\mathfrak{t}_1 \times \mathfrak{g}_2$ and $\mathfrak{sl}_2 \times \mathfrak{g}_2$.
- The only nontrivial self-dual representation of $\mathfrak{sl}_3$ of dimension at most 6 is the sum of the standard 3-dimensional representation and its dual. This rules out $\mathfrak{sl}_2 \times \mathfrak{sl}_3$.

- The only nontrivial self-dual representation of $\mathfrak{so}_6$ is the standard representation, which is not symplectic.
- The smallest dimension of a nontrivial representation of $\mathfrak{so}_7$ is $7 > 6$.

The above considerations, together with an additional argument to rule out the group U(2) (see [**16**, Rem. 2.3]), allow us to conclude that G^0 must correspond to one of the following groups (the subscripts in the notation are defined below):

$\dim(T)$	Connected groups satisfying (ST1) and (ST2)
1	$\mathrm{U}(1)_3$, $\mathrm{SU}(2)_3$, $\mathrm{SO}(3)$
2	$\mathrm{U}(1){\times}\mathrm{U}(1)_2$, $\mathrm{U}(1){\times}\mathrm{SU}(2)_2$, $\mathrm{SU}(2){\times}\mathrm{U}(1)_2$, $\mathrm{SU}(2){\times}\mathrm{SU}(2)_2$, $\mathrm{SU}(3)$
3	$\mathrm{U}(1){\times}\mathrm{U}(1){\times}\mathrm{U}(1)$, $\mathrm{U}(1){\times}\mathrm{U}(1){\times}\mathrm{SU}(2)$, $\mathrm{U}(1){\times}\mathrm{SU}(2){\times}\mathrm{SU}(2)$, $\mathrm{U}(3)$, $\mathrm{U}(1){\times}\mathrm{USp}(4)$, $\mathrm{SU}(2){\times}\mathrm{SU}(2){\times}\mathrm{SU}(2)$, $\mathrm{SU}(2){\times}\mathrm{USp}(4)$, $\mathrm{USp}(6)$

Each group comes equipped with a 6-dimensional representation, which may be inferred from the preceding table using the following considerations:

- The groups SU(2), USp(4), USp(6) carry their standard representations.
- For $d = 2, 3$, the notations $\mathrm{U}(1)_d, \mathrm{SU}(2)_d$ refer to the d-fold diagonal representations of U(1), SU(2).
- The groups U(1), U(3) are embedded into SU(2), USp(6) by the formula

$$A \mapsto \begin{pmatrix} A & 0 \\ 0 & \overline{A} \end{pmatrix}. \tag{3.1}$$

- The groups SO(3) and SU(3) are viewed in USp(6) via the above embedding of U(3).
- Note that the abstract group $\mathrm{U}(1) \times \mathrm{SU}(2)$ is considered twice: once as a subgroup of $\mathrm{U}(1) \times \mathrm{U}(1) \times \mathrm{SU}(2)$ via the diagonal embedding of U(1), and once as a subgroup of $\mathrm{U}(1) \times \mathrm{SU}(2) \times \mathrm{SU}(2)$ via the diagonal embedding of SU(2).

In the case USp(6) and all of the product cases, one easily verifies that (ST4) is satisfied by following the methods of [**15**, §4]. This leaves the cases of SO(3), SU(3), and U(3), which we work out by hand. Recall that U(3) is embedded into USp(6) via (3.1). As a result, we see that the endomorphisms of $\mathbb{C}^6$ commuting with SO(3) or SU(3) are those given by block scalar matrices, all of which also commute with U(3). It follows that SO(3) and SU(3) do not satisfy (ST4).

The possible values for a connected G satisfying the Sato–Tate axioms are listed in the fourth column of Table 1.

3.2. Geometric endomorphism algebras. In this section we recover the possible geometric $\mathbb{R}$-algebras of endomorphisms of abelian threefolds from our classification of identity components of Sato–Tate groups. We then describe the possible geometric $\mathbb{Q}$-algebras of endomorphisms they correspond, by using Shimura's refinement of Albert's classification of division algebras equipped with a positive involution.

3.2.1. *Geometric endomorphism $\mathbb{R}$-algebras.* For an abelian threefold A defined over a number field k, we can recover the geometric endomorphism $\mathbb{R}$-algebra $\mathrm{End}(A_{\overline{k}})_{\mathbb{R}} := \mathrm{End}(A_{\overline{k}}) \otimes \mathbb{R}$ by first finding the subalgebra of $\mathrm{End}(\mathbb{C}^6)$ commuting with $\mathrm{ST}(A)^0$, then picking out the maximal positive definite subspace of this algebra under the Rosati form [**15**, Prop. 2.19].

In the case of USp(6) and all the other product cases, the above computation is easily carried out with the methods of [**15**, §4]. In the case of U(3) the Rosati form is given up to scalars by

$$\begin{pmatrix} uI_3 & 0 \\ 0 & vI_3 \end{pmatrix} \mapsto 6uv = \frac{3}{2}((u+v)^2 - (u-v)^2),$$

so the maximal positive definite subspace consists of those matrices for which $u+v \in \mathbb{R}, u-v \in i\mathbb{R}$, or equivalently $v = \overline{u}$. Therefore $\mathrm{End}(A_{\overline{k}})_{\mathbb{R}} \simeq \mathbb{C}$. We assign letters **A**, **B**, ..., **N** to each of the possibilities for $\mathrm{End}(A_{\overline{k}})_{\mathbb{R}}$. We refer to this letter as the *absolute type* of A. The absolute types together with the possible values for $\mathrm{End}(A_{\overline{k}})_{\mathbb{R}}$ are listed in the first and second columns of Table 1.

3.2.2. *Geometric endomorphism $\mathbb{Q}$-algebras.* For a simple abelian variety A of dimension $g \leq 3$, the work of Albert and Shimura leaves the following possibilities for the geometric $\mathbb{Q}$-algebra of endomorphisms $\mathrm{End}(A_{\overline{k}})_{\mathbb{Q}} := \mathrm{End}(A_{\overline{k}}) \otimes \mathbb{Q}$. Note that for a fixed $g \leq 3$, each possibility is distinguished by the $\mathbb{Q}$-dimension d of $\mathrm{End}(A_{\overline{k}})_{\mathbb{Q}}$ (equal to the $\mathbb{R}$-dimension of $\mathrm{End}(A_{\overline{k}})_{\mathbb{R}}$). (See for example [**31**].)

g	d	Albert type	$\mathbb{Q}$-algebra
1	1	I(1)	$\mathbb{Q}$
	2	IV(1,1)	imaginary quadratic field
2	2	I(1)	real quadratic field
	4	I(2)	indefinite quaternion algebra
	6	IV(2,1)	quartic CM field
3	1	I(1)	$\mathbb{Q}$
	2	IV(1,1)	imaginary quadratic field
	3	I(3)	totally real cubic field
	6	IV(3,1)	sextic CM field

From the list of geometric $\mathbb{Q}$-algebra endomorphism types for simple abelian varieties of dimension $g \leq 3$ one can determine all possibilities for $g = 3$ by taking all possible products (including self products) that yield an abelian variety of dimension 3; there are 22 ways to do this. Each yields a distinct type of $\mathbb{Q}$-algebra, but only 14 distinct $\mathbb{R}$-algebras (for example, taking three non-isogenous elliptic curves of type I(1) yields the same $\mathbb{R}$-algebra $\mathbb{R} \times \mathbb{R} \times \mathbb{R}$ as a single RM abelian threefold of type I(3)). This leads to the following description of absolute types in terms of the isogeny decomposition of $A_{\overline{k}}$ (compare with [**35**, §4, Table 2]):

(**A**) $A_{\overline{k}}$ is simple of type I(1);
(**B**) $A_{\overline{k}}$ is simple of type IV(1,1);
(**C**) $A_{\overline{k}}$ is isogenous to the product of an elliptic curve without CM and a simple abelian surface of type I(1);
(**D**) $A_{\overline{k}}$ is isogenous to the product of an elliptic curve with CM and a simple abelian surface of type I(1);
(**E**) $A_{\overline{k}}$ is either
- simple of type I(3), or
- isogenous to the product of an elliptic curve without CM and a simple abelian surface of type I(2), or
- isogenous to the product of three pairwise non-isogenous elliptic curves without CM;

(**F**) $A_{\overline{k}}$ is either

- isogenous to the product of an elliptic curve with CM and two non-isogenous elliptic curves without CM, or
- isogenous to the product of an elliptic curve with CM and a simple abelian surface of type I(2);

(G) $A_{\overline{k}}$ is either
- isogenous to the product of an elliptic curve without CM and two non-isogenous elliptic curves with CM, or
- isogenous to the product of an elliptic curve without CM and a simple abelian surface of type IV(2,1);

(H) $A_{\overline{k}}$ is either
- simple of type IV(3,1), or
- isogenous to the product of an elliptic curve with CM and a simple abelian surface of type IV(2,1), or
- isogenous to the product of three pairwise non-isogenous elliptic curves with CM;

(I) $A_{\overline{k}}$ is either
- isogenous to the product of an elliptic curve without CM and a simple abelian surface of type II(1), or
- isogenous to the product of an elliptic curve E without CM and the square of an elliptic curve without CM non-isogenous to E;

(J) $A_{\overline{k}}$ is either
- isogenous to the product of an elliptic curve with CM and the square of an elliptic curve without CM, or
- isogenous to the product of an elliptic curve with CM and a simple abelian surface of type II(1);

(K) $A_{\overline{k}}$ is isogenous to the product of an elliptic curve without CM and the square of an elliptic curve with CM;

(L) $A_{\overline{k}}$ is isogenous to the product of an elliptic curve E with CM and the square of an elliptic curve with CM non-isogenous to E;

(M) $A_{\overline{k}}$ is isogenous to the cube of an elliptic curve without CM;

(N) $A_{\overline{k}}$ is isogenous to the cube of an elliptic curve with CM.

REMARK 3.3. The third column of Table 1 summarizes the discussion above. In the table one can see that absolute types are ordered according to the lexicographic ordering of pairs (r, d) where r is the rank of the Néron-Severi group of $A_{\overline{k}}$ and d is the $\mathbb{R}$-dimension of $\operatorname{End}(A_{\overline{k}})_{\mathbb{R}}$. See [**8**] for an interpretation of r and d in terms of the moments of the Sato–Tate group.

3.3. Extensions. Let G^0 be one of the connected groups obtained in §3.1. Let N and Z denote the normalizer and centralizer of G^0 within USp(6). The general strategy in [**17**] to prove Theorem 3.1 exploits the bijection between:

- The set of (conjugacy classes of) closed subgroups G of USp(6) with identity component G^0 that satisfy (ST3);
- The set of (conjugacy classes of) finite subgroups H of N/G^0 such that HG^0 satisfies (ST3).

In this section we report on some aspects of this proof of Theorem 3.1. We will organize the discussion into four cases depending on the structure of the normalizer N. We say that G^0 is *decomposable* if there is a subgroup G_1^0 of SU(2) and a

Type	$\mathrm{End}(A_{\overline{k}})_{\mathbb{R}}$	Splitting of $A_{\overline{k}}$	$\mathrm{ST}(A)^0$
A	$\mathbb{R}$	T	$\mathrm{USp}(6)$
B	$\mathbb{C}$	S_{IM}	$\mathrm{U}(3)$
C	$\mathbb{R}\times\mathbb{R}$	$E\times S$	$\mathrm{SU}(2)\times\mathrm{USp}(4)$
D	$\mathbb{C}\times\mathbb{R}$	$E_{\mathrm{CM}}\times S$	$\mathrm{U}(1)\times\mathrm{USp}(4)$
E	$\mathbb{R}\times\mathbb{R}\times\mathbb{R}$	T_{RM} $E\times S_{\mathrm{RM}}$ $E\times E'\times E''$	$\mathrm{SU}(2)\times\mathrm{SU}(2)\times\mathrm{SU}(2)$
F	$\mathbb{C}\times\mathbb{R}\times\mathbb{R}$	$E_{\mathrm{CM}}\times E'\times E''$ $E_{\mathrm{CM}}\times S_{\mathrm{RM}}$	$\mathrm{U}(1)\times\mathrm{SU}(2)\times\mathrm{SU}(2)$
G	$\mathbb{C}\times\mathbb{C}\times\mathbb{R}$	$E_{\mathrm{CM}}\times E'_{\mathrm{CM}}\times E''$ $S_{\mathrm{CM}}\times E$	$\mathrm{U}(1)\times\mathrm{U}(1)\times\mathrm{SU}(2)$
H	$\mathbb{C}\times\mathbb{C}\times\mathbb{C}$	T_{CM} $E_{\mathrm{CM}}\times S_{\mathrm{CM}}$ $E_{\mathrm{CM}}\times E'_{\mathrm{CM}}\times E''_{\mathrm{CM}}$	$\mathrm{U}(1)\times\mathrm{U}(1)\times\mathrm{U}(1)$
I	$\mathbb{R}\times\mathrm{M}_2(\mathbb{R})$	$E\times(E')^2$ $E\times S_{\mathrm{QM}}$	$\mathrm{SU}(2)\times\mathrm{SU}(2)_2$
J	$\mathbb{C}\times\mathrm{M}_2(\mathbb{R})$	$E_{\mathrm{CM}}\times(E')^2$ $E_{\mathrm{CM}}\times S_{\mathrm{QM}}$	$\mathrm{U}(1)\times\mathrm{SU}(2)_2$
K	$\mathbb{R}\times\mathrm{M}_2(\mathbb{C})$	$E\times(E'_{\mathrm{CM}})^2$	$\mathrm{SU}(2)\times\mathrm{U}(1)_2$
L	$\mathbb{C}\times\mathrm{M}_2(\mathbb{C})$	$E_{\mathrm{CM}}\times(E'_{\mathrm{CM}})^2$	$\mathrm{U}(1)\times\mathrm{U}(1)_2$
M	$\mathrm{M}_3(\mathbb{R})$	E^3	$\mathrm{SU}(2)_3$
N	$\mathrm{M}_3(\mathbb{C})$	E^3_{CM}	$\mathrm{U}(1)_3$

TABLE 1. Real endomorphism algebras and connected parts of Sato–Tate groups of abelian threefolds. We describe the decomposition up to isogeny of $A_{\overline{k}}$; when doing so, we denote by E, S, T simple abelian varieties over $\overline{k}$ of respective dimensions 1, 2, 3. Given E, we denote by E' an elliptic curve not isogenous to E (a similar convention applies to E''). For $B\in\{E,S,T\}$, we write B_{CM} to indicate that B has complex multiplication by a CM field of degree $2\dim(B)$ and B_{RM} to indicate that B has real multiplication by a field of degree $\dim(B)$. The lack of a subindex indicates that B has trivial endomorphism ring. We write S_{IM} (resp. S_{QM}) to indicate that S has multiplication by an imaginary quadratic field (resp. by an indefinite quaternion algebra over $\mathbb{Q}$).

subgroup G_2^0 of $\mathrm{USp}(4)$ such that

$$G^0 = G_1^0\times G_2^0\,. \tag{3.2}$$

The indecomposable cases correspond to $G^0=\mathrm{USp}(6)$, $\mathrm{U}(3)$. Suppose that G^0 is decomposable and let N_1 (resp. N_2) denote the normalizer of G_1^0 (resp. G_2^0) in $\mathrm{SU}(2)$ (resp. $\mathrm{USp}(4)$). We say that that G^0 is a *split product* (resp. a *non-split product*) if

$$N = N_1\times N_2\,,\qquad (\text{resp. } N\supsetneq N_1\times N_2).$$

Eight of the possibilities for G^0 happen to be split products. The non-split product cases include the *triple products* $\mathrm{SU}(2) \times \mathrm{SU}(2) \times \mathrm{SU}(2)$ and $\mathrm{U}(1) \times \mathrm{U}(1) \times \mathrm{U}(1)$, and also the *diagonal products* $\mathrm{SU}(2)_3$ and $\mathrm{U}(1)_3$.

REMARK 3.4. We will only impose (ST3) explicitly in the case of diagonal products (and, in fact, only a relaxed version of (ST3)). However, it should be noted that (ST3) will be imposed in an implicit way for the split products. We will postpone to §4 the verification that (ST3) is satisfied for all the extensions described in this section. In particular, it will follow from the discussion in §4 that, when G^0 is indecomposable or a triple product, any closed subgroup G of N with identity component G^0 automatically satisfies (ST3).

The main purpose of this section is to describe the 30 group extensions that potentially satisfy the Sato–Tate axioms and are maximal with respect to finite index inclusions (up to conjugacy). The distribution of these groups among the possibilities for G^0 will be discussed below and it is summarized in Table 3.

3.3.1. *Indecomposable cases.* In case $G^0 = \mathrm{USp}(6)$ (type **A**), it is clear that $G = G^0$, which is obviously maximal. In case $G^0 = \mathrm{U}(3)$ (type **B**), we have $Z = \mathrm{U}(1)_3 \subset G^0$ and $N = G^0 \cup JG^0$. Consequently, $N/G^0 \simeq \mathrm{C}_2$, so G equals either G^0 or N, which is the only maximal group for the identity component $G^0 = \mathrm{U}(3)$.

3.3.2. *Split products.* Note that in this case

$$G = G_1 \times G_2 ,$$

where G_1 is a subgroup of N_1 with identity component G_1^0 and G_2 is a subgroup of N_2 with identity component G_2^0. Here G_i^0 is as in (3.2). We start with the cases where we must have $G_i = G_i^0$ for one i, in which case the options come directly from the other factor.

- In the case $G^0 = \mathrm{SU}(2) \times \mathrm{USp}(4)$ (type **C**), we must have $G_i = G_i^0$ for $i = 1, 2$, so $G = G^0$. This yields a maximal group.
- In the case $G^0 = \mathrm{U}(1) \times \mathrm{USp}(4)$ (type **D**), we must have $G_2 = G_2^0$, so $G = G_1 \times \mathrm{USp}(4)$ for $G_1 \in \{\mathrm{U}(1), N(\mathrm{U}(1))\}$. Then $G_1 = N(\mathrm{U}(1))$ yields a maximal group.
- In the case $G^0 = \mathrm{SU}(2) \times \mathrm{SU}(2)_2$ (type **I**), we must have $G_1 = G_1^0$, so $G = \mathrm{SU}(2) \times G_2$ for some G_2 among the groups of dimension $d = 3$ in [**15**, Table 8]. Among these groups, the maximal ones are $J(E_4)$ and $J(E_6)$. For this identity component we find 10 extensions, among which two are maximal, namely $\mathrm{SU}(2) \times J(E_4)$ and $\mathrm{SU}(2) \times J(E_6)$.
- In the case $G^0 = \mathrm{SU}(2) \times \mathrm{U}(1)_2$ (type **K**), we must have $G_1 = G_1^0$, so $G = \mathrm{SU}(2) \times G_2$ for some G_2 among the groups of dimension $d = 1$ in [**15**, Table 8]. Among these groups, the maximal ones are $J(D_6)$ and $J(O)$. For this identity component we find 32 extensions, among which two are maximal, namely $\mathrm{SU}(2) \times J(D_6)$ and $\mathrm{SU}(2) \times J(O)$.
- In the case $G^0 = \mathrm{U}(1) \times \mathrm{U}(1) \times \mathrm{SU}(2)$ (type **G**), take $G_1^0 = \mathrm{SU}(2)$ and $G_2^0 = \mathrm{U}(1) \times \mathrm{U}(1)$; we must have $G_1 = G_1^0$, so $G = \mathrm{SU}(2) \times G_2$ for some G_2 among the groups with dimension $d = 2$ in [**15**, Table 8]. Among these groups, only $F_{a,b,c}$ is maximal. For this identity component we find 8 extensions, of which one is maximal: $\mathrm{SU}(2) \times F_{a,b,c}$ (we will return to this case in §4.1).

REMARK 3.5. As indicated by the asterisks that appear in [**15**, Table 8], the subgroup $F_{a,b,c}$ of $\mathrm{USp}(4)$ is one of three subgroups that satisfy the Sato–Tate axioms for abelian surfaces but do not arise as the Sato–Tate group of an abelian surface. This explains the discrepancy between the counts 55 and 52 in [**15**] of groups that satisfy the Sato–Tate axioms and groups that arise as Sato–Tate groups, and it increases the number of maximal Sato–Tate groups with identity component $\mathrm{U}(1) \times \mathrm{U}(1)$ (two of the maximal subgroups of $F_{a,b,c}$ arise as Sato–Tate groups). A similar phenomenon occurs here, as explained in Remark 4.6 below.

We next consider cases where $G_1^0 = \mathrm{U}(1)$. In such cases, $G_1 \in \{\mathrm{U}(1), N(\mathrm{U}(1))\}$, so for any given G there exists a group G_2 with connected part G_2^0 such that either $G = \mathrm{U}(1) \times G_2$, $G = N(\mathrm{U}(1)) \times G_2$, or G is the fiber product $N(\mathrm{U}(1)) \times_{\mathrm{C}_2} G_2$ with respect to the unique nontrivial homomorphism $N(\mathrm{U}(1)) \to \mathrm{C}_2$ and some nontrivial homomorphism $G_2 \to \mathrm{C}_2$. The latter may be characterized by the group G_2 and the index-2 subgroup K which is the kernel of the homomorphism.

- In the case $G^0 = \mathrm{U}(1) \times \mathrm{SU}(2) \times \mathrm{SU}(2)$ (type **F**), take $G_1^0 = \mathrm{U}(1)$ and $G_2^0 = \mathrm{SU}(2) \times \mathrm{SU}(2)$; then $G_2 \in \{\mathrm{SU}(2) \times \mathrm{SU}(2), N(\mathrm{SU}(2) \times \mathrm{SU}(2))\}$ and there is a unique fiber product in the latter case. For this identity component we find 5 extensions, of which one is maximal: $N(\mathrm{U}(1)) \times N(\mathrm{SU}(2) \times \mathrm{SU}(2))$.
- In the case $G^0 = \mathrm{U}(1) \times \mathrm{SU}(2)_2$ (type **J**), G_2 must be taken from the groups with $d = 3$ in [**15**, Table 8]. Unique fiber products occur for $G_2 = E_2, E_4, E_6, J(E_1), J(E_3)$. Multiple fiber products occur for $G_2 = J(E_{2n})$ for $n = 1, 2, 3$, as we may take the kernel to be either E_{2n} or one of the two copies of $J(E_n)$. (The latter are conjugate to each other via the normalizer of $J(E_{2n})$, and so give rise to conjugate fiber products.) For this identity component we find 31 extensions, of which two are maximal: $N(\mathrm{U}(1)) \times J(E_4)$ and $N(\mathrm{U}(1)) \times J(E_6)$.
- In the case $G^0 = \mathrm{U}(1) \times \mathrm{U}(1)_2$ (type **L**), we find the maximal extensions $N(\mathrm{U}(1)) \times J(D_6)$ and $N(\mathrm{U}(1)) \times J(O)$. We defer to [**17**] the discussion yielding 122 extensions in this case.

3.3.3. *Triple products.* We next classify the options for G assuming that $G^0 \simeq \mathrm{SU}(2) \times \mathrm{SU}(2) \times \mathrm{SU}(2)$ (type **E**). In this case, $N/G^0 \simeq \mathrm{S}_3$; more precisely, N is generated by G^0 plus the permutation matrices. We find four options for G in this case, corresponding to the subgroups of S_3 up to conjugacy, which we may identify with $\mathrm{S}_1, \mathrm{S}_2, \mathrm{A}_3, \mathrm{S}_3$. We obtain a unique maximal extension in this case.

We next classify the options for G assuming that $G^0 \simeq \mathrm{U}(1) \times \mathrm{U}(1) \times \mathrm{U}(1)$ (type **H**). In this case, $Z = G^0$ and N/G^0 is isomorphic to the wreath product $\mathrm{C}_2 \wr \mathrm{S}_3$. We find 33 options for G in this case, all contained in a unique maximal extension (we will return to this case in §4.1).

3.3.4. *The diagonal product* $\mathrm{SU}(2)_3$. We next classify the options for G assuming that $G^0 \simeq \mathrm{SU}(2)_3$ (type **M**). In this case, Z equals the group $\mathrm{O}(3)$ realized as a block matrix with scalar entries, $Z \cap G^0 = \{\pm 1\}$, and $N = ZG^0$, so $N/G^0 \simeq \mathrm{SO}(3)$. We first bound the order of a cyclic subgroup of $\mathrm{SO}(3)$ satisfying (ST3). Any cyclic subgroup of $\mathrm{SO}(3)$ of order n is conjugate to

$$\langle A \rangle, \qquad A = \begin{pmatrix} 1 & 0 & 0 \\ 0 & \cos 2\pi/n & \sin 2\pi/n \\ 0 & -\sin 2\pi/n & \cos 2\pi/n \end{pmatrix}.$$

By (ST3) the average of the trace of the representation $\wedge^2\mathbb{C}^6$ on the coset of G^0 containing A, which is given by

$$\mathrm{E}[\mathrm{Trace}(\gamma, \wedge^2\mathbb{C}^6) : \gamma \in AG^0] = |\,\mathrm{Trace}(A)|^2 = (1 + 2\cos 2\pi/n)^2,$$

must be an integer; this happens if and only if $n \in \{1,2,3,4,6\}$. Since every finite subgroup of SO(3) is either cyclic, dihedral, or one of the three exceptional groups (tetrahedral, octahedral, icosahedral), we obtain component groups isomorphic to

$$\mathrm{C}_1,\ \mathrm{C}_2,\ \mathrm{C}_3,\ \mathrm{C}_4,\ \mathrm{C}_6,\ \mathrm{D}_2,\ \mathrm{D}_3,\ \mathrm{D}_4,\ \mathrm{D}_6,\ \mathrm{A}_4,\ \mathrm{S}_4.$$

For this identity component we find 11 extensions, among which two are maximal, namely those with component groups D_6 and S_4, which we denote by $\mathrm{SU}(2)_3[\mathrm{D}_6]$ and $\mathrm{SU}(2)_3[\mathrm{S}_4]$.

3.3.5. *The diagonal product* $\mathrm{U}(1)_3$. We finally report on the classification for the options for G achieved in [**17**] assuming that $G^0 \simeq \mathrm{U}(1)_3$ (type **N**). In this case, $Z = \mathrm{U}(3)$ embedded as in §3.1, $G^0 \subset Z$, and $N = Z \cup JZ$, so $N/G^0 \simeq \mathrm{PSU}(3) \rtimes \mathrm{C}_2$ for the action of C_2 on PSU(3) via complex conjugation.

As in the case $G^0 \simeq \mathrm{SU}(2)_3$, if the cyclic subgroup of PSU(3) generated by $A \in Z$ satisfies (ST3), then $|\mathrm{Trace}(A)|^2$ is an integer (and similarly for every power of A). It will suffice to apply (ST3) in this form, which we denote by (ST3′), as we will show *a posteriori* in §4 that every group we obtain also satisfies (ST3) in full.

We first focus on subgroups of PSU(3). Since SU(3) surjects onto PSU(3) with kernel μ_3, we may identify finite subgroups of PSU(3) with finite subgroups of SU(3) containing μ_3; for the latter, we use the classification by Blichfeldt–Dickson–Miller [**29**, Chapter XII].

LEMMA 3.6. *Every finite subgroup of* SU(3) *containing* μ_3 *is conjugate to a subgroup which is either*

- *contained in the diagonal torus,*
- *contained in* $\mathrm{SU}(3) \cap (\mathrm{U}(1) \times \mathrm{U}(2))$,
- *a semidirect product of* A_3 *acting on a diagonal abelian group,*
- *a (not necessarily split) extension of* S_3 *by a diagonal abelian group, or*
- *one of seven exceptional cases whose images in* PSU(3) *have orders* 36, 72, 216 *(the Hessian group),* 60 *(the group* A_5*),* 360 *(the group* A_6*), or* 168 *(the group* PSL(2,7)*).*

To reduce the classification to a finite problem, we study the abelian case in detail. In what follows, let $D(u,v,w)$ denote the 3×3 diagonal matrix with diagonal entries $e^{2\pi i u}, e^{2\pi i v}, e^{2\pi i w}$.

LEMMA 3.7. *Let* H *be a finite cyclic subgroup of* SU(3) *containing* μ_3 *and satisfying (ST3′). Then* $\#(H/\mu_3)$ *divides one of* 7, 8, 12.

PROOF. By conjugating, we may ensure that H is generated by some matrix of the form $D(u,v,w)$ with $u,v,w \in \mathbb{Q}$ and $u+v+w \in \mathbb{Z}$, and the problem is to bound the least common denominator of u,v,w subject to the condition that

$$e^{2\pi i(u-v)} + e^{2\pi i(u-w)} + e^{2\pi i(v-u)} + e^{2\pi i(v-w)} + e^{2\pi i(w-u)} + e^{2\pi i(w-v)} \in \mathbb{Z}.$$

This amounts to finding the cyclotomic points on a certain algebraic curve (i.e., solving a multiplicative Manin-Mumford problem), which can be done by following either Conway–Jones [**9**] or Beukers–Smyth [**7**]; we defer to [**17**] for further details. □

REMARK 3.8. By Lemma 3.7, the only primes that can divide the order of a finite subgroup of PSU(3) satisfying (ST3′) are 2, 3, 7; in particular, the exceptional subgroups $\mathrm{A}_5, \mathrm{A}_6$ are excluded. We may moreover classify all of the cases in which 7 divides the order as follows.

Up to conjugation, there is a unique copy H_0 of C_7 in PSU(3) satisfying (ST3′); it is generated by the image of $D(\frac{1}{7}, \frac{2}{7}, \frac{4}{7})$. By Lemma 3.7, H_0 is not contained in any larger abelian group satisfying (ST3′); it is also not contained in any conjugate of SU(2). By Lemma 3.6 plus a direct verification that PSL(2, 7) satisfies (ST3′), it follows (after checking that these cases do all work) that up to conjugacy, the finite subgroups of PSU(3) satisfying (ST3′) of order divisible by 7 are

$$H_0, \; H_0 \rtimes \mathrm{A}_3, \; \mathrm{PSL}(2,7).$$

Having addressed the subgroups of PSU(3) of order divisible by 7, we henceforth restrict our attention to finite subgroups of PSU(3) with 3-smooth order.

REMARK 3.9. Let H be a finite subgroup of SU(3) of 3-smooth order containing μ_3 and satisfying (ST3′). If H is a diagonal abelian group, then it must be isomorphic to $\mathrm{C}_m \times \mathrm{C}_n$ for some integers m, n with $\mathrm{lcm}(3, m)$ dividing n. It is straightforward to exhaustively consider all choices of m, n consistent with Lemma 3.7 to identify cases satisfying (ST3′); we find that in all cases, $m \in \{1, 2, 3, 4, 6\}$.

If H is an extension of A_3 by a diagonal abelian group H_0, then H_0 cannot be cyclic (because no cyclic group of order dividing 8 or 12 has a faithful automorphism of order 3). It must thus have the form $\mathrm{C}_m \times \mathrm{C}_n$ with $m \in \{2, 3, 4, 6\}$. It is straightforward to identify the cases from the previous paragraph which are stable under A_3; each of these cases is in fact stable under S_3.

REMARK 3.10. Using Lemma 3.7, Remark 3.8, and Remark 3.9, the classification of finite subgroups of SU(3) containing μ_3 and satisfying (ST3′) reduces to the case of a nonabelian subgroup H of $\mathrm{SU}(3) \cap (\mathrm{U}(1) \times \mathrm{U}(2))$. The latter group admits $\mathrm{U}(1) \times \mathrm{SU}(2)$ as a double cover via the map $(u, A) \mapsto (u^2, u^{-1}A)$. For $\tilde{H}$ the preimage of H in this cover, the projection of $\tilde{H}$ to SU(2) cannot be an abelian group of odd order; it must therefore be the inverse image of a dihedral, tetrahedral, or octahedral subgroup of SO(3). Meanwhile, the projection of $\tilde{H}$ to U(1) has order dividing 16 or 24. One may thus obtain a bound on such groups by testing (ST3′) for each finite subgroup (up to conjugacy) of each of the subgroups

$$\mathrm{C}_{96} \times 2\mathrm{D}_{48}, \; \mathrm{C}_{96} \times 2O.$$

We were not able to completely automate this calculation; the classification in [**17**] is human-readable, although we did cross-check several steps by numerical computation.

It remains to classify finite subgroups of $\mathrm{PSU}(3) \rtimes \mathrm{C}_2$ not contained in PSU(3) whose intersection with PSU(3) equals a fixed group H known to satisfy (ST3′); we call each such subgroup a *double cover* of H. (In general, not all double covers are semidirect products of H by C_2, but the maximal groups all are.) In most cases, we work with a conjugacy representative of H which is stable under the action $g \mapsto \overline{g}$ of complex conjugation; under this condition, the double covers of H are the groups of the form $\langle H, Jg \rangle$ for some $g \in N_H$ satisfying $\overline{g}g \in H$. The last condition means that the image of g in N_H/H maps to its inverse under complex conjugation; the values in N_H/H corresponding to the same double cover as g are the images of $\overline{h}^{-1}gh$ for all $h \in N_H$.

REMARK 3.11. One complicating feature is that not every double cover of a given H extends to every overgroup. In particular, the intersection of a maximal finite subgroup of $\mathrm{PSU}(3) \rtimes \mathrm{C}_2$ satisfying (ST3′) with PSU(3) need not be a *maximal* finite subgroup of PSU(3) satisfying (ST3′).

THEOREM 3.12. *Up to conjugacy, there are* 12 *maximal subgroups of* $\mathrm{PSU}(3) \rtimes \mathrm{C}_2$ *satisfying (ST3′), as listed in Table 2. (See below for explanation of the table entries.)*

The maximal subgroups in Theorem 3.12 are described in Table 2. In each row, M is an imaginary quadratic field of class number 1, $\tilde{H}$ is a subgroup of $\mathrm{GL}_3(M)$, H is the projective image of $\tilde{H}$, and $H \rtimes \mathrm{C}_2$ is a particular choice of semidirect product (to be discussed in more detail in Lemma 4.10). The notation for $\tilde{H}$ is to be interpreted as follows.

- The labels $G(m, n, p)$ and G_i refer to the Shephard-Todd notation [**32**] for complex reflection groups. In these cases, M is the field of traces of $\tilde{H}$ as per [**28**, Table 1].
- The notation $[G_8 : 4]$ refers to a non-normal index-4 subgroup of $G(1,1,1) \times G_8$. It is not a complex reflection group.
- The notation G_4' refers to a copy of G_4 embedded via its two-dimensional irreducible representation with rational traces (a cubic twist of the reflection representation). That representation has a nontrivial Schur index: it can only be realized over fields that split the quaternion algebra $(-1,-1)_{\mathbb{Q}}$ (compare [**14**, Proposition 3.5]). By convention, we do not regard G_4' as a complex reflection group.

The action of C_2 on H is induced by an action of C_2 on $\tilde{H}$ specified as follows.

- For $c \in \mathrm{C}_2$ and $c_M \in \mathrm{Gal}(M/\mathbb{Q})$ generators and $\rho : \tilde{H} \to \mathrm{GL}_3(M)$ the implied representation, we have $\rho \cong c_M \circ \rho \circ c$.
- In the cases where $\tilde{H}$ is written as a product, the action of c respects this decomposition.
- The actions on $G(n,1,1), G(4,4,3), G(6,2,3), G_4, G_8, G_{12}, G_{24}, G_{25}$ are all outer, represented by the action of c_M on a c_M-stable presentation [**28**, Theorem 1.7].
- The action on $[G_8 : 4]$ is outer, but this leaves two possibilities for the semidirect product. One of these extends to G_8; we take the other one.
- The action on $G(3,3,2)$ is trivial.
- The actions on $G(2,1,2), G(6,6,2)$ are outer.
- The action on G_4' is outer, but this leaves two possibilities for the semidirect product. One of these recovers $\langle 144, 125\rangle$; we take the other one.

REMARK 3.13. Since we are only interested in H rather than $\tilde{H}$, some variation in the choice of complex reflection groups is possible in Table 2. For the most part, we have made choices to minimize $\#\tilde{H}/\#H$.

REMARK 3.14. By choosing explicit presentations of the groups in Table 2, it is possible to verify by computer calculations that the subgroups of these groups account for 171 distinct extensions of $\mathrm{U}(1)_3$, and that there are no containment relations among the groups in the table. The difficult part of the classification, which we defer to [**17**], is to check that every possible extension is accounted for in this manner; in other words, that there are no maximal subgroups missing from the table below.

H	$H \rtimes \mathrm{C}_2$	$\tilde{H}$	$\#\tilde{H}/\#H$	M
$\langle 24,1\rangle$	$\langle 48,15\rangle$	$G(1,1,1)\times[G_8:4]$	1	$\mathbb{Q}(i)$
$\langle 24,10\rangle$	$\langle 48,15\rangle$	$G(3,1,1)\times G(2,1,2)$	1	$\mathbb{Q}(\zeta_3)$
$\langle 24,5\rangle$	$\langle 48,38\rangle$	$G(4,1,1)\times G(3,3,2)$	1	$\mathbb{Q}(i)$
$\langle 24,5\rangle$	$\langle 48,41\rangle$	$G(4,1,1)\times G(6,6,2)$	2	$\mathbb{Q}(i)$
$\langle 48,29\rangle$	$\langle 96,193\rangle$	$G(1,1,1)\times G_{12}$	1	$\mathbb{Q}(\sqrt{-2})$
$\langle 72,25\rangle$	$\langle 144,125\rangle$	$G(3,1,1)\times G_4$	1	$\mathbb{Q}(\zeta_3)$
$\langle 72,25\rangle$	$\langle 144,127\rangle$	$G(3,1,1)\times G_4'$	1	$\mathbb{Q}(\zeta_3)$
$\langle 96,67\rangle$	$\langle 192,988\rangle$	$G(1,1,1)\times G_8$	1	$\mathbb{Q}(i)$
$\langle 96,64\rangle$	$\langle 192,956\rangle$	$G(4,4,3)$	1	$\mathbb{Q}(i)$
$\langle 168,42\rangle$	$\langle 336,208\rangle$	G_{24}	2	$\mathbb{Q}(\sqrt{-7})$
$\langle 216,92\rangle$	$\langle 432,523\rangle$	$G(6,2,3)$	3	$\mathbb{Q}(\zeta_3)$
$\langle 216,153\rangle$	$\langle 432,734\rangle$	G_{25}	3	$\mathbb{Q}(\zeta_3)$

TABLE 2. Maximal extensions of $\mathrm{U}(1)_3$.

$\mathrm{ST}(A)^0$	$\dim \mathrm{ST}(A)^0$	Extensions	Maximal
$\mathrm{USp}(6)$	21	1	1
$\mathrm{U}(3)$	9	2	1
$\mathrm{SU}(2)\times\mathrm{USp}(4)$	13	1	1
$\mathrm{U}(1)\times\mathrm{USp}(4)$	11	2	1
$\mathrm{SU}(2)\times\mathrm{SU}(2)\times\mathrm{SU}(2)$	9	4	1
$\mathrm{U}(1)\times\mathrm{SU}(2)\times\mathrm{SU}(2)$	7	5	1
$\mathrm{U}(1)\times\mathrm{U}(1)\times\mathrm{SU}(2)$	5	8	1
$\mathrm{U}(1)\times\mathrm{U}(1)\times\mathrm{U}(1)$	3	33	1
$\mathrm{SU}(2)\times\mathrm{SU}(2)_2$	6	10	2
$\mathrm{U}(1)\times\mathrm{SU}(2)_2$	4	31	2
$\mathrm{SU}(2)\times\mathrm{U}(1)_2$	4	32	2
$\mathrm{U}(1)\times\mathrm{U}(1)_2$	2	122	2
$\mathrm{SU}(2)_3$	3	11	2
$\mathrm{U}(1)_3$	1	171	12

TABLE 3. For each connected Sato–Tate group, we list the number of extensions satisfying the Sato–Tate axioms and the number of those which are maximal.

4. Realization

The goal of this section is to provide a proof of the following result, assuming Theorem 3.1.

THEOREM 4.1. *Up to conjugacy, there are* 410 *closed subgroups of* USp(6) *that occur as Sato–Tate groups of abelian threefolds over number fields. Among these, the* 33 *that are maximal with respect to finite index inclusions arise as Sato–Tate groups of abelian threefolds. (Hence by base extension, all* 410 *groups occur for abelian threefolds over number fields.)*

The proof of the theorem has two parts. On the one hand, when the identity component contains $\mathrm{U}(1)\times\mathrm{U}(1)$ as a factor, we refine the group-theoretic argument by making a further analysis of CM types that rules out 23 of the groups satisfying the Sato–Tate axioms (§4.1). In particular, the set of maximal groups among those satisfying the refined analysis of CM types differs from the set of maximal groups satisfying the Sato–Tate axioms (see Remark 4.6).

On the other hand, we realize the remaining 410 groups as Sato–Tate groups of certain abelian threefolds (§4.2). Crucially, it will suffice to realize the 33 maximal groups among these, as then their subgroups of finite index can be achieved by considering base extensions (by Lemma 2.8). As an aside, we note that at present we do not know exactly which groups can occur over $\mathbb{Q}$ (certainly not all of them: no connected group with a unique factor isomorphic to $\mathrm{U}(1)$ can arise over $\mathbb{Q}$, since this would require a CM elliptic curve to have all its endomorphisms defined over $\mathbb{Q}$).

We note in passing that all of the examples of types other than **N** will carry a principal polarization. The analysis of polarizations in the type **N** case is more subtle, and is not included here.

4.1. Upper bound. Let A be an abelian threefold with $\mathrm{ST}(A) = G$ and suppose that G^0 contains $\mathrm{U}(1) \times \mathrm{U}(1)$ as a factor.

Suppose first that $G^0 = \mathrm{SU}(2) \times \mathrm{U}(1) \times \mathrm{U}(1)$. Then A must split up to isogeny as the product of an elliptic curve without CM and an abelian surface whose Sato–Tate group has identity component $\mathrm{U}(1)\times\mathrm{U}(1)$. By [**15**, §4.3, §4.4], of the 8 options for G described in §3.3.2, we may omit the 3 for which G_2 cannot occur for abelian surfaces.

Suppose next that $G^0 = \mathrm{U}(1) \times \mathrm{U}(1) \times \mathrm{U}(1)$ so that N/G^0 is isomorphic to the wreath product $\mathrm{C}_2 \wr \mathrm{S}_3$. Let a, b, c be representatives of the nontrivial cosets in the three factors of $N(\mathrm{U}(1))$, and set t, s correspond to the permutations $(12), (123)$ in S_3.

If A is not simple, then it splits as a product of a CM elliptic curve and a CM abelian surface; this means that up to conjugacy, $\mathrm{ST}(A)$ is contained in the product of a subgroup of $\langle a, b, t\rangle$ and a subgroup of $\langle c\rangle$. By the analysis from [**15**, §4.3], the former must be a subgroup of either $\langle a, b\rangle$ or $\langle at\rangle$.

To handle the case where A is simple, we recall some basic facts about abelian varieties with complex multiplication, following [**12**, §1] and [**24**, §1] (and ultimately [**33**, §5.5]). For future reference, we temporarily allow A to have arbitrary dimension.

DEFINITION 4.2. A *CM field* is a totally imaginary quadratic extension of a totally real number field. A *CM type* is a pair (M, Φ) consisting of a CM field M and a section Φ of the map grouping the complex embeddings of M into conjugate pairs.

Given a CM type (M, Φ), for any $\mathbb{Z}$-lattice $\mathfrak{m}$ in M, we may use Φ to specify a complex structure on $M \otimes_{\mathbb{Q}} \mathbb{R}$, and thus view $(M \otimes_{\mathbb{Q}} \mathbb{R})/\mathfrak{m}$ as a complex torus. Changing the choice of $\mathfrak{m}$ gives rise to an isogenous torus [**33**, Proposition 5.13]; consequently, we may associate a polarization to this torus using the trace pairing on M, and thus obtain an abelian variety A over $\mathbb{C}$ with complex multiplication by some order in M. Note that if M has degree $2n$, then A has dimension n.

DEFINITION 4.3. Given a CM type (M, Φ), let L be the Galois closure of M in $\mathbb{C}$, which is again a CM field. The action of $\mathrm{Gal}(L/\mathbb{Q})$ on the complex embeddings of M by postcomposition acts on the CM types for M. The *reflex field* of (M, Φ) is the fixed field M^* of the stabilizer of Φ in $\mathrm{Gal}(L/\mathbb{Q})$; note that M^* is defined as a subfield of L (and hence of $\mathbb{C}$), whereas M does not come with a distinguished embedding into L. (There is also a CM type associated to M^*, the *reflex type*, which we will not discuss here.)

LEMMA 4.4. *Fix an embedding of k into $\mathbb{C}$. Let A be a simple polarized abelian variety of dimension n over k with CM type (M, Φ).*

(a) *The field M^* is Galois over $k \cap M^*$.*
(b) *The field kM^* is the endomorphism field of A.*
(c) *The group $\mathrm{Gal}(kM^*/k)$ is canonically isomorphic to a subgroup of $\mathrm{Aut}(M/\mathbb{Q})$.*

PROOF. For (a), (b), see [**33**, Theorem 5.15, Proposition 5.17]. For (c), the action from (b) gives an action of $\mathrm{Gal}(kM^*/k)$ on $\mathrm{End}(A_{\mathbb{C}})_{\mathbb{Q}} = M$. □

REMARK 4.5. To trace the effect of the previous discussion on Sato–Tate groups, we identify $\mathrm{ST}(A)/\mathrm{ST}(A)^0$ with a subgroup of $\mathrm{C}_2 \wr \mathrm{S}_n$ using the embedding of M into $\mathrm{End}(A_K)_{\mathbb{R}}$ to fix the identification of $\mathrm{U}(1) \times \cdots \times \mathrm{U}(1)$ with $\prod_{\tau \in \Phi} \mathbb{C}^\times$. Then the subgroup $\mathrm{ST}(A)/\mathrm{ST}(A)^0$ is contained in $\mathrm{Gal}(L/(k \cap M^*))$ and the composition

$$\mathrm{ST}(A)/\mathrm{ST}(A)^0 \to \mathrm{Gal}(L/(k \cap M^*)) \to \mathrm{Gal}(M^*/(k \cap M^*)) \cong \mathrm{Gal}(kM^*/k)$$

is the canonical map (see Remark 2.9).

We now specialize back to the case where A is a simple CM abelian threefold. In this case, as per [**12**, §5.1.1], the sextic field M must be one of the following.

- A cyclic Galois extension of $\mathbb{Q}$. In this case, we must have $G/G^0 \subseteq \langle abc, s\rangle$ as this is the unique copy of C_6 in $\mathrm{C}_2 \wr \mathrm{S}_3$.
- A compositum of an imaginary quadratic field M_0 and a non-Galois totally real cubic field M_1. In this case, $\mathrm{Aut}(M/\mathbb{Q}) = \mathrm{Gal}(M_0/\mathbb{Q})$ and so $G/G^0 \subseteq \langle abc\rangle$.
- A non-Galois extension of $\mathbb{Q}$ whose Galois closure has group $\mathrm{C}_2 \wr \mathrm{A}_3$ or $\mathrm{C}_2 \wr \mathrm{S}_3$. In this case, $\mathrm{Aut}(M/\mathbb{Q})$ is trivial, as then is G/G^0.

To summarize, we list the 33 conjugacy classes of subgroups of N/G^0 (as may be verified using GAP or MAGMA) in Table 4. In the table, $\star$ indicates a normal subgroup, while an underline denotes a group that is permitted by the previous analysis to occur as a Sato–Tate group of an abelian threefold; we will see later that these groups all do occur (see §4.2.3).

REMARK 4.6. The above discussion eliminates, in particular, the groups $\mathrm{SU}(2) \times F_{a,b,c}$ and $N(\mathrm{U}(1) \times \mathrm{U}(1) \times \mathrm{U}(1))$. These are the maximal groups among those satisfying the Sato–Tate axioms for the absolute types **G** and **H** as seen in §3.3.2 and §3.3.3, respectively. In their place, five groups become maximal among the list of groups satisfying the above refined analysis. The groups $\mathrm{SU}(2) \times F_{a,b}$ and $\mathrm{SU}(2) \times F_{ac}$ become maximal for the absolute type **G**, and the groups with groups of connected components $\langle a, b, c\rangle$, $\langle at, c\rangle$, and $\langle abc, s\rangle$ become maximal for the absolute type **H**. The refined analysis thus yields 33 maximal groups in total.

Order	Isom type	Groups	Total
1	C_1	$\underline{\langle e\rangle}$	1
2	C_2	$\underline{\langle a\rangle}$, $\underline{\langle ab\rangle}$, $\underline{\langle abc\rangle}\star$, $\langle t\rangle$, $\langle ct\rangle$	5
3	C_3	$\underline{\langle s\rangle}$	1
4	C_4	$\underline{\langle at\rangle}$, $\underline{\langle act\rangle}$	2
4	C_2^2	$\underline{\langle a,b\rangle}$, $\underline{\langle a,bc\rangle}$, $\underline{\langle ab,bc\rangle}\star$, $\langle c,t\rangle$, $\langle ab,t\rangle$, $\langle ab,ct\rangle$, $\langle abc,t\rangle$	7
6	C_6	$\underline{\langle abc,s\rangle}$	1
6	S_3	$\langle s,t\rangle$, $\langle abct,s\rangle$	2
8	$\mathrm{C}_2\times\mathrm{C}_4$	$\underline{\langle c,at\rangle}$	1
8	C_2^3	$\underline{\langle a,b,c\rangle}\star$, $\langle ab,c,t\rangle$	2
8	D_4	$\langle a,b,t\rangle$, $\langle ab,bc,t\rangle$, $\langle a,b,ct\rangle$, $\langle ab,bc,ct\rangle$	4
12	D_6	$\langle abc,s,t\rangle$	1
12	A_4	$\langle ab,bc,s\rangle\star$	1
16	$\mathrm{C}_2\times\mathrm{D}_4$	$\langle a,b,c,t\rangle$	1
24	$\mathrm{C}_2\wr\mathrm{A}_3$	$\langle a,b,c,s\rangle\star$	1
24	S_4	$\langle ab,bc,s,t\rangle\star$, $\langle ab,bc,at,s\rangle\star$	2
48	$\mathrm{C}_2\wr\mathrm{S}_3$	$\langle a,b,c,s,t\rangle\star$	1

TABLE 4. Conjugacy classes of subgroups of N/G^0 for $G^0 \simeq \mathrm{U}(1)\times\mathrm{U}(1)\times\mathrm{U}(1)$. The starred groups are normal; the underlined groups can be realized by Sato–Tate groups.

REMARK 4.7. Recall from Remark 3.2 that Theorem 3.1 gives only an upper bound on the set of conjugacy classes of subgroups of $\mathrm{GL}_6(\mathbb{C})$ satisfying the Sato–Tate axioms for an abelian threefold. The lower bound is not obtained in this argument because we only verify (ST3) for some choices of the character χ.

However, it can be shown that all 433 conjugacy classes that occur do in fact satisfy (ST3) in full. For the 410 groups that occur in Theorem 4.1, the realizations of these groups as Sato–Tate groups of abelian threefolds will *a posteriori* confirm (ST3). To settle the remaining cases, it suffices to check (ST3) for the full normalizers of $\mathrm{U}(1)\times\mathrm{U}(1)\times\mathrm{SU}(2)$ and $\mathrm{U}(1)\times\mathrm{U}(1)\times\mathrm{U}(1)$ in $\mathrm{USp}(6)$. For G equal to one of these groups, we can write G as $G_0 \rtimes K$ where K is a finite subgroup of $\mathrm{USp}(6)\cap\mathrm{GL}_6(\mathbb{Q})$; this means that (ST3) holds for every χ which is the base extension of a $\mathbb{Q}$-linear representation of $\mathrm{GL}_6(\mathbb{Q})$, and this covers all χ by virtue of the usual description of irreducible representations of $\mathrm{GL}_6(\mathbb{C})$ via Schur functors.

4.2. Lower bound. We next exhibit examples of abelian threefolds defined over $\mathbb{Q}$ realizing each of the 33 maximal groups described in Remark 4.6.

4.2.1. *Indecomposable cases.* The unique group of absolute type **A** is realized by any abelian threefold with trivial endomorphism ring, and in particular by the Jacobian of a generic genus-3 curve. For a concrete example of absolute type **A**, by a theorem of Zarhin [**40**, Theorem 2.1], the Jacobian of the hyperelliptic curve $y^2 = x^7 - x + 1$ has trivial endomorphism ring.

The unique maximal group of absolute type **B** is realized by the Jacobian of a generic Picard curve. For a concrete example of absolute type **B**, we take an example of Upton [**38**, §6.1.3]: the Jacobian of the Picard curve $C : y^3 = x^4 + x + 1$ has maximal 67-adic Galois image. Note that Upton's result uses a calculation of

Frobenius traces plus a big image criterion in the manner of [**39**]; in general, we do not know of a good way to compute directly with the mod-ℓ Galois representation for any prime ℓ other than 2, for which the Galois action on bitangents provides a computable model.

4.2.2. *Split products.* Note that the maximal split products are all of the form $G = G_1 \times G_2$, where $G_1 = N(\mathrm{U}(1))$ or $\mathrm{SU}(2)$, and

$$G_2 = \mathrm{USp}(4),\ N(\mathrm{SU}(2)\times \mathrm{SU}(2)),\ F_{ac},\ F_{a,b},\ J(E_4),\ J(E_6),\ J(D_6),\ \text{ or } J(O).$$

Fix one such G_2. As shown by [**15**, Table 11], there exists a principally polarized abelian surface S defined over $\mathbb{Q}$ with $\mathrm{ST}(S) \simeq G_2$. If $G_1 = \mathrm{SU}(2)$, then G is realized by the product of S with an elliptic curve without CM defined over $\mathbb{Q}$; if $G_1 = N(\mathrm{U}(1))$, then G is realized by the product of S with an elliptic curve defined over $\mathbb{Q}$ with CM by a quadratic imaginary field not contained in the endomorphism field of S. One readily verifies that examples of elliptic curves satisfying this condition can be easily found.

4.2.3. *Triple products.* The unique maximal group for the absolute type **E** can be realized in the following manner. Let E be an elliptic curve without CM defined over a non Galois cubic extension $F/\mathbb{Q}$ which is not a $\mathbb{Q}$-curve (that is, there exists $\sigma \in G_{\mathbb{Q}}$ such that E and ${}^{\sigma}E$ are not $\overline{\mathbb{Q}}$-isogenous). It immediately follows that if σ and σ' are different embeddings of F into $\overline{\mathbb{Q}}$, then ${}^{\sigma}E$ and ${}^{\sigma'}E$ are not $\overline{\mathbb{Q}}$-isogenous, and therefore the maximal group for the absolute type **E** is realized by taking the Weil restriction of scalars of E from F down to $\mathbb{Q}$.

We next realize the maximal groups of absolute type **H**:

- The product of three elliptic curves defined over $\mathbb{Q}$ with CM by pair-wise non-isomorphic imaginary quadratic fields realizes the maximal group with group of components $\langle a, b, c\rangle$.
- Let S be a principally polarized abelian surface over $\mathbb{Q}$ with CM by a quartic CM field (as the example given in [**15**, Table 11] realizing the group F_{ac}, for example). Let E be an elliptic curve defined over $\mathbb{Q}$ with CM by a quadratic imaginary field not contained in the endomorphism field of S. Then $E \times S$ realizes the maximal group with group of components $\langle at, c\rangle$.
- As discussed in §4.1, the maximal group with group of components $\langle abc, s\rangle$ is realized by a principally polarized abelian threefold with CM by a cyclic CM field of degree 6. As a concrete example, we may take the Jacobian of the curve $y^2 = x^7 + 1$, which has CM by the field $\mathbb{Q}(\zeta_7)$.

4.2.4. *The diagonal product* $\mathrm{SU}(2)_3$. We next realize the two maximal groups of absolute type **M**. We will use the construction of Example 2.11, but this does not yield examples carrying a principal polarization; we thus supplement the discussion with some explicit examples that are principally polarized.

In order to realize $\mathrm{SU}(2)_3[\mathrm{S}_4]$, let $L/\mathbb{Q}$ denote a Galois extension with Galois group S_4 and let

$$\xi\colon \mathrm{Gal}(L/\mathbb{Q}) \to \mathrm{GL}_3(\mathbb{Z})$$

denote a faithful integral degree 3 representation. Let E be an elliptic curve without complex multiplication defined over $\mathbb{Q}$ and let A be the twist of E^3 by ξ. From the fact that ξ has projective image S_4, we easily deduce that A has Sato–Tate group $\mathrm{SU}(2)_3[\mathrm{S}_4]$.

Alternatively, note that this group occurs for the Jacobian of a generic twist of the curve $y^2 = x^8 + 14x^4 + 1$, which has an automorphism group of order 48. See for example [**2**, Table 4, row 11].

In order to realize $\mathrm{SU}(2)_3[\mathrm{D}_6]$, let $L/\mathbb{Q}$ denote a Galois extension with Galois group D_6 and let

$$\xi\colon \mathrm{Gal}(L/\mathbb{Q}) \to \mathrm{GL}_2(\mathbb{Z})$$

denote a faithful integral degree 2 representation. Let E be an elliptic curve without complex multiplication defined over $\mathbb{Q}$, set $S = E^2$, and let S_ξ be the twist of S by ξ. Since the projective image of ξ is D_3, the endomorphism field of S_ξ has Galois group D_3. Then $\mathrm{SU}(2)_3[\mathrm{D}_6]$ is realized by the product of S_ξ and a nontrivial quadratic twist E' of E over a quadratic field not contained in the endomorphism field of S.

Alternatively, let C_2 be the curve $y^2 = x^6 + x^3 + 4$, which is given in [**15**, Table 11] as an example with Sato-Tate group $J(E_3)$. The curve C_2 is a twist of $y^2 = x^6 + x^{3/2} + 1$, whose quotient by the involution

$$(x, y) \mapsto (1/x, y/x^3)$$

may be identified with the curve $C_1 : y^2 = (x+2)(x^3-3x+1/2)$ via the map $(x, y) \mapsto (x+1/x, y(x+1)/x^2)$; taking the product of $\mathrm{Jac}(C_2)$ with a generic quadratic twist of $\mathrm{Jac}(C_1)$ yields an abelian threefold with Sato-Tate group $\mathrm{SU}(2)_3[\mathrm{D}_6]$.

REMARK 4.8. Note that in both of the constructions realizing $\mathrm{SU}(2)_3[\mathrm{D}_6]$, we use an abelian surface with Sato-Tate group $J(E_3)$. One may wonder whether it is possible to construct an abelian threefold A defined over $\mathbb{Q}$ with $\mathrm{ST}(A) = \mathrm{SU}(2)_3[\mathrm{D}_6]$ such that A is the product of an elliptic curve E over $\mathbb{Q}$ and an abelian surface S over $\mathbb{Q}$ with $\mathrm{ST}(S) = J(E_6)$. This seemingly natural construction is ruled out by [**18**, §3D], which shows that an elliptic factor of such an S never admits a model up to isogeny defined over $\mathbb{Q}$.

4.2.5. *The diagonal product* $\mathrm{U}(1)_3$. We finally realize the twelve maximal groups of absolute type **N**. We do this in a uniform fashion using the presentations given in Table 2. Hereafter, let $H, \tilde{H}$ correspond to a row of the table.

We first verify that the construction of Example 2.12 is applicable. In the cases where $\tilde{H}$ is written as a product, we apply Remark 2.13 to the two-dimensional factor. Using the results of [**13**], we see the following.

- For the factors G_{12}, G_8, the integral representation ρ_0 is uniquely determined by ρ, so point (i) of Remark 2.13 applies.
- For the factor $G(3, 3, 2)$, point (ii) of Remark 2.13 applies.

To handle the remaining factors, take $\alpha = 3$ if $M = \mathbb{Q}(i)$ or $\alpha = 2$ if $M = \mathbb{Q}(\zeta_3)$. In Table 5, we give explicit generators for the image of one choice of ρ_0, presented so that the reduction modulo α becomes lower triangular. One may check easily that while the reduction modulo α is manifestly reducible, the reduction modulo α^2 admits no invariant vector not divisible by α. We may thus apply [**13**, Corollary 1.5] to see that there are exactly two choices for ρ_0, and that these two are exchanged by conjugation by $\begin{pmatrix} \alpha & 0 \\ 0 & 1 \end{pmatrix}$. Consequently, either these two choices are preserved by $\rho_0 \mapsto c_M \circ \rho_0 \circ c$, in which case Example 2.12 applies directly; or they are exchanged and Remark 2.14 applies.

Factor	Generators
$[G_8:4]$	$\begin{pmatrix} -1-i & -3 \\ i & 1+i \end{pmatrix}, \begin{pmatrix} 1 & 3 \\ -1 & -2 \end{pmatrix}$
$G(2,1,2)$	$\begin{pmatrix} -1 & -2 \\ 0 & 1 \end{pmatrix}, \begin{pmatrix} -1 & 0 \\ 1 & 1 \end{pmatrix}$
$G(6,6,2)$	$\begin{pmatrix} 1 & 0 \\ -1 & -1 \end{pmatrix}, \begin{pmatrix} -1 & -3 \\ 1 & 2 \end{pmatrix}$
G_4	$\begin{pmatrix} -1 & -2 \\ 1 & 1 \end{pmatrix}, \begin{pmatrix} -1 & -2 \\ 0 & \zeta_3 \end{pmatrix}$
G_4'	$\begin{pmatrix} -1 & -2 \\ 1 & 1 \end{pmatrix}, \begin{pmatrix} -\zeta_3 & -2\zeta_3 \\ 0 & \zeta_3^2 \end{pmatrix}$

TABLE 5. Presentations of two-dimensional factors of $\tilde{H}$.

REMARK 4.9. Although it is not necessary for our present discussion, for future reference we resolve the ambiguity in the previous discussion.

- For the factors $[G_8:4], G(2,1,2), G(6,6,2)$, we are in a situation similar to point (ii) of Remark 2.13: the choices for ρ_0 are fixed by c_M and exchanged by c, so we must apply Remark 2.14 rather than Example 2.12.
- For the factor G_4', we argue as in point (ii) of Remark 2.13. The group $G_4' \cong \langle 24, 3\rangle$ has a unique 2-Sylow subgroup

$$P = \left\langle \begin{pmatrix} -1 & -2 \\ 1 & 1 \end{pmatrix}, \begin{pmatrix} 1 & -2\zeta_3^3 \\ \zeta_3 & -1 \end{pmatrix} \right\rangle .$$

  The two choices of ρ_0, restricted to P, have reductions modulo 2 which factor through the quotient $\mathrm{C}_2 \times \mathrm{C}_2$ of P; by labeling the elements of P and of $\mathbb{F}_4$ and keeping track of signs of permutations, we see that the reductions are interchanged by c_M.

  For the extension by $\mathrm{Gal}(M/\mathbb{Q})$ in question, the 2-Sylow subgroup is $\langle 16, 13\rangle$. The quotient of this group by the center of P is abelian, so the reductions are preserved by c.

  Putting this together, we see that the two choices for ρ_0 must be exchanged by $\rho_0 \mapsto c_M \circ \rho_0 \circ c$, so we must apply Remark 2.14 rather than Example 2.12.
- For the factor G_4, recall from the explanation of Table 2 that we get the same projective image by taking the factor G_4' and modifying the action of c; we may thus reuse the previous analysis. In this case, the 2-Sylow subgroup is $\langle 16, 8\rangle$. The quotient of this group by the center of P is D_4, so the reductions are interchanged by c. Consequently, the two choices for ρ_0 are preserved by $\rho_0 \mapsto c_M \circ \rho_0 \circ c$, so we may apply Example 2.12 directly.

In the cases where $\tilde{H}$ is three-dimensional, we again apply the results of **[13]**. The only case where ρ_0 is not uniquely determined by ρ is $\tilde{H} = G_{25}$, in which case there are two choices for ρ_0 which are reducible modulo $1-\zeta_3$. However, the two reductions have invariant subspaces of different dimensions, so the map $\rho_0 \mapsto c_M \circ \rho_0 \circ c$ fixes them both and Example 2.12 applies directly.

From the previous analysis, it follows that we can apply the construction of Example 2.12 to obtain an abelian threefold A over $\mathbb{Q}$ with $\mathrm{ST}(A)^0 \simeq \mathrm{U}(1)_3$ and $\mathrm{ST}(A)/\mathrm{ST}(A)^0 \simeq H \rtimes \mathrm{C}_2$ *provided* that we can solve the Galois embedding problem for the pair $(\tilde{H} \rtimes \mathrm{C}_2, \tilde{H})$ and the extension $M/\mathbb{Q}$. We first give a uniform solution for the complex reflection groups.

LEMMA 4.10. *Consider any row of Table 2 for which $\tilde{H}$ is a complex reflection group (that is, $\tilde{H} \neq G(1,1,1) \times [G_8 : 4], G(3,1,1) \times G_4'$). Let $V = M^3$ be the representation space on which $\mathrm{GL}_3(M)$ acts.*

(a) *The invariant ring $R = (\mathrm{Sym}\, V^\vee)^{\tilde{H}}$ is a polynomial ring in 3 variables over M.*
(b) *The ring R^c is a polynomial ring in 3 variables over $\mathbb{Q}$.*

PROOF. To check (a), apply the Chevalley–Shephard–Todd theorem, which is valid for arbitrary complex reflection groups [**32**, §7], [**6**, Theorem 7.2.1]. We may deduce (b) from (a) by Galois descent. □

For those rows of Table 2 to which Lemma 4.10 applies, the affine space corresponding to R^c contains a Zariski-dense open subspace every point of which defines a solution of the desired embedding problem.

REMARK 4.11. For the various choices of $\tilde{H}$ subsumed in the previous argument, one can also give more direct solutions of the Galois embedding problem that may be better suited for generating small examples. This is perhaps least obvious when H is the Hessian; in this case, one has (as pointed out by Noam Elkies) an isomorphism $\tilde{H} \cong \mathrm{U}(3, \mathbb{F}_4)$ via which $\tilde{H}$ occurs as the 2-division field of a generic Picard curve over $\mathbb{Q}$.

To deal with the remaining cases, we give more direct solutions of the Galois embedding problems in question.

- For the case $\tilde{H} = G(1,1,1) \times [G_8 : 4]$, we find a Galois extension L of $\mathbb{Q}$ containing $\mathbb{Q}(i)$ with $\mathrm{Gal}(L/\mathbb{Q}) \cong \langle 48, 15\rangle$, $\mathrm{Gal}(L/\mathbb{Q}(i)) \cong \langle 24, 1\rangle$. By way of motivation, note that the second row of Table 2 also represents the abstract group $\langle 48, 15\rangle$ but with a different presentation: in that case, one takes the semidirect product of $\mathrm{C}_3 \times \mathrm{D}_4$ by C_2 acting by an outer automorphism of each factor. It therefore suffices to take L to be the compositum of two Galois extensions $L_1/\mathbb{Q}, L_2/\mathbb{Q}$ in which $L_1/\mathbb{Q}$ is an S_3-extension, $L_2/\mathbb{Q}$ is a D_8-extension containing $\mathbb{Q}(i)$ as the fixed subfield of one copy of D_4, and the fixed field of the other copy of D_4 coincides with the quadratic subfield of $L_1/\mathbb{Q}$.

  To confirm that this can be achieved, we give an explicit example. The degree-8 number field `8.2.85525504.1` contains $\mathbb{Q}(\sqrt{17})$, its discriminant is -1 times a square, and its Galois group is D_8; we may thus take L_2 to be the Galois closure of this field. The degree-3 number field `3.3.2057.1` has discriminant which is 17 times a square and its Galois group is S_3; we may thus take L_1 to be the Galois closure of this field.
- For the case $\tilde{H} = G(3,1,1) \times G_4'$, we find a Galois extension L of $\mathbb{Q}$ containing $\mathbb{Q}(\zeta_3)$ with $\mathrm{Gal}(L/\mathbb{Q}) \cong \langle 144, 125\rangle$, $\mathrm{Gal}(L/\mathbb{Q}(\zeta_3)) \cong \langle 72, 25\rangle$. It suffices to take L to be the compositum of two Galois extensions $L_1/\mathbb{Q}, L_2/\mathbb{Q}$ in which $L_1/\mathbb{Q}$ is an S_3-extension containing $\mathbb{Q}(\zeta_3)$ and $L_2/\mathbb{Q}$ is an extension containing $\mathbb{Q}(\zeta_3)$ with $\mathrm{Gal}(L/\mathbb{Q}) \cong \langle 48, 33\rangle$, $\mathrm{Gal}(L/\mathbb{Q}(\zeta_3)) \cong \langle 24, 3\rangle$.

To confirm that this can be achieved, we give an explicit example. We take $L_1 = \mathbb{Q}(\zeta_3, 2^{1/3})$. To choose L_2, we consult the Klüners-Malle database **[25]**, which includes the polynomial

$$x^{16} - x^{15} - 2x^{14} + 8x^{13} + 10x^{12} - 19x^{11} - 4x^{10} + 64x^9 + \\ 12x^8 - 94x^7 + 38x^6 + 119x^5 - 64x^4 - 97x^3 + 48x^2 + 11x + 1$$

as an entry for the permutation representation 16T60 (in GAP notation). Using MAGMA, one may confirm that this polynomial is irreducible with Galois group $\langle 48, 33\rangle$, and that the unique quadratic subfield of the splitting field is $\mathbb{Q}(\zeta_3)$.

References

[1] P.B. Allen, F. Calegari, A. Caraiani, T. Gee, D. Helm, B.V. Le Hung, J. Newton, P. Scholze, R. Taylor, and J.A. Thorne, *Potential automorphy over CM fields*, arXiv:1812.09999v1 (2018).

[2] Sonny Arora, Victoria Cantoral-Farfán, Aaron Landesman, Davide Lombardo, and Jackson S. Morrow, *The twisting Sato-Tate group of the curve $y^2 = x^8 - 14x^4 + 1$*, Math. Z. **290** (2018), no. 3-4, 991–1022, DOI 10.1007/s00209-018-2049-6. MR3856841

[3] Grzegorz Banaszak and Kiran S. Kedlaya, *An algebraic Sato-Tate group and Sato-Tate conjecture*, Indiana Univ. Math. J. **64** (2015), no. 1, 245–274, DOI 10.1512/iumj.2015.64.5438. MR3320526

[4] Grzegorz Banaszak and Kiran S. Kedlaya, *Motivic Serre group, algebraic Sato-Tate group and Sato-Tate conjecture*, Frobenius distributions: Lang-Trotter and Sato-Tate conjectures, Contemp. Math., vol. 663, Amer. Math. Soc., Providence, RI, 2016, pp. 11–44, DOI 10.1090/conm/663/13348. MR3502937

[5] Thomas Barnet-Lamb, Toby Gee, and David Geraghty, *The Sato-Tate conjecture for Hilbert modular forms*, J. Amer. Math. Soc. **24** (2011), no. 2, 411–469, DOI 10.1090/S0894-0347-2010-00689-3. MR2748398

[6] D. J. Benson, *Polynomial invariants of finite groups*, London Mathematical Society Lecture Note Series, vol. 190, Cambridge University Press, Cambridge, 1993, DOI 10.1017/CBO9780511565809. MR1249931

[7] F. Beukers and C. J. Smyth, *Cyclotomic points on curves*, Number theory for the millennium, I (Urbana, IL, 2000), A K Peters, Natick, MA, 2002, pp. 67–85. MR1956219

[8] Edgar Costa, Francesc Fité, Andrew V. Sutherland, *Arithmetic invariants from Sato–Tate moments*, C. R. Math. Acad. Sci. Paris **357** (2019), 823–826.

[9] J. H. Conway and A. J. Jones, *Trigonometric Diophantine equations (On vanishing sums of roots of unity)*, Acta Arith. **30** (1976), no. 3, 229–240, DOI 10.4064/aa-30-3-229-240. MR422149

[10] Edgar Costa, Nicolas Mascot, Jeroen Sijsling, and John Voight, *Rigorous computation of the endomorphism ring of a Jacobian*, Math. Comp. **88** (2019), no. 317, 1303–1339, DOI 10.1090/mcom/3373. MR3904148

[11] P. Deligne and J.S. Milne, *Tannakian Categories*, Lecture Notes in Math. 900, Springer, New York, 1982, 100–228.

[12] B. Dodson, *The structure of Galois groups of* CM*-fields*, Trans. Amer. Math. Soc. **283** (1984), no. 1, 1–32, DOI 10.2307/1999987. MR735406

[13] Walter Feit, *Some integral representations of complex reflection groups*, J. Algebra **260** (2003), no. 1, 138–153, DOI 10.1016/S0021-8693(02)00629-4. Special issue celebrating the 80th birthday of Robert Steinberg. MR1973580

[14] Francesc Fité and Xavier Guitart, *Fields of definition of elliptic k-curves and the realizability of all genus 2 Sato-Tate groups over a number field*, Trans. Amer. Math. Soc. **370** (2018), no. 7, 4623–4659, DOI 10.1090/tran/7074. MR3812090

[15] Francesc Fité, Kiran S. Kedlaya, Víctor Rotger, and Andrew V. Sutherland, *Sato-Tate distributions and Galois endomorphism modules in genus 2*, Compos. Math. **148** (2012), no. 5, 1390–1442, DOI 10.1112/S0010437X12000279. MR2982436

[16] Francesc Fité, Kiran S. Kedlaya, and Andrew V. Sutherland, *Sato-Tate groups of some weight 3 motives*, Frobenius distributions: Lang-Trotter and Sato-Tate conjectures, Contemp. Math., vol. 663, Amer. Math. Soc., Providence, RI, 2016, pp. 57–101, DOI 10.1090/conm/663/13350. MR3502939

[17] Francesc Fité, Kiran S. Kedlaya, Andrew V. Sutherland, *Sato–Tate groups of abelian threefolds*, preprint.

[18] Francesc Fité and Andrew V. Sutherland, *Sato-Tate distributions of twists of $y^2 = x^5 - x$ and $y^2 = x^6 + 1$*, Algebra Number Theory **8** (2014), no. 3, 543–585, DOI 10.2140/ant.2014.8.543. MR3218802

[19] Robert Guralnick and Kiran S. Kedlaya, *Endomorphism fields of abelian varieties*, Res. Number Theory **3** (2017), Paper No. 22, 10, DOI 10.1007/s40993-017-0088-4. MR3720508

[20] Everett W. Howe, Franck Leprévost, and Bjorn Poonen, *Large torsion subgroups of split Jacobians of curves of genus two or three*, Forum Math. **12** (2000), no. 3, 315–364, DOI 10.1515/form.2000.008. MR1748483

[21] Ernst Kani, *Products of CM elliptic curves*, Collect. Math. **62** (2011), no. 3, 297–339, DOI 10.1007/s13348-010-0029-1. MR2825715

[22] Kiran S. Kedlaya and Andrew V. Sutherland, *Computing L-series of hyperelliptic curves*, Algorithmic number theory, Lecture Notes in Comput. Sci., vol. 5011, Springer, Berlin, 2008, pp. 312–326, DOI 10.1007/978-3-540-79456-1_21. MR2467855

[23] Kiran S. Kedlaya and Andrew V. Sutherland, *Hyperelliptic curves, L-polynomials, and random matrices*, Arithmetic, geometry, cryptography and coding theory, Contemp. Math., vol. 487, Amer. Math. Soc., Providence, RI, 2009, pp. 119–162, DOI 10.1090/conm/487/09529. MR2555991

[24] P. Kılıçer, *The CM class number one problem for curves*, thesis, Université de Bordeaux, 2016, `https://openaccess.leidenuniv.nl/handle/1887/41145`.

[25] Jürgen Klüners and Gunter Malle, *A database for field extensions of the rationals*, LMS J. Comput. Math. **4** (2001), 182–196, DOI 10.1112/S1461157000000851. MR1901356

[26] The LMFDB Collaboration, *The L-functions and Modular Forms Database*, `https://www.lmfdb.org`, 2018 (accessed Oct 2019).

[27] Shouhei Ma, *Decompositions of an Abelian surface and quadratic forms* (English, with English and French summaries), Ann. Inst. Fourier (Grenoble) **61** (2011), no. 2, 717–743, DOI 10.5802/aif.2627. MR2895071

[28] I. Marin and J. Michel, *Automorphisms of complex reflection groups*, Represent. Theory **14** (2010), 747–788, DOI 10.1090/S1088-4165-2010-00380-5. MR2746138

[29] G. A. Miller, H. F. Blichfeldt, and L. E. Dickson, *Theory and applications of finite groups*, Dover Publications, Inc., New York, 1961. MR0123600

[30] D. Mumford, *A note of Shimura's paper "Discontinuous groups and abelian varieties"*, Math. Ann. **181** (1969), 345–351, DOI 10.1007/BF01350672. MR248146

[31] Frans Oort, *Endomorphism algebras of abelian varieties*, Algebraic geometry and commutative algebra, Vol. II, Kinokuniya, Tokyo, 1988, pp. 469–502, DOI 10.1016/B978-0-12-348032-3.50007-4. MR977774

[32] G. C. Shephard and J. A. Todd, *Finite unitary reflection groups*, Canad. J. Math. **6** (1954), 274–304, DOI 10.4153/cjm-1954-028-3. MR59914

[33] Goro Shimura, *On the zeta-function of an abelian variety with complex multiplication*, Ann. of Math. (2) **94** (1971), 504–533, DOI 10.2307/1970768. MR288089

[34] Tetsuji Shioda and Naoki Mitani, *Singular abelian surfaces and binary quadratic forms*, Classification of algebraic varieties and compact complex manifolds, Springer, Berlin, 1974, pp. 259–287. Lecture Notes in Math., Vol. 412, DOI 10.1007/BFb0066163. MR0382289

[35] Andrew V. Sutherland, Sato–Tate distributions, in *Analytic Methods in Arithmetic Geometry: Arizona Winter School, March 12–16, 2016, The University of Arizona, Tucson, AZ*, Contemp. Math. 740, Amer. Math.Soc., Providence 2019, 197–248, DOI: 10.1090/conm/740, MR4033732

[36] Andrew V. Sutherland, *Fast Jacobian arithmetic for hyperelliptic curves of genus 3*, Proceedings of the Thirteenth Algorithmic Number Theory Symposium, Open Book Ser., vol. 2, Math. Sci. Publ., Berkeley, CA, 2019, pp. 425–442, DOI 10.2140/obs.2019.2.425. MR3952026

[37] Andrew V. Sutherland, *A database of nonhyperelliptic genus-3 curves over $\mathbb{Q}$*, Proceedings of the Thirteenth Algorithmic Number Theory Symposium, Open Book Ser., vol. 2, Math. Sci. Publ., Berkeley, CA, 2019, pp. 443–459, DOI 10.2140/obs.2019.2.443. MR3952027

[38] M. G. Upton, *Galois representations attached to Picard curves*, J. Algebra **322** (2009), no. 4, 1038–1059, DOI 10.1016/j.jalgebra.2009.04.021. MR2537671

[39] Adrian Vasiu, *Surjectivity criteria for p-adic representations. II*, Manuscripta Math. **114** (2004), no. 4, 399–422, DOI 10.1007/s00229-004-0465-x. MR2081941

[40] Yuri G. Zarhin, *Hyperelliptic Jacobians without complex multiplication*, Math. Res. Lett. **7** (2000), no. 1, 123–132, DOI 10.4310/MRL.2000.v7.n1.a11. MR1748293

Francesc Fité, Department of Mathematics, Massachusetts Institute of Technology, 77 Massachusetts Ave., Cambridge, MA 02139, USA
Email address: ffite@mit.edu

Kiran S. Kedlaya, Department of Mathematics, University of California, San Diego, 9500 Gilman Drive #0112, La Jolla, CA 92093, USA
Email address: kedlaya@ucsd.edu

Andrew V. Sutherland, Department of Mathematics, Massachusetts Institute of Technology, 77 Massachusetts Ave., Cambridge, MA 02139, USA
Email address: drew@math.mit.edu

Contemporary Mathematics
Volume **770**, 2021
https://doi.org/10.1090/conm/770/15433

Arithmetic, Geometry, and Coding Theory: Homage to Gilles Lachaud

Sudhir R. Ghorpade, Christophe Ritzenthaler, François Rodier, and Michael A. Tsfasman

ABSTRACT. We give an overview of several of the mathematical works of Gilles Lachaud and provide a historical context. This is interspersed with some personal anecdotes highlighting many facets of his personality.

The AGC^2T Conference at CIRM[1] in June 2019 was dedicated to Gilles Lachaud (26 July 1946 – 21 February 2018) who was a founder of this series of conferences which began in 1987. Two speakers spoke exclusively about the work and persona of Gilles Lachaud, while many others paid rich tributes to him during their talks[2].

This article is a homage to Gilles and it attempts to give an outline of his mathematical journey and glimpses of his work and personality. It is divided in four parts. The first part, written by Rodier, gives an account of Gilles' early works including his Ph.D. thesis related to automorphic forms, works on Waring's problem and real quadratic fields, and some of his earlier articles on coding theory. Next, Tsfasman recounts the beginnings of algebraic geometric codes, his interactions with Gilles Lachaud, and some of their joint work. The third part, by Ghorpade, describes some of Gilles' work on linear codes and on counting the number of rational points of algebraic varieties over finite fields, as well as some related developments. Finally, Ritzenthaler gives a personalized account of Gilles' work that deals mainly with non-hyperelliptic curves and abelian varieties.

1. Gilles Lachaud's early works

François Rodier

I met Gilles Lachaud in 1969 when he had just been appointed to what was still called the University of Paris, as Roger Godement's assistant. He was a pure

Sudhir Ghorpade is partially supported by DST-RCN grant INT/NOR/RCN/ICT/P-03/2018 from the Dept. of Science & Technology, Govt. of India, MATRICS grant MTR/2018/000369 from the Science & Engg. Research Board, and IRCC award grant 12IRAWD009 from IIT Bombay.

Michael Tsfasman is supported in part by ANR project FLAIR (ANR-17-CE40-0012).

[1]CIRM = Centre International de Rencontres Mathématiques = International Centre for Mathematics Meetings (https://www.cirm-math.fr) is based in Luminy, near Marseille, France.

[2]The slides of most of the talks in AGC^2T-2019, and videos of some of them, are available at: https://conferences.cirm-math.fr/1921.html

product of the university. He did not want to go to the so-called "Grandes Écoles". Gilles had always refused this royal road.

During the academic year 1969-70, Roger Godement had left Paris for a six-month stay at the Institute for Advanced Study in Princeton, USA, to work with, among others, Robert Langlands. Meanwhile, Godement's assistants organized the same DEA (Masters degree) course that he offered each year. With Gilles, there were Miss Chambon, Jean-Pierre Labesse, and Michel Duflo, assisted by François Bruhat. It was by attending these classes on *Representations of Locally Compact Groups* that I first met Gilles. He had lectured on spectral theory, a field in which he specialized.

Having started a thesis myself under the direction of Godement, I met Gilles quite often, especially during the weekly sessions of seminars. He came to the symposium in 1971 in Budapest where we had the opportunity to cross the Iron Curtain to meet the Soviet School of Group Theory, and first Israel Gelfand, who had been authorized to cross the border of the USSR to come to Hungary. Then, in the evening, Gilles organized some parties with the younger people in the conference.

In 1972, an International Summer School was organized by the University of Antwerp, on *Modular functions of one variable*. But it was financed by NATO. The dislike of the army by Godement which was well known, made him prefer a symposium organized by the AMS in Williamston, in the United States on *Harmonic analysis on homogeneous spaces*. Gilles was one of the mathematicians who followed Godement instead of attending the symposium in Antwerp. This was the first time that I did my presentation in English in front of an amphitheater filled with people that I considered more capable than me. So Gilles offered me to make me rehearse it. Gilles was more sure of himself in English, but he still asked me to assist in exchange for his own rehearsal.

After publishing two articles in French in the journal Inventiones: *Analyse spectrale des formes automorphes et séries d'Eisenstein* [**19**] in 1978, and *Variations sur un thème de Mahler* [**21**] in 1979, Gilles defended his thesis of "Doctorat d'État" at Paris Diderot University - Paris 7 in June 1979 on *Analyse spectrale et prolongement analytique: Séries d'Eisenstein, Fonctions Zêta et nombre de solutions d'équations diophantiennes* [**22**] under the direction of Roger Godement. In this thesis, Gilles reproves the results of Selberg on the meromorphic prolongation of the Eisenstein series by a method of perturbing the continuous spectra of self-adjoint operators. He also obtains the analytical extension of a type of generalized zeta functions, and applies these results to diophantine equations. The Rivoire prize from the University of Clermont-Ferrand was awarded to him in 1979 for his thesis.

More specifically, in the article *Spectral analysis of automorphic forms and Eisenstein series*, Gilles is concerned with the spectral decomposition of the left regular representation π of G on $K\backslash G/\Gamma$ for G a real, linear, connected simple Lie group of rank 1, K a maximal compact subgroup, Γ a discrete subgroup such that G/Γ has finite invariant volume. Then, by a general principle of functional analysis, there exists a spectral measure m on the spectrum of the algebra $L^1(G,K)$ of complex integrable functions on G bi-invariant under K. This spectrum is a closed subset of the complex plane, such that we have

$$(\pi(F)f,g) = \int \tilde{F}(\lambda)dm_{f,g}(\lambda)$$

by noting $\tilde{F}$ the Gel'fand transform of the function $F \in L^1(G, K)$, and $dm_{f,g}$ the complex measure associated with the spectral measure m and the elements f, g of $L^2(K\backslash G/\Gamma)$. Some eigenfunctions of $\pi(F)$ are given by Eisenstein's series. In the case of $G = SL(2, \mathbb{R})$, $\Gamma = SL(2, \mathbb{Z})$, $K = SO(2)$ they are written

$$E_s(z) = \frac{1}{2} y^s \sum_{(c,d)=1} |cz + d|^{-2s}, \qquad \Re e\ s > 1, z = x + iy, y > 0,$$

since one can identify the symmetric space $K\backslash G$ with the upper half-plane. They extend analytically to the entire complex plane, and their value on the spectrum of the algebra $L^1(G, K)$ allows to calculate the continuous part of the measures $dm_{f,g}$. It was A. Selberg who stated for the first time these results in the case the group G is of rank one on the field of rationals, following his own method.

The aim of Gilles is to use another method based on the theory of perturbations of the continuous spectra of self-adjoint operators. It consist in finding a representation π' of G in $L^2(K\backslash G/\Gamma)$ fulfilling two properties. First its spectral analysis is known, second it is close to the representation π in a suitable sense, namely that the difference $\pi(F) - \pi'(F)$ must satisfy a certain compactness criterion for sufficiently numerous $F \in L^1(G; K)$. This method, known from Selberg and Gel'fand, was developed by Faddeev in the case of the group $SL(2, R)$. Hence, the ideas used by Gilles in proving his main theorems are not completely different from known ones. But, after using many general theorems, the theory has been given a smooth and systematic look.

In the article *Variations on a theme of Mahler*, he studied the function

$$N(t) = \#\{x \in \mathbb{Z}^n : |F(x)| \leq t\}$$

for multivariate polynomials F with n variables on $\mathbb{Z}$. More precisely let $S = \{p_1, \ldots, p_s\}$ be a finite set of prime numbers, $P = p_1 \ldots p_s$ and $\mathbb{Z}^n(S)$ the set of vectors $x = (x_1, \ldots, x_n)$ of $\mathbb{Z}^n$ whose coordinates are prime to P as a whole. For any prime p, we denote $|x|_p$ the absolute value of $x \in \mathbb{Q}$ associated with the place defined by p. Denote $|x|_0$ the Archimedean absolute value of $\mathbb{Q}$. If $x \in \mathbb{Q}^n$, we write:

$$F_S(x) = |F(x)|_0 \prod_{p \in S} |F(x)|_p.$$

Gilles proves the following theorem about $N_S(t) = \#\{x \in \mathbb{Z}^n : |F_S(x)| \leq t\}$.

THEOREM 1.1. *With the previous notations, let us suppose the form F is anisotropic on $\mathbb{R}$ and on $\mathbb{Q}_p$ when $p \in S$. Then*

$$N_S(t) = V_S t^{n/d} + O(t^{(n-\theta)/d}),$$

where V_S is a constant (which he computes) and where $\theta > 1$.

He deduced the analytic continuation of certain Zeta functions.

Strangely enough, because the theory of automorphic forms has developed strongly, under the impetus of Langlands among others, these publications have had little echo. But Gilles' interest in this theme has not disappeared since, as we can see in several of his later publications, and perhaps one of the last *The distribution of the trace in the compact group of type G_2* **[36]**, where he used the theory of representations of Lie groups. This article responds to a request from J.-P. Serre concerning the calculation of a certain exponential sum of degree 7, involving the G_2 group as in the theory of Sato-Tate.

Gilles was interested in other things than the theory of automorphic forms. He was also concerned about number theory and history of mathematics.

He gave lectures at various universities about some points of history of mathematics, the foundations of arithmetic and the connections between mathematics and philosophy, mysticism and magic. In the article *Nombres et archétypes: du vieux et du neuf [Number and archetypes: something old, something new]* [**20**], he suggested: mathematics and philosophy have been estranged since the time of Diophantos, whose number theory can be viewed as a succession of relations between objects whose status is determined axiomatically and mathematical truth is determined internally, without reference to external reality; there is another tradition of mathematics working under mystical influences, e.g., Euclid's Pythagorean emphasis on perfect numbers and solids. This led Gilles to speculate on the connections between mathematics and magic with the philosophers like Plato, Iamblichus and Ferdinand Gonseth.

He contributed to the publication of *Les Arithmétiques* by Diophantus in 1984 (*R. Rashed, Diophante. Les Arithmétiques, t. III : Livre IV, ccvi + 162 p. ; Paris, Les Belles Lettres, 1984*). In [**6**, p. 367], Karine Chemla writes about this book: "R. Rashed in the publication of third book of Diophantus makes in his commentary, after having gathered in an abstract of algebraic geometry, written in conjunction with the number theorist Lachaud, the set of techniques that the translation of the text of Diophantus supposes. This abstract thus presents the double virtue of being a well-made mathematical introduction and of being adapted to the text of Diophantus. We have here a most remarkable example of what the combination of mathematics and history can produce." Indeed, Gilles wrote about fifty pages of introduction to modern algebraic geometry, including Bézout theorem, a discussion of birationnal equivalence, and the quartic associated to the double equations of degree 2. He gives some examples of applications of elementary algebraic geometry to diophantine analysis avoiding the megamachines often needed in these kind of results.

In Number Theory, he worked on Waring's problem. He wrote an article *Une présentation adélique de la série singulière et du problème de Waring [An adelic presentation of the singular series and the Waring problem]* [**23**] to consider Waring's problem and the corresponding singular series in the adelic setting from ideas of Igusa, Ono and Weil. The local singular series in the place p is expressed in the form of counting points. Later it is this expression that applies in the demonstration of some examples of Yuri Manin's conjectures about the asymptotic behaviour of the number of points of bounded height on an algebraic variety defined over a number field when the height goes to infinity by some people who tackled that subject such as Antoine Chambert-Loir, Yuri Tschinkel, Régis de la Bretèche, Tim Browning, Emmanuel Peyre.

He worked also on real quadratic fields. In the article *On real quadratic fields* [**25**], he studies the caliber of a real quadratic field $Q(\sqrt{d})$ (the number $k(d)$ of reduced primitive binary quadratic forms of discriminant d). A theorem of Siegel says there are but a finite number of real quadratic fields with a given caliber but it is not effective. Gilles shows that under GRH, the only real quadratic fields with caliber one are the seven fields $Q(\sqrt{t})$ with $t = 2, 5, 13, 29, 53, 173, 293$. Also he shows that under GRH, the only principal real quadratic fields with discriminant $d = r^2 + 1 \equiv 1 \pmod 4$ are the six fields $Q(\sqrt{t})$ with $t = 5, 17, 37, 101, 197, 677$.

For that he proved a formula involving the class number $h(d)$ and the fundamental unit e_0: $h(d)\log e_0(d) < c\cdot k(d)$ where $c < 4.230$. In 2010, this has been proved and generalised by Byungheup Jun and Jungyun Lee, in the article *Caliber number of real quadratic fields* [**17**], where they find, without GRH, the fields with caliber 1 and also, the fields of caliber 2 and discriminant $\not\equiv 5 \pmod 8$.

We did not see each other so often anymore when he was appointed professor in Nice in 1981. But an article of Tsfasman and Vlăduţ dealing with application of modular functions to the theory of Goppa codes led Gilles to be interested in Information Theory. In 1985, he made a presentation at the prestigious Bourbaki seminar on *the Geometric Goppa codes* [**24**] where he explained these works. I was there, especially since at that time, I wanted to change my field of research from the Theory of Group Representations to the Theory of Codes.

It was also in 1985 that Gilles took up the position of *Directeur de Recherches CNRS* and moved from Nice to Luminy. He became the second director of CIRM, from 1986 to 1991. The CIRM was inaugurated in 1991, and Gilles knew it already, as vice-president of SMF, i.e., *Société Mathématique de France* (French Mathematical Society), during 1982–1984, and as the president of the scientific council of CIRM. His stewardship was around the beginning of CIRM. With the technical team, he had a pioneering spirit. During that period, there was a rise in scientific activities: from 20 symposia in 1986, it reached 37 in 1991.

Already, at that time, a certain number of symposia foreshadowed in part the major currents of mathematical research. There were some big conferences, such as the 300-person symposium organized by Mohammed Mebkhout, or the *Journées Arithmétiques* in 1989, with a similar number of participants, organised by Gilles. The highlight of this period was the construction of the new library. The former was housed in the premises of the University library, it was open at the same hours and it was far from CIRM. Obtaining the necessary funding to build new library for CIRM was not easy, but finally he was able to launch an architectural competition. Of course there was some complication during the works. For instance he wanted to build a beautiful amphitheater in the new building, but while digging, they fell on the rock of a creek (that's why the conference room at that time was horizontal). The library was finally inaugurated in 1991, at the time of the 10th anniversary of CIRM. It had been a really short time (5 years) since the granting of the credits until the inauguration. Gilles showed his ecological feeling and ended this period with a long-term wish: CIRM has 4 hectares of a forest that is quite exceptional, since it is a northern forest in the middle of the southern forest of creeks. The flora is well supplied, but he complained that the forest areas are not restored at all. He wished those who follow him to achieve preservation and enrichment of this precious ecological heritage.

In 1987, while Gilles was the head of CIRM, he organised with Jacques Wolfmann from the University of Toulon, a conference named Algebraic Geometry and Coding Theory (AGCT). At that time, one did not imagine that this conference would still exist now under a slightly modified name because of the renewal of the subjects: Arithmetic, Geometry, Cryptography, and Coding Theory (AGC^{2}T).

He did several works concerning popularization of Mathematics. He wrote several articles in the general public journal *La Recherche* (Les codes correcteurs d'erreurs) or for the *Encyclopedia Universalis* (Qu'est ce qu'une équation).

In the article *The weights of the orthogonals of the extended quadratic binary Goppa codes* [**46**] with Jacques Wolfmann they started from results on elliptic curves and Kloosterman sums over the finite field $\mathbb{F}_{2^t}$, and they determined the weights of the orthogonals of some binary linear codes. The function $f(x) = x^{-1}$ on $\mathbb{F}_{2^t}$ resists to differential cryptanalysis and it is used in the "Rijndael" protocol retained in the call for proposals for the Advanced Encryption Standard (AES) launched by the National Institute of Standards and Technology (NIST). They could compute the Walsh spectrum of that function from their result giving a relation between the number of points of ordinary elliptic curve and Kloosterman sums. This article has become the most quoted article of his.

A little before 1990, Gilles created a working group on the Theory of Information that he set up at CIRM, that included Guy Chassé, Yves Driencourt, Dominique Le Brigand, Jean-Francis Michon, Marc Perret, Robert Rolland, and myself. This group was very active and organised a meeting every month.

At that time, he was also working with Gérard Rauzy and Jean-Yves Girard and with the head of CNRS to create a CNRS proper laboratory dedicated to Discrete Mathematics. With CIRM, CNRS wanted to create a "mathematics campus" in Luminy. The LMD (Discrete Mathematics Laboratory) was created in Marseille in October 1992 with a team ATI (Arithmetic and Theory of Information) that he led and that I joined naturally.

Gilles was generous, always willing to help others. He helped me a lot to make my moving easier, because I was disabled at the time. In order that I get used to giving talks to students he left me a course of DEA/Master2 to teach in his place.

He became the Director of IML (Mathematics Institute of Luminy) from 2006 to 2011, and then gave me the place of Director. At that time, I saw the investment in time and energy that this appointment required, and wondered how he had managed to occupy this position twice while he kept on doing mathematics.

He was optimistic as shows the last letter I received from him, one month before he died.

> *"Pour ma santé, ça va bien, je suis normal, j'ai un excellent traitement, [...] Le résultat est que je deviens casanier et paresseux. Donc je fais des maths à la maison, sur les statistiques de traces de représentations, ce qui est une activité qui se fait assis.*
>
> *J'arrive quand même à sortir dans le quartier, et j'ai pu aller à Besançon présider la thèse de Philippe Lebacque.*
>
> *Il faudrait tout de même qu'on se voie un de ces jours. Quel est ton jour? Le jeudi?"* [3]

Besides mathematics, Gilles' interests included philosophy, literature, art, history, and so on. We miss Gilles a lot. He breathed tranquility. I retain his vast culture and his humanism. He leaves a great void.

---

[3] *"For my health, I'm fine, I'm normal, I have an excellent treatment, [...] The result is that I become a homebody and lazy. So I do maths at home, on statistics of traces of representations, which is an activity that is done sitting.*

I can still go out in the neighborhood, and I went to Besançon to preside over the thesis of Philippe Lebacque.

We should still see each other one of these days. What is your day? Thursday?"

2. Gilles Lachaud: friend and mathematician

Michael A. Tsfasman

Algebraic geometry codes were discovered by Goppa at the beginning of 1980s and became widely known to mathematicians after 1981 when our preprint of [**52**]—with the proof that there exist codes over the Gilbert–Varshamov bound—was smuggled through the iron curtain. In 1985, Gilles Lachaud gave a talk on the subject [**24**] at the Bourbaki seminar. Most likely he was attracted by the subject because our proof was based on modular curves, and his basic education was in modular and automorphic forms. In 1987, he organised the first AGCT meeting at CIRM and invited me to come. Strong mutual dislike between me and the Soviet regime naturally led to the refusal of the leave permit needed at the time to cross the border. Two years later—the regime being much weakened by perestroika—I came for AGCT-2, and we saw each other for the first time. Gilles met me at the entrance to CIRM, and we became close friends ever after.

I was impressed by the scope of his interests with quite an intersection with mine: literature, philosophy, history, religion, art, and science, not to speak about mathematics. I appreciated it even more when I learned French enough to switch to it in our communication.

Gilles was the director of CIRM for 6 years and his influence is felt there even now. He built the library that should bear his name. And he reorganized CIRM in such a way as to make it maximally friendly for the mathematical community in everything from perfect blackboards to wonderful cuisine. Then he organised the LMD (Discrete Mathematics Laboratory) that later became the Institute of Mathematics of Luminy.

It is because of him that among the countries to work outside Russia I have chosen France.

Now let me describe the part of his legacy that is close to my interests. I mean mostly geometry over finite fields and a little bit coding theory.

Gilles' work on codes and curves over finite fields started with questions where he could use his expertise in exponential sums, Kloostermann sums, Eisenstein sums, and so on [**26**, **28**, **29**, **31**, **32**, **45**].

He also did some nice work, joint with J. Stern, on purely coding theory questions [**41**, **42**]. In the first of these papers they prove that there are families of asymptotically good codes over $\mathbb{F}_q$ with polynomial complexity of construction whose relative weights are as close to $(q-1)/q$ as we want. In the second one they prove a nice asymptotic lower bound for polynomially constructed spherical codes and for the (polynomial) kissing number.

How many $\mathbb{F}_q$-points can there be on an abelian variety? On the Jacobian of a curve? Since $h_X = |J_X(\mathbb{F}_q)| = P_X(1)$ the obvious bound is

$$(\sqrt{q}-1)^{2g} \leq h \leq (\sqrt{q}+1)^{2g}.$$

In his paper with Martin–Deschamps [**37**], they prove much better bounds:

THEOREM. *For a smooth geometrically irreducible algebraic curve X of genus g, the number $h_X = |J_X(\mathbb{F}_q)|$ satisfies the lower bounds*

$$h_X \geq q^{g-1} \frac{(q-1)^2}{(q+1)(g+1)}$$

and

$$h_X \geq (\sqrt{q}-1)^2\ \frac{q^{g-1}-1}{g}\ \frac{|X(\mathbb{F}_q)|+q-1}{q-1}.$$

If $g > \frac{\sqrt{q}}{2}$ *and if* $X(\mathbb{F}_q)$ *is nonempty, then*

$$h_X \geq (q^g-1)\ \frac{q-1}{q+g+gq}.$$

Much later, in joint papers with Aubry and Haloui [**2, 3**], they ameliorated these bounds. In the same paper they proved an analogue of Serre's bound for an arbitrary abelian variety.

His paper with S. Vlăduţ [**44**] treats the function field analogue of the famous Gauss problem asking whether the number of real quadratic fields with class number 1 is infinite, or not. Its weak version asks whether the number of principal algebraic number fields (i.e., with class number 1) is infinite. Both are wide open. In the function field case we have two types of their analogues, the obvious one asks just for infinity of such fields and is easy to solve in the affirmative, the more reasonable one asks for infinity of such fields when the constant field $\mathbb{F}_q$ is fixed.

Here is an illustrious specimen of their results:

THEOREM. *Assume* $q = 4, 9, 25, 49$, *or* 169. *Let* $P \in \mathbb{P}^1(\mathbb{F}_q)$. *Then there are infinitely many extensions* (K, S) *of the pair* $(\mathbb{F}_q(T), \{P\})$ *such that:*

(1) *The place* P *completely splits in* K.
(2) *The field* K *is a Galois extension of* $\mathbb{F}_q(T)$.
(3) *The ring* A_S *is principal.*

Now let me discuss in more details our only joint paper *Formules explicites pour le nombre de points des variétés sur un corps fini [Explicit formulas for the number of points of varieties over a finite field]* [**43**]. In mid-nineties, I realised that though there are plenty of mathematicians working on curves over finite fields and their number of points, actually no-one had ever studied the number of points on surfaces. After a while I had decided to write down my first results on surfaces and multi-dimensional varieties. It was the moment of my intensive attempts to study French (Gilles' influence, of course). So I started to write my only French language paper [**53**]. Quite naturally, the French of it was awful. So I asked Gilles to correct it. He did it perfectly and little by little we started to think on the subject together.

For curves the asymptotic problem is that of the behaviour of the ratio $|X(\mathbb{F}_q)|/g_X$. For multidimensional varieties I suggested to divide by the sum of Betti numbers. The problem becomes mutidimensional, say for surfaces we have two parameters instead of one, $x = b_1/(b_1+b_2)$ and $y = N/(b_1+b_2)$, here $N = |X(\mathbb{F}_q)|$. Therefore, the answer is not a number as in the case of curves (where, at least when q is a square, it is equal to $\sqrt{q}-1$), but a curve on the (x, y)-plane.

One of the results of that paper of mine was the generalization of the Drinfeld–Vlăduţ asymptotic upper bound to the multi-dimensional case. I did that using the elementary approach of Drinfeld–Vlăduţ. It was clear for both of us that the explicit formulae approach could be better.

The process of our work showed the difference between the Russian and French mathematical schools, as well as Gilles' style of doing mathematics. I would give a vague and often incorrect idea and try to explain it to him. Next day he would come with a couple of pages: precise definitions and clear results. I would read and criticize it, and then in the morning a better version was given to me.

After you write down explicit formulae, to get bounds you need to choose good test functions. These test functions should be doubly positive, meaning that both their values and their Fourier coefficients are positive. Gilles noticed that these functions happen to be related with (doubly) positive kernels used in sophisticated analysis starting from 19th century. After handing the previous night results to me to read, he would go to CIRM library to look through classical books to find new and new ones for us to use.

Classics mainly considered doubly positive kernels (densities of the corresponding measures) such as, for example, Fejér, Johnson, and de la Vallée-Poussin kernels, and the Jacobi kernel.

There is also a more recent source of doubly positive sequences. Namely, J. Oesterlé, who studied for the case of curves a more subtle question on the minimum genus of a curve having a prescribed number N of points, used explicit formulas and substituted into them doubly positive sequences optimal for precisely this problem, which involves two free parameters, N and g. However, in the asymptotic problem, two parameters appear not for curves but for surfaces, so Oesterlé sequences yield good bounds for the case of surfaces too.

Each of the above-mentioned doubly positive kernels or sequences gives a specific bound on the number of points. There are countably many such bounds, their graphs intersect each other, and the best known bound is given by their envelope curve.

The result (one of, the paper being large enough) is depicted on the following figures for the case of $q = 2$ (on these plots, the abscissa axis corresponds to the parameter $\frac{b_1}{b_1+b_2}$, and the ordinate axis, to the parameter $\frac{N}{b_1+b_2}$): Fig. 1 shows bounds that correspond to the above-described doubly positive objects and their envelope; Fig. 2, besides the envelope, shows the Weil bound W, Serre bound S, generalized elementary bound L, and the existence bound M from [**53**].

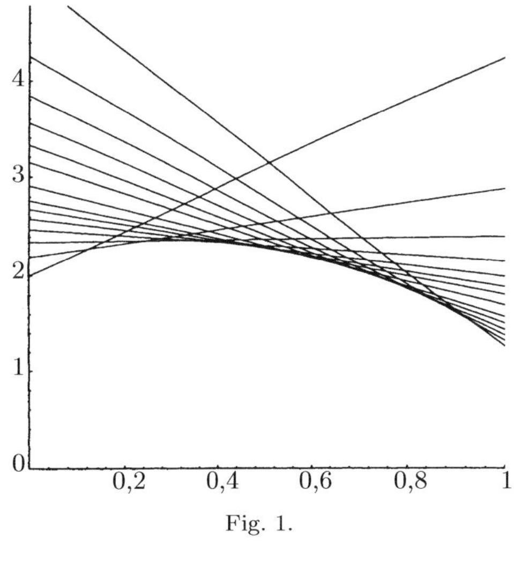

Fig. 1.

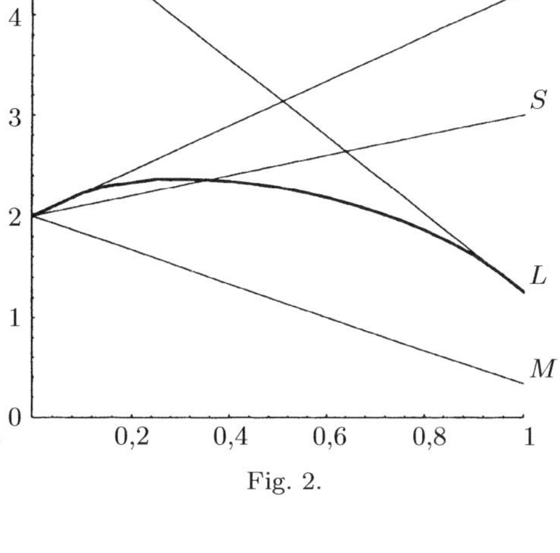

Fig. 2.

3. A Tribute to Gilles Lachaud

Sudhir R. Ghorpade

During August 1993, there was a 3-week CIMPA[4] School on Discrete Mathematics at Lanzhou, China. The speakers included Gilles Lachaud, Maurice Mignotte, Robert Rolland, and Wu Wen-Tsun. I was one of the participants in this school, and it was here that I first met Gilles Lachaud. As it turned out, that was the beginning of a long and cherished mathematical association and close friendship. In what follows, we will discuss some of Gilles' work on topics such as counting points of algebraic varieties over finite fields, and linear error correcting codes.

In the previous two sections, the Bourbaki seminar [**24**] and some of the earlier works of Lachaud on coding theory, including his much cited paper with Wolfman [**46**], have been mentioned. I will begin with an innocuous little paper on projective Reed-Muller codes that first appeared in the Cachan proceedings [**27**]. Recall that given any positive integers d, m, the (generalized) Reed-Muller code $\mathrm{RM}_q(d,m)$ of order d and length $n = q^m$ can be defined as the image of the evaluation map

$$\mathrm{Ev} : \mathbb{F}_q[x_1, \dots, x_m]_{\le d} \to \mathbb{F}_q^n \quad \text{given by} \quad \mathrm{Ev}(f) = (f(P_1), \dots, f(P_n)),$$

where $\mathbb{F}_q[x_1, \dots, x_m]_{\le d}$ denotes the $\mathbb{F}_q$-vector space of all polynomials of (total) degree $\le d$ in m variables $x_1, \dots, x_m$ with coefficients in the finite field $\mathbb{F}_q$, and where $P_1, \dots, P_n$ is an ordered listing of distinct points in $\mathbb{A}^m(\mathbb{F}_q)$. This has been widely studied in coding theory, and we cite in particular, the work of Delsarte, Goethals, and MacWilliams [**10**]. Gilles proposes a natural generalization of this where the affine space $\mathbb{A}^m$ is replaced by the projective space $\mathbb{P}^m$ and quite naturally, $\mathbb{F}_q[x_1, \dots, x_m]_{\le d}$ is replaced by the space $\mathbb{F}_q[x_0, x_1, \dots, x_m]_d$ of homogeneous polynomials of degree d in $m+1$ variables. Further, he takes fixed representatives in $\mathbb{F}_q^{m+1}$ of the points of $\mathbb{P}^m(\mathbb{F}_q)$ (for instance, those where the first nonzero coordinate is 1). One can then define a similar evaluation map and its image is the projective Reed-Muller code $\mathrm{PRM}_q(d,m)$ of order d. It has length

$$p_m := |\mathbb{P}^m(\mathbb{F}_q)| = q^m + q^{m-1} + \cdots + q + 1. \tag{3.1}$$

Moreover, if $d \le q$, then the evaluation map is injective and so the dimension of $\mathrm{PRM}_q(d,m)$ is $\binom{m+d}{d}$. In [**27**], Gilles gives a lower bound on the minimum distance of $\mathrm{PRM}_q(d,m)$, and computes the exact value in the special cases $d = 1$ (where the Plotkin bound is attained) and $d = 2$. The latter uses classical facts about zeros of quadrics in finite projective spaces. For the general case, one needs tight bounds on the maximum number of $\mathbb{F}_q$-rational points on a hypersurface in $\mathbb{P}^m$ or in other words, the maximum number, say N_d, of zeros in $\mathbb{P}^m(\mathbb{F}_q)$ that a homogeneous polynomial of degree d in $m+1$ variables with coefficients in $\mathbb{F}_q$ can have. Tsfasman conjectured that if $d \le q+1$, then $N_d = dq^{m-1} + p_{m-2}$, where p_j is as in (3.1) if j is a nonnegative integer and $p_j := 0$ if $j < 0$. Tsfasman's conjecture was quickly proved in the affirmative by J.-P. Serre, who communicated it to Tsfasman in a letter which was published some two years later [**50**]. Gilles was then in a position to complete the results of [**27**] and give a formula for the minimum distance in the general case in [**30**], where he argues that from coding

[4]CIMPA = *Centre International de Mathématiques Pures et Appliquées* = *International Centre of Pure and Applied Mathematics*. It is an organization based in France that conducts instructional schools on mathematics in "developing countries". For more information, see: https://www.cimpa.info/en

theoretic viewpoint, projective Reed-Muller codes are somewhat better than (affine, or generalized) Reed-Muller codes, and as such, they deserve to be studied further. A beginning was already made by Sorensen [**51**] whose paper, published in the same year as [**50**], and gives, in effect, an alternative proof of Tsfasman's conjecture and additional results of particular interest to coding theory.

The introduction of projective Reed-Muller codes by Gilles and the question about the minimum distance of these codes, which is essentially equivalent to Tsfasman's conjecture mentioned above, has had many offshoots, and we will briefly describe some of them.

Tsfasman's conjecture is about projective hypersurfaces defined over $\mathbb{F}_q$ of degree $d \leq q+1$. Now a hypersurface is the simplest example of a complete intersection in $\mathbb{P}^m$ of high dimension (vis-à-vis m). Keeping this in view, Gilles formulated the following conjecture (which appeared in print much later, in our joint paper [**13**]).

CONJECTURE. *If X is a complete intersection in $\mathbb{P}^m$ defined over $\mathbb{F}_q$ of dimension $n \geq m/2$ and degree $d \leq q+1$, then*

$$|X(\mathbb{F}_q)| \leq dp_n - (d-1)p_{2n-m} = d(p_n - p_{2n-m}) + p_{2n-m}.$$

Observe that if X is a hypersurface (so that $n = m-1$), then this is Tsfasman's Conjecture. On the other hand, Tsfasman together with Boguslavsky, generalized his conjecture about hypersurfaces of a given degree $d \leq q+1$ to intersection, say $S_{d,r}$ of r hypersurfaces, each defined by a homogeneous polynomial of degree $d < q$, such that these r polynomials are linearly independent. Tsfasman-Boguslavsky Conjecture proposes an intricate, but explicit, formula $T_r(d,m)$ for the maximum number of $\mathbb{F}_q$-rational points of algebraic sets such as $S_{d,r}$. Again, if $r = 1$, then this reduces to Tsfasman's Conjecture about hypersurfaces.

In 2015-2017, that is, roughly about two decades after these conjecture were made, two interesting developments took place almost simultaneously. On the one hand, Alain Couvreur [**7**] showed that Lachaud's Conjecture, and in fact, a more general result, holds in the affirmative. On the other hand, Mrinmoy Datta and I showed that the Tsfasman-Boguslavsky Conjecture holds if $r \leq m+1$, but it is false, in general (cf. [**8**], [**9**]). I think Gilles was rather happy and proud that his guess turned out to be correct. Shortly before these developments took place, Gilles himself returned to the topic and together with Robert Rolland, he wrote a nice article [**40**] giving various bounds for the number of $\mathbb{F}_q$-rational points of affine as well as projective varieties over finite fields. In AGC^{2}T-2015, Lachaud, Couvreur and I gave talks in a single session to explain these developments, and I remember how Gilles provided a nice perspective on the topic in the first talk of that session. I might mention in passing that a refined version of Tsfasman-Boguslavsky Conjecture has been proposed in [**9**] and [**5**]. The newer conjectures remain open in general although they are known to be valid in many cases (cf. [**4**] and [**5**]).

Another interesting offshoot of the work of Gilles on projective Reed-Muller code and the inequality Serre proved to answer Tsfasman's Conjecture is a development where Gilles himself was involved. It is the subject matter of his paper [**1**], which was published barely six months before he passed away. Perhaps it is worthwhile to mention a background to this. In February 2016, Everett Howe, Kristin Lauter, and Judy Walker organized a workshop on Algebraic Geometry for Coding Theory and Cryptography at the Institute for Pure and Applied Mathematics (IPAM) on the campus of the University of California, Los Angeles. They said that this will not be a standard "five talks per day with a break for lunch" type of conference, but instead an activity focused on research projects that will be started during the workshop with a hope that it will result in ongoing collaboration among participants. Gilles and I were asked to be co-leaders of a working group. Both of us were apprehensive about this experiment. Nonetheless, we put together a group of six participants including Yves Aubry, Wouter Castryck, Mike O'Sullivan, and Samrith Ram. The broad topic for our group was "Number of points of algebraic sets over finite fields". After some introductory talks, we discussed every day of the week (and sometimes late in the evenings as well) and bounced ideas off each other. As a result of this brainstorming, and continued correspondence thereafter, we wrote an article concerning an analogue of Serre's inequality for hypersurfaces in weighted projective space, and weighted projective Reed-Muller codes. Gilles was a very active participant in this, and wrote almost single handedly a compendium on the (geometry of) weighted projective spaces as a handy manual for us, which formed an appendix to our article [**1**]. Although there are many partial results in this article, the problem of determining the maximum number of $\mathbb{F}_q$-rational points on a hypersurface in weighted projective spaces is open, in general, and a conjecture stated here could be a pointer for further research on the topic. Before ending the discussion of projective Reed-Muller codes introudced by Gilles, I would like to mention an article of well-known finite geometers and group theorists Bill Kantor and Ernie Shult [**18**], which was published in 2013. Here they remark that their first main theorem can be viewed as a statement about a certain code C having a check matrix whose columns consist of one nonzero vector in each Veronesean point. Then they write: *We have not been able to find any reference to this code in the literature. It is probably worth studying, at least from a geometric perspective.* In fact, the code C that they talk about is nothing but the dual of the projective Reed-Muller code studied by Gilles about 25 years ago!

I would now like to go back to my first meeting with Gilles at the 1993 CIMPA workshop in China, which was also the first time I learned about error correcting codes from the lecture courses of Robert Rolland and Gilles Lachaud. In the course of lectures or possibly, during the discussions over meals, I learned about the seemingly intractable open problem of determining an explicit formula for the number, say $\gamma(q;k,n)$, of q-ary MDS codes of length n and dimension k. The answer was known for $k \leq 2$ and also for $k = 3$ and $n \leq 8$. I could see that the problem was equivalent to determining the number of $\mathbb{F}_q$-rational points of the open subset $U_{k,n}$ of the Grassmannian $G_{k,n}$ consisting of points of $G_{k,n}$ for which *all* the Plücker coordinates are nonzero. Based on this observation, I could make some observations, which interested Gilles and he encouraged me to think further. The activity picked up when Gilles arranged for me to get a fellowship from CNRS to visit him Marseille for a month in June 1996. This was the first of my numerous visits to Marseille in subsequent years. Using the connection with Grassmannians together with the work of Nogin [**49**] on Grassmann codes, and various combinatorial inputs, we could obtain explicit lower and upper bounds for $\gamma(q;k,n)$ and an asymptotic formula in the general case. Further, we could use some of the auxiliary results together with Grothendieck-Lefschetz trace formula to derive some geometric properties of certain linear sections of Grassmannians. These results are given in our article *Hyperplane sections of Grassmannians and the number of MDS linear codes* [**12**]. Originally, there was to be a remark in this paper indicating how some results of Nogin on higher weights of Grassmann codes could also be derived as a consequence of the combinatorial results we had in preparation for our results on $\gamma(q; k, n)$. But then there was to be a conference in Guanajuato, Mexico in 1998, which Gilles encouraged me to attend. We decided to expand this remark and used the opportunity to introduce what seemed a natural generalization of Grassmann codes, called Schubert codes. This paper *Higher weights of Grassmann codes* [**11**] was published in the proceedings of the Guanajuato conference and it contained a conjecture about the minimum distance of Schubert codes. As it turned out, [**11**] led to a lot of further research and it has many more citations than [**12**]. The moral of the story for me (and I think Gilles agreed with that) was: A good problem is sometimes more valuable than a good theorem!

In the course of our work on MDS codes, we thought it would be nice to have concrete estimates for the number of $\mathbb{F}_q$-rational points of arbitrary algebraic varieties, especially when the variety is not necessarily smooth. The best general estimate we had at our disposal was the following inequality proved by Lang and Weil in 1954:

LANG–WEIL INEQUALITY. *Let X be an (absolutely) irreducible projective variety in $\mathbb{P}^N$ defined over $\mathbb{F}_q$. If X has dimension n and degree d, then*

$$\big||X(\mathbb{F}_q)| - p_n\big| \leq (d-1)(d-2)q^{n-(1/2)} + Cq^{n-1},$$

where C is a constant depending only on N, n, and d.

When the variety is smooth, and better still, a complete intersection, much sharper estimates are available such as the one obtained by Deligne in 1973 as a consequence of his seminal work on Weil conjectures.

DELIGNE'S INEQUALITY FOR SMOOTH COMPLETE INTERSECTIONS. *If X is a nonsingular complete intersection in $\mathbb{P}^N$ over $\mathbb{F}_q$ of dimension $n = N - r$, then*

$$\big||X(\mathbb{F}_q)| - p_n\big| \leq b'_n\, q^{n/2}.$$

Here $b'_n = b_n - \epsilon_n$ is the primitive nth Betti number of X (where $\epsilon_n = 1$ if n is even and $\epsilon_n = 0$ if n is odd), and $p_n := |\mathbb{P}^n(\mathbb{F}_q)| = q^n + q^{n-1} + \cdots + q + 1$.

We remark that if X has multidegree $\mathbf{d} = (d_1, \ldots, d_r)$, then a formula of Hirzebruch shows that $b'_n = b'_n(N, \mathbf{d})$ equals

$$(-1)^{n+1}(n+1) + \sum_{c=r}^{N}(-1)^{N+c}\binom{N+1}{c+1} \sum_{\substack{\nu_1+\cdots+\nu_r=c \\ \nu_i \geq 1\ \forall i}} d_1^{\nu_1}\cdots d_r^{\nu_r}.$$

From a practical viewpoint, one likes to have some control on the constant C appearing in Lang-Weil inequality and have a Deligne-like inequality for complete intersections (in particular, hypersurfaces) that are not necessarily smooth. We had an idea how this might be achieved, and we worked for years together to prove the following result ([**13**, Theorem 6.1]), which is one of the main results of our paper *Étale cohomology, Lefschetz theorems and number of points of singular varieties over finite fields* [**13**], published in a special volume of *Moscow Mathematical Journal* dedicated to Yuri I. Manin on the occasion of his 65th birthday.

DELIGNE-TYPE INEQUALITY FOR ARBITRARY COMPLETE INTERSECTIONS. *Let X be an irreducible complete intersection of dimension n in $\mathbb{P}^N_{\mathbb{F}_q}$, defined by $r =$*

$N-n$ equations, with multidegree $\mathbf{d}=(d_1,\ldots,d_r)$, and let s be the dimension of the singular locus of X. Then

$$||X(\mathbb{F}_q)|-p_n| \leq b'_{n-s-1}(N-s-1,\mathbf{d})\, q^{(n+s+1)/2}+C_s(X)q^{(n+s)/2},$$

where $C_s(X)$ is a constant independent of q. If X is nonsingular, then $C_{-1}(X)=0$. If $s\geq 0$, then

$$C_s(X)\leq 9\times 2^r\times (r\delta+3)^{N+1} \quad \textit{where} \quad \delta=\max\{d_1,\ldots,d_r\}.$$

For normal complete intersections, this may be viewed as a common refinement of Deligne's inequality and the Lang-Weil inequality. Corollaries include previous results of Aubry and Perret (1996), Shparlinskiĭ and Skorobogatov (1990), as well as Hooley and Katz (1991).

We also used the power of Weil conjectures and some estimates of Katz (2001) to arrive at a version of Lang-Weil inequality with an explicit bound on the constant C appearing therein ([**13**, Theorem 11.1]). A precise statement is given below. Here, given any m-tuple $\mathbf{d}=(d_1,\ldots,d_m)$ of positive integers, we say that an affine (resp. projective) variety in $\mathbb{A}^N_{\mathbb{F}_q}$ (resp. $\mathbb{P}^N_{\mathbb{F}_q}$) is of *type* $(m,N,\mathbf{d})$ if it can be defined by the vanishing of m polynomials in N variables (resp. m homogeneous polynomials in $N+1$ variables) of degrees $d_1,\ldots,d_m$ with coefficients in $\mathbb{F}_q$.

Effective Lang-Weil inequality. *Suppose X is a projective variety in $\mathbb{P}^N_{\mathbb{F}_q}$ or an affine variety in $\mathbb{A}^N_{\mathbb{F}_q}$ defined over $\mathbb{F}_q$. Let $n=\dim X$ and $d=\deg X$. Then*

$$||X(\mathbb{F}_q)|-p_n|\leq (d-1)(d-2)q^{n-(1/2)}+C_+(\bar{X})\,q^{n-1},$$

where $C_+(\bar{X})$ is independent of q. Moreover, if X is of type $(m,N,\mathbf{d})$, with $\mathbf{d}=(d_1,\ldots,d_m)$, and if $\delta=\max\{d_1,\ldots,d_m\}$, then

$$C_+(\bar{X})\leq \begin{cases} 9\times 2^m\times (m\delta+3)^{N+1} & \textit{if } X \textit{ is projective}\\ 6\times 2^m\times (m\delta+3)^{N+1} & \textit{if } X \textit{ is affine.}\end{cases}$$

As a corollary, one obtains an analogue of a result of Schmidt (1974) on a lower bound for the number of points of irreducible hypersurfaces over $\mathbb{F}_q$.

Finally, we showed that certain *conjectural statements* of Lang and Weil (1954), relating the Weil zeta function of a smooth projective variety X defined over $\mathbb{F}_q$ and the "characteristic polynomial" of its Picard variety, hold true provided one uses the "correct" Picard variety. For more details, one may refer to our paper [**13**] or the expository article *Number of solutions of equations over finite fields and a conjecture of Lang and Weil* [**14**] that appeared in the proceedings of an international conference on Number Theory and Discrete Mathematics, held at Chandigarh, India, which Gilles attended.

Although it took a long time to complete this work, it was quite satisfying in the end. What was particularly gratifying was to see many seemingly disparate topics to which our results found applications (references to some of such applications as well as to extensions and generalizations of our results, especially by Cafure and Matera, can be found in the corrigenda and addenda [**15**] to [**13**]). For me, it was a great learning experience to work with Gilles on this topic and to witness firsthand his tenacity and technical skills. At the beginning of this paper, there is a *Sanskrit* verse from *Rg Veda* that can roughly be translated as "Their cord was extended across". Now one might be tempted to think that this was put in there by the Indian coauthor. But the fact is it was Gilles who insisted that we insert it. To him, it conveyed somehow that we are merely extending the key ideas of our illustrious predecessors. Also, he felt it was pertinent because we frequently use Bertini-type theorem to cut the variety by suitable linear sections!

I would like to end by reproducing the concluding remarks in my talk on Gilles at AGC^{2}T-2019:

> Gilles Lachaud has made important and lasting contributions to mathematics, especially in the study of algebraic varieties over finite fields and linear codes. His knowledge and interests were deep and wide. When he became interested in some topic, he would usually delve deeper and spend considerable time learning many aspects of it. As far as I have seen, he would never be in a rush to publish quickly, but would prefer to take time and be thorough. Besides his contributions to mathematics, Gilles was an institution builder. He helped nurture an institution like CIRM. Also, the continuing success of the AGCT conferences owes largely to his vision and efforts.
>
> Other than scientific institutes and conferences, Gilles served as the President of the French Pavilion at Auroville, near Pondicherry, India. He had read or had at least browsed through significant amount of ancient and modern Sanskrit works, including the *Vedas, Upanishadas*, and the scholarly treatises of *Sri Aurobindo.*
>
> Above all, Gilles was a wonderful human being, generous, warmhearted, kind, and always willing to help others, especially students and younger colleagues. His untimely demise last year is a great loss to our subject and the community. Personally, it has been a pleasure and honor to have known him. He will certainly be missed...

4. Gilles Lachaud's work

Christophe Ritzenthaler

Among the four contributors, I am the one who met Gilles the latest, precisely in 2006. He was then director of IML (Institut de Mathématiques de Luminy) in Marseille and I was starting as Maître de Conférences there in the ATI (Arithmétique et Théorie de l'Information) team. I think I owe to our common interest in genus 3 non-hyperelliptic curves that he invited me to apply for this position.

In 2005, Gilles had indeed published two articles [**33, 34**] on the Klein quartic (the smooth quartic $X : x^3y+y^3z+z^3x = 0$) in *Moscow Mathematical Journal*[5]. As many other people (a full book "The eightfold way" [**47**] was published a few years earlier on this single curve), Gilles was fascinated by the conjunction of beautiful analytic, algebraic and arithmetic geometry and group and representation theory to address this particular genus 3 non-hyperelliptic curve. In the first and longest paper *Ramanujan modular forms and the Klein quartic*, he conjugates his passion for history of mathematics and research, unravelling Ramanujan's contributions to some identities between modular forms of level 7. The relation with modular forms boils down to the fact that X is isomorphic over $\mathbb{C}$ to the modular curve $X(7)$,

[5]From some manuscript notes I saw, it was part of a broader program to study the geometry and arithmetic of more general family of quartics with non-trivial automorphism groups.

since they are, up to isomorphisms, the unique genus 3 curve with automorphism group isomorphic to $\mathrm{PSL}_2(\mathbb{F}_7)$. He then gives four different expressions for an L-series which encodes the number of points of the Klein quartic over finite fields, and derives some relations between the Weil polynomial of $X \otimes \mathbb{F}_p$ and the one of an elliptic curve E with complex multiplication by $\mathbb{Z}[(1+\sqrt{-7})/2]$. In the second article *The Klein quartic as a cyclic group generator*, Gilles comes back to this last problem and gives new expressions for the Weil polynomial. He does it not only for the Klein quartic but also for some of its twists $ax^3y + by^3z + cz^3x = 0$ with $a, b, c \in \mathbb{F}_p^\times$. This is realized as a special case of a more general result on Faddeev curves that he proves using Davenport-Hasse's method. These results have been expanded later by Meagher and Top in [**48**] to all twists of the Klein quartic.

When I arrived in Marseille, it was then natural that we started a collaboration on a tantalizing question of Serre on genus 3 curves. In his lectures *Rational points on curves over finite fields* at Harvard in 1985, Serre explains that a principally polarized abelian variety A of dimension $g > 2$ defined over a perfect field k, which is geometrically a Jacobian, is not necessarily a Jacobian over k (unlike in dimension 1 or 2). The obstruction is given by a quadratic character of $\mathrm{Gal}(\bar{k}/k)$ and is called *Serre's obstruction.* When $k \subset \mathbb{C}$ and $g = 3$, he speculates in his lectures (and gives more details in a letter to Top in 2003), that this obstruction can be computed in terms of the value of a certain Siegel modular form called χ_{18}. At that time, it was pure magic to us as this Siegel modular form was a product of so-called theta constants that are purely transcendental objects and which only "see" the curve over $\mathbb{C}$. We could not imagine them containing arithmetic information. In order to understand Serre's intuition, we decided to work out the formulae in the particular case of Ciani quartic. They form a dimension 3 family of quartics with automorphism group containing a group isomorphic to $(\mathbb{Z}/2\mathbb{Z})^2$. This structure helps decomposing their Jacobians into a product of three explicit elliptic curves up to isogeny. In 2000, Howe, Leprevost and Poonen in [**16**] had shown how, conversely, starting from elliptic curves E_1, E_2 and E_3 defined over a field k and a "good" k-rational maximal isotropic subgroup $G \subset (E_1 \times E_2 \times E_3)[2]$, the abelian threefold $A = (E_1 \times E_2 \times E_3)/G$ is the Jacobian over k of an explicit Ciani quartic if and only if a certain expression in terms of the coefficients of the E_i is a non-zero square in k. They had found an algebraic expression of Serre's obstruction for A in this particular case! In *On some questions of Serre on abelian threefolds* [**38**], we worked out this expression in terms of the theta constants associated to the E_i, then to A and eventually in terms of χ_{18} as required. This first collaboration with Gilles went very smoothly: I had the tools to manipulate theta functions, and he knew very well the geometry of this family of quartics and useful decompositions of the symplectic group from his earlier works. This result was presented at the conference organized for Gilles' 60th birthday in Tahiti in 2007.

More than confirming Serre's intuition for a dimension 3 family, this work made us realize the correct normalization of the Siegel modular form we should look for. We used this knowledge to work out a complete proof of Serre's formula in our next article *Jacobians among abelian threefolds: a formula of Klein and a question of Serre* [**39**], in collaboration with Zykin (simultaneously, Meagher got a similar result in his PhD). In order to achieve that, the best formalism was to use the theory of algebraic stacks and geometric modular forms. It sounded scary at first but Gilles was confident and calm as usual and we managed to get our way through.

The cherry on top of the cake was that we could make precise a formula due to Klein linking this modular form to the discriminant of plane quartics. This formula is the first beachhead that enables connecting very precisely the world of modular forms and invariants as it was done previously in genus 1 and 2. We used it recently with Lercier to give the complete dictionary in genus 3.

Three years later, I was leaving for Rennes, and we did not get the chance to collaborate more. Driven by his tireless curiosity and maybe triggered by a conference and a winter school at CIRM in 2014 and discussions with Kohel, Gilles' interest moved towards arithmetic statistics. In his first article on the topic [**35**], *On the distribution of the trace in the unitary symplectic group and the distribution of Frobenius*, he considers the group USp_{2g} which corresponds to the "generic case" for curves and abelian varieties in Katz-Sarnak theory. He recalls the link between the distribution associated to the trace on USp_{2g} and the distribution of the number of points for curves or abelian varieties over finite fields. Although this is well-known to experts, this is done with great care and with the purpose of making it explicit enough for anybody who would need to do actual computations. When $g = 2$ and 3, he then gives some useful expressions for the distribution law in terms of hypergeometric series and related functions and draws some inspiring graphics representing the density of curves with a certain Weil polynomial. In 2018, in *The distribution of the trace in the compact group of type G_2* [**36**], he works out similar formulas for the case of a group (the only compact simple Lie group of type G_2) related to exponential sums and introduced by Katz. In the present volume, one can find his last work on the case of $SU(3)$.

During my time in Marseille, Gilles, as director, managed to create such a pleasant working atmosphere that the years I spent there were extraordinary. Although, as I found out later, such a position is really time-consuming, his door was always open, and he never made anybody felt indebted for the facilities he provided. I learned a lot from working with him and I still meditate about his way of addressing new problems. When I consider them as "enemies" that should surrender after you have attacked them from every angles, Gilles never got upset by their resistance and approached them peacefully. He was also a true humanist as François, Michael and Sudhir mentioned, with many passions outside mathematics. In his car, stuck in the traffic on our way back from Luminy some evenings, I was always happy to listen to his insights on Japanese culture, his collaborative projects with India or lately his study with his wife Patricia of the secrets of the painting *The Tempest* by Giorgione. I guess we will all miss his friendly attitude towards life.

References

[1] Yves Aubry, Wouter Castryck, Sudhir R. Ghorpade, Gilles Lachaud, Michael E. O'Sullivan, and Samrith Ram, *Hypersurfaces in weighted projective spaces over finite fields with applications to coding theory*, Algebraic geometry for coding theory and cryptography, Assoc. Women Math. Ser., vol. 9, Springer, Cham, 2017, pp. 25–61, DOI 10.1007/978-3-319-63931-4_2. MR3775424

[2] Yves Aubry, Safia Haloui, and Gilles Lachaud, *Sur le nombre de points rationnels des variétés abéliennes et des Jacobiennes sur les corps finis* (French, with English and French summaries), C. R. Math. Acad. Sci. Paris **350** (2012), no. 19-20, 907–910, DOI 10.1016/j.crma.2012.10.001. MR2990901

[3] Yves Aubry, Safia Haloui, and Gilles Lachaud, *On the number of points on abelian and Jacobian varieties over finite fields*, Acta Arith. **160** (2013), no. 3, 201–241, DOI 10.4064/aa160-3-1. MR3106095

[4] Peter Beelen, Mrinmoy Datta, and Sudhir R. Ghorpade, *A combinatorial approach to the number of solutions of systems of homogeneous polynomial equations over finite fields*, 2018, arXiv:math.AG/1807.01683.

[5] Peter Beelen, Mrinmoy Datta, and Sudhir R. Ghorpade, *Maximum number of common zeros of homogeneous polynomials over finite fields*, Proc. Amer. Math. Soc. **146** (2018), no. 4, 1451–1468, DOI 10.1090/proc/13863. MR3754333

[6] Karine Chemla, Régis Morelon, and André Allard, *La tradition arabe de diophante d'alexandrie : à propos de quatre livres des arithmétiques perdus en grec retrouvés en arabe*, L'antiquité classique **55** (1986), 351–375.

[7] Alain Couvreur, *An upper bound on the number of rational points of arbitrary projective varieties over finite fields*, Proc. Amer. Math. Soc. **144** (2016), no. 9, 3671–3685, DOI 10.1090/proc/13015. MR3513530

[8] Mrinmoy Datta and Sudhir R. Ghorpade, *On a conjecture of Tsfasman and an inequality of Serre for the number of points of hypersurfaces over finite fields*, Mosc. Math. J. **15** (2015), no. 4, 715–725, DOI 10.17323/1609-4514-2015-15-4-715-725. MR3438829

[9] Mrinmoy Datta and Sudhir R. Ghorpade, *Number of solutions of systems of homogeneous polynomial equations over finite fields*, Proc. Amer. Math. Soc. **145** (2017), no. 2, 525–541, DOI 10.1090/proc/13239. MR3577858

[10] P. Delsarte, J.-M. Goethals, and F. J. MacWilliams, *On generalized Reed-Muller codes and their relatives*, Information and Control **16** (1970), 403–442. MR274186

[11] Sudhir R. Ghorpade and Gilles Lachaud, *Higher weights of Grassmann codes*, Coding theory, cryptography and related areas (Guanajuato, 1998), Springer, Berlin, 2000, pp. 122–131. MR1749453

[12] Sudhir R. Ghorpade and Gilles Lachaud, *Hyperplane sections of Grassmannians and the number of MDS linear codes*, Finite Fields Appl. **7** (2001), no. 4, 468–506, DOI 10.1006/ffta.2000.0299. MR1866340

[13] Sudhir R. Ghorpade and Gilles Lachaud, *Étale cohomology, Lefschetz theorems and number of points of singular varieties over finite fields*, Mosc. Math. J. **2** (2002), no. 3, 589–631, Dedicated to Yuri I. Manin on the occasion of his 65th birthday. MR1988974

[14] Sudhir R. Ghorpade and Gilles Lachaud, *Number of solutions of equations over finite fields and a conjecture of Lang and Weil*, Number theory and discrete mathematics (Chandigarh, 2000), Trends Math., Birkhäuser, Basel, 2002, pp. 269–291. MR1962145

[15] Sudhir R. Ghorpade and Gilles Lachaud, *Corrigenda and addenda: Étale cohomology, Lefschetz theorems and number of points of singular varieties over finite fields [MR1988974]*, Mosc. Math. J. **9** (2009), no. 2, 431–438, DOI 10.17323/1609-4514-2009-9-2-431-438. MR2568444

[16] Everett W. Howe, Franck Leprévost, and Bjorn Poonen, *Large torsion subgroups of split Jacobians of curves of genus two or three*, Forum Math. **12** (2000), no. 3, 315–364, DOI 10.1515/form.2000.008. MR1748483

[17] Byungheup Jun and Jungyun Lee, *Caliber number of real quadratic fields*, J. Number Theory **130** (2010), no. 11, 2586–2595, DOI 10.1016/j.jnt.2010.05.010. MR2678863

[18] W. M. Kantor and E. E. Shult, *Veroneseans, power subspaces and independence*, Adv. Geom. **13** (2013), no. 3, 511–531, DOI 10.1515/advgeom-2012-0042. MR3100924

[19] Gilles Lachaud, *Analyse spectrale des formes automorphes et séries d'Eisenstein* (French), Invent. Math. **46** (1978), no. 1, 39–79, DOI 10.1007/BF01390102. MR506041

[20] Gilles Lachaud, *Nombres et archétypes: du vieux et du neuf*, Eleutheria (1979), no. 2, 133–140. MR622276

[21] Gilles Lachaud, *Variations sur un thème de Mahler* (French), Invent. Math. **52** (1979), no. 2, 149–162, DOI 10.1007/BF01403062. MR536078

[22] Gilles Lachaud, *Analyse spectrale et prolongement analytique: séries d'Eisenstein, fonctions zéta et nombre de solutions d'équations diophantiennes*, Ann. Sci. Univ. Clermont-Ferrand II Math. (1982), no. 21, 1–14. MR706119

[23] Gilles Lachaud, *Une présentation adélique de la série singulière et du problème de Waring* (French), Enseign. Math. (2) **28** (1982), no. 1-2, 139–169. MR670501

[24] Gilles Lachaud, *Les codes géométriques de Goppa*, Seminar Bourbaki, Vol. 1984/85, no. 133-134, 1986, pp. 189–207. MR837220

[25] Gilles Lachaud, *On real quadratic fields*, Bull. Amer. Math. Soc. (N.S.) **17** (1987), no. 2, 307–311, DOI 10.1090/S0273-0979-1987-15571-6. MR903739

[26] Gilles Lachaud, *Sommes d'Eisenstein et nombre de points de certaines courbes algébriques sur les corps finis* (French, with English summary), C. R. Acad. Sci. Paris Sér. I Math. **305** (1987), no. 16, 729–732. MR920053

[27] Gilles Lachaud, *Projective Reed-Muller codes* (English, with French summary), Coding theory and applications (Cachan, 1986), Lecture Notes in Comput. Sci., vol. 311, Springer, Berlin, 1988, pp. 125–129, DOI 10.1007/3-540-19368-5_13. MR960714

[28] Gilles Lachaud, *Exponential sums and the Carlitz-Uchiyama bound*, Coding theory and applications (Toulon, 1988), Lecture Notes in Comput. Sci., vol. 388, Springer, New York, 1989, pp. 63–75, DOI 10.1007/BFb0019847. MR1023680

[29] Caterina De Simone, *The cut polytope and the Boolean quadric polytope*, Discrete Math. **79** (1989/90), no. 1, 71–75, DOI 10.1016/0012-365X(90)90056-N. MR1032635

[30] Gilles Lachaud, *The parameters of projective Reed-Muller codes* (English, with French summary), Discrete Math. **81** (1990), no. 2, 217–221, DOI 10.1016/0012-365X(90)90155-B. MR1054981

[31] Gilles Lachaud, *Artin-Schreier curves, exponential sums, and the Carlitz-Uchiyama bound for geometric codes*, J. Number Theory **39** (1991), no. 1, 18–40, DOI 10.1016/0022-314X(91)90031-6. MR1123166

[32] Gilles Lachaud, *Artin-Schreier curves, exponential sums, and coding theory*, Theoret. Comput. Sci. **94** (1992), no. 2, 295–310, DOI 10.1016/0304-3975(92)90040-M. Discrete mathematics and applications to computer science (Marseille, 1989). MR1157861

[33] Gilles Lachaud, *The Klein quartic as a cyclic group generator*, Mosc. Math. J. **5** (2005), no. 4, 857–868, 973, DOI 10.17323/1609-4514-2005-5-4-857-868. MR2266462

[34] Gilles Lachaud, *Ramanujan modular forms and the Klein quartic* (English, with English and Russian summaries), Mosc. Math. J. **5** (2005), no. 4, 829–856, 972–973, DOI 10.17323/1609-4514-2005-5-4-829-856. MR2266461

[35] Kevin James, *Variants of the Sato-Tate and Lang-Trotter conjectures*, Frobenius distributions: Lang-Trotter and Sato-Tate conjectures, Contemp. Math., vol. 663, Amer. Math. Soc., Providence, RI, 2016, pp. 175–184, DOI 10.1090/conm/663/13354. MR3502943

[36] Gilles Lachaud, *The distribution of the trace in the compact group of type G_2*, Arithmetic geometry: computation and applications, Contemp. Math., vol. 722, Amer. Math. Soc., Providence, RI, 2019, pp. 79–103, DOI 10.1090/conm/722/14536. MR3896850

[37] Gilles Lachaud and Mireille Martin-Deschamps, *Nombre de points des jacobiennes sur un corps fini* (French), Acta Arith. **56** (1990), no. 4, 329–340, DOI 10.4064/aa-56-4-329-340. MR1096346

[38] Gilles Lachaud and Christophe Ritzenthaler, *On some questions of Serre on abelian threefolds*, Algebraic geometry and its applications, Ser. Number Theory Appl., vol. 5, World Sci. Publ., Hackensack, NJ, 2008, pp. 88–115, DOI 10.1142/9789812793430_0005. MR2484050

[39] Gilles Lachaud, Christophe Ritzenthaler, and Alexey Zykin, *Jacobians among abelian threefolds: a formula of Klein and a question of Serre*, Math. Res. Lett. **17** (2010), no. 2, 323–333, DOI 10.4310/MRL.2010.v17.n2.a11. MR2644379

[40] Gilles Lachaud and Robert Rolland, *On the number of points of algebraic sets over finite fields* (English, with English and French summaries), J. Pure Appl. Algebra **219** (2015), no. 11, 5117–5136, DOI 10.1016/j.jpaa.2015.05.008. MR3351576

[41] Gilles Lachaud and Jacques Stern, *Polynomial-time construction of codes. I. Linear codes with almost equal weights*, Appl. Algebra Engrg. Comm. Comput. **3** (1992), no. 2, 151–161, DOI 10.1007/BF01387197. MR1325753

[42] Gilles Lachaud and Jacques Stern, *Polynomial-time construction of codes. II. Spherical codes and the kissing number of spheres*, IEEE Trans. Inform. Theory **40** (1994), no. 4, 1140–1146, DOI 10.1109/18.335961. MR1301424

[43] Gilles Lachaud and Michael A. Tsfasman, *Formules explicites pour le nombre de points des variétés sur un corps fini* (French), J. Reine Angew. Math. **493** (1997), 1–60, DOI 10.1515/crll.1997.493.1. MR1491808

[44] Gilles Lachaud and Serge Vlăduţ, *Gauss problem for function fields*, J. Number Theory **85** (2000), no. 2, 109–129, DOI 10.1006/jnth.2000.2523. MR1802707

[45] Gilles Lachaud and Jacques Wolfmann, *Sommes de Kloosterman, courbes elliptiques et codes cycliques en caractéristique* 2 (French, with English summary), C. R. Acad. Sci. Paris Sér. I Math. **305** (1987), no. 20, 881–883. MR925289

[46] Gilles Lachaud and Jacques Wolfmann, *The weights of the orthogonals of the extended quadratic binary Goppa codes*, IEEE Trans. Inform. Theory **36** (1990), no. 3, 686–692, DOI 10.1109/18.54892. MR1053865

[47] Silvio Levy (ed.), *The eightfold way*, Mathematical Sciences Research Institute Publications, vol. 35, Cambridge University Press, Cambridge, 1999. The beauty of Klein's quartic curve. MR1722410

[48] Stephen Meagher and Jaap Top, *Twists of genus three curves over finite fields*, Finite Fields Appl. **16** (2010), no. 5, 347–368, DOI 10.1016/j.ffa.2010.06.001. MR2678623

[49] D. Yu. Nogin, *Codes associated to Grassmannians*, Arithmetic, geometry and coding theory (Luminy, 1993), de Gruyter, Berlin, 1996, pp. 145–154. MR1394931

[50] Jean-Pierre Serre, *Lettre à M. Tsfasman* (French, with English summary), Astérisque **198-200** (1991), 11, 351–353 (1992). Journées Arithmétiques, 1989 (Luminy, 1989). MR1144337

[51] Anders Bjært Sørensen, *Projective Reed-Muller codes*, IEEE Trans. Inform. Theory **37** (1991), no. 6, 1567–1576, DOI 10.1109/18.104317. MR1134296

[52] M. A. Tsfasman, S. G. Vlădu ţ, and Th. Zink, *Modular curves, Shimura curves, and Goppa codes, better than Varshamov-Gilbert bound*, Math. Nachr. **109** (1982), 21–28, DOI 10.1002/mana.19821090103. MR705893

[53] Michael A. Tsfasman, *Nombre de points des surfaces sur un corps fini* (French, with English and French summaries), Arithmetic, geometry and coding theory (Luminy, 1993), de Gruyter, Berlin, 1996, pp. 209–224. MR1394936

Sudhir R. Ghorpade, Department of Mathematics, Indian Institute of Technology Bombay, Powai, Mumbai 400076, India

Email address: srg@math.iitb.ac.in

Christophe Ritzenthaler, Univ Rennes, IRMAR, Campus de Beaulieu, 35042 Rennes, France

Email address: christophe.ritzenthaler@univ-rennes1.fr

François Rodier, Aix Marseille Université, CNRS, Centrale Marseille, Institut de Mathématiques de Marseille, UMR 7373, 13288 Marseille, France

Email address: francois.rodier@univ-amu.fr

Michael A. Tsfasman, CNRS, Laboratoire de Mathematiques de Versailles (UMR 8100), France; Institute for Information Transmission Problems, Moscow, Russia; Independent University of Moscow, Russia

Email address: mtsfasman@yandex.ru

Contemporary Mathematics
Volume **770**, 2021
https://doi.org/10.1090/conm/770/15434

Elliptic curves with large Tate–Shafarevich groups over $\mathbb{F}_q(t)$

Richard Griffon and Guus de Wit

ABSTRACT. Let $\mathbb{F}_q$ be a finite field of odd characteristic p. We exhibit elliptic curves over the rational function field $K = \mathbb{F}_q(t)$ whose Tate-Shafarevich groups are large. Precisely, we consider certain infinite sequences of explicit elliptic curves E the Tate–Shafarevich group $Ш(E)$ of which, we prove, is finite and satisfies $|Ш(E)| = H(E)^{1+o(1)}$ as $H(E) \to \infty$, where $H(E)$ denotes the exponential differential height of E. We further show that, despite being large, these Tate–Shafarevich groups have trivial p-primary part.

The proof involves explicitly computing the L-functions of these elliptic curves, proving the BSD conjecture for them, and obtaining estimates on the size of the central value of their L-function.

Introduction

0.1. The size of Tate–Shafarevich groups of elliptic curves. Fix a finite field $\mathbb{F}_q$ of odd characteristic p and let $K := \mathbb{F}_q(t)$ denote the rational function field over $\mathbb{F}_q$. Let E be an elliptic curve over K which, we assume, is non-isotrivial (*i.e.*, its j-invariant $j(E) \in K$ does not lie in $\mathbb{F}_q$). One attaches to E its Tate–Shafarevich group, denoted by $Ш(E)$. Among other reasons, the arithmetic significance of $Ш(E)$ stems from its measuring, in a certain sense, how badly the local–global principle fails for E. The Tate–Shafarevich group remains a mysterious object; for instance, the finiteness of $Ш(E)$ is still conjectural in general, even though it has been proved in a number of cases.

Let us assume for now that $Ш(E)$ is indeed finite: what can then be said about its size? There are several natural choices of numerical invariants of E to compare $|Ш(E)|$ to. Here, we choose the exponential differential height $H(E)$ and the conductor $N(E)$: these are defined by

$$H(E) := q^{\frac{1}{12}\deg \Delta_{\min}(E)} \qquad \text{and} \qquad N(E) := q^{\deg \mathcal{N}(E)},$$

where $\Delta_{\min}(E)$ and $\mathcal{N}(E)$ denote the minimal discriminant and the conductor divisors of E, respectively. Goldfeld and Szpiro [**1**] have proven upper bounds on the order of $Ш(E)$ in terms of these two invariants.

Let us quote a special case of their result:

THEOREM 1 (Goldfeld–Szpiro [**1**]). *Let E be a non-isotrivial elliptic curve over K. Assume that E has finite Tate–Shafarevich group.*

2020 *Mathematics Subject Classification.* Primary 11G05, 11G40; Secondary 14G10, 11L05.
Key words and phrases. Elliptic curves over function fields, Tate-Shafarevich groups, Explicit computation of L-functions, BSD conjecture, Gauss and Kloosterman sums.

(*i*) *Then, for all* $\varepsilon > 0$, *one has* $|\text{Ш}(E)| \ll_{q,\varepsilon} H(E)^{1+\varepsilon}$.
(*ii*) *If moreover,* $j(E) \in K$ *is not a* p*-th power in* K, *then, for all* $\varepsilon > 0$, *one has* $|\text{Ш}(E)| \ll_{q,\varepsilon} N(E)^{1/2+\varepsilon}$.

In this statement[1], item *(ii)* follows from item *(i)* by applying Szpiro's inequality for elliptic curves with separable j-invariants. It is then natural to wonder about the optimality of *(i)* and *(ii)*: apart from the ε's, are the exponents of the height (1) and of the conductor (1/2) in these upper bounds best possible? In other words, as E ranges over all elliptic curves over K, what is the largest power of $H(E)$ – or $N(E)$ – that does appear in $|\text{Ш}(E)|$, up to a $\pm\varepsilon$ for all $\varepsilon > 0$?

In the analogous setting of elliptic curves over $\mathbb{Q}$, the analogue of Theorem 1 is known to be a consequence of the ABC conjecture (see §1–3 in [**1**]). Moreover, de Weger [**21**] conjectures that the exponents 1 and 1/2 should indeed be optimal in that context (see Conjectures 2 and 4 in [**21**]). Following the analogy between the arithmetics of elliptic curves over $\mathbb{Q}$ and over K, one can translate the statement of de Weger's conjecture (the quantities $H(E)$ and $N(E)$ here correspond to the 1/12-th power of the minimal discriminant and to the conductor of an elliptic curve over $\mathbb{Q}$, respectively). This translation results in the following:

CONJECTURE 2 (de Weger [**21**]). *Assuming finiteness of the relevant Tate–Shafarevich groups,*

(i) *For any* $\varepsilon > 0$, *there are infinitely many elliptic curves* E/K *such that*
$$|\text{Ш}(E)| \gg_{q,\varepsilon} H(E)^{1-\varepsilon}.$$
(ii) *For any* $\varepsilon > 0$, *there are infinitely many elliptic curves* E/K *such that*
$$|\text{Ш}(E)| \gg_{q,\varepsilon} N(E)^{1/2-\varepsilon}.$$

In the same paper, de Weger proves, conditionally to the Birch and Swinnerton-Dyer (BSD) conjecture, that (the analogue of) Conjecture 2(i) holds for elliptic curves over $\mathbb{Q}$. Prior to [**21**], Mai and Murty [**10**, Theorem 2] had shown, again conditionally to the BSD conjecture, a weaker version of Conjecture 2(ii) for elliptic curves over $\mathbb{Q}$, with the exponent 1/2 replaced by 1/4. Both of these results rely on considering sequences of well-chosen quadratic twists of a given elliptic curve.

In the context of elliptic curves over K, the above-stated Conjecture 2(i) is not difficult to prove. One can indeed take advantage of the existence of inseparable isogenies of large degree to construct sequences of elliptic curves over K with large Tate–Shafarevich groups (see §1.5, where we build one such example). In such sequences, the inseparability degree of the j-invariant grows to infinity. Moreover, by construction of these sequences, the elliptic curves therein are K-isogenous: one cannot, therefore, hope to deduce from these any result towards Conjecture 2(ii). One also notices that the p-primary parts of the Tate–Shafarevich groups of the elliptic curves belonging to these sequences are "large"; and that, actually, the order of the p-primary part already accounts for the observed "large Ш" phenomenon.

We are therefore led to ask the following questions:

QUESTION 1 *For a given* $\varepsilon > 0$, *are there infinitely many* pairwise non K-isogenous *elliptic curves* E/K *such that* $|\text{Ш}(E)| \gg_{q,\varepsilon} H(E)^{1-\varepsilon}$*?*

[1] For two functions $f(x)$, $g(x)$ defined on $[0,\infty)$, we use Vinogradov's notation "$f(x) \ll_a g(x)$" to mean that there exists a constant $C > 0$, depending at most on the parameters a, such that $|f(x)| \leq Cg(x)$ for $x \to \infty$. This is synonymous with Landau's notation "$f(x) = O_a(g(x))$".

QUESTION 2 *If so, is the fact that their Tate–Shafarevich groups are "large" always explained by a "large p-primary part of* Ш*" phenomenon?*

QUESTION 3 *In the first question, can one also require that the j-invariants of the involved elliptic curves have bounded inseparability degree?*

In this paper, we give positive answers to the first and third questions:

THEOREM A. *For all* $\varepsilon > 0$*, there are infinitely many* pairwise non K-isogenous *elliptic curves* E/K *with separable* j*-invariant and finite Tate–Shafarevich group, such that*

$$|Ш(E)| \geq H(E)^{1-\varepsilon}.$$

We also prove a result in direction of Conjecture 2(ii), with the same exponent 1/4 as in [**10**]:

THEOREM B. *For all* $\varepsilon > 0$*, there are infinitely many* pairwise non $\overline{K}$-isomorphic *elliptic curves* E/K *with separable* j*-invariant and finite Tate–Shafarevich group, such that*

$$|Ш(E)| \geq N(E)^{1/4-\varepsilon}.$$

In contrast to the aforementioned results concerning elliptic curves over $\mathbb{Q}$, Theorems A and B are unconditional, and the involved elliptic curves are not quadratic twists of each other. Our proof is constructive and effective in that we exhibit sequences of elliptic curves over K satisfying these properties and we provide explicit bounds on the order of their Tate–Sharafarevich groups.

Finally, for all the elliptic curves in these sequences, we prove that the p-primary parts of their Tate–Shafarevich groups are trivial, thus answering in the negative the second question raised above.

0.2. Elliptic curves with large Tate–Shafarevich groups. Theorems A and B both follow from our main theorem, which we now state. Let $\mathbb{F}_q$ be a finite field of odd characteristic p. For any parameter $\gamma \in \mathbb{F}_q^\times$ and any integer $a \geq 1$, we write $\wp_a(t) := t^{q^a} - t \in \mathbb{F}_q[t]$ and consider the elliptic curve $E_{\gamma,a}$ defined over K by the Weierstrass model:

$$E_{\gamma,a}: \qquad y^2 = x\big(x^2 + \wp_a(t)x + \gamma\big). \tag{1}$$

The main result of this paper is the following:

THEOREM C. *In the above setting,*

(*1*) *As* $\gamma \in \mathbb{F}_q^\times$ *and* $a \geq 1$ *vary, the curves* $E_{\gamma,a}$ *are pairwise neither* $\overline{K}$*-isomorphic nor* K*-isogenous. Moreover, the* j*-invariant* $j(E_{\gamma,a})$ *is separable.*

(*2*) *For any* $\gamma \in \mathbb{F}_q^\times$ *and* $a \geq 1$*, the Tate–Shafarevich group* $Ш(E_{\gamma,a})$ *is finite.*

(*3*) *Given* $\gamma \in \mathbb{F}_q^\times$*, as* $a \to \infty$*, we have*

$$|Ш(E_{\gamma,a})| = H(E_{\gamma,a})^{1+o(1)}.$$

(*4*) *For any* $\gamma \in \mathbb{F}_q^\times$ *and* $a \geq 1$*, the* p*-primary part of* $Ш(E_{\gamma,a})$ *is trivial.*

One can reformulate *(3)* as follows: given $\varepsilon > 0$, for any $\gamma \in \mathbb{F}_q^\times$ and any large enough integer $a \geq 1$ (depending on ε), we have

$$H(E_{\gamma,a})^{1-\varepsilon} \leq |Ш(E_{\gamma,a})| \leq H(E_{\gamma,a})^{1+\varepsilon}.$$

It is then clear that, for a given $\varepsilon > 0$, all but finitely many of the $E_{a,\gamma}$'s satisfy $|\text{Ш}(E_{\gamma,a})| \geq H(E_{\gamma,a})^{1-\varepsilon}$. Further, we will see that $H(E_{\gamma,a}) = N(E_{\gamma,a})^{1/4}$, so that $|\text{Ш}(E_{\gamma,a})| \geq N(E_{\gamma,a})^{1/4-\varepsilon}$ holds for all but finitely many of the $E_{\gamma,a}$'s.

In particular, Theorem C does imply Theorems A and B.

0.3. Organisation of the paper. We now explain the overall strategy of the proof of Theorem C as we describe the plan of the paper. The first section briefly recalls the definitions and main properties of the objects that will be used throughout the paper. In the last subsection §1.5, we show how the Frobenius isogenies allow one to construct sequences of elliptic curves with large Tate–Shafarevich groups (thus proving Conjecture 2(i) above). We found it worthwhile to include this construction since the argument in §1.5 illustrates on a simple example the general structure of the proof of Theorem C.

The elliptic curves $E_{\gamma,a}$ are introduced in section 2, where we also calculate some of their more easily accessible invariants: their height and conductor, their Tamagawa number, as well as the torsion subgroup of $E_{\gamma,a}(K)$. The curves $E_{\gamma,a}$ were first studied by Pries and Ulmer in [**13**]: we recall in §2.2 how they were constructed there, and some of the results of [**13**]. Item *(1)* of Theorem C is proved in Proposition 2.4.

A large part of the proof of Theorem C is analytic, in that we rely on a detailed study of the relevant L-functions. Our first main goal is therefore to obtain an explicit expression for the L-function of $E_{\gamma,a}$. To that end, we introduce some notation in section 3. We define a certain finite set $P_q(a)$ of places of K and, to each place $v \in P_q(a)$ we attach in §3.3 two character sums over the residue field of K at v: a Gauss sum $\mathbf{g}(v)$ and a Kloosterman sum $\mathbf{Kl}_\gamma(v)$. We then show (see Theorem 4.1) that, for any integer $a \geq 1$ and any $\gamma \in \mathbb{F}_q^\times$, the L-function $L(E_{\gamma,a}, T)$ of $E_{\gamma,a}$ admits the following expression:

$$L(E_{\gamma,a}, T) = \prod_{v \in P_q(a)} \left(1 - \mathbf{g}(v)\mathbf{Kl}_\gamma(v)T^{\deg v} + \mathbf{g}(v)^2 q^{\deg v} T^{2\deg v}\right). \tag{2}$$

This identity is proved in section 4 by an elementary method based on the definition of $L(E_{\gamma,a}, T)$ and manipulation of character sums. This result, which is instrumental in the proof of our main theorem, may be of independent interest.

Using (2) and arithmetic properties of Gauss and Kloosterman sums, we elucidate in Theorem 5.2 the p-adic valuations of the zeros of $L(E_{\gamma,a}, T)$. As a consequence, we deduce that $L(E_{\gamma,a}, T)$ does not vanish at the central point $T = q^{-1}$ for its functional equation (see Theorem 5.3). This non-vanishing result is enough to ensure that the BSD conjecture holds for $E_{\gamma,a}$ (Corollary 5.4) which, in turn, has several important corollaries for our study. First, the Tate–Shafarevich group $\text{Ш}(E_{\gamma,a})$ is indeed (unconditionally) finite, as claimed in Theorem C*(2)*. Secondly, we derive from the BSD formula and from our computations in section 2 that

$$|\text{Ш}(E_{\gamma,a})| = q^{-1} H(E_{\gamma,a}) \cdot L(E_{\gamma,a}, q^{-1}). \tag{3}$$

Lastly – even though this is less central to our point – the Mordell–Weil group $E_{\gamma,a}(K)$ is finite and, taking into account our description of its torsion subgroup, we conclude that $E_{\gamma,a}(K) = \{\mathcal{O}, (0,0)\}$.

In light of the link (3) between the order of the Tate–Shafarevich group and the central value of $L(E_{\gamma,a}, T)$, we estimate the size of $|\text{Ш}(E_{\gamma,a})|$ in terms of $H(E_{\gamma,a})$

by proving adequate upper and lower bounds on $L(E_{\gamma,a}, q^{-1})$. Specifically, in order to show that Theorem C*(3)* holds, we need to prove that

$$-o(1) \leq \frac{\log L(E_{\gamma,a}, q^{-1})}{\log H(E_{\gamma,a})} \leq o(1) \qquad (\text{as } a \to \infty). \tag{4}$$

The proof of these inequalities is carried out in section 7. Proving the lower bound in (4) is the crucial step. After evaluating expression (2) for $L(E_{\gamma,a}, T)$ at $T = q^{-1}$, straightforward analytical considerations yield that $L(E_{\gamma,a}, q^{-1})$ can be bounded from below by

$$\log L(E_{\gamma,a}, q^{-1}) \geq \sum_{v \in P_q(a)} w\left(\frac{\mathsf{Kl}_\gamma(v)}{2q^{\deg v/2}}\right), \qquad \text{where } w(x) := \log|x^2(1-x^2)|.$$

For any place $v \in P_q(a)$, the algebraic number $\mathsf{Kl}_\gamma(v)/2q^{\deg v/2}$ is known to be totally real; moreover, its image in any complex embedding of $\overline{\mathbb{Q}}$ lies in the interval $[-1, 1]$ by the Weil bound for Kloosterman sums. It is apparent that the size of the right-hand side of the previous display depends on how the set

$$\Theta_{\gamma,a} := \{\mathsf{Kl}_\gamma(v)/2q^{\deg v/2}\}_{v \in P_q(a)}$$

is distributed in $[-1, 1]$. We study the relevant aspects of the distribution of $\Theta_{\gamma,a}$ in section 6. Namely, we first recall from [**3**] an effective version of an asymptotic equidistribution statement, the qualitative form of which says that, asymptotically as $a \to \infty$, the set $\Theta_{\gamma,a}$ becomes equidistributed in $[-1, 1]$ with respect to the measure $\frac{2}{\pi}\sqrt{1-x^2}\,dx$. Next we prove a quantitative version of the fact that $\Theta_{\gamma,a}$ "avoids" the points $-1, 0$ and 1, which are the poles of $x \mapsto w(x)$ on $[-1, 1]$.

We then conclude the proof of the lower bound in (4) by coupling these two facts about $\Theta_{\gamma,a}$ with a suitable approximation of w on $[-1, 1]$. The resulting estimates (4) constitute Theorem 7.1. We combine these to (3) in §8.2 to complete the proof of Theorem C*(3)*.

Finally, in order to prove Theorem C*(4)*, we show that $|\text{Ш}(E_{\gamma,a})|$ is relatively prime to p (see Theorem 8.1). To do so, we start from the BSD formula (3): instead of estimating the "archimedean size" of the central value $L(E_{\gamma,a}, q^{-1})$ as we did above, we now estimate its "p-adic size". The p-adic valuation of $L(E_{\gamma,a}, q^{-1})$ visibly depends on the p-adic valuations of the zeros of $L(E_{\gamma,a}, T)$. The above-mentioned Theorem 5.2 allows us to obtain the desired result, in §8.1.

1. Invariants of elliptic curves over function fields

Let $\mathbb{F}_q$ be a finite field of characteristic p and $K := \mathbb{F}_q(t)$ denote the field of rational functions on the projective line $\mathbb{P}^1_{/\mathbb{F}_q}$. For simplicity, and since we will not need the general case, we assume throughout that $p \neq 2$. Let M_K denote the set of all places v of K, which may be identified with the set of closed points of $\mathbb{P}^1_{/\mathbb{F}_q}$. Among those is the degree-1 closed point $\infty = [0 : 1]$; we write M_K° for the set $M_K \setminus \{\infty\}$ of closed points of $\mathbb{A}^1_{/\mathbb{F}_q} = \mathbb{P}^1_{/\mathbb{F}_q} \setminus \{\infty\}$. There is a 1-to-1 correspondence between M_K° and the set of monic irreducible elements of $\mathbb{F}_q[t]$.

In this section, we introduce the relevant invariants of elliptic curves over K. For a more detailed introduction to the arithmetic of elliptic curves over function fields and their invariants, the reader is referred to [**19**]. Some of the results quoted in this section are valid more generally for elliptic curves (or even abelian varieties) over a function field of positive characteristic, but we only state the special cases that are required for our purpose. We do not mention K in the notation used for the invariants of elliptic curves over K which we consider: unless explicitly noted otherwise, these invariants are relative to K.

1.1. Conductor and height. Let E be a non-isotrivial elliptic curve over K (*i.e.*, its j-invariant $j(E)$ is not a constant rational function). We denote by $\Delta_{\min}(E)$ its minimal discriminant divisor, which is a divisor on $\mathbb{P}^1_{/\mathbb{F}_q}$. The *exponential differential height* $H(E)$ is defined by

$$H(E) := q^{\frac{1}{12}\deg \Delta_{\min}(E)}.$$

The conductor divisor $\mathcal{N}(E)$ of E is also a divisor on $\mathbb{P}^1_{/\mathbb{F}_q}$ (see [**16**, Chap. IV, §10] for the precise definition). We then let

$$N(E) := q^{\deg \mathcal{N}(E)}.$$

We will call $N(E)$ the "*(numerical) conductor*" of E. An easy inequality between the conductor and the minimal discriminant divisors yields that $N(E)^{1/12} \leq H(E)$. In the other direction, Szpiro's inequality states that

$$H(E) \leq N(E)^{1/2},$$

provided that $j(E) \in K$ is not a p-th power in K (see [**1**, Thm. 3] and the references in that article). Note that two K-isogenous elliptic curves have the same numerical conductor (see §1.3).

1.2. The Tate–Shafarevich group. Fix a separable closure K^{sep} of K and consider the Galois cohomology group $\mathrm{H}^1(K, E) := \mathrm{H}^1(\mathrm{Gal}(K^{\text{sep}}/K), E(K^{\text{sep}}))$. For any place $v \in M_K$, there is a similarly defined group $\mathrm{H}^1(K_v, E)$, where K_v denotes the completion of K at v. Recall that the *Tate–Shafarevich group of* E is defined by:

$$Ш(E) := \ker\left(\mathrm{H}^1(K, E) \longrightarrow \prod_{v \in M_K} \mathrm{H}^1(K_v, E)\right),$$

where the product runs over all places v of K and the arrow is the product of the canonical restriction maps $\mathrm{H}^1(K, E) \to \mathrm{H}^1(K_v, E)$.

Perhaps a more illuminating way of thinking about $Ш(E)$ is to observe that it is in bijection with the set of K-isomorphism classes of pairs (C, ϕ), where C/K is a curve of genus 1 which has at least one K_v-rational point for all places $v \in M_K$ and ϕ is a $\overline{K}$-isomorphism between E and $\mathrm{Jac}(C)$ (see [**17**, Chap. X, §3-4] for more details). One can show that the class of such a pair (C, ϕ) is trivial in $Ш(E)$ if and only if C has a K-rational point. This point of view makes it clearer that $Ш(E)$ measures a local-global obstruction.

By construction, the Tate–Shafarevich group $Ш(E)$ is a torsion abelian group. It is conjectured, but not known in general, that $Ш(E)$ is finite (in the cases we study, we will show that it is indeed finite).

1.3. The L-function. For any place $v \in M_K$, we denote by $\mathbb{F}_v$ the residue field of K at v, which is a finite extension of $\mathbb{F}_q$, and by $\deg v = [\mathbb{F}_v : \mathbb{F}_q]$ the degree of v. For such a v, let $(\widetilde{E})_v$ denote the reduction of E at v. This plane cubic curve over $\mathbb{F}_v$ is not necessarily smooth: we say that E has *good reduction at* v if it is, and that E has *bad reduction at* v otherwise. In both cases, it makes sense to count

$\mathbb{F}_v$-rational points on $(\widetilde{E})_v$. For any $v \in M_K$, put $a_v(E) := |\mathbb{F}_v| + 1 - |(\widetilde{E})_v(\mathbb{F}_v)|$ and define the *L-function of E* by the formal Euler product

$$L(E,T) := \prod_{v\in M_K} L_v(E,T)^{-1}, \tag{1.1}$$

where the product runs over all places of K and where

$$L_v(E,T) = \begin{cases} 1 - a_v(E)T^{\deg v} + q^{\deg v}T^{2\deg v} & \text{if } E \text{ has good reduction at } v, \\ 1 - a_v(E)T^{\deg v} & \text{if } E \text{ has bad reduction at } v. \end{cases} \tag{1.2}$$

By the Hasse bound, the Euler product in (1.1) *a priori* converges on the complex disk $\{T \in \mathbb{C} : |T| < q^{-3/2}\}$. However, deep results of Grothendieck and Deligne (among others) actually prove that $L(E,T)$ is a polynomial with integral coefficients in T, with constant coefficient 1. The degree $b(E)$ of that polynomial is given by the Grothendieck–Ogg–Shafarevich formula, which states that $\deg L(E,T) = \deg \mathcal{N}(E) - 4$. Moreover, the L-function satisfies the expected functional equation: $L(E,T) = \pm 1 \cdot (qT)^{b(E)} \cdot L(E, 1/q^2T)$. Lastly, $L(E,T)$ satisfies the Riemann Hypothesis *i.e.*, the complex zeros of $z \mapsto L(E,z)$ have magnitude q^{-1}.

In particular, the value of $L(E,T)$ at the central point for its functional equation (*viz.* $T = q^{-1}$) is a rational number, and the Riemann Hypothesis allows one to see that $L(E,q^{-1})$ is non-negative.

Recall that two K-isogenous elliptic curves E and E' have the same L-function: see [**5**, App. C] for a proof. In particular, their L-functions have the same degree and the Grothendieck–Ogg–Shafarevich formula implies that $\deg \mathcal{N}(E) = \deg \mathcal{N}(E')$, so that the numerical conductors $N(E)$ and $N(E')$ coincide.

By the results recalled in the previous paragraph, the L-function of E can be written as a product of the form

$$L(E,T) = \prod_{k=1}^{b(E)} (1 - z_k T),$$

for some algebraic integers z_k. These z_k's are the inverse zeros of the polynomial $L(E,T)$. We choose a prime ideal $\mathfrak{P}$ of $\overline{\mathbb{Q}}$ above p and denote by $\operatorname{ord}_{\mathfrak{P}} : \overline{\mathbb{Q}}^\times \to \mathbb{Q}$ the corresponding discrete valuation, so normalised that $\operatorname{ord}_{\mathfrak{P}}(q) = 1$. For all $k \in \{1, \dots, b(E)\}$, we let $\lambda_k = \operatorname{ord}_{\mathfrak{P}}(z_k) \in \mathbb{Q}$. The functional equation of $L(E,T)$ implies that both z_k and q^2/z_k are algebraic integers, so that $0 \leq \lambda_k \leq 2$ for all k. Renumbering the z_k's if necessary, we can assume that $\lambda_1 \leq \lambda_2 \leq \cdots \leq \lambda_{b(E)}$. Thus ordered, the sequence $\{\lambda_1, \lambda_2, \dots, \lambda_{b(E)}\}$ will be called the *p-adic slope sequence of* $L(E,T)$. It then follows from the functional equation satisfied by $L(E,T)$ that one has $\lambda_{b(E)-k} = 2 - \lambda_k$ for any $k \in \{1, \dots, b(E)\}$.

1.4. The BSD conjecture. Birch and Swinnerton-Dyer (and Tate, in this context) conjectured that the analytic behaviour of the L-function $z \mapsto L(E,z)$ around $z = q^{-1}$ encodes arithmetic information about E. The reader is referred to [**18**], [**1**, §4] or [**19**] for more detailed accounts of the BSD conjecture.

In the case that $L(E,T)$ does not vanish at $T = q^{-1}$, which is the most relevant for our purpose, their conjecture is entirely proved. We state the end result as follows:

THEOREM 1.1 (BSD conjecture for elliptic curves of analytic rank 0). *Let E be a non-isotrivial elliptic curve over K. Assume that $L(E,q^{-1}) \neq 0$. Then the following statements hold:*

(1) *The Mordell–Weil group $E(K)$ is finite (i.e., $E(K)$ is torsion).*
(2) *The Tate–Shafarevich group $Ш(E)$ is finite.*
(3) *Moreover, one has*

$$L(E,q^{-1}) = \frac{|Ш(E)|}{H(E)} \cdot \frac{q \cdot \tau(E)}{|E(K)|^2}, \tag{1.3}$$

where $\tau(E)$ denotes the Tamagawa number of E (i.e., the product over all places v of K of the number of connected components of the fiber over v of the Néron model of E).

We sketch a proof of this theorem. Write $\rho_E = \operatorname{ord}_{T=q^{-1}} L(E,T)$ for the *analytic rank of E i.e.*, the multiplicity of q^{-1} as a root of $L(E,T)$. Tate [**18**] has proved that $0 \leq \operatorname{rank}_{\mathbb{Z}} E(K) \leq \rho_E$. By assumption we have $\rho_E = 0$ here, so that $\operatorname{rank}_{\mathbb{Z}} E(K)$ vanishes too: hence the finiteness of $E(K)$. Moreover, we have $\rho_E = \operatorname{rank}_{\mathbb{Z}} E(K)$ (this equality is the so-called "weak BSD conjecture").

On the other hand, deep results by Tate [**18**] and Milne [**12**] show that the "weak BSD conjecture" implies the whole BSD conjecture. More precisely, they prove the equivalence:

$$\begin{aligned}\rho_E = \operatorname{rank}_{\mathbb{Z}} E(K) \iff \exists\, \ell \text{ prime} : Ш(E)[\ell^\infty] \text{ is finite}\\ \iff \text{the BSD conjecture holds for } E/K.\end{aligned}$$

Milne [**12**] assumes that $p \neq 2$, but note that the equivalence is now known to hold even in even characteristic, by subsequent work of Kato and Trihan [**8**].

Therefore, the equality $\rho_E = \operatorname{rank}_{\mathbb{Z}} E(K)$ implies *(2)* and *(3)*.

1.5. Large Ш via Frobenius isogenies. This section is inspired by [**20**, §5]: our main purpose here is to illustrate, on a very simple example, the steps that we shall follow to prove that some elliptic curves have a large Tate–Shafarevich group.

Let E be a non-isotrivial elliptic curve over $K = \mathbb{F}_q(t)$ and, for any integer $n \geq 1$, let E_n denote the base change of E/K under the p^n-th power Frobenius morphism $\mathrm{Fr}_{p^n} : K \to K$. In other words, we put $E_n = E \times_K K$ where the underlying map $K \to K$ is Fr_{p^n}. The induced map $F_{p^n} : E \to E_n$ is a purely inseparable isogeny of degree p^n (see [**19**, p. 225]). We obtain in this manner a sequence $(E_n)_{n\geq 1}$ of elliptic curves over K. One obviously has $j(E_n) = j(E)^{p^n}$, so that the inseparability degree of $j(E_n)$ is unbounded as n varies. Moreover, we have $N(E_n) = N(E)$ since the curves are K-isogenous (see §1.1). For the same reason, we have $L(E_n,T) = L(E,T)$ for all $n \geq 1$. On the other hand, the proof of Theorem 5.1 in [**20**] shows that the height $H(E_n)$ tends to infinity as $n \to \infty$.

Now let us *assume that E is chosen so that $L(E,q^{-1})$ is non-zero.* Clearly, the central value $L(E_n,q^{-1})$ is then also non-zero for any $n \geq 1$. This implies that the full BSD conjecture holds for all the elliptic curves E_n in the sequence (see Theorem 1.1). In particular, the Tate–Shafarevich groups $Ш(E_n)$ are all finite. Furthermore, the BSD formula (1.3) states that

$$L(E_n,q^{-1}) = \frac{|Ш(E_n)|}{H(E_n)} \cdot \frac{q \cdot \tau(E_n)}{|E_n(K)_{\mathrm{tors}}|^2}.$$

The growth of $\tau(E_n)$ can be estimated in terms of $H(E_n)$: [**6**, Thm. 1.22] or [**2**, Thm. 1.5.4] yield that $\log \tau(E_n) = o\big(\log H(E_n)\big)$ as $n \to \infty$. Further, there is a uniform bound on torsion subgroups of elliptic curves over K (see Proposition 7.1 in [**19**, Lect. 1] for instance): here, this shows that $|E_n(K)_{\mathrm{tors}}|$ remains bounded as $n \to \infty$. From the last displayed identity and these two bounds, we deduce that

$$\frac{\log |\text{Ш}(E_n)|}{\log H(E_n)} = 1 + \frac{\log L(E_n, q^{-1})}{\log H(E_n)} + o(1) \qquad (\text{as } n \to \infty).$$

Since $L(E_n, q^{-1}) = L(E, q^{-1}) = O(1)$, we therefore obtain that

$$\log |\text{Ш}(E_n)| \sim \log H(E_n) \qquad (\text{as } n \to \infty).$$

In other words, we have $|\text{Ш}(E_n)| = H(E_n)^{1+o(1)}$ as $n \to \infty$, so that the elliptic curves E_n do have "large Tate–Shafarevich groups".

Let us denote by $v_p : \mathbb{Q}^\times \to \mathbb{Q}$ the p-adic valuation on $\mathbb{Q}$, normalised so that $v_p(q) = 1$. We now evaluate the p-adic valuations of both sides of the BSD formula (instead of their logarithm, as we just did). The above mentioned upper bound on $\tau(E_n)$ also yields that $v_p(\tau(E_n)) = O\big(v_p(H(E_n))\big)$ as $n \to \infty$. Hence, we have

$$\frac{v_p(|\text{Ш}(E_n)|)}{v_p(H(E_n))} \geq 1 + \frac{v_p(L(E_n, q^{-1}))}{v_p(H(E_n))} + o(1) = 1 + o(1).$$

Besides, the order of $\text{Ш}(E_n)[p^\infty]$ equals $q^{v_p(|\text{Ш}(E_n)|)}$. Since $v_p(H(E_n))$ is none other than $\log H(E_n)/\log q$, we deduce from the above that

$$|\text{Ш}(E_n)[p^\infty]| \geq H(E_n)^{1+o(1)} \qquad (\text{as } n \to \infty).$$

Therefore, the "large Ш" phenomenon for the elliptic curves in the sequence $(E_n)_{n\geq 1}$ is "explained" by a large p-primary part of Ш.

The set of elliptic curves E to which this construction applies is non-empty. For instance, consider the Legendre elliptic curve E/K given by $y^2 = x(x-1)(x-t)$. This curve is non-isotrivial, and a straightforward application of Tate's algorithm shows that the conductor divisor $\mathcal{N}(E)$ has degree 4. The Grothendieck–Ogg–Shafarevich formula then predicts that $\deg L(E, T) = 0$. This forces the L-function to be trivial (*i.e.*, $L(E, T) = 1$); in particular, $L(E, q^{-1})$ is non-zero.

The previous paragraph shows that we have:

PROPOSITION 1.2. *For any integer $n \geq 1$, consider the elliptic curve E_n defined over K by*

$$E_n : \qquad y^2 = x\big(x-1\big)\big(x - t^{p^n}\big).$$

For all $n \geq 1$, the Tate–Shafarevich group $\text{Ш}(E_n)$ is finite and, as $n \to \infty$, one has

$$|\text{Ш}(E_n)| = H(E_n)^{1+o(1)}.$$

This proposition proves that Conjecture 2(i) in the introduction is true. To prove Theorem C, we rely on a somewhat similar strategy: the various steps we follow in the rest of the paper are but more elaborate versions of the ones in the proof of Proposition 1.2.

REMARK 1.3. In case the characteristic p of K is congruent to 1 modulo 6, the authors of [**4**] have very recently produced a sequence of elliptic curves over K with large Tate–Shafarevich groups, the p-primary parts of which are trivial. In contrast to the elliptic curves studied in the present paper, though, the elliptic curves forming the sequence of [**4**], being sextic twists of $y^2 = x^3 + 1$ over K, are all isotrivial and are all $\overline{K}$-isomorphic (they have j-invariant 0). Theorem C is also independent of the congruence of p modulo 6.

2. The elliptic curves $E_{\gamma,a}$

Let $\mathbb{F}_q$ be a finite field of characteristic $p \geq 3$, $K = \mathbb{F}_q(t)$ and M_K be the set of places of K. Recall that we identify, as we may, the set $M_K^\circ = M_K \setminus \{\infty\}$ both with the set of monic irreducible polynomials in $\mathbb{F}_q[t]$ and with the set of closed points of $\mathbb{A}^1_{/\mathbb{F}_q} = \mathbb{P}^1_{/\mathbb{F}_q} \setminus \{\infty\}$.

For any integer $a \geq 1$, we let $\wp_a(t) := t^{q^a} - t \in \mathbb{F}_q[t]$.

2.1. Definition. Let γ be a nonzero element of $\mathbb{F}_q$ and $a \geq 1$ be an integer. We let $E_{\gamma,a}$ be the elliptic curve over K given by the affine Weierstrass model

$$E_{\gamma,a}: \qquad y^2 = x^3 + \wp_a(t)x^2 + \gamma x. \tag{2.1}$$

Its j-invariant is easily computed to be

$$j(E_{\gamma,a}) = 2^8 \cdot \frac{(\wp_a(t)^2 - 3\gamma)^3}{\gamma^2(\wp_a(t)^2 - 4\gamma)}, \tag{2.2}$$

and the discriminant of the model (2.1) is $\Delta_{\gamma,a} = 2^4\gamma^2(\wp_a(t)^2 - 4\gamma)$. Note that the j-invariant is not constant, hence the curve $E_{\gamma,a}$ is non-isotrivial. We also remark that $j(E_{\gamma,a})$ is not a p-th power in K *i.e.*, $j(E_{\gamma,a})$ is separable.

Even though the model (2.1) still makes sense when $p = 2$ or $\gamma = 0$, the expression of its discriminant $\Delta_{\gamma,a}$ shows that the curve $E_{\gamma,a}$ defined by (2.1) is not smooth in these cases. Hence our assuming that $p \neq 2$ and $\gamma \neq 0$ to ensure that $E_{\gamma,a}$ is a *bona fide* elliptic curve.

2.2. Construction of $E_{\gamma,a}$. The sequences $\{E_{\gamma,a}\}_{a\geq 1}$ are essentially some of the titular Artin–Schreier families of elliptic curves studied in [**13**] (see §6.2 there). We briefly give more details about how the sequences $\{E_{\gamma,a}\}_{a\geq 1}$ were constructed in *loc. cit.* and "compare" them to other similarly constructed sequences.

The basic input of the construction in [**13**] is a pair (f, g) of rational functions $\mathbb{P}^1_{/\mathbb{F}_q} \to \mathbb{P}^1_{/\mathbb{F}_q}$ whose divisors satisfy mild conditions. The output is a sequence, indexed by powers Q of q, of smooth projective surfaces X_Q over $\mathbb{F}_q$, each equipped with a natural fibration $\pi_Q : X_Q \to \mathbb{P}^1_{/\mathbb{F}_q}$ and admitting a dominant rational map $C_{f,Q} \times C_{g,Q} \dashrightarrow X_Q$ from the product of two curves $C_{f,Q}$, $C_{g,Q}$ defined over $\mathbb{F}_q$ in terms of f and g. One of the main results of [**13**] is that the Jacobian variety of the generic fiber of π_Q satisfies the BSD conjecture.

Pries and Ulmer further classify (see Proposition 3.1.5 and §4.2 in [**13**]) the various polar behaviours of f and g for which the resulting sequence $(X_Q)_Q$ is a sequence of elliptic surfaces (*i.e.*, for which the generic fiber of π_Q is an elliptic curve over K). This classification results in seven "types" of pairs (f, g).

The elliptic curve $E_{\gamma,a}$ defined by (2.1) corresponds to the case $(2, 1+1)$ of their construction. Specifically, one starts with $f(u) = u^2$ (one double pole at $\infty \in \mathbb{P}^1_{/\mathbb{F}_q}$) and $g(v) = v + \gamma/v$ (one simple pole at 0 and one simple pole at ∞). One obtains, for any power Q of q, a smooth projective surface X_Q over $\mathbb{F}_q$ which is birational to the affine surface Y_Q defined in affine coordinates (u, v, t) by

$$Y_Q: \qquad f(u) - g(v) = t^Q - t.$$

The surface X_Q is equipped with a surjective morphism $\pi_Q : X_Q \to \mathbb{P}^1_{/\mathbb{F}_q}$, which extends the natural projection $Y_Q \to \mathbb{A}^1_{/\mathbb{F}_q}$ given by $(u, v, t) \mapsto t$. Writing $Q = q^a$ for some integer $a \geq 1$, the elliptic curve $E_{\gamma,a}$ over K then arises as the generic fiber of π_Q. The model (2.1) for $E_{\gamma,a}/K$ indeed appears naturally by putting the equation for the generic fiber of π_Q into Weierstrass form.

We wish to remark that one other case of the Pries–Ulmer construction is, for a good choice of parameters, "isogenous" to the case $(2, 1+1)$. Given an element $\gamma \in \mathbb{F}_q^\times$, consider the two rational maps $f, g : \mathbb{P}^1_{/\mathbb{F}_q} \to \mathbb{P}^1_{/\mathbb{F}_q}$ given by $f(u) = u^2$ and $g(v) = v^2 + \gamma/v^2$. In the terminology of [**13**], this input data has type $(2, 2+2)$ since f has a single double pole at ∞, and g has double poles at 0 and at ∞. For any power Q of q, let X_Q be the smooth projective surface over $\mathbb{F}_q$, fibered over $\mathbb{P}^1_{/\mathbb{F}_q}$ *via* $\pi_Q : X_Q \to \mathbb{P}^1_{/\mathbb{F}_q}$, which is birational to the affine surface Y_Q over $\mathbb{F}_q$ given by

$$Y_Q : \qquad f(u) - g(v) = t^Q - t$$

and such that the morphism $\pi_Q : X_Q \to \mathbb{P}^1_{/\mathbb{F}_q}$ extending $(u, v, t) \in Y_Q \mapsto t \in \mathbb{A}^1_{/\mathbb{F}_q}$ is minimal. Writing $Q = q^a$ for some integer $a \geq 1$, we denote by $E'_{\gamma,a}/K$ the generic fiber of π_Q. By construction, after clearing denominators, the curve $E'_{\gamma,a}$ is given in affine coordinates $(u, v) \in \mathbb{A}^2_{/K}$ by

$$E'_{\gamma,a} : \qquad u^2v^2 = v^4 + \wp_a(t)v^2 + \gamma.$$

The change of coordinates $(u, v) \mapsto (x, y) = \big(2v(u+v), 4v\big(uv + v^2 + \wp_a(t)\big)\big)$ then provides the following affine Weierstrass model for $E'_{\gamma,a}$ over K:

$$E'_{\gamma,a} : \qquad y^2 = \big(x + \wp_a(t)\big)\big(x^2 - 4\gamma\big).$$

Our main observation is:

PROPOSITION 2.1. *For any $\gamma \in \mathbb{F}_q^\times$ and any $a \geq 1$, the elliptic curve $E'_{\gamma,a}$ defined over K by the affine Weierstrass model*

$$E'_{\gamma,a} : \qquad y^2 = \big(x + \wp_a(t)\big)\big(x^2 - 4\gamma\big)$$

is 2-isogenous to $E_{\gamma,a}$ over K.

PROOF. Using the formulae in [**17**, Chap. III, Ex. 4.5], we find that the map

$$\phi : (x, y) \in E_{\gamma,a} \mapsto \big(y^2/x^2 - \wp_a(t), y(1 - \gamma/x^2)\big) \in E'_{\gamma,a},$$

which is clearly defined over K, provides the desired 2-isogeny. □

Since they are K-isogenous, the elliptic curves $E_{\gamma,a}$ and $E'_{\gamma,a}$ have the same conductor, they share the same L-function, and their Mordell-Weil ranks are equal. Therefore, our main results about the elliptic curves $E_{\gamma,a}$ (such as the explicit expression for their L-function, and the fact that they have large Tate-Shafarevich groups) are also valid for the curves $E'_{\gamma,a}$.

Furthermore, we know by [**13**, Coro. 3.1.4] that, for all $\gamma \in \mathbb{F}_q^\times$ and $a \geq 1$, the elliptic curves $E_{\gamma,a}$ and $E'_{\gamma,a}$ satisfy the BSD conjecture. The "geometric" calculations of [**13**, §6.2] also show that the Mordell-Weil group $E_{\gamma,a}(K)$ has rank 0. We will recover these two facts *via* a different method, as corollaries of our explicit expression for the L-function of $E_{\gamma,a}$ (see Theorem 5.3 and Corollary 5.4).

2.3. Bad reduction and invariants. We let $B_{\gamma,a} \subset M_K^\circ$ denote the subset corresponding to monic irreducible polynomials in $\mathbb{F}_q[t]$ which divide $\wp_a(t)^2 - 4\gamma$. Equivalently, $B_{\gamma,a}$ is the set of places $v \in M_K^\circ$ dividing the discriminant of the Weierstrass model (2.1) for $E_{\gamma,a}$.

PROPOSITION 2.2. *For any $\gamma \in \mathbb{F}_q^\times$ and any integer $a \geq 1$, the elliptic curve $E_{\gamma,a}$ has good reduction outside $B_{\gamma,a} \cup \{\infty\}$. Moreover, its places of bad reduction are as follows:*

Place v	*Reduction at* v	$\delta_v(E_{\gamma,a})$	$\nu_v(E_{\gamma,a})$
$v \in B_{\gamma,a}$	*Multiplicative (fiber of type* $\mathbf{I}_1$*)*	1	1
∞	*Additive (fiber of type* $\mathbf{I}^*_{4q^a}$*)*	$4q^a + 6$	2

where, for any place $v \in M_K$, we have denoted by $\delta_v(E_{\gamma,a})$ (resp. by $\nu_v(E_{\gamma,a})$) the order at v of the minimal discriminant divisor of $E_{\gamma,a}$ (resp. of its conductor divisor).

PROOF. This follows from a routine application of Tate's algorithm to compute the type of the fibers of bad reduction (as explained in [**16**, §IV.9]). □

From the above proposition, we deduce explicit expressions for the exponential differential height $H(E_{\gamma,a})$ and the "numerical" conductor $N(E_{\gamma,a})$ of $E_{\gamma,a}$ (as defined in §1.1):

$$H(E_{\gamma,a}) = q^{\frac{1}{12}\deg \Delta_{\min}(E_{\gamma,a})} = q^{(q^a+1)/2}, \quad \text{and} \quad N(E_{\gamma,a}) = q^{\deg \mathcal{N}(E_{\gamma,a})} = q^{2(q^a+1)}. \tag{2.3}$$

Indeed, since $p \neq 2$ and $\gamma \neq 0$, the polynomial $\wp_a(t)^2 - 4\gamma \in \mathbb{F}_q[t]$ is squarefree, and therefore we have

$$\sum_{v \in B_{\gamma,a}} \deg v = \sum_{v \mid \wp_a(t)^2 - 4\gamma} \deg v = \deg\left(\wp_a(t)^2 - 4\gamma\right) = 2q^a.$$

It is clear from (2.3) that we have $H(E_{\gamma,a}) = N(E_{\gamma,a})^{1/4}$.

The *Tamagawa number* $\tau(E_{\gamma,a})$ of $E_{\gamma,a}$ is the product over all places $v \in M_K$ of the number of components in the special fiber of the Néron model of $E_{\gamma,a}$ at v. It is also readily calculated from the above proposition with the help of the table on p. 365 of [**16**]. We find that the only contribution to $\tau(E_{\gamma,a})$ comes from the fiber above ∞: since the special fiber of $E_{\gamma,a}$ at ∞ has Kodaira type $\mathbf{I}^*_{4q^a}$, its group of components is isomorphic to $(\mathbb{Z}/2\mathbb{Z})^2$. We obtain that

$$\tau(E_{\gamma,a}) = 4. \tag{2.4}$$

REMARK 2.3. For the computation of the L-function in section 4, it will be useful to have minimal integral models of $E_{\gamma,a}$ at places in M_K° at our disposal. For a place $v \in M_K^\circ$, we find by comparing the valuation at v of the discriminant $\Delta_{\gamma,a}$ of the model (2.1) with the valuation at v of the minimal discriminant in the above table, that the Weierstrass model (2.1) is minimal and integral at v.

This computation of the invariants of $E_{\gamma,a}$ yields the following:

PROPOSITION 2.4. *As γ varies in $\mathbb{F}_q^\times$ and as $a \geq 1$ varies among integers, the elliptic curves $E_{\gamma,a}$ are pairwise non $\overline{K}$-isomorphic and pairwise non K-isogenous.*

PROOF. Two elliptic curves over K are $\overline{K}$-isomorphic if and only if they have the same j-invariant. It is obvious from (2.2) that $\deg j(E_{\gamma,a}) = 4a$ is strictly increasing when $a \geq 1$ grows. Also apparent on (2.2) is the fact that, for a fixed a, the position of the poles of $j(E_{\gamma,a})$ varies with $\gamma \in \mathbb{F}_q^{\times}$. Hence the first assertion.

In a similar vein, it follows from (2.3) that $\deg \mathcal{N}(E_{\gamma,a})$ increases with $a \geq 1$. Since two K-isogenous elliptic curves have equal conductor divisors, we conclude that $E_{\gamma,a}$ and $E_{\gamma,a'}$ cannot be K-isogenous for any $a \neq a'$. For a given $a \geq 1$ and distinct $\gamma, \gamma' \in \mathbb{F}_q^{\times}$, the curves $E_{\gamma,a}$ and $E_{\gamma',a}$ are not K-isogenous either, since the sets $B_{\gamma,a} \cup \{\infty\}$ and $B_{\gamma',a} \cup \{\infty\}$ of places where they have bad reduction differ (*i.e.*, their conductor divisors have different supports). □

2.4. Torsion subgroup. We conclude this section by elucidating the structure of the torsion subgroup of the Mordell-Weil group $E_{\gamma,a}(K)$:

THEOREM 2.5. *For any $\gamma \in \mathbb{F}_q^{\times}$ and any integer $a \geq 1$, let $P_0 \in E_{\gamma,a}(K)$ be the point with coordinates $(0,0)$ in the Weierstrass model (2.1).*
Then we have $E_{\gamma,a}(K)_{\mathrm{tors}} = \{\mathcal{O}, P_0\}$.

We will show later on (Corollary 5.4) that the whole group $E_{\gamma,a}(K)$ is torsion, so that the above result provides a complete list of the K-rational points on $E_{\gamma,a}$.

PROOF. We see on (2.2) that the j-invariant of $E_{\gamma,a}$ is not a p-th power in K, so that Proposition 7.1 in [**19**, Lect. I] ensures that $E_{\gamma,a}(K)_{\mathrm{tors}}$ contains no point with p-th power order.

As was shown in the previous subsection, $E_{\gamma,a}$ has additive reduction at ∞. We may thus make use of Lemma 7.8 of [**15**], which asserts that the prime-to-p part of $E_{\gamma,a}(K)_{\mathrm{tors}}$ embeds into the group G_∞ of components of the special fiber at ∞ of the Néron model of $E_{\gamma,a}$. Since the reduction at ∞ is of type $\mathbf{I}^*_{4q^a}$, the table in §7.2 of [**15**] tells us that $G_\infty \simeq (\mathbb{Z}/2\mathbb{Z})^2$.

Combining the last two paragraphs, we deduce that $E_{\gamma,a}(K)_{\mathrm{tors}}$ is isomorphic to a subgroup of $(\mathbb{Z}/2\mathbb{Z})^2$. In particular, the torsion subgroup of $E_{\gamma,a}(K)$ consists only of 2-torsion points and we infer that $E_{\gamma,a}(K)_{\mathrm{tors}} = E_{\gamma,a}(K)[2]$.

The 2-torsion subgroup of $E_{\gamma,a}(\overline{K})$ is readily computed: it consists of 4 points, given in the homogenised version of (2.1) by

$$\mathcal{O} = [0:1:0], \quad P_0 = [0:0:1],$$
$$P_+ = \left[\wp_a(t) + \sqrt{\wp_a(t)^2 - 4\gamma} : 0 : -2\right], \quad P_- = \left[\wp_a(t) - \sqrt{\wp_a(t)^2 - 4\gamma} : 0 : -2\right].$$

Since $p \neq 2$ and $\gamma \neq 0$, the polynomial $\wp_a(t)^2 - 4\gamma \in \mathbb{F}_q[t]$ is squarefree, so that only the first two points are K-rational (the latter two are rational over the quadratic extension $K(\sqrt{\wp_a(t)^2 - 4\gamma})$ of K). We deduce that $E_{\gamma,a}(K)[2] = \{\mathcal{O}, P_0\}$. □

3. Preliminaries on character sums

In the next section (see Theorem 4.1), we will compute an explicit expression for the L-function of the curve $E_{\gamma,a}$ introduced above. To carry out this computation, we first need to set up some notation and conventions. These will be in force for the rest of the paper.

3.1. Gauss sums and Kloosterman sums. Let $\mathbb{F}$ be a finite field of odd characteristic p. Any additive character ψ on $\mathbb{F}$ may and will be assumed to take values in the p-th cyclotomic field $\mathbb{Q}(\zeta_p)$. For any finite extension $\mathbb{F}'/\mathbb{F}$, we denote the trace map by $\mathrm{Tr}_{\mathbb{F}'/\mathbb{F}} : \mathbb{F}' \to \mathbb{F}$. If ψ is an additive character on $\mathbb{F}$, then the composition $\psi \circ \mathrm{Tr}_{\mathbb{F}'/\mathbb{F}}$ is an additive character on $\mathbb{F}'$.

We denote by $\lambda : \mathbb{F}^\times \to \overline{\mathbb{Q}}^\times$ (or $\lambda_{\mathbb{F}}$ if confusion is likely to arise) the unique nontrivial multiplicative character of order 2 on $\mathbb{F}^\times$. We extend λ to the whole of $\mathbb{F}$ by setting $\lambda(0) := 0$.

DEFINITION 3.1. For any additive character ψ on $\mathbb{F}$, define the *quadratic Gauss sum* $\mathrm{G}_{\mathbb{F}}(\psi, \lambda)$ by

$$\mathrm{G}_{\mathbb{F}}(\psi, \lambda) := -\sum_{x\in\mathbb{F}} \lambda(x)\psi(x).$$

Note our choice of normalising $\mathrm{G}_{\mathbb{F}}(\psi, \lambda)$ by multiplying the sum by -1. By construction, the sum $\mathrm{G}_{\mathbb{F}}(\psi, \lambda)$ is an algebraic integer in the cyclotomic field $\mathbb{Q}(\zeta_p)$. Recall the following facts about Gauss sums:

(Ga 1) For any nontrivial additive character ψ on $\mathbb{F}$, any $\alpha \in \mathbb{F}^\times$ and any finite extension $\mathbb{F}'/\mathbb{F}$, define $\psi_{\mathbb{F}'}^{(\alpha)}$ by $x \mapsto \psi \circ \mathrm{Tr}_{\mathbb{F}'/\mathbb{F}}(\alpha x)$. (This map $\psi_{\mathbb{F}'}^{(\alpha)}$ is a nontrivial additive character on $\mathbb{F}'$.) Letting $\alpha' := \alpha^{|\mathbb{F}|}$, one has

$$\mathrm{G}_{\mathbb{F}'}(\psi_{\mathbb{F}'}^{(\alpha')}, \lambda_{\mathbb{F}'}) = \mathrm{G}_{\mathbb{F}'}(\psi_{\mathbb{F}'}^{(\alpha)}, \lambda_{\mathbb{F}'}).$$

(Ga 2) For any nontrivial additive character ψ on $\mathbb{F}$ and any finite extension $\mathbb{F}'/\mathbb{F}$, one has

$$\mathrm{G}_{\mathbb{F}'}(\psi \circ \mathrm{Tr}_{\mathbb{F}'/\mathbb{F}}, \lambda_{\mathbb{F}'}) = \mathrm{G}_{\mathbb{F}}(\psi, \lambda_{\mathbb{F}})^{[\mathbb{F}':\mathbb{F}]}.$$

(Ga 3) For any nontrivial additive character ψ on $\mathbb{F}$, one has $|\mathrm{G}_{\mathbb{F}}(\psi, \lambda)| = |\mathbb{F}|^{1/2}$ in any complex embedding of $\mathbb{Q}(\zeta_p)$.

(Ga 4) For any nontrivial additive character ψ on $\mathbb{F}$, the quotient $\mathrm{G}_{\mathbb{F}}(\psi, \lambda)/|\mathbb{F}|^{1/2}$ is a 4th root of unity (which may be explicitly determined).

These results are quite classical, and the reader is referred to [**9**, Chap. V, §2] for their proofs.

DEFINITION 3.2. For $\alpha \in \mathbb{F}$ and an additive character ψ on $\mathbb{F}$, define the *Kloosterman sum* $\mathrm{Kl}_{\mathbb{F}}(\psi; \alpha)$ by

$$\mathrm{Kl}_{\mathbb{F}}(\psi; \alpha) := -\sum_{x\in\mathbb{F}^\times} \psi\left(x + \frac{\alpha}{x}\right). \tag{3.1}$$

Again, we point out our choice of normalising the sum by multiplying it by -1. One can show that the Kloosterman sum $\mathrm{Kl}_{\mathbb{F}}(\psi; \alpha)$ is a totally real algebraic integer in $\mathbb{Q}(\zeta_p)$ *i.e.*, $\mathrm{Kl}_{\mathbb{F}}(\psi; \alpha) \in \mathbb{Z}[\zeta_p + \zeta_p^{-1}]$. We remind the reader of the following facts about Kloosterman sums:

(Kl 1) For any nontrivial additive character ψ on $\mathbb{F}$ and any $\alpha \in \mathbb{F}^\times$, one has the identity (sometimes called Salié's formula):

$$\mathrm{Kl}_{\mathbb{F}}(\psi; \alpha) = -\sum_{y\in\mathbb{F}} \lambda(y^2 - 4\alpha)\psi(y).$$

(Kl 2) For any nontrivial additive character ψ on $\mathbb{F}$, any $\alpha \in \mathbb{F}$ and any finite extension $\mathbb{F}'/\mathbb{F}$, letting $\alpha' = \alpha^{|\mathbb{F}|}$, one has

$$\mathrm{Kl}_{\mathbb{F}'}(\psi \circ \mathrm{Tr}_{\mathbb{F}'/\mathbb{F}}; \alpha') = \mathrm{Kl}_{\mathbb{F}'}(\psi \circ \mathrm{Tr}_{\mathbb{F}'/\mathbb{F}}; \alpha).$$

(Kl 3) For any nontrivial additive character ψ on $\mathbb{F}$ and any $\alpha \in \mathbb{F}^\times$, there is a unique pair $\{\mathsf{kl}_{\mathbb{F}}(\psi;\alpha), \mathsf{kl}'_{\mathbb{F}}(\psi;\alpha)\}$ of algebraic integers, whose product is $|\mathbb{F}|$ and such that, for any finite extension $\mathbb{F}'/\mathbb{F}$, one has

$$\mathrm{Kl}_{\mathbb{F}'}(\psi \circ \mathrm{Tr}_{\mathbb{F}'/\mathbb{F}};\alpha) = \mathsf{kl}_{\mathbb{F}}(\psi;\alpha)^{[\mathbb{F}':\mathbb{F}]} + \mathsf{kl}'_{\mathbb{F}}(\psi;\alpha)^{[\mathbb{F}':\mathbb{F}]}.$$

(Kl 4) For any nontrivial additive character ψ on $\mathbb{F}$ and any $\alpha \in \mathbb{F}^\times$, one has $|\mathsf{kl}_{\mathbb{F}}(\psi;\alpha)| = |\mathsf{kl}'_{\mathbb{F}}(\psi;\alpha)| = |\mathbb{F}|^{1/2}$ in any complex embedding of $\overline{\mathbb{Q}}$. Hence, one has $|\mathrm{Kl}_{\mathbb{F}}(\psi;\alpha)| \leq 2|\mathbb{F}|^{1/2}$ in any complex embedding of $\mathbb{Q}(\zeta_p)$.

These results are classical: see [**9**, Chap. V, §5] for proofs thereof. For convenience, we also state here a fact that will only be proved later on (see Lemma 5.1 and Remark 6.2):

(Kl 5) For any nontrivial additive character ψ on $\mathbb{F}$ and any $\alpha \in \mathbb{F}^\times$, $\mathrm{Kl}_{\mathbb{F}}(\psi;\alpha)$ is a p-adic unit in $\mathbb{Q}(\zeta_p)$. In particular, one has $0 < |\mathrm{Kl}_{\mathbb{F}}(\psi;\alpha)| < 2|\mathbb{F}|^{1/2}$ in any complex embedding of $\mathbb{Q}(\zeta_p)$.

3.2. Places of degree dividing a. Let $\mathbb{F}_q$ be a finite field of odd characteristic, and let $K := \mathbb{F}_q(t)$ denote the rational function field over $\mathbb{F}_q$.

DEFINITION 3.3. For any integer $a \geq 1$, let $P_q(a)$ denote the set of places $v \in M_K$ such that $v \neq 0, \infty$ and $\deg v \mid a$. That is to say, $P_q(a)$ is the set of closed points of the multiplicative group $\mathbb{G}_{m/\mathbb{F}_q} = \mathbb{P}^1_{/\mathbb{F}_q} \setminus \{0,\infty\}$ whose degree divides a. In the usual identification between places in M_K° and monic irreducible elements of $\mathbb{F}_q[t]$, the set $P_q(a)$ corresponds to the set of monic irreducible polynomials $B \in \mathbb{F}_q[t]$ such that $B \neq t$ and $\deg B \mid a$.

The latter interpretation allows for an easy estimation of the cardinality $|P_q(a)|$. Indeed the Prime Number Theorem for $\mathbb{F}_q[t]$ states that, for any integer $n \geq 1$, the number $\pi_q(n)$ of monic irreducible polynomials in $\mathbb{F}_q[t]$ of degree n satisfies: $q^n/n - q^{n/2} \leq \pi_q(n) \leq q^n/n$ (see, for instance, [**14**, Thm. 2.2] and its proof). Noting that $|P_q(a)| = -1 + \sum_{n|a} \pi_q(n)$, we deduce the existence of constants $c_q, c'_q > 0$, depending at most on q, such that

$$\forall a \geq 1, \qquad c'_q \cdot \frac{q^a}{a} \leq |P_q(a)| \leq c_q \cdot \frac{q^a}{a}. \tag{3.2}$$

3.3. The sums $\mathbf{g}(v)$ and $\mathsf{Kl}_\gamma(v)$. Let $\mathbb{F}_q$ be a finite field of odd characteristic p, and endow $\mathbb{F}_q$ with a nontrivial additive character ψ_q taking values in the p-th cyclotomic field $\mathbb{Q}(\zeta_p)$. For any finite extension $\mathbb{F}/\mathbb{F}_q$ we "lift" ψ_q to a nontrivial character $\psi_{\mathbb{F}}$ on $\mathbb{F}$ by composing ψ_q with the relative trace map; in other words, we let $\psi_{\mathbb{F}} := \psi_q \circ \mathrm{Tr}_{\mathbb{F}/\mathbb{F}_q}$.

DEFINITION 3.4. Let $\gamma \in \mathbb{F}_q^\times$ and $a \geq 1$ be an integer. For any place $v \in P_q(a)$, we denote by $\mathbb{F}_v$ the residue field of K at v, and by $d_v := [\mathbb{F}_v : \mathbb{F}_q]$ the degree of v.

Viewing v as the $\mathrm{Gal}(\overline{\mathbb{F}_q}/\mathbb{F}_q)$-orbit of an $\overline{\mathbb{F}_q}$-rational point of $\mathbb{G}_{m/\mathbb{F}_q}$, we may pick an element $\beta_v \in \mathbb{F}_v^\times \subset \overline{\mathbb{F}_q}^\times$ representing that orbit v. (The choices of representatives of v in $\overline{\mathbb{F}_q}^\times$ are then $\beta_v, \beta_v^q, \beta_v^{q^2}, \ldots, \beta_v^{q^{d_v-1}}$.) The map $\psi_{\mathbb{F}_v}^{(\beta_v)} : x \mapsto \psi_q \circ \mathrm{Tr}_{\mathbb{F}_v/\mathbb{F}_q}(\beta_v x)$ defines a nontrivial additive character on $\mathbb{F}_v$.

- We denote the Gauss sum $\mathrm{G}_{\mathbb{F}_v}(\psi_{\mathbb{F}_v}^{(\beta_v)}, \lambda_{\mathbb{F}_v})$ by $\mathbf{g}(v)$. A repeated application of (Ga 1) shows that the definition of $\mathbf{g}(v)$ makes sense, in that the value of the sum $\mathrm{G}_{\mathbb{F}_v}(\psi_{\mathbb{F}_v}^{(\beta_v)}, \lambda_{\mathbb{F}_v})$ does not depend on a particular choice of representative β_v for the orbit v.

- We denote the Kloosterman sum $\mathrm{Kl}_{\mathbb{F}_v}(\psi_{\mathbb{F}_v}^{(\beta_v)};\gamma)$ by $\mathbf{Kl}_\gamma(v)$, and we let $\{\mathbf{kl}_\gamma(v), \mathbf{kl}'_\gamma(v)\}$ be the pair of algebraic integers attached to the Kloosterman sum $\mathbf{Kl}_\gamma(v)$ as in (Kl 3). Again, these definitions make sense: repeated applications of (Kl 2) imply that the value of $\mathrm{Kl}_{\mathbb{F}_v}(\psi_{\mathbb{F}_v}^{(\beta_v)};\gamma)$, and hence the pair of algebraic integers attached to it by (Kl 3), do not depend on the choice of β_v in v.

4. The L-function of $E_{\gamma,a}$

The main objective of this section is to provide an explicit expression for the L-function of the elliptic curve $E_{\gamma,a}$. The resulting formula will prove instrumental for the proof of our main theorem. In the notation set up in the previous section, the result is as follows:

THEOREM 4.1. *Let $\mathbb{F}_q$ be a finite field of odd characteristic. For any $\gamma \in \mathbb{F}_q^\times$ and any $a \geq 1$, consider the elliptic curve $E_{\gamma,a}$ over $K = \mathbb{F}_q(t)$ defined by (2.1). The L-function $L(E_{\gamma,a}, T) \in \mathbb{Z}[T]$ of $E_{\gamma,a}$ is given by:*

$$L(E_{\gamma,a},T) = \prod_{v\in P_q(a)} \left(1 - \mathbf{g}(v)\mathbf{kl}_\gamma(v)T^{\deg v}\right)\left(1 - \mathbf{g}(v)\mathbf{kl}'_\gamma(v)T^{\deg v}\right), \tag{4.1}$$

where $P_q(a)$ denotes the set of places defined in §3.2, and $\mathbf{g}(v)$, $\mathbf{kl}_\gamma(v)$, $\mathbf{kl}'_\gamma(v)$ denote the algebraic integers attached to any $v \in P_q(a)$ in §3.3.

The proof of this theorem occupies the rest of the section: the next subsection proves a useful identity between character sums, and the following subsection contains the computation leading to Theorem 4.1.

REMARK 4.2.

(1) Given the definition of $\{\mathbf{kl}_\gamma(v), \mathbf{kl}'_\gamma(v)\}$ and formula (Kl 3), an equivalent way of formulating (4.1) is:

$$L(E_{\gamma,a},T) = \prod_{v\in P_q(a)} \left(1 - \mathbf{g}(v)\mathbf{Kl}_\gamma(v)T^{\deg v} + \mathbf{g}(v)^2 q^{\deg v}T^{2\deg v}\right). \tag{4.2}$$

(2) To define the sums $\mathbf{g}(v)$ and $\mathbf{Kl}_\gamma(v)$ for a place $v \in P_q(a)$, we started by fixing a nontrivial additive character ψ_q on $\mathbb{F}_q$. We remark that the expression for the L-function of $E_{\gamma,a}$ obtained in Theorem 4.1 is independent of this choice. Indeed, a different choice of ψ_q has the sole effect of permuting the factors in (4.1), for the L-function $L(E_{\gamma,a}, T)$ has integral coefficients.

(3) By definition, we have $\sum_{v\in P_q(a)} \deg v = |\mathbb{G}_m(\mathbb{F}_{q^a})| = q^a - 1$. Hence, as a polynomial in T, the L-function of $E_{\gamma,a}$ has degree

$$b(E_{\gamma,a}) = \deg L(E_{\gamma,a},T) = 2(q^a - 1).$$

This is compatible with the Grothendieck–Ogg–Shafarevich formula and the value of $\deg \mathcal{N}(E_{\gamma,a})$ found in (2.3). Note that $b(E_{\gamma,a})$ is even.

4.1. An identity between character sums. For this subsection, we let $\mathbb{F}$ be a finite field of odd characteristic, which we equip with an additive character ψ.

Denote by $\lambda_{\mathbb{F}} = \lambda : \mathbb{F}^\times \to \{\pm 1\}$ the unique quadratic character on $\mathbb{F}^\times$, extended by $\lambda(0) := 0$. For any $\gamma \in \mathbb{F}^\times$, define the double character sum

$$S(\mathbb{F}, \psi, \gamma) := \sum_{z\in\mathbb{F}} \sum_{x\in\mathbb{F}} \lambda(x^3 + zx^2 + \gamma x)\psi(z).$$

Note that the terms with $x = 0$ do not contribute to the sum since $\lambda(0) = 0$. We may therefore disregard these terms and use the multiplicativity of λ to get:

$$S(\mathbb{F}, \psi, \gamma) = \sum_{x\neq 0} \lambda(x^2) \left\{ \sum_{z\in\mathbb{F}} \lambda\left(z + \frac{x^2+\gamma}{x}\right)\psi(z) \right\}.$$

For a given $x \in \mathbb{F}^\times$, we put $u = z + (x^2 + \gamma)/x$ in the sum displayed between brackets: using the additivity of ψ and the definition of $\mathrm{G}_{\mathbb{F}}(\psi, \lambda)$, we obtain that

$$\sum_{z\in\mathbb{F}} \lambda\left(z + \frac{x^2+\gamma}{x}\right)\psi(z) = \psi\left(-x - \frac{\gamma}{x}\right) \sum_{u\in\mathbb{F}} \lambda(u)\psi(u) = -\psi\left(-x - \frac{\gamma}{x}\right) \cdot \mathrm{G}_{\mathbb{F}}(\psi, \lambda).$$

Multiplying this identity by $\lambda(x^2)$, and summing over all $x \in \mathbb{F}^\times$ then yields that

$$S(\mathbb{F}, \psi, \gamma) = -\mathrm{G}_{\mathbb{F}}(\psi, \lambda) \cdot \sum_{x\in\mathbb{F}^\times} \lambda(x)^2 \psi\left(-x - \frac{\gamma}{x}\right).$$

Noting that $\lambda(x)^2 = 1$ for all $x \neq 0$ (because λ has order 2) and reindexing the sum by setting $y = -x$ entails that

$$S(\mathbb{F}, \psi, \gamma) = -\mathrm{G}_{\mathbb{F}}(\psi, \lambda) \cdot \sum_{y\in\mathbb{F}^\times} \psi\left(y + \frac{\gamma}{y}\right) = \mathrm{G}_{\mathbb{F}}(\psi, \lambda) \cdot \mathrm{Kl}_{\mathbb{F}}(\psi; \gamma).$$

We have thus proved:

LEMMA 4.3. *In the above setting, one has $S(\mathbb{F}, \psi, \gamma) = \mathrm{G}_{\mathbb{F}}(\psi, \lambda) \cdot \mathrm{Kl}_{\mathbb{F}}(\psi; \gamma)$, where the Gauss and Kloosterman sums are as defined in* §3.1.

Additionally, it is immediate to check that all three sums in the above identity vanish when ψ is trivial.

4.2. Proof of Theorem 4.1. Fix a parameter $\gamma \in \mathbb{F}_q^\times$ and an integer $a \geq 1$ as in the statement of the theorem. Our starting point for the computation is expression (4.3) below for the L-function of $E_{\gamma,a}$.

For any $\tau \in \overline{\mathbb{F}_q}$, we denote by $v_\tau \in M_K^\circ$ the corresponding place of K; we pick a minimal integral Weierstrass (affine) model of $E_{\gamma,a}/K$ at v_τ of the form $y^2 = f_\tau(x,t)$ where $f_\tau(x,t)$ is a monic cubic polynomial in x, with coefficients in $\mathbb{F}_q[t]$. Recall that $\lambda_{\mathbb{F}_{q^n}} : \mathbb{F}_{q^n} \to \{\pm 1\}$ denotes the unique character of exact order 2 on $\mathbb{F}_{q^n}^\times$, extended by $\lambda_{\mathbb{F}_{q^n}}(0) := 0$ to the whole of $\mathbb{F}_{q^n}$. With this notation, one has the equality of formal power series:

$$\log L(E_{\gamma,a}, T) = -\sum_{n=1}^{\infty} \left(\sum_{\tau\in\mathbb{F}_{q^n}} \sum_{x\in\mathbb{F}_{q^n}} \lambda_{\mathbb{F}_{q^n}}\left(f_\tau(x,\tau)\right) \right) \frac{T^n}{n}. \tag{4.3}$$

This expression can be derived from the definition (1.1)-(1.2) of the L-function, just as in [**3**] (see Lemma 4.6 and the following paragraph there) or [**4**, §4]. Here, we have implicitly used the fact that $E_{\gamma,a}$ has additive reduction at the place ∞ (see Proposition 2.2) to ignore the local term corresponding to this place. At a place of additive reduction, the local Euler factor of $L(E_{\gamma,a}, T)$ in (1.2) is indeed trivial.

We next aim at giving a more explicit expression of the inner double sums in (4.3). As was pointed out in Remark 2.3, one can choose, for any $\tau \in \overline{\mathbb{F}_q}$,

$$f_\tau(x,t) = x^3 + \wp_a(t)x^2 + \gamma x.$$

For any integer $n \geq 1$, the inner double sum in (4.3) can thus be rewritten as

$$\sum_{\tau\in\mathbb{F}_{q^n}} \sum_{x\in\mathbb{F}_{q^n}} \lambda_{\mathbb{F}_{q^n}}(f_\tau(x,\tau)) = \sum_{z\in\mathbb{F}_{q^n}} \sum_{x\in\mathbb{F}_{q^n}} \big|\{\tau \in \mathbb{F}_{q^n} : \wp_a(\tau) = z\}\big| \cdot \lambda_{\mathbb{F}_{q^n}}(x^3 + zx^2 + \gamma x).$$

Moreover, Lemma 4.5 in [**3**] asserts that, for any $z \in \mathbb{F}_{q^n}$,

$$\big|\{\tau \in \mathbb{F}_{q^n} : \wp_a(\tau) = z\}\big| = \sum_{\beta\in\mathbb{F}_{q^n}\cap\mathbb{F}_{q^a}} \psi_q \circ \mathrm{Tr}_{\mathbb{F}_{q^n}/\mathbb{F}_q}(\beta x).$$

For all $\beta \in \mathbb{F}_{q^n}$, the map $z \mapsto \psi_q \circ \mathrm{Tr}_{\mathbb{F}_{q^n}/\mathbb{F}_q}(\beta z)$ is an additive character on $\mathbb{F}_{q^n}$, which we denote by $\psi_{q^n}^{(\beta)}$. Hence, for any integer $n \geq 1$, we have

$$\sum_{\tau\in\mathbb{F}_{q^n}} \sum_{x\in\mathbb{F}_{q^n}} \lambda_{\mathbb{F}_{q^n}}(f_\tau(x)) = \sum_{\beta\in\mathbb{F}_{q^n}\cap\mathbb{F}_{q^a}} S(\mathbb{F}_{q^n}, \psi_{q^n}^{(\beta)}, \gamma),$$

$$\text{where } S(\mathbb{F}_{q^n}, \psi_{q^n}^{(\beta)}, \gamma) = \sum_{z\in\mathbb{F}_{q^n}} \sum_{x\in\mathbb{F}_{q^n}} \lambda_{\mathbb{F}_{q^n}}(x^3 + zx^2 + \gamma x)\psi_{q^n}^{(\beta)}(z).$$

Lemma 4.3 above now yields that $S(\mathbb{F}_{q^n}, \psi_{q^n}^{(\beta)}, \gamma) = \mathrm{G}_{\mathbb{F}_{q^n}}(\psi_{q^n}^{(\beta)}, \lambda) \cdot \mathrm{Kl}_{\mathbb{F}_{q^n}}(\psi_{q^n}^{(\beta)}; \gamma)$ for all $\beta \in \mathbb{F}_{q^n}$, where the Gauss and Kloosterman sums are as in §3.1. Combining the above equalities, we obtain that

$$(4.4) \qquad -\log L(E_{\gamma,a}, T) = \sum_{n=1}^{\infty} \left(\sum_{\beta\in\mathbb{F}_{q^n}\cap\mathbb{F}_{q^a}} \mathrm{G}_{\mathbb{F}_{q^n}}(\psi_{q^n}^{(\beta)}, \lambda) \cdot \mathrm{Kl}_{\mathbb{F}_{q^n}}(\psi_{q^n}^{(\beta)}; \gamma) \right) \frac{T^n}{n}.$$

When $\beta = 0$, the additive character $\psi_{q^n}^{(\beta)}$ is trivial, so both $\mathrm{G}_{\mathbb{F}_{q^n}}(\psi_{q^n}^{(\beta)}, \lambda)$ and $\mathrm{Kl}_{\mathbb{F}_{q^n}}(\psi_{q^n}^{(\beta)}; \gamma)$ vanish. For any $\beta \in (\mathbb{F}_{q^a} \cap \mathbb{F}_{q^n}) \setminus \{0\}$, we denote the place of K containing β by $v_\beta \in M_K^\circ$ (equivalently, v_β is the $\mathrm{Gal}(\overline{\mathbb{F}_q}/\mathbb{F}_q)$-orbit of β). The place v_β is not $0 \in M_K^\circ$ and has degree $\deg v_\beta = [\mathbb{F}_q(\beta) : \mathbb{F}_q]$, which divides both a and n. In particular, v_β belongs to $P_q(a)$.

LEMMA 4.4. *In the notation of* §3.3, *one has* $\mathrm{G}_{\mathbb{F}_{q^n}}(\psi_{q^n}^{(\beta)}, \lambda) = \mathbf{g}(v_\beta)^{n/\deg v_\beta}$ *and* $\mathrm{Kl}_{\mathbb{F}_{q^n}}(\psi_{q^n}^{(\beta)}; \gamma) = \mathsf{kl}_\gamma(v_\beta)^{n/\deg v_\beta} + \mathsf{kl}'_\gamma(v_\beta)^{n/\deg v_\beta}$.

PROOF. For brevity, we write $d = \deg v_\beta$ (recall that d divides n). By multiplicativity of the trace in towers of extensions and because $\beta \in \mathbb{F}_{q^d}$, we have $\psi_{\mathbb{F}_{q^n}}^{(\beta)} = \psi_{\mathbb{F}_{q^d}}^{(\beta)} \circ \mathrm{Tr}_{\mathbb{F}_{q^n}/\mathbb{F}_{q^d}}$. Moreover, $\lambda = \lambda_{\mathbb{F}_{q^n}}$ coincides with $\lambda_{\mathbb{F}_{q^d}} \circ \mathbf{N}_{\mathbb{F}_{q^n}/\mathbb{F}_{q^d}}$. Hence, the Hasse–Davenport relation for Gauss sums (Ga 2) implies that

$$\mathrm{G}_{\mathbb{F}_{q^n}}(\psi_{q^n}^{(\beta)}, \lambda) = \mathrm{G}_{\mathbb{F}_{q^n}}(\psi_{q^d}^{(\beta)} \circ \mathrm{Tr}_{\mathbb{F}_{q^n}/\mathbb{F}_{q^d}}, \lambda_{\mathbb{F}_{q^d}} \circ \mathbf{N}_{\mathbb{F}_{q^n}/\mathbb{F}_{q^d}}) = \mathrm{G}_{\mathbb{F}_{q^d}}(\psi_{q^d}^{(\beta)}, \lambda_{\mathbb{F}_{q^d}})^{[\mathbb{F}_{q^n}:\mathbb{F}_{q^d}]}.$$

Since, by definition, we have $\mathbf{g}(v_\beta) = \mathrm{G}_{\mathbb{F}_{q^d}}(\psi_{q^d}^{(\beta)}, \lambda_{\mathbb{F}_{q^d}})$, the first identity is proved. The second identity is proved in a similar fashion, using the Hasse–Davenport relation for Kloosterman sums (Kl 3). □

Plugging the identities of Lemma 4.4 into (4.4) and exchanging the order of summation yields that

$$-\log L(E_{\gamma,a},T) = \sum_{\beta\in\mathbb{F}_{q^a}\setminus\{0\}} \sum_{\substack{n\geq 1\\ \deg v_\beta | n}} \left(\left(\mathbf{g}(v_\beta)\mathsf{kl}_\gamma(v_\beta)\right)^{n/\deg v_\beta} + \left(\mathbf{g}(v_\beta)\mathsf{kl}'_\gamma(v_\beta)\right)^{n/\deg v_\beta}\right)\frac{T^n}{n}.$$

Upon reindexing the second sum (by setting $m = n/\deg v_\beta$), we obtain that

$$-\log L(E_{\gamma,a},T) = \sum_{\beta\in\mathbb{F}_{q^a}^\times} \frac{1}{\deg v_\beta}\left\{ \sum_{m=1}^{\infty} \frac{(\mathbf{g}(v_\beta)\mathsf{kl}_\gamma(v_\beta)T^{\deg v_\beta})^m}{m} + \sum_{m=1}^{\infty} \frac{(\mathbf{g}(v_\beta)\mathsf{kl}'_\gamma(v_\beta)T^{\deg v_\beta})^m}{m}\right\}.$$

Whence we have, by the formal power series identity $-\log(1-u) = \sum_{m=1}^{\infty}\frac{u^m}{m}$,

$$\log L(E_{\gamma,a},T) = \sum_{\beta\in\mathbb{F}_{q^a}^\times} \frac{1}{\deg v_\beta} \log\left((1-\mathbf{g}(v_\beta)\mathsf{kl}_\gamma(v_\beta)T^{\deg v_\beta})(1-\mathbf{g}(v_\beta)\mathsf{kl}'_\gamma(v_\beta)T^{\deg v_\beta})\right).$$

Finally we group the indices $\beta\in\mathbb{F}_{q^a}^\times$ corresponding to the same $v\in P_q(a)$, and get:

$$\begin{aligned}\log L(E_{\gamma,a},T) &= \sum_{v\in P_q(a)}\sum_{\beta\in v} \frac{1}{\deg v}\log\left((1-\mathbf{g}(v)\mathsf{kl}_\gamma(v)T^{\deg v})(1-\mathbf{g}(v)\mathsf{kl}'_\gamma(v)T^{\deg v})\right)\\ &= \sum_{v\in P_q(a)} \log\left((1-\mathbf{g}(v)\mathsf{kl}_\gamma(v)T^{\deg v})(1-\mathbf{g}(v)\mathsf{kl}'_\gamma(v)T^{\deg v})\right).\end{aligned}$$

To conclude the proof of Theorem 4.1, there only remains to exponentiate this identity. □

5. Non-vanishing of $L(E_{\gamma,a},T)$ at the central point and consequences

As before, let $\mathbb{F}_q$ be a finite field of odd characteristic p and $K := \mathbb{F}_q(t)$. For any parameter $\gamma\in\mathbb{F}_q^\times$ and any integer $a\geq 1$, we consider the elliptic curve $E_{\gamma,a}$ defined over K by (2.1). In this section, we describe the behaviour of the L-function $z\mapsto L(E_{\gamma,a},z)$ around $z=q^{-1}$ (see Theorem 5.3). We first gather some information about its p-adic slope sequence from Theorem 4.1 and results about Gauss and Kloosterman sums. We then derive an important corollary of this analytic study, namely the BSD conjecture for the elliptic curves $E_{\gamma,a}$ (see Corollary 5.4).

5.1. p-adic slopes of $L(E_{\gamma,a},T)$. We choose, once and for all, a prime ideal $\mathfrak{P}$ of $\overline{\mathbb{Q}}$ above p. We denote by $\mathrm{ord}_{\mathfrak{P}}:\overline{\mathbb{Q}}^\times\to\mathbb{Q}$ the corresponding discrete valuation, so normalised that $\mathrm{ord}_{\mathfrak{P}}(q)=1$. The goal of this subsection is to compute the p-adic slope sequence of $L(E_{\gamma,a},T)$ explicitly (see §1.3).

We first prove a probably well-known lemma for which we could not find a proof in the literature:

LEMMA 5.1. *Let $\mathbb{F}$ be a finite extension of $\mathbb{F}_q$, and ψ be a nontrivial additive character on $\mathbb{F}$. For any $\alpha \in \mathbb{F}^\times$, consider the Kloosterman sum $\mathrm{Kl}_{\mathbb{F}}(\psi;\alpha)$ and the pair $\{\mathsf{kl}_{\mathbb{F}}(\psi;\alpha), \mathsf{kl}'_{\mathbb{F}}(\psi;\alpha)\}$ of algebraic integers associated to it by (Kl 3). Then we have*

$$\{\mathrm{ord}_{\mathfrak{P}}\, \mathsf{kl}_{\mathbb{F}}(\psi;\alpha), \mathrm{ord}_{\mathfrak{P}}\, \mathsf{kl}'_{\mathbb{F}}(\psi;\alpha)\} = \big\{0, [\mathbb{F} : \mathbb{F}_q]\big\}.$$

Equivalently, one has $\mathrm{ord}_{\mathfrak{P}}\, \mathrm{Kl}_{\mathbb{F}}(\psi;\alpha) = 0$.

The last statement is equivalent to the first assertion of (Kl 5). In the same setting, the $\mathfrak{P}$-adic valuation of the Gauss sum $\mathrm{G}_{\mathbb{F}}(\psi,\lambda)$ is easily determined: (Ga 4) shows that the sums $\mathrm{G}_{\mathbb{F}}(\psi,\lambda)$ has the same $\mathfrak{P}$-adic valuation as $|\mathbb{F}|^{1/2}$, so that

$$\mathrm{ord}_{\mathfrak{P}}\, \mathrm{G}_{\mathbb{F}}(\psi,\lambda) = \frac{[\mathbb{F} : \mathbb{F}_q]}{2}. \tag{5.1}$$

PROOF. By construction, $\mathrm{Kl}_{\mathbb{F}}(\psi;\alpha)$ is an element of $\mathbb{Q}(\zeta_p)$. The unique prime ideal of $\mathbb{Q}(\zeta_p)$ above p is the principal ideal $I = (\zeta_p - 1)$, so that $p \cdot \mathbb{Z}[\zeta_p] = I^{p-1}$. Since $\psi(y)$ is a power of ζ_p for all $y \in \mathbb{F}$, we have $\psi(y) \equiv 1 \bmod I$ and we find that

$$-\mathrm{Kl}_{\mathbb{F}}(\psi;\alpha) \equiv \sum_{x\in\mathbb{F}^\times} 1 \equiv |\mathbb{F}| - 1 \equiv -1 \bmod I,$$

because $|\mathbb{F}|$ is a power of p. In particular, $\mathrm{Kl}_{\mathbb{F}}(\psi;\alpha) \not\equiv 0 \bmod p$ in $\overline{\mathbb{Q}}$, and we indeed have $\mathrm{ord}_{\mathfrak{P}}\, \mathrm{Kl}_{\mathbb{F}}(\psi;\alpha) = 0$.

On the other hand, (Kl 3) implies that the algebraic integers $\mathsf{kl}_{\mathbb{F}}(\psi;\alpha)$ and $\mathsf{kl}'_{\mathbb{F}}(\psi;\alpha)$ satisfy:

$$\mathsf{kl}_{\mathbb{F}}(\psi;\alpha) \cdot \mathsf{kl}'_{\mathbb{F}}(\psi;\alpha) = |\mathbb{F}| \quad \text{and} \quad \mathsf{kl}_{\mathbb{F}}(\psi;\alpha) + \mathsf{kl}'_{\mathbb{F}}(\psi;\alpha) = \mathrm{Kl}_{\mathbb{F}}(\psi;\alpha).$$

The pair $\{v, v'\}$ formed by their $\mathfrak{P}$-adic valuations thus satisfies the following constraints: $v, v' \geq 0$, $v + v' = \mathrm{ord}_{\mathfrak{P}}\, |\mathbb{F}| = [\mathbb{F} : \mathbb{F}_q]$, and $\min\{v, v'\} \leq \mathrm{ord}_{\mathfrak{P}}\, \mathrm{Kl}_{\mathbb{F}}(\psi;\alpha) = 0$. The only pair $\{v, v'\}$ fulfilling these requirements is $\{0, [\mathbb{F} : \mathbb{F}_q]\}$.

Hence the result. ☐

We can now state and prove the following:

THEOREM 5.2. *Let $\gamma \in \mathbb{F}_q^\times$ and $a \geq 1$. Consider the elliptic curve $E_{\gamma,a}/K$ defined by (2.1), and write $b = \deg L(E_{\gamma,a}, T)$. Recall from Remark 4.2(3) that $b = 2(q^a - 1)$ is even. The L-function $L(E_{\gamma,a}, T)$ has p-adic slope sequence*

$$\lambda_1 = \frac{1}{2},\ \lambda_2 = \frac{1}{2},\ \ldots,\ \lambda_{b/2} = \frac{1}{2},\ \lambda_{b/2+1} = \frac{3}{2},\ \lambda_{b/2+2} = \frac{3}{2},\ \ldots,\ \lambda_b = \frac{3}{2}.$$

PROOF. By definition (see §1.3), the p-adic slope sequence of $L(E_{\gamma,a}, T)$ is the (suitably indexed) multiset of $\mathfrak{P}$-adic valuations of the inverses of zeros of the L-function $z \mapsto L(E_{\gamma,a}, z)$.

Upon staring at the expression for $L(E_{\gamma,a}, T)$ obtained in Theorem 4.1, one immediately sees that an algebraic number $z \in \overline{\mathbb{Q}}$ is the inverse of a zero of $L(E_{\gamma,a}, T)$ if and only if there exists a place $v \in P_q(a)$ with $z^{\deg v} \in \{\mathbf{g}(v)\mathbf{kl}_\gamma(v), \mathbf{g}(v)\mathbf{kl}'_\gamma(v)\}$. In particular, if z is the inverse of a zero of $L(E_{\gamma,a}, T)$, there exists $v \in P_q(a)$ such that $\mathrm{ord}_{\mathfrak{P}}(z^{\deg v})$ equals one of $\mathrm{ord}_{\mathfrak{P}}(\mathbf{g}(v)\mathbf{kl}_\gamma(v))$ or $\mathrm{ord}_{\mathfrak{P}}(\mathbf{g}(v)\mathbf{kl}'_\gamma(v))$.

As can be seen from applying formula (5.1) in the case where $\mathrm{G}_{\mathbb{F}}(\psi,\lambda) = \mathbf{g}(v)$, we have $\mathrm{ord}_{\mathfrak{P}}\, \mathbf{g}(v) = \deg v/2$. The previous lemma applied to $\mathrm{Kl}_{\mathbb{F}}(\psi;\alpha) = \mathbf{Kl}_\gamma(v)$

further yields that $\{\mathrm{ord}_{\mathfrak{P}} \mathbf{kl}_\gamma(v), \mathrm{ord}_{\mathfrak{P}} \mathbf{kl}'_\gamma(v)\} = \{0, \deg v\}$. We therefore have

$$\begin{aligned}\mathrm{ord}_{\mathfrak{P}} z &= \frac{\mathrm{ord}_{\mathfrak{P}}(z^{\deg v})}{\deg v} \\ &\in \left\{\frac{\mathrm{ord}_{\mathfrak{P}} \mathbf{g}(v) + \mathrm{ord}_{\mathfrak{P}} \mathbf{kl}_\gamma(v)}{\deg v}, \frac{\mathrm{ord}_{\mathfrak{P}} \mathbf{g}(v) + \mathrm{ord}_{\mathfrak{P}} \mathbf{kl}'_\gamma(v)}{\deg v}\right\} = \left\{\frac{1}{2}, \frac{3}{2}\right\}.\end{aligned}$$

As was recalled in §1.3, the p-adic slope sequence $\{\lambda_k\}_{k=1}^b$ of $L(E_{\gamma,a}, T)$ admits the following symmetry: for a given $s \in [0,2]$, there are as many indices k such that $\lambda_k = s$ as there are indices k' such that $\lambda_{k'} = 2 - s$. From this observation and from the above display, we immediately conclude that, among the b slopes of the L-function $L(E_{\gamma,a}, T)$, half of them equal $1/2$ and the other half equal $3/2$. The result follows by choosing a suitable numbering of these slopes. □

5.2. Non-vanishing at the central point. The following is a direct consequence of Theorem 5.2:

THEOREM 5.3. *For any $\gamma \in \mathbb{F}_q^\times$ and any $a \geq 1$, the L-function $L(E_{\gamma,a}, T)$ does not vanish at $T = q^{-1}$. In other words, one has $\mathrm{ord}_{T=q^{-1}} L(E_{\gamma,a}, T) = 0$.*

PROOF. Given the expression for $L(E_{\gamma,a}, T)$ of Theorem 4.1, it suffices to show that, for any $v \in P_q(a)$, the two factors

$$1 - \mathbf{g}(v)\mathbf{kl}_\gamma(v)q^{-\deg v} \qquad \text{and} \qquad 1 - \mathbf{g}(v)\mathbf{kl}'_\gamma(v)q^{-\deg v}$$

of $L(E_{\gamma,a}, q^{-1})$ are nonzero. As we have seen in the previous subsection, for all places $v \in P_q(a)$, we have

$$\{\mathrm{ord}_{\mathfrak{P}}(\mathbf{g}(v)\mathbf{kl}_\gamma(v)), \mathrm{ord}_{\mathfrak{P}}(\mathbf{g}(v)\mathbf{kl}'_\gamma(v))\} = \left\{\frac{\deg v}{2}, \frac{3\deg v}{2}\right\}.$$

Since $\mathrm{ord}_{\mathfrak{P}}(q^{\deg v}) = \deg v$, neither of $\mathbf{g}(v)\mathbf{kl}_\gamma(v)$ or $\mathbf{g}(v)\mathbf{kl}'_\gamma(v)$ can equal $q^{\deg v}$, and we are done. □

Remark 6.2 outlines an alternative proof of Theorem 5.3: there, we obtain the non-vanishing by relying on the "angular distribution" of the Gauss and Kloosterman sums.

5.3. The BSD conjecture for $E_{\gamma,a}$. Given the non-vanishing of $L(E_{\gamma,a}, T)$ at $T = q^{-1}$, we deduce from the BSD result (see Theorem 1.1) that the BSD conjecture holds for all the elliptic curves $E_{\gamma,a}$.

COROLLARY 5.4. *For all $\gamma \in \mathbb{F}_q^\times$ and all integers $a \geq 1$, one has:*

1. *The Mordell–Weil group $E_{\gamma,a}(K)$ consists of the two points $\mathcal{O}$ and $P_0 = (0,0)$.*
2. *The Tate–Shafarevich group $\text{Ш}(E_{\gamma,a})$ is finite.*
3. *The following identity holds:*

$$L(E_{\gamma,a}, q^{-1}) = \frac{|\text{Ш}(E_{\gamma,a})|}{q^{-1} \cdot H(E_{\gamma,a})}. \tag{5.2}$$

That the elliptic curves $E_{\gamma,a}$ satisfy the BSD conjecture is not new: as was already mentioned in §2.2, it was first proved by Pries and Ulmer in [**13**], where the proof relies on the geometry of the minimal regular model of $E_{\gamma,a}$. Our proof appears to be largely independent of theirs. Pries and Ulmer have further shown (also by a geometric argument) that $E_{\gamma,a}(K)$ has rank 0 (see §6.2 in [**13**]).

PROOF. By Theorem 1.1*(1)*, the Mordell–Weil group $E_{\gamma,a}(K)$ is finite, hence coincides with its torsion subgroup. Theorem 2.5 then entails the first assertion. The second assertion is copied verbatim from the second item of Theorem 1.1. As for identity (5.2), it suffices to plug the values $|E_{\gamma,a}(K)| = 2$ and $\tau(E_{\gamma,a}) = 4$ (both calculated in §2.3) into the BSD formula (1.3). □

6. Distribution of the Kloosterman sums $\mathsf{Kl}_\gamma(v)$

In this section, we work in the following setting: we fix a finite field $\mathbb{F}_q$ of odd characteristic p, endowed with a nontrivial additive character ψ_q; we also choose, once and for all, an embedding $\iota : \overline{\mathbb{Q}} \hookrightarrow \mathbb{C}$. For any $\gamma \in \mathbb{F}_q^\times$ and any $a \geq 1$, we consider the elliptic curves $E_{\gamma,a}$ defined by (2.1).

In the next section, we will describe the asymptotic behaviour of $L(E_{\gamma,a}, q^{-1})$ as $a \to \infty$. Doing so will require some knowledge about the "angular distribution" in the complex plane of (the images under ι of) the algebraic integers $\mathsf{kl}_\gamma(v), \mathsf{kl}'_\gamma(v)$, for $v \in P_q(a)$. We therefore lay out, in this section, the relevant facts about that distribution.

6.1. Angles of Gauss and Kloosterman sums. For a place $v \in P_q(a)$, the Weil bounds (Ga 3) and (Kl 4) for Gauss and Kloosterman sums reveal that the images under ι of the algebraic integers $\mathbf{g}(v)$, $\mathsf{kl}_\gamma(v)$ and $\mathsf{kl}'_\gamma(v)$ lie on the complex circle $\{z \in \mathbb{C} : |z| = q^{\deg v/2}\}$. We may therefore introduce the following angles:

DEFINITION 6.1. For a place $v \in P_q(a)$,

- We know from (Ga 4) that the algebraic number $\mathbf{g}(v)/q^{\deg v/2}$ is a 4th root of unity. Thus, there is a unique $\varepsilon(v) \in \{0, \pi/2, \pi, 3\pi/2\}$ such that
$$\iota(\mathbf{g}(v)) = q^{\deg v/2} \mathrm{e}^{i\varepsilon(v)}.$$
- By (Kl 4), the complex number $\iota(\mathsf{Kl}_\gamma(v))$ is real with $|\iota(\mathsf{Kl}_\gamma(v))| \leq 2q^{\deg v/2}$. Hence there exists a unique angle $\boldsymbol{\theta}_\gamma(v) \in [0, \pi]$ such that
$$\iota(\mathsf{Kl}_\gamma(v)) = 2q^{\deg v/2} \cos \boldsymbol{\theta}_\gamma(v).$$

For all $v \in P_q(a)$, we remark that one has
$$\{\iota(\mathsf{kl}_\gamma(v)), \iota(\mathsf{kl}'_\gamma(v))\} = \left\{q^{\deg v/2}\mathrm{e}^{i\boldsymbol{\theta}_\gamma(v)}, q^{\deg v/2}\mathrm{e}^{-i\boldsymbol{\theta}_\gamma(v)}\right\}.$$

With this new notation, the expression for $L(E_{\gamma,a}, T)$ obtained in Theorem 4.1 becomes:

(6.1)
$$L(E_{\gamma,a}, T) = \prod_{v \in P_q(a)} \left(1 - \mathrm{e}^{i(\varepsilon(v)+\boldsymbol{\theta}_\gamma(v))}(qT)^{\deg v}\right)\left(1 - \mathrm{e}^{i(\varepsilon(v)-\boldsymbol{\theta}_\gamma(v))}(qT)^{\deg v}\right).$$

Note that, whilst the angles $\boldsymbol{\theta}_\gamma(v)$ individually depend on the choice of ι, the set $\{\boldsymbol{\theta}_\gamma(v)\}_{v \in P_q(a)}$ does not: a different choice of ι only permutes the various $\boldsymbol{\theta}_\gamma(v)$, since $L(E_{\gamma,a}, T)$ has integral coefficients.

Evaluating both sides of the above equality at $T = q^{-1}$ yields that

$$L(E_{\gamma,a}, q^{-1}) = \prod_{v \in P_q(a)} \left(1 + \mathrm{e}^{2i\varepsilon(v)} - 2\mathrm{e}^{i\varepsilon(v)} \cos \boldsymbol{\theta}_\gamma(v)\right). \tag{6.2}$$

REMARK 6.2. (1) For any $v \in P_q(a)$, Lemma 5.1 implies that the angle $\boldsymbol{\theta}_\gamma(v)$ does not lie in $\{0, \pi/2, \pi\}$. Indeed, were $\boldsymbol{\theta}_\gamma(v)$ to hit one of $0, \pi/2$ or π, then the sum $\mathsf{Kl}_\gamma(v)$ would equal $+2q^{\deg v/2}, 0$ or $-2q^{\deg v/2}$, respectively. This is incompatible with Lemma 5.1 since $\mathsf{Kl}_\gamma(v)$ would then not be a p-adic unit.

More generally, with the same construction as in Definition 6.1, one can associate an angle $\theta_{\mathbb{F},\psi,\alpha} \in [0, \pi]$ to any of the Kloosterman sums $\mathrm{Kl}_{\mathbb{F}}(\psi; \alpha)$ introduced in §3.1. A straightforward adaptation of the above argument shows that $\theta_{\mathbb{F},\psi,\alpha} \notin \{0, \pi/2, \pi\}$, which directly implies the second assertion in (Kl 5).

(2) The previous item actually provides a second proof of the non-vanishing of the L-function $L(E_{\gamma,a}, T)$ at $T = q^{-1}$, as follows. It is clear that, for a place $v \in P_q(a)$, the factor indexed by v in the product (6.1) vanishes at $T = q^{-1}$ if and only if $\boldsymbol{\varepsilon}(v) \equiv \pm\boldsymbol{\theta}_\gamma(v) \bmod 2\pi$. By Definition 6.1 above, we know that $\boldsymbol{\varepsilon}(v) \in \{0, \pi/2, 3\pi/2, 2\pi\}$. What we have shown in item (1) proves that the v-th factor in (6.1) does not vanish at $T = q^{-1}$. Hence $L(E_{\gamma,a}, q^{-1})$ is nonzero.

6.2. Angular distribution of the sums $\mathsf{Kl}_\gamma(v)$. Recall that the Sato–Tate measure μ_∞ on $[0, \pi]$ is defined by $\mathrm{d}\mu_\infty := \frac{2}{\pi}\sin^2\theta\,\mathrm{d}\theta$. We denote by μ_a the discrete probability measure on $[0, \pi]$ supported on the set $\{\boldsymbol{\theta}_\gamma(v)\}_{v \in P_q(a)}$ of angles introduced in the previous subsection. In other words, we put

$$\mu_a := \frac{1}{|P_q(a)|}\sum_{v \in P_q(a)} \delta\{\boldsymbol{\theta}_\gamma(v)\},$$

where $\delta\{x\}$ denotes the Dirac delta measure at $x \in [0, \pi]$. Note that μ_a depends neither on the choice of a nontrivial additive character ψ_q on $\mathbb{F}_q$ nor on that of an embedding $\iota : \overline{\mathbb{Q}} \hookrightarrow \mathbb{C}$. The measure μ_a does depend on $\mathbb{F}_q$ and γ though, but we chose not to reflect this in the notation for brevity.

In a previous work, the first-named author has proved that, as $a \to \infty$, the sequence of measures $(\mu_a)_{a \geq 1}$ converges weak-$*$ to μ_∞ in a quantitative way. More specifically, Theorem 6.6 in [**3**] states that:

THEOREM 6.3. *In the above setting, given any continuously differentiable function $g : [0, \pi] \to \mathbb{C}$ and any $\gamma \in \mathbb{F}_q^\times$, as $a \geq 1$ tends to $+\infty$, one has*

$$\left|\int_{[0,\pi]} g\,\mathrm{d}\mu_a - \int_{[0,\pi]} g\,\mathrm{d}\mu_\infty\right| \ll_q \frac{a^{1/2}}{q^{a/4}} \cdot \int_0^\pi |g'(t)|\,\mathrm{d}t, \tag{6.3}$$

where the implicit constant is effective and depends at most on q.

The reader is referred to section 6 of [**3**] for a detailed proof of that statement. Note that μ_a is the same measure as the one denoted by ν_a there.

6.3. 'Small' angles of Kloosterman sums. As we have seen in Remark 6.2, the fact that $L(E_{\gamma,a}, T)$ does not vanish at $T = q^{-1}$ is equivalent to the statement that none of the angles $\boldsymbol{\theta}_\gamma(v)$ lies in $\{0, \pi/2, \pi\}$. In this subsection, we obtain a quantitative version of the fact that the set $\{\boldsymbol{\theta}_\gamma(v)\}_{v \in P_q(a)}$ "avoids" $\{0, \pi/2, \pi\}$.

THEOREM 6.4. *Let p be an odd prime number. There exists a constant $\sigma_p > 0$, depending at most on p, such that the following holds. Let $\mathbb{F}_q$ be a finite field* *of characteristic* p *endowed with a nontrivial additive character* ψ_q*, and fix an embedding* $\iota : \overline{\mathbb{Q}} \hookrightarrow \mathbb{C}$*. For any parameter* $\gamma \in \mathbb{F}_q^\times$ *and any integer* $a \geq 1$*, one has*

$$\{\boldsymbol{\theta}_\gamma(v)\}_{v\in P_q(a)} \subset \left[\varepsilon_a, \tfrac{\pi}{2} - \varepsilon_a\right] \cup \left[\tfrac{\pi}{2} + \varepsilon_a, \pi - \varepsilon_a\right] \quad \text{where } \varepsilon_a = (q^a)^{-\sigma_p}.$$

In other words, the measure μ_a *is supported on* $\left[\varepsilon_a, \frac{\pi}{2} - \varepsilon_a\right] \cup \left[\frac{\pi}{2} + \varepsilon_a, \pi - \varepsilon_a\right]$*.*

Corollary 5.5 in [**3**] already shows that $\boldsymbol{\theta}_\gamma(v) \in [\varepsilon_a, \pi - \varepsilon_a]$ for all $v \in P_q(a)$; so we actually only need to prove that $|\boldsymbol{\theta}_\gamma(v) - \pi/2| \geq \varepsilon_a$. However, for convenience, we give a complete proof of Theorem 6.4.

The main tool to prove this result is the following version of Liouville's inequality, which is taken from the introduction of [**11**]. Let $P \in \mathbb{Z}[X]$ be a polynomial of degree N. For any algebraic number $\kappa \in \overline{\mathbb{Q}}$, let $ht(\kappa)$ denote its absolute logarithmic Weil height. If $P(\kappa) \neq 0$, then

$$\frac{\log|P(\kappa)|}{[\mathbb{Q}(\kappa):\mathbb{Q}]} \geq -\log\|P\|_1 - N \cdot ht(\kappa), \tag{6.4}$$

in any complex embedding of $\mathbb{Q}(\kappa)$. Here we have denoted by $\|P\|_1$ the sum of the absolute values of the coefficients of P. The reader is referred to [**11**] for this statement and its proof.

Proof of Theorem 6.4. In the setting of the theorem, we claim that there exists a constant $\sigma_p > 0$ such that, for all $v \in P_q(a)$, one has

$$\min\{\boldsymbol{\theta}_\gamma(v), |\boldsymbol{\theta}_\gamma(v) - \pi/2|, \pi - \boldsymbol{\theta}_\gamma(v)\} \geq (q^{\deg v})^{-\sigma_p}.$$

Since $\deg v$ divides a for all $v \in P_q(a)$, the theorem clearly follows from this claim.

In order to prove the claim, consider the algebraic number $\kappa := \mathsf{kl}_\gamma(v) q^{-\deg v/2}$. Up to replacing $\mathsf{kl}_\gamma(v)$ by $\mathsf{kl}'_\gamma(v)$, we can assume that $\iota(\kappa) = e^{i\boldsymbol{\theta}_\gamma(v)}$. It is relatively easy to show that $[\mathbb{Q}(\kappa):\mathbb{Q}] \leq 2(p-1)$ and that $ht(\kappa) \leq \log(q^{\deg v/2})$, see [**3**, Lem. 5.3] for details. Let us now apply Liouville's inequality to κ with the following three polynomials: $P_1 = X - 1$, $P_2 = X + 1$ and $P_3 = X^4 - 1$.

By (Kl5) or Remark 6.2, we know that $\kappa \notin \{1, i, -1\}$, so that $P_j(\kappa) \neq 0$ for $j \in \{1, 2, 3\}$. Combined with our estimates of the degree and of the height of κ, Liouville's inequality (6.4) yields that, for $j \in \{1, 2, 3\}$,

$$\begin{aligned} \log|\iota(P_j(\kappa))| &\geq -[\mathbb{Q}(\kappa):\mathbb{Q}](\log 2 + 4ht(\kappa)) \geq -2(p-1)(\log 2 + 2\log q^{\deg v}) \\ &\geq -6(p-1)\log q^{\deg v}. \end{aligned} \tag{6.5}$$

Besides, a quick analysis shows that, for any $\theta \in [0, \pi]$, one has

$$\begin{aligned} |P_1(e^{i\theta})| &= |e^{i\theta} - 1| \leq |\theta| = \theta, \\ |P_2(e^{i\theta})| &= |e^{i\theta} + 1| = |e^{i(\theta-\pi)} - 1| \leq |\theta - \pi| = \pi - \theta, \\ \text{and} \quad |P_3(e^{i\theta})| &= |e^{4i\theta} - 1| = |e^{i\theta} - i| \cdot |e^{i\theta} + i| \cdot |e^{2i\theta} - 1| \\ &\leq 4|e^{i\theta} - i| = 4|e^{i(\theta-\pi/2)} - 1| \leq 4|\theta - \pi/2|. \end{aligned}$$

Applying these inequalities to $\theta = \boldsymbol{\theta}_\gamma(v)$ and using (6.5), we directly obtain that

$$\boldsymbol{\theta}_\gamma(v) \geq |P_1(\kappa)| \geq (q^{\deg v})^{-6(p-1)},$$

$$\text{and that} \quad \pi - \boldsymbol{\theta}_\gamma(v) \geq |P_2(\kappa)| \geq (q^{\deg v})^{-6(p-1)}.$$

By further noting that $q^{2\deg v} \geq 4$, we also get that

$$|\pi/2 - \boldsymbol{\theta}_\gamma(v)| \geq |P_3(\kappa)|/4 \geq (q^{\deg v})^{-6(p-1)}/4 \geq (q^{\deg v})^{-(6p-4)}.$$

The claim now follows immediately, with $\sigma_p = 6p - 4 > 0$ being a suitable choice of constant. □

7. Estimates of the central value $L(E_{\gamma,a}, q^{-1})$

The goal of this section is to prove a precise asymptotic estimate on the central value $L(E_{\gamma,a}, q^{-1})$ as $a \to \infty$. This will provide the crucial input for our bounds on the size of $Ш(E_{\gamma,a})$ in the next section.

The result is as follows:

THEOREM 7.1. *Let $\mathbb{F}_q$ be a finite field of odd characteristic, and $K = \mathbb{F}_q(t)$. There are positive constants c_1, c_2 (depending at most on q) such that: for any $\gamma \in \mathbb{F}_q^\times$ and any $a \geq 1$, the central value $L(E_{\gamma,a}, q^{-1})$ of the L-function of the elliptic curve $E_{\gamma,a}$ satisfies:*

$$-\frac{c_1}{a} \leq \frac{\log |L(E_{\gamma,a}, q^{-1})|}{\log H(E_{\gamma,a})} \leq \frac{c_2}{a}. \tag{7.1}$$

The proof of this theorem will be given in §7.2 after proving an intermediate result in the following subsection. The upper bound in (7.1) provides a slight improvement on the generic upper bound on the central value: for any non-isotrivial elliptic curve E over K with $L(E, q^{-1}) \neq 0$, it is known that

$$\frac{\log L(E, q^{-1})}{\log H(E)} \leq c_3 \cdot \frac{\log \log \log H(E)}{\log \log H(E)} \qquad (\text{as } H(E) \to \infty),$$

for some constant $c_3 > 0$. This can be proved by using the fact that the zeros of $L(E, T)$ become uniformly equidistributed on the circle $\{z \in \mathbb{C} : |z| = q^{-1}\}$ as $H(E) \to \infty$ (see [**6**, Thm. 7.5]). For easier comparison, note that $\log \log H(E_{\gamma,a})$ has the same order of magnitude as a, as was shown in (2.3).

The lower bound in (7.1) is, on the other hand, much stronger than the generic lower bound on the central value. For a non-isotrivial elliptic curve E with $L(E, q^{-1}) \neq 0$, the latter only yields that

$$-1 + \frac{c_4}{\log \log H(E)} \leq \frac{\log L(E, q^{-1})}{\log H(E)} \qquad (\text{as } H(E) \to \infty),$$

for some $c_4 < 0$. *Via* the BSD formula, this translates into a lower bound on the order of $Ш(E)$ which is weaker than the one we require to prove Theorem C*(3)*.

7.1. Convergence of a certain sequence. Let $a \in \mathbb{Z}_{\geq 1}$ and $\gamma \in \mathbb{F}_q^\times$. We keep the notation introduced in section 6. Consider the non-negative function $W : [0, \pi] \to \mathbb{R}$ defined by

$$W(\theta) := \begin{cases} -\log\left(\sin^2\theta \cdot \cos^2\theta\right) & \text{if } \theta \in [0,\pi] \setminus \{0, \pi/2, \pi\}, \\ 0 & \text{if } \theta \in \{0, \pi/2, \pi\}. \end{cases}$$

The function $\theta \mapsto W(\theta)$ is continuously differentiable on $(0, \pi/2) \cup (\pi/2, \pi)$ and, for all $\theta \in [0, \pi]$, it satisfies $W(\pi - \theta) = W(\theta)$. Moreover, a routine check shows that W is integrable on $[0, \pi]$ for the Lebesgue measure, as well as for the Sato–Tate measure μ_∞.

By Remark 6.2, the function W is also integrable for the measure μ_a introduced in §6.2: it thus makes sense to consider the sequence $I_W = \left(\int_{[0,\pi]} W \,\mathrm{d}\mu_a\right)_{a\geq 1}$. Even though we know, by Theorem 6.3, that $(\mu_a)_{a\geq 1}$ converges weak-$*$ to μ_∞, we cannot directly conclude that the sequence I_W converges since W is not continuous on $[0,\pi]$. Nonetheless, the goal of this subsection is to show the following:

Proposition 7.2. *In the above setting, the sequence* $\left(\int_{[0,\pi]} W \,\mathrm{d}\mu_a\right)_{a\geq 1}$ *converges to* $\int_{[0,\pi]} W \,\mathrm{d}\mu_\infty$. *More precisely, one has*

$$\left| \int_{[0,\pi]} W \,\mathrm{d}\mu_a - \int_{[0,\pi]} W \,\mathrm{d}\mu_\infty \right| \ll_q \frac{a^{3/2}}{q^{a/4}} \qquad (\textit{as } a \to \infty),$$

where the implicit constant is effective and depends at most on q.

Even though the exact value of the limit is of little importance to us, one can actually compute that $\int_{[0,\pi]} W \,\mathrm{d}\mu_\infty = \log(16)$. Using the definition of the measure μ_a, the above result can thus be rewritten in a more explicit form:

$$\frac{1}{|P_q(a)|} \sum_{v\in P_q(a)} \log\left(\sin^2 \boldsymbol{\theta}_\gamma(v) \cdot \cos^2 \boldsymbol{\theta}_\gamma(v)\right) = -\log(16) + O_q\left(\frac{a^{3/2}}{q^{a/4}}\right) \qquad (\text{as } a \to \infty).$$

Proof. We pick, once and for all, a nondecreasing continuously differentiable function $\beta_0 : [0,1] \to [0,1]$ such that $\beta_0(x) = 0$ for all $x \in [0,1/3]$, and $\beta_0(x) = 1$ for all $x \in [2/3, 1]$. For a small enough $\varepsilon > 0$, we define a "smoothing" function $\beta_\varepsilon : [0,\pi] \to [0,1]$ as follows: define β_ε on $[0,\pi/2]$ by

$$\beta_\varepsilon(\theta) := \begin{cases} \beta_0(\theta/\varepsilon) & \text{if } \theta \in [0,\varepsilon], \\ 1 & \text{if } \theta \in [\varepsilon, \pi/2 - \varepsilon], \\ \beta_0((\pi/2-\theta)/\varepsilon) & \text{if } \theta \in [\pi/2-\varepsilon, \pi/2], \end{cases}$$

and extend it to $[0,\pi]$ by requiring that $\beta_\varepsilon(\pi-\theta) = \beta_\varepsilon(\theta)$ for all $\theta \in [0,\pi]$.

For any $\varepsilon > 0$, we put $W_\varepsilon := W \cdot \beta_\varepsilon$. The function $W_\varepsilon : [0,\pi] \to \mathbb{R}$ is continuously differentiable on the whole interval $[0,\pi]$, it coincides with W on $[\varepsilon, \pi/2-\varepsilon] \cup [\pi/2+\varepsilon, \pi-\varepsilon]$, and it satisfies $W_\varepsilon(\pi-\theta) = W_\varepsilon(\theta)$ for all $\theta \in [0,\pi]$. Let us also record the following estimates:

Lemma 7.3. *For all* $\varepsilon \in (0, 1/4)$, *one has*

$$(a) \int_0^\pi |W_\varepsilon'(t)| \,\mathrm{d}t \ll |\log\varepsilon|, \qquad (b) \int_{[0,\pi]} |W - W_\varepsilon| \,\mathrm{d}\mu_\infty \ll \varepsilon|\log\varepsilon|.$$

We postpone proving these two bounds until the end of the section, and carry on with the proof of Proposition 7.2. In the notation introduced above, the triangle inequality yields that

$$\left|\int_{[0,\pi]} W\,\mathrm{d}\mu_a - \int_{[0,\pi]} W\,\mathrm{d}\mu_\infty\right| \leq \int_{[0,\pi]} |W - W_\varepsilon|\,\mathrm{d}\mu_a + \left|\int_{[0,\pi]} W_\varepsilon\,\mathrm{d}\mu_a - \int_{[0,\pi]} W_\varepsilon\,\mathrm{d}\mu_\infty\right| + \int_{[0,\pi]} |W - W_\varepsilon|\,\mathrm{d}\mu_\infty. \tag{7.2}$$

We treat each of the three terms appearing in the right-hand side of this inequality separately.

The third term $\int_{[0,\pi]} |W - W_\varepsilon|\,\mathrm{d}\mu_\infty$ can be directly bounded with the help of Lemma 7.3*(b)*.

To show that the middle term is small when $a \to \infty$, we apply Theorem 6.3 to W_ε, which is permissible since W_ε is continuously differentiable on $[0,\pi]$. We obtain that

$$\left|\int_{[0,\pi]} W_\varepsilon\,\mathrm{d}\mu_a - \int_{[0,\pi]} W_\varepsilon\,\mathrm{d}\mu_\infty\right| \ll_q \frac{a^{1/2}}{q^{a/4}} \cdot \int_0^\pi |W_\varepsilon'(t)|\,\mathrm{d}t \ll_q \frac{a^{1/2}}{q^{a/4}} \cdot |\log\varepsilon|,$$

where the last inequality follows from Lemma 7.3*(a)*.

Finally, we claim that the first term $\int_{[0,\pi]} |W - W_\varepsilon|\,\mathrm{d}\mu_a$ on the right-hand side of (7.2) vanishes provided that $\varepsilon > 0$ is small enough. Indeed, we know by Theorem 6.4 that the support of μ_a is contained in $[\varepsilon_a, \pi/2 - \varepsilon_a] \cup [\pi/2 + \varepsilon_a, \pi - \varepsilon_a]$ where $\varepsilon_a = (q^a)^{-\sigma_p}$ for some constant $\sigma_p > 0$. By the proof of that theorem, one can take $\sigma_p = 6p - 4$. On the other hand, as was noted in a previous paragraph, W and W_ε coincide on $[\varepsilon, \pi/2 - \varepsilon] \cup [\pi/2 + \varepsilon, \pi - \varepsilon]$. Hence, choosing $\varepsilon < \varepsilon_a$, we have $\int_{[0,\pi]} |W - W_\varepsilon|\,\mathrm{d}\mu_a = 0$.

Summing up these three contributions, inequality (7.2) yields that

$$\left|\int_{[0,\pi]} W\,\mathrm{d}\mu_a - \int_{[0,\pi]} W\,\mathrm{d}\mu_\infty\right| \ll_q 0 + \frac{a^{1/2}}{q^{a/4}} \cdot |\log\varepsilon| + \varepsilon|\log\varepsilon|,$$

for all $\varepsilon < \varepsilon_a$. Note that $(q^a)^{-6q} < \varepsilon_a$ because $6q > 6p - 4 = \sigma_p$. We may therefore take $\varepsilon = (q^a)^{-6q}$, and this choice provides the desired bound. □

Proof of Lemma 7.3. Both W and W_ε are periodic of period $\pi/2$ and are symmetric around $\pi/4$ (*i.e.*, $W(\pi/2 - \theta) = W(\theta)$ for all $\theta \in [0, \pi/2]$, and similarly for W_ε). The same holds for $|W_\varepsilon'|$ and $|W - W_\varepsilon|$. We are thus reduced to proving the following two bounds:

(a') $\displaystyle\int_0^{\pi/4} |W_\varepsilon'(t)|\,\mathrm{d}t \ll |\log\varepsilon|$, *(b')* $\displaystyle\int_{[0,\pi/2]} |W - W_\varepsilon|\,\mathrm{d}\mu_\infty \ll \varepsilon|\log\varepsilon|$.

To that end, we first gather a few useful remarks:

(i) Since $W_\varepsilon = W \cdot \beta_\varepsilon$ and since $0 \leq \beta_\varepsilon(t) \leq 1$, we have
$$\forall t \in [0,\pi], \qquad |W_\varepsilon'(t)| \leq |W'(t)| + |\beta_\varepsilon'(t)|W(t).$$

(ii) By construction of β_ε, we have $|\beta_\varepsilon'(t)| = 0$ for all $t \in [\varepsilon, \pi/4]$. We also have $|\beta_\varepsilon'(t)| \leq M/\varepsilon$ for $t \in [0, \varepsilon]$, where $M := \sup_{x\in[0,1]} |\beta_0'(x)|$.

(iii) The function W_ε is constant (equal to 0) on $[0, \varepsilon/3]$, so that $W_\varepsilon'(t) = 0$ for all $t \in (0, \varepsilon/3)$.

(iv) The functions W and W_ε are symmetric around $\pi/4$, and they coincide on $[\varepsilon, \pi/2-\varepsilon]$. Moreover, we have

$$\forall t \in [0,\pi/2], \qquad |W(t) - W_\varepsilon(t)| = (1-\beta_\varepsilon(t))W(t) \le W(t).$$

(v) Classical convexity inequalities imply that $(\sin t)^{-1} \le \pi/(2t)$ and $(\cos t)^{-1} \le \pi/(\pi - 2t)$ for all $t \in (0, \pi/4]$. We deduce that

$$W(t) = 2\log\left((\sin t\cdot \cos t)^{-1}\right) \le 2\log\left(\frac{\pi^2}{2t(\pi-2t)}\right) \le 2\log\frac{\pi}{t},$$

for all t in this interval. In particular, this yields that

$$\int_0^\varepsilon W(t)\,\mathrm{d}t \le -2\int_0^\varepsilon \log\frac{t}{\pi}\,\mathrm{d}t \ll \varepsilon|\log\varepsilon|.$$

(vi) For all $t \in (0,\pi/4]$, a simple calculation shows that

$$W'(t) = 2\left(\frac{\sin t}{\cos t} - \frac{\cos t}{\sin t}\right) = -\frac{2\cos(2t)}{\sin t\cdot\cos t}.$$

Thus, using the same classical inequalities as in the previous item, we obtain that, for all $t \in (0,\pi/4]$,

$$|W'(t)| \le \frac{2}{\sin t\cdot \cos t} \le \frac{2\pi^2}{2t(\pi-2t)} \le \frac{2\pi}{t}.$$

We now combine the above items to derive the desired inequalities. First, we have

$$\begin{aligned}\int_0^\varepsilon |W_\varepsilon'(t)|\,\mathrm{d}t &\overset{\text{by (iii)}}{=} \int_{\varepsilon/3}^\varepsilon |W_\varepsilon'(t)|\,\mathrm{d}t \overset{\text{by (i)}}{\le} \int_{\varepsilon/3}^\varepsilon |W'(t)|\,\mathrm{d}t + \int_{\varepsilon/3}^\varepsilon W(t)|\beta_\varepsilon'(t)|\,\mathrm{d}t \\ &\overset{\text{by (vi), (ii)}}{\le} \int_{\varepsilon/3}^\varepsilon \frac{2\pi}{t}\,\mathrm{d}t + \int_{\varepsilon/3}^\varepsilon W(t)\frac{M}{\varepsilon}\,\mathrm{d}t \overset{\text{by (v)}}{\ll} |\log\varepsilon|.\end{aligned}$$

Furthermore, we have

$$\begin{aligned}\int_\varepsilon^{\pi/4} |W_\varepsilon'(t)|\,\mathrm{d}t &\overset{\text{by (i)}}{\le} \int_\varepsilon^{\pi/4} |W'(t)|\,\mathrm{d}t + \int_\varepsilon^{\pi/4} W(t)|\beta_\varepsilon'(t)|\,\mathrm{d}t \\ &\overset{\text{by (vi), (ii)}}{\le} \int_\varepsilon^{\pi/4} \frac{2\pi}{t}\,\mathrm{d}t \ll |\log\varepsilon|.\end{aligned}$$

Summing the last two displays proves *(a')*. Finally, we notice that

$$\begin{aligned}\int_{[0,\pi/2]} |W - W_\varepsilon|\,\mathrm{d}\mu_\infty &= \frac{2}{\pi}\int_0^{\pi/2} |W(t) - W_\varepsilon(t)|\sin^2(t)\,\mathrm{d}t \\ &\le \frac{2}{\pi}\int_0^{\pi/2} |W(t) - W_\varepsilon(t)|\,\mathrm{d}t \overset{\text{by (iv)}}{\le} \frac{4}{\pi}\int_0^\varepsilon W(t)\,\mathrm{d}t \overset{\text{by (v)}}{\ll} \varepsilon|\log\varepsilon|,\end{aligned}$$

which proves *(b')* and concludes the proof of the lemma. □

7.2. Proof of Theorem 7.1. With Proposition 7.2 at hand, we may now proceed to prove Theorem 7.1. In the notation introduced in the previous sections, we begin by taking the logarithm of identity (6.2):

$$\begin{aligned}\log L(E_{\gamma,a}, q^{-1}) &= \log |L(E_{\gamma,a}, q^{-1})| \\ &= \sum_{v\in P_q(a)} \log\left|1 + \mathrm{e}^{2i\varepsilon(v)} - 2\mathrm{e}^{i\varepsilon(v)}\cos\boldsymbol{\theta}_\gamma(v)\right| \\ &= \sum_{v\in P_q(a)} \log |F_v(\boldsymbol{\theta}_\gamma(v))|,\end{aligned} \tag{7.3}$$

where the functions $F_v : [0, \pi] \to \mathbb{R}$ are defined as follows:

$$\forall v \in P_q(a), \qquad F_v : \theta \mapsto \begin{cases} 2 - 2\cos\theta & \text{if } \varepsilon(v) = 0, \\ -2i\cos\theta & \text{if } \varepsilon(v) = \pi/2, \\ 2 + 2\cos\theta & \text{if } \varepsilon(v) = \pi, \\ 2i\cos\theta & \text{if } \varepsilon(v) = 3\pi/2. \end{cases}$$

Note that $|F_v(\boldsymbol{\theta}_\gamma(v))| > 0$ for all $v \in P_q(a)$ because $\boldsymbol{\theta}_\gamma(v) \notin \{0, \pi/2, \pi\}$ by (Kl5) (see Remark 6.2). Straightforward analytic estimates show that, for any place $v \in P_q(a)$, one has

$$\forall\theta \in [0, \pi], \qquad \sin^2\theta \cdot \cos^2\theta \le |F_v(\theta)| \le 4.$$

In particular, writing $W(\theta) = -\log\left(\sin^2\theta\cdot\cos^2\theta\right)$ as in §7.1, we obtain that

$$\forall v \in P_q(a), \qquad -W(\boldsymbol{\theta}_\gamma(v)) \le \log |F_v(\boldsymbol{\theta}_\gamma(v))| \le \log 4.$$

Summing this chain of inequalities over all $v \in P_q(a)$ and using equality (7.3) yields:

$$\frac{1}{\log H(E_{\gamma,a})} \cdot \sum_{v\in P_q(a)} -W(\boldsymbol{\theta}_\gamma(v)) \le \frac{\log L(E_{\gamma,a}, q^{-1})}{\log H(E_{\gamma,a})} \le \log 4 \cdot \frac{|P_q(a)|}{\log H(E_{\gamma,a})}.$$

It is clear from these inequalities that Theorem 7.1 will be proved once we show the following two assertions:

(C_1) There exists a constant $c_1 > 0$ such that

$$\frac{1}{\log H(E_{\gamma,a})} \sum_{v\in P_q(a)} W(\boldsymbol{\theta}_\gamma(v)) \le \frac{c_1}{a},$$

(C_2) There exists a constant $c_2 > 0$ such that $\dfrac{|P_q(a)|}{\log H(E_{\gamma,a})} \le \dfrac{c_2}{a}$.

We thus now prove these two claims. The second one (C_2) is a direct consequence of the following two estimates: we know from (2.3) that $\log H(E_{\gamma,a}) \gg_q q^a$, and from (3.2) that $|P_q(a)| \ll_q q^a/a$.

Next we turn to the proof of (C_1): by Proposition 7.2, we have

$$\frac{1}{|P_q(a)|} \sum_{v\in P_q(a)} W(\boldsymbol{\theta}_\gamma(v)) = \int_{[0,\pi]} W\,\mathrm{d}\mu_\infty + O_q\left(\frac{a^{3/2}}{q^{a/4}}\right),$$

as $a \to \infty$, where the implicit constant depends at most on q. Therefore, making use of the estimate in (C_2) which we have just proved, we obtain that

$$0 \leq \frac{1}{\log H(E_{\gamma,a})} \sum_{v \in P_q(a)} W(\boldsymbol{\theta}_\gamma(v)) = \frac{|P_q(a)|}{\log H(E_{\gamma,a})} \cdot \frac{1}{|P_q(a)|} \sum_{v \in P_q(a)} W(\boldsymbol{\theta}_\gamma(v))$$
$$\leq \frac{c_2}{a}\left(\int_{[0,\pi]} W \,\mathrm{d}\mu_\infty + O_q\left(\frac{a^{3/2}}{q^{a/4}}\right)\right) \leq \frac{c_1}{a},$$

for some constant $c_1 > 0$, depending at most on q. This proves the first claim (C_1) and concludes the proof of Theorem 7.1. □

8. Proof of Theorem C

In this section, we gather our results so far to prove our main result (Theorem C in the introduction). We have already proved assertions *(1)* and *(2)* of that theorem: see Proposition 2.4 and Corollary 5.4*(2)*, respectively. In the following two subsections, we prove assertions *(4)* and *(3)* of Theorem C, in this order.

8.1. The p-primary part of $\text{Ш}(E_{\gamma,a})$. Let us first prove assertion *(4)* of Theorem C, concerning the p-primary part of $\text{Ш}(E_{\gamma,a})$:

THEOREM 8.1. *Let $\mathbb{F}_q$ be a finite field of odd characteristic p, and $K = \mathbb{F}_q(t)$. For any $\gamma \in \mathbb{F}_q^\times$ and any integer $a \geq 1$, consider the elliptic curve $E_{\gamma,a}$ over K defined by* (2.1). *Then the p-primary part of $\text{Ш}(E_{\gamma,a})$ is trivial. In other words, the integer $|\text{Ш}(E_{\gamma,a})|$ is relatively prime to p.*

PROOF. Recall from §5.1 that $\operatorname{ord}_{\mathfrak{P}} : \overline{\mathbb{Q}}^\times \to \mathbb{Q}$ denotes an extension of the p-adic value to $\overline{\mathbb{Q}}$, normalised so that $\operatorname{ord}_{\mathfrak{P}}(q) = 1$. Taking $\mathfrak{P}$-adic valuation of both sides of the BSD formula (5.2), we obtain that

$$\operatorname{ord}_{\mathfrak{P}} |\text{Ш}(E_{\gamma,a})| = \operatorname{ord}_{\mathfrak{P}} L(E_{\gamma,a}, q^{-1}) + \operatorname{ord}_{\mathfrak{P}} H(E_{\gamma,a}) - 1. \tag{8.1}$$

Our formula (2.3) for the height of $E_{\gamma,a}$ implies that $\operatorname{ord}_{\mathfrak{P}} H(E_{\gamma,a}) - 1 = (q^a - 1)/2$. To conclude the proof, it thus suffices to see that $\operatorname{ord}_{\mathfrak{P}} L(E_{\gamma,a}, q^{-1}) = -(q^a - 1)/2$. Indeed, one would then deduce from (8.1) that $\operatorname{ord}_{\mathfrak{P}} |\text{Ш}(E_{\gamma,a})| = 0$, and this would directly show that $\text{Ш}(E_{\gamma,a})$ has trivial p-primary part (by the structure theorem for finite abelian groups).

We now proceed to compute $\operatorname{ord}_{\mathfrak{P}} L(E_{\gamma,a}, q^{-1})$. Evaluating at $T = q^{-1}$ the expression for $L(E_{\gamma,a}, T)$ obtained in Theorem 4.1, and taking $\mathfrak{P}$-adic valuations on both sides of the resulting identity yields that

$$\operatorname{ord}_{\mathfrak{P}} L(E_{\gamma,a}, q^{-1}) = \sum_{v \in P_q(a)} \operatorname{ord}_{\mathfrak{P}}\left(1 - \frac{\mathbf{g}(v)\mathbf{kl}_\gamma(v)}{q^{\deg v}}\right) + \operatorname{ord}_{\mathfrak{P}}\left(1 - \frac{\mathbf{g}(v)\mathbf{kl}'_\gamma(v)}{q^{\deg v}}\right).$$

The results proved in §5.1 imply that, for all places $v \in P_q(a)$, we have

$$\left\{\operatorname{ord}_{\mathfrak{P}}\left(\mathbf{g}(v)\mathbf{kl}_\gamma(v)q^{-\deg v}\right), \operatorname{ord}_{\mathfrak{P}}\left(\mathbf{g}(v)\mathbf{kl}'_\gamma(v)q^{-\deg v}\right)\right\} = \left\{-\frac{\deg v}{2}, \frac{\deg v}{2}\right\}.$$

Using the cases of equality in the non-archimedean triangle inequality, we then obtain that

$$\operatorname{ord}_{\mathfrak{P}} L(E_{\gamma,a}, q^{-1}) = \sum_{v \in P_q(a)} \min\left\{0, -\frac{\deg v}{2}\right\} + \min\left\{0, \frac{\deg v}{2}\right\} = -\frac{1}{2} \sum_{v \in P_q(a)} \deg v.$$

As has already been observed, we have $\sum_{v\in P_q(a)} \deg v = |\mathbb{G}_m(\mathbb{F}_{q^a})| = q^a - 1$. We therefore conclude that $\mathrm{ord}_{\mathfrak{P}} L(E_{\gamma,a}, q^{-1}) = -(q^a-1)/2$, as was to be shown. □

REMARK 8.2. (1) In a very recent paper [**20**], Ulmer introduces the notion of *dimension of* Ш for abelian varieties over $\mathbb{F}_q(t)$ whose Tate–Shafarevich group is finite. In the special case of $E_{\gamma,a}$, this invariant is defined as follows: for all integers $n \geq 1$, let $K_n := \mathbb{F}_{q^n}(t)$ and consider

$$\dim Ш(E_{\gamma,a}) := \lim_{n\to\infty} \frac{\log\left|Ш(E_{\gamma,a}\times_K K_n/K_n)[p^\infty]\right|}{\log q^n}.$$

Proposition 4.1 of [**20**] shows that the limit exists and is a nonnegative integer, called the *dimension of* Ш *of* $E_{\gamma,a}$. Further, [**20**, Prop. 4.2] provides an expression for $\dim Ш(E_{\gamma,a})$ in terms of the p-adic slope sequence of the L-function of $E_{\gamma,a}$. Using that expression and our Theorem 5.2, an easy calculation yields that $\dim Ш(E_{\gamma,a}) = 0$.

(2) The previous item shows that the order of the p-primary part of the Tate–Shafarevich group of the base-changed elliptic curve $E_{\gamma,a}\times_K K_n$ over K_n grows slowly with n (its log is $o(n)$). Replacing q by q^n in the proof of Theorem 8.1, one actually proves the stronger assertion that, for all $n\geq 1$,

$$|Ш(E_{\gamma,a}\times_K K_n/K_n)[p^\infty]| = 1.$$

8.2. The size of $Ш(E_{\gamma,a})$. Finally, we prove assertion *(3)* of Theorem C about the size of $Ш(E_{\gamma,a})$. We actually show a slightly more precise result:

THEOREM 8.3. *Let $\mathbb{F}_q$ be a finite field of odd characteristic and $K = \mathbb{F}_q(t)$. For any $\gamma\in\mathbb{F}_q^\times$ and any integer $a\geq 1$, consider the elliptic curve $E_{\gamma,a}$ over K defined by* (2.1). *Then, as $a\to\infty$, we have*

$$|Ш(E_{\gamma,a})| = H(E_{\gamma,a})^{1+O_q(1/a)},$$

where the implicit constant is effective and depends at most on q.

Recall from (2.3) that $H(E_{\gamma,a}) = N(E_{\gamma,a})^{1/4}$. Hence, the above implies:

COROLLARY 8.4. *In the same setting, as $a\to\infty$, we have*

$$|Ш(E_{\gamma,a})| = N(E_{\gamma,a})^{1/4+O_q(1/a)}.$$

PROOF (OF THEOREM 8.3). We note that, by (2.3), $\log\log H(E_{\gamma,a})$ and a have the same order of magnitude when $a\to\infty$, the involved constants depending at most on q. The elliptic curve $E_{\gamma,a}$ satisfies the BSD conjecture (see Corollary 5.4): taking the logarithm of both sides of the BSD formula (5.2) and reordering terms yields that

$$\frac{\log|Ш(E_{\gamma,a})|}{\log H(E_{\gamma,a})} = 1 - \frac{\log q}{\log H(E_{\gamma,a})} + \frac{\log L(E_{\gamma,a}, q^{-1})}{\log H(E_{\gamma,a})}. \tag{8.2}$$

The term $\log q/\log H(E_{\gamma,a})$ is clearly $o\big(1/\log\log H(E_{\gamma,a})\big) = o(a^{-1})$ as $a\to\infty$. To control the right-most term, we use our bound on the central value $L(E_{\gamma,a}, q^{-1})$: we deduce from Theorem 7.1 that

$$\frac{\left|\log L(E_{\gamma,a}, q^{-1})\right|}{\log H(E_{\gamma,a})} = O\big(1/\log\log H(E_{\gamma,a})\big) = O(1/a),$$

as $a\to\infty$, where the implicit constant depends at most on q. Plugging these two estimates into (8.2) completes the proof. □

REMARK 8.5. Theorem 8.3 can be interpreted as an analogue of the Brauer–Siegel theorem for the sequences $\{E_{\gamma,a}\}_{a\geq 1}$. We refer the reader to [**6**, **7**] for a detailed description of the analogue we have in mind: let us simply recall that Hindry and Pacheco have introduced the *Brauer–Siegel ratio* of an abelian variety with finite Tate–Shafarevich group and that, in the case at hand, the Brauer–Siegel ratio is given by $\mathfrak{Bs}(E_{\gamma,a}) = \log|\text{Ш}(E_{\gamma,a})|/\log H(E_{\gamma,a})$. (For an elliptic curve with positive Mordell–Weil rank, the Brauer–Siegel ratio also includes the Néron–Tate regulator.) With this notation, Theorem 8.3 can be rewritten in a compact form:

$$\mathfrak{Bs}(E_{\gamma,a}) = 1 + O_q\big(1/a\big) \qquad (\text{as } a \to \infty). \tag{8.3}$$

There is only a handful of sequences of elliptic curves over K for which one can prove that the Brauer–Siegel ratio has a limit and that this limit is 1 (see [**2**, **3**, **6**, **7**] and references therein). The families studied in the present paper therefore provide further examples of that behaviour. However, it seems interesting to remark that the sequences $\{E_{\gamma,a}\}_{a\geq 1}$ consist of rank 0 elliptic curves, whereas previous articles considered sequences with unbounded rank.

Acknowledgments

This article is partly based on GdW's Leiden Master's thesis [**22**], written under the supervision of RG. The authors would like to thank Bas Edixhoven, Marc Hindry and Douglas Ulmer for fruitful conversations about this work and useful comments on earlier versions thereof.

GdW received funding from the ALGANT Master Programme during his studies at Università degli studi di Milano and Universiteit Leiden. Work on this project was started at Universiteit Leiden. RG now works at Universität Basel, with funding from the Swiss National Science Foundation (SNSF Professorship #170565 awarded to Pierre Le Boudec). These institutions are gratefully thanked for creating great working conditions, and for their financial support. RG also acknowledges funding from Agence Nationale de la Recherche (Grant ANR-17-CE40-0012 FLAIR).

References

[1] Dorian Goldfeld and Lucien Szpiro, *Bounds for the order of the Tate-Shafarevich group*, Compositio Math. **97** (1995), no. 1-2, 71–87. Special issue in honour of Frans Oort. MR1355118

[2] Richard Griffon. *Analogues du théorème de Brauer-Siegel pour quelques familles de courbes elliptiques.* PhD thesis, Université Paris Diderot, July 2016.

[3] Richard Griffon, *Bounds on special values of L-functions of elliptic curves in an Artin-Schreier family*, Eur. J. Math. **5** (2019), no. 2, 476–517, DOI 10.1007/s40879-018-0284-3. MR3946589

[4] Richard Griffon and Douglas Ulmer, *On the arithmetic of a family of twisted constant elliptic curves*, Pacific J. Math. **305** (2020), no. 2, 597–640, DOI 10.2140/pjm.2020.305.597. MR4099310

[5] Benedict H. Gross, *Lectures on the conjecture of Birch and Swinnerton-Dyer*, Arithmetic of L-functions, IAS/Park City Math. Ser., vol. 18, Amer. Math. Soc., Providence, RI, 2011, pp. 169–209, DOI 10.1090/pcms/018/08. MR2882691

[6] Marc Hindry and Amílcar Pacheco, *An analogue of the Brauer-Siegel theorem for abelian varieties in positive characteristic*, Mosc. Math. J. **16** (2016), no. 1, 45–93, DOI 10.17323/1609-4514-2016-16-1-45-93. MR3470576

[7] Marc Hindry, *Analogues of Brauer-Siegel theorem in arithmetic geometry*, Arithmetic geometry: computation and applications, Contemp. Math., vol. 722, Amer. Math. Soc., Providence, RI, 2019, pp. 19–41, DOI 10.1090/conm/722/14531. MR3896847

[8] Kazuya Kato and Fabien Trihan, *On the conjectures of Birch and Swinnerton-Dyer in characteristic* $p > 0$, Invent. Math. **153** (2003), no. 3, 537–592, DOI 10.1007/s00222-003-0299-2. MR2000469

[9] Rudolf Lidl and Harald Niederreiter, *Finite fields*, 2nd ed., Encyclopedia of Mathematics and its Applications, vol. 20, Cambridge University Press, Cambridge, 1997. With a foreword by P. M. Cohn. MR1429394

[10] Liem Mai and M. Ram Murty, *A note on quadratic twists of an elliptic curve*, Elliptic curves and related topics, CRM Proc. Lecture Notes, vol. 4, Amer. Math. Soc., Providence, RI, 1994, pp. 121–124, DOI 10.1090/crmp/004/09. MR1260959

[11] Maurice Mignotte and Michel Waldschmidt, *On algebraic numbers of small height: linear forms in one logarithm*, J. Number Theory **47** (1994), no. 1, 43–62, DOI 10.1006/jnth.1994.1025. MR1273455

[12] J. S. Milne, *On a conjecture of Artin and Tate*, Ann. of Math. (2) **102** (1975), no. 3, 517–533, DOI 10.2307/1971042. MR414558

[13] Rachel Pries and Douglas Ulmer, *Arithmetic of abelian varieties in Artin-Schreier extensions*, Trans. Amer. Math. Soc. **368** (2016), no. 12, 8553–8595, DOI 10.1090/tran6641. MR3551581

[14] Michael Rosen, *Number theory in function fields*, Graduate Texts in Mathematics, vol. 210, Springer-Verlag, New York, 2002, DOI 10.1007/978-1-4757-6046-0. MR1876657

[15] Matthias Schütt and Tetsuji Shioda, *Elliptic surfaces*, Algebraic geometry in East Asia—Seoul 2008, Adv. Stud. Pure Math., vol. 60, Math. Soc. Japan, Tokyo, 2010, pp. 51–160, DOI 10.2969/aspm/06010051. MR2732092

[16] Joseph H. Silverman, *Advanced topics in the arithmetic of elliptic curves*, Graduate Texts in Mathematics, vol. 151, Springer-Verlag, New York, 1994, DOI 10.1007/978-1-4612-0851-8. MR1312368

[17] Joseph H. Silverman, *The arithmetic of elliptic curves*, 2nd ed., Graduate Texts in Mathematics, vol. 106, Springer, Dordrecht, 2009, DOI 10.1007/978-0-387-09494-6. MR2514094

[18] John Tate, *On the conjectures of Birch and Swinnerton-Dyer and a geometric analog*, Séminaire Bourbaki, Vol. 9, Soc. Math. France, Paris, 1995, Exp. No. 306, pp. 415–440. MR1610977

[19] Douglas Ulmer, *Elliptic curves over function fields*, Arithmetic of L-functions, IAS/Park City Math. Ser., vol. 18, Amer. Math. Soc., Providence, RI, 2011, pp. 211–280, DOI 10.1090/pcms/018/09. MR2882692

[20] Douglas Ulmer, *On the Brauer-Siegel ratio for abelian varieties over function fields*, Algebra Number Theory **13** (2019), no. 5, 1069–1120, DOI 10.2140/ant.2019.13.1069. MR3981314

[21] Benjamin M. M. de Weger, *$A + B = C$ and big Sha's*, Quart. J. Math. Oxford Ser. (2) **49** (1998), no. 193, 105–128, DOI 10.1093/qjmath/49.193.105. MR1617347

[22] Guus de Wit. *Elliptic curves over function fields with large Tate–Shafarevich groups.* Master's thesis, Universiteit Leiden, July 2018. (available at `www.math.u-bordeaux.fr/~ybilu/algant/algant_theses.php`).

Richard Griffon, Departement Mathematik und Informatik, Universität Basel, Spiegelgasse 1, CH-4051 Basel, Switzerland

Email address: `richard.griffon@gmail.com`

Guus de Wit, Mathematisch Instituut, Universiteit Leiden, PO Box 9512, 2300RA Leiden, The Netherlands

Email address: `dewit.guus@gmail.com`

Contemporary Mathematics
Volume **770**, 2021
https://doi.org/10.1090/conm/770/15435

On Sato–Tate distributions, extremal traces, and real multiplication in genus 2

David Kohel and Yih-Dar Shieh

ABSTRACT. The vertical Sato–Tate conjectures gives expected trace distributions for for families of curves. We develop exact expression for the distribution associated to degree-4 representations of USp(4), SU(2) × SU(2) and SU(2) in the neighborhood of the extremities of the Weil bound. As a consequence we derive qualitative distinctions between the extremal traces arising from generic genus-2 curves and genus-2 curves with real or quaternionic multiplication. In particular we show, in a specific sense, to what extent curves with real multiplication dominate the contribution to extremal traces.

1. Introduction

Let $C/\mathbb{F}_q$ be a curve of genus g over the field $\mathbb{F}_q$ of q elements and let J be its Jacobian. The characteristic polynomial of the Frobenius endomorphism $\pi \in \mathrm{End}(J)$, is of the form

$$\chi_C(x) = x^{2g} - a_1 x^{2g-1} + a_2 x^{2g-2} + \cdots + a_2 q^{g-2} x^2 - a_1 q^{g-1} x + q^g.$$

We define the normalized Weil polynomial to be the characteristic polynomial of $\pi \otimes \sqrt{q}^{-1}$ in $\mathrm{End}(J) \otimes_{\mathbb{Z}} \mathbb{R}$:

$$\tilde{\chi}_C(x) = x^{2g} - \tilde{a}_1 x^{2g-1} + \tilde{a}_2 x^{2g-2} + \cdots + \tilde{a}_2 x^2 - \tilde{a}_1 x + 1,$$

with normalized roots $\tilde{\alpha}_1, \ldots, \tilde{\alpha}_g, \tilde{\alpha}_{g+1}, \ldots, \tilde{\alpha}_{2g}$ satisfying $\tilde{\alpha}_j \tilde{\alpha}_{j+g} = 1$ where $\tilde{\alpha}_j = e^{i\theta_j}$. We denote the real numbers $\tilde{\alpha}_j + \tilde{\alpha}_{j+g}$ by t_j, and define $(s_1, \ldots, s_g)$ to be the symmetric polynomials in $(t_1, \ldots, t_g)$. We call the polynomial

$$\prod_{i=1}^{g} (x - t_i) = x^g - s_1 x^{g-1} + s_2 x^{g-2} + \cdots + (-1)^g s_g$$

the normalized real Weil polynomial of $C/\mathbb{F}_q$, noting in particular that $s_1 = \tilde{a}_1$ and $s_2 = \tilde{a}_2 - g$.

The Sato–Tate conjecture for an non-CM elliptic curve $C/\mathbb{Q}$ concerns the equidistribution with respect to a Haar measure induced by the group SU(2) of the Frobenius angles θ_1 or normalized Frobenius traces t_1 as q varies over the reductions of C over finite fields $\mathbb{F}_q$, and has been generalized to curves of higher genus or abelian varieties of higher dimension. In higher degree, this is expressed as an equidistribution of the tuples of Frobenius angles $(\theta_1, \ldots, \theta_g)$, of real traces

2020 *Mathematics Subject Classification.* Primary 11F03, 11G20, 11G25, 14G10, 14D05.

$(t_1, \dots, t_g)$ or of the symmetric sums $(s_1, \dots, s_g)$ with respect to a induced Haar measure of an associated compact Lie group, the Sato–Tate group.

These generalizations are typically grouped together as horizontal Sato–Tate conjectures. Katz introduced a vertical Sato–Tate conjecture concerning the variation of a family $\mathcal{C}/S$ of curves, abelian varieties or exponential sums over a base scheme S. Specifically, the vertical Sato–Tate conjectures concerns the limit of the sets of Frobenius traces associated to the fibers point in $S(\mathbb{F}_q)$, as q goes to infinity. One can view the horizontal Sato–Tate conjecture as a statement that Frobenius distribution for a fiber over a point S follows the equidistribution of the vertical Sato–Tate distribution. Whereas the horizontal Sato–Tate conjectures are, in many instances, still open, the vertical Sato–Tate conjectures are more amenable to proof. In this work we assume the expected vertical Sato–Tate conjecture and as a consequence derive a qualitative distinction between extremal traces for quaternionic multiplication (QM), real multiplication (RM) and generic families, and when referring to the Sato–Tate group over a finite field we are implicitly considering the vertical aspect of these conjectures.

In what follows we consider the case of genus-2 curves, for which the generic Sato–Tate group is USp(4). When the Jacobian is split or admits RM over the base field, the Frobenius endomorphism commutes with this structure, and the Sato–Tate group is restricted to $\mathrm{SU}(2) \times \mathrm{SU}(2)$. Such families of curves arise when the base is a cover of a Humbert surface $\mathcal{H}_D$ over which the RM ring is defined. The Humbert surface is a moduli space, contained in the 3-fold $\mathcal{M}_2$ of moduli for genus-2 curves, classifying isomorphism classes of curves whose Jacobians admit endomorphisms by a real quadratic order of discriminant D. A special case is when D is a square, corresponding to orders in the real quadratic ring $\mathbb{Z}[x]/(x^2 - x) \cong \mathbb{Z} \times \mathbb{Z}$ of discriminant 1. The embedding of the Sato–Tate group $\mathrm{SU}(2) \times \mathrm{SU}(2)$ in USp(4) depends on the choice of Humbert surface (in particular on its discriminant), but the induced Haar measure on Frobenius traces remains invariant of this embedding. Over a finite field, every curve $C/\mathbb{F}_q$ lies in the image of some Humbert surface, and the objective of this work is to understand how this stratification of the space $\mathcal{M}_2$ by Humbert surfaces $\mathcal{H}_D$ lets us understand the distributions of normalized Frobenius traces s_1 near the extremities of the Weil interval $[-2g, 2g] = [-4, 4]$. To complete the picture of this stratification, we consider the Frobenius distributions of families with quaternionic multiplication (QM). These arise as families over Shimura curves X^N, admitting QM by an indefinite order in a quaternion algebra of discriminant N. Such curves appear as components of the intersections of Humbert surfaces, and give rise to the Sato–Tate group SU(2) embedded diagonally (up to conjugation) as a subgroup $\Delta \subset \mathrm{SU}(2) \times \mathrm{SU}(2)$ in a degree-4 representation.

The study of extremal traces (in particular curves with many points) has a long history, motivated by applications to coding theory and the rich mathematical structure going into their study. In the next section we introduce the notation for this study before turning to the Weyl integration formulas in Section 3, which gives an explicit form for the Haar measures induced by the symplectic groups SU(2) and USp(4). In the following Section 4, we develop precise Taylor series expressions for the Haar measure induced on the trace function s_1, and devote Section 5 to certain global expressions for the continuous distribution functions for USp(4) and $\mathrm{SU}(2) \times \mathrm{SU}(2)$ on s_1 obtained by Gilles Lachaud. As an application to the explicit Taylor series expansions, in Section 6, we give a qualitative comparison of the expected contribution of a generic family over $\mathcal{M}_2$, of an RM family over some Humbert surface $\mathcal{H}_D$, and of a QM family over a Shimura curve X^N. In particular we show that a Shimura curve X^N and associated Sato–Tate group Δ gives the greatest density of extremal traces of these groups, but that a Humbert surface of small discriminant dominates the contribution to extremal traces of genus-2 curves.

2. Background and notation

We consider Galois representations in USp(4) and its subgroups SU(2)×SU(2), and SU(2) arising from families of curves of genus 2. We assume a generic normalized Weil polynomial takes the form

$$\widetilde{\chi}(x) = x^4 - s_1x^3 + (s_2+2)x^2 - s_1x + 1,$$

in particular, s_1 represents the normalized trace of a general element. Under a splitting of the generic normalized Weil polynomial as

$$(x - e^{i\theta_1})(x - e^{-i\theta_1})(x - e^{i\theta_2})(x - e^{-i\theta_2}) = (x^2 - t_1x + 1)(x^2 - t_2x + 1),$$

we consider the transformation of Haar measures induced by the changes of variables between local parameters (θ_1, θ_2), (t_1, t_2), and (s_1, s_2) on the respective domains of support Θ_2, I_2 and Σ_2. In order to focus on the role of the discriminant of the normalized real polynomial:

$$(x - 2\cos(\theta_1))(x - 2\cos(\theta_2)) = (x - t_1)(x - t_2) = x^2 - s_1x + s_2,$$

we consider $D_0 = (t_1-t_2)^2 = s_1^2-4s_2$, and its square root $\delta_0 = t_2-t_1$. Similarly, for a root π of $\chi(x)$, with conjugate $\bar{\pi} = \pi^{-1}$, the relative ring extension $\mathbb{Z}[\pi,\bar{\pi}]/\mathbb{Z}[\pi+\bar{\pi}]$ has norm discriminant

$$D_1 = (4-t_1^2)(4-t_2^2) = (4+s_2)^2 - 4s_1^2 = (4-2s_1+s_2)(4+2s_1+s_2).$$

As in the above construction, we fix the relations such as $(s_1, s_2) = (t_1+t_2, t_1t_2)$, and identify D_0 and D_1 in $\mathbb{Z}[t_1,t_2]$ and $\mathbb{Z}[s_1,s_2]$, irrespective of the ring parameters over which they are defined.

The goal of this work is to investigate the asymptotic trace distributions in the neighborhood of $s_1 = -4$ (or by symmetry, in the neighborhood of $s_1 = 4$). In particular we investigate the interpretation of the Haar measure on conjugacy classes via a stratification of the moduli space $\mathcal{M}_2$, with generic Sato–Tate group USp(4), by the Humbert surfaces $\mathcal{H}_D$, with generic Sato–Tate SU(2) × SU(2), and the contributions from Shimura curves, whose associated Sato–Tate group is a diagonal image of SU(2). We recall that the embeddings of these groups determined by the particular Humbert surface or Shimura curve varies by a conjugation, but the traces and integration formulas remain invariant.

3. Weyl integration formula

The Weyl integration formula gives an integral expression for the Haar measure on the space of Frobenius angles $(\theta_1,\dots,\theta_g) \in \Theta_g = [0,\pi]^g$ or of the space of real traces:

$$(t_1,\dots,t_g) = (2\cos(\theta_1),\dots,2\cos(\theta_g)) \in I_g = [-2,2]^g.$$

In particular we focus on $g = 2$ and the distributions induced by subgroups of USp(4). We recall the form of the Weyl integration formula for the unitary symplectic group from Weyl [**6**, Theorem 7.8B] (see Katz and Sarnak [**2**, §5.0.4]).

THEOREM 1 (Weyl). The Haar measure induced by $\mathrm{USp}(2g)$ on the angle space Θ_g is given by the formula

$$\mu_G(\theta_1,\ldots,\theta_g) = \frac{2^{g^2}}{g!\pi^g}\prod_{i<j}\big(\cos(\theta_i)-\cos(\theta_j)\big)^2\prod_{i=1}^{g}\sin^2(\theta_i)\,d\theta_1\cdots d\theta_g.$$

Degree 2. As a corollary, we specialize to the case $g = 1$, where we find the classic distributions for the degree-2 representations of $\mathrm{SU}(2) = \mathrm{USp}(2)$ arising in the Sato–Tate conjecture for elliptic curves. To simplify notation we note $K = \mathrm{SU}(2)$ and use the same notation μ_K for the measure on the spaces $\Theta = [0,\pi]$ and $I = [-2,2]$, distinguishing the domain by the variable name.

COROLLARY 2. The Haar measure induced by $K = \mathrm{SU}(2)$ on the angle space $\Theta = [0,\pi]$ is given by the formula

$$\mu_K(\theta) = \frac{2}{\pi}\sin^2(\theta)d\theta,$$

and, in terms of the trace $t = 2\cos(\theta)$ in $I = [-2,2]$, the Haar measure takes the form

$$\mu_K(t) = \frac{1}{2\pi}\sqrt{4-t^2}\,dt.$$

We define the associated probability density functions on Θ and I by

$$f_K(\theta) = \frac{\mu_K(\theta)}{d\theta} = \frac{2}{\pi}\sin^2(\theta) \text{ and } f_K(t) = \frac{\mu_K(t)}{dt} = \frac{1}{2\pi}\sqrt{4-t^2}.$$

giving well-known density functions of the Sato–Tate conjectures:

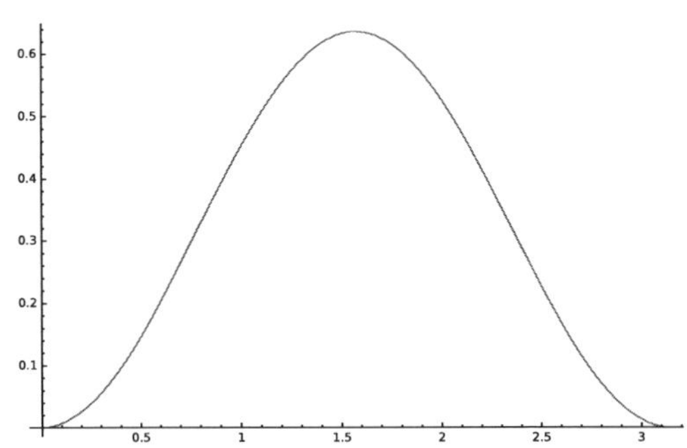

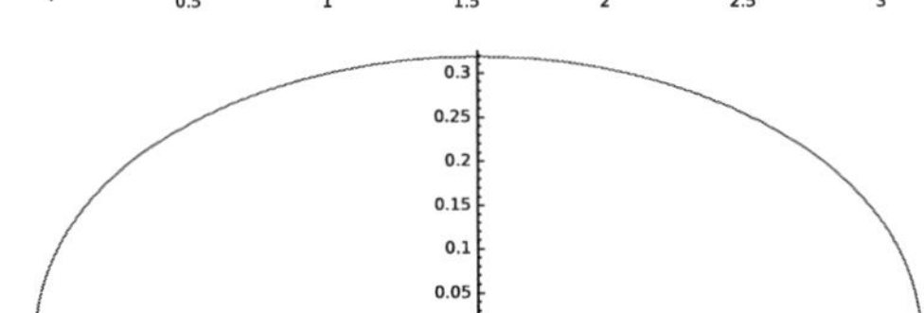

and associated cumulative density functions

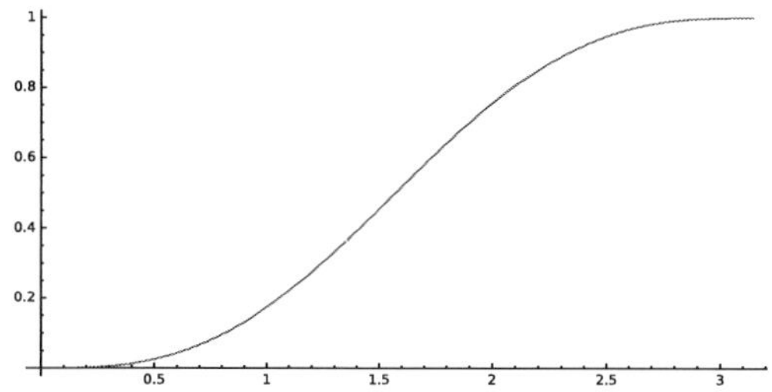

$$F_K(\theta) = \int_0^\theta \mu_K(\theta)$$

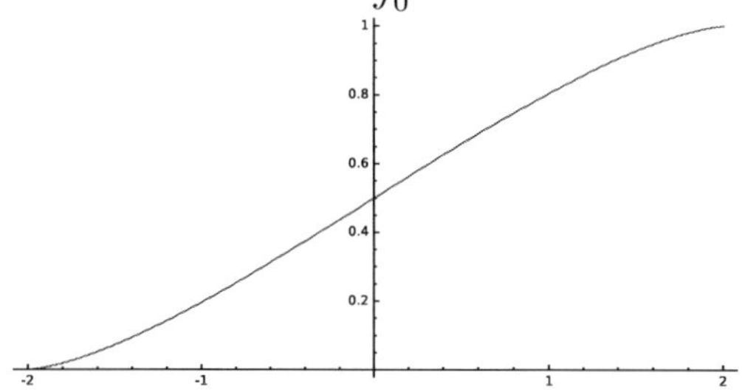

$$F_K(t) = \int_{-2}^t \mu_K(t)$$

In particular, the latter distribution, on the trace space I, the cumulative distribution function measures the contribution of extremal traces in a neighborhood of $t = -2$, corresponding to elliptic curves with many points.

Degree 4. From the perspective of Sato–Tate distributions of genus-2 curves, we are interested in the degree-4 representations of USp(4), of a subgroup SU(2) $\times$ SU(2), and of the image of the diagonal map $\Delta : \mathrm{SU}(2) \to \mathrm{SU}(2) \times \mathrm{SU}(2)$.

COROLLARY 3. The Haar measure induced by $G = \mathrm{USp}(4)$ on the angle space Θ_2 is given by the formula

$$\mu_G(\theta_1, \theta_2) = \frac{8}{\pi^2} (\cos(\theta_1) - \cos(\theta_2))^2 \sin^2(\theta_1) \sin^2(\theta_2) d\theta_1 d\theta_2.$$

The Haar measure for SU(2) $\times$ SU(2) is obtained as the product measure for the case $g = 1$.

COROLLARY 4. The Haar measure induced by $H = \mathrm{SU}(2) \times \mathrm{SU}(2)$ on the angle space Θ_2 is given by the formula

$$\mu_H(\theta_1, \theta_2) = \frac{4}{\pi^2} \sin^2(\theta_1) \sin^2(\theta_2) d\theta_1 d\theta_2.$$

In what follows we continue to use the notation G for USp(4), H for SU(2) $\times$ SU(2), and write Δ for the diagonal image of SU(2) in H. Making the change of variables $t_j = 2\cos\theta_j$, such that $\sin(\theta_1)\sin(\theta_2)d\theta_1 d\theta_2 = dt_1 dt_2$, we obtain the expression in (t_1, t_2) for the Haar measure induced by G and H:

$$\mu_G(t_1, t_2) = \frac{1}{8\pi^2}(t_1 - t_2)^2 \sqrt{(4 - t_1^2)(4 - t_2^2)}\, dt_1 dt_2 = \frac{1}{8\pi^2} D_0 \sqrt{D_1}\, dt_1 dt_2,$$

$$\mu_H(t_1, t_2) = \frac{1}{4\pi^2} \sqrt{(4 - t_1^2)(4 - t_2^2)}\, dt_1 dt_2 = \frac{1}{4\pi^2} \sqrt{D_1}\, dt_1 dt_2.$$

where $D_0 = (t_1 - t_2)^2$ and $D_1 = (4 - t_1^2)(4 - t_2^2)$. These groups of rank 2 have measure supported on the whole domain I_2. The measure for Δ is supported on the closed domain $t_1 = t_2$ (or $\theta_1 = \theta_2$ in Θ_2). We defer the discussion of the trace distribution for Δ until after the introduction of suitable transformations of the domain.

Domains of integration. In order to focus on the behavior of the normalized trace function $s_1 = t_1 + t_2$ of the representation in USp(4), and relate the measure to the coefficients of the normalized Weil polynomial, we first consider the transformation $I_2 \to \Sigma_2$, where

$$\Sigma_2 = \{(s_1, s_2) \in \mathbb{R}^2 \mid 2|s_1| \le s_2 + 4, 4s_2 \le s_1^2\},$$

sending (t_1, t_2) to $(s_1, s_2) = (t_1 + t_2, t_1 t_2)$.

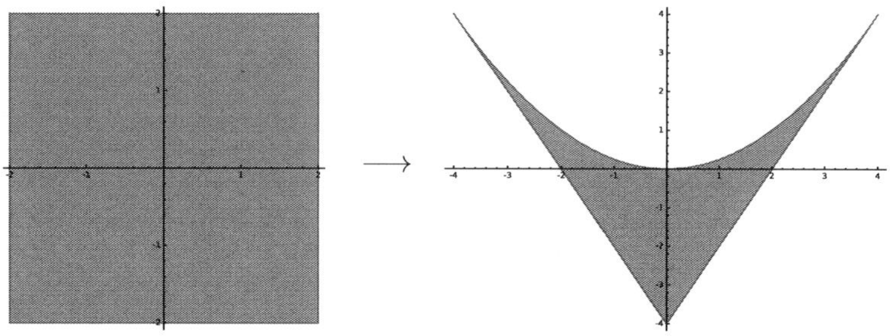

As for the discriminants D_i, we continue to use the same names for the measures μ_G and μ_H on the spaces Θ_2, I_2 and Σ_2, distinguishing the domain by the variables names. In particular, we recall the previously determined expressions for the discriminants

$$D_0 = s_1^2 - 4s_2 \text{ and } D_1 = (4 + s_2)^2 - 4s_1^2$$

and, in view of the equivalence of alternating volume forms,

$$ds_1 ds_2 = (dt_1 + dt_2)(t_1 dt_2 + t_2 dt_1) = (t_1 - t_2) dt_1 dt_2 = \sqrt{D_0}\, dt_1 dt_2,$$

and a factor of 2 from the double cover $I_2 \to \Sigma_2$, the induced Haar measure becomes

$$\mu_G(s_1, s_2) = \frac{1}{4\pi^2}\sqrt{D_0 D_1}\, ds_1 ds_2, \text{ and } \mu_H(s_1, s_2) = \frac{1}{2\pi^2}\sqrt{\frac{D_1}{D_0}}\, ds_1 ds_2.$$

In order to identify the role of the normalized discriminant D_0 of the real subring $\mathbb{Z}[\pi + \bar{\pi}]$ and with the view of obtaining a simple domain of integration for integrating over the fibers above s_1, we set $\delta_0^2 = D_0 = s_1^2 - 4s_2$ and apply the transformation $(s_1, s_2) \mapsto (s_1, \delta_0)$ from Σ_2 to the domain

$$\mathfrak{D}_2 = \{(s_1, \delta_0) \in [-4, 4] \times [0, 4] \mid \delta_0 \pm s_1 \le 4\}.$$

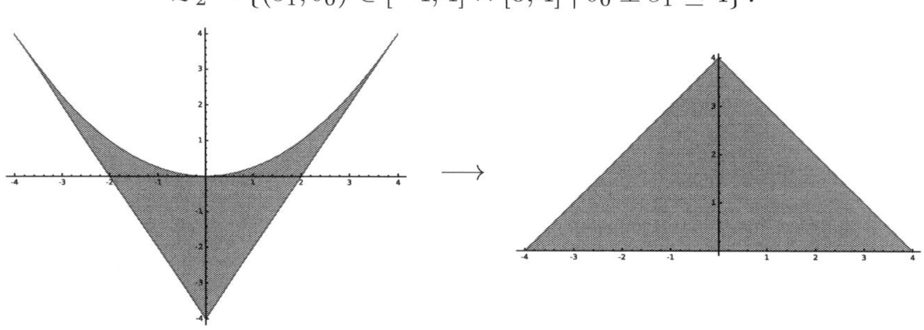

By taking the positive branch of $\delta_0 = \sqrt{D_0} \geq 0$, this map is an isomorphism. Moreover, from $s_1^2 - 4s_2 = \delta_0^2$, the volume forms satisfy

$$ds_1 ds_2 = \frac{1}{2}\delta_0 \, d\delta_0 ds_1,$$

from which we write

$$\mu_G(s_1, \delta_0) = \frac{1}{8\pi^2}\delta_0^2\sqrt{D_1}\, d\delta_0 ds_1, \text{ and } \mu_H(s_1, \delta_0) = \frac{1}{4\pi^2}\sqrt{D_1}\, d\delta_0 ds_1,$$

noting that D_1 takes the form $D_1 = \dfrac{((4+s_1)^2 - \delta_0^2)((4-s_1)^2 - \delta_0^2)}{16}$.

REMARK. The formal agreement of the expressions for μ_G on I_2 and $\mathfrak{D}_2$,

$$\mu_G(t_1, t_2) = \frac{1}{8\pi^2}D_0\sqrt{D_1}\, dt_1 dt_2 \text{ and } \mu_G(s_1, \delta_0) = \frac{1}{8\pi^2}D_0\sqrt{D_1}\, d\delta_0 ds_1,$$

and similarly for μ_H,

$$\mu_H(t_1, t_2) = \frac{1}{4\pi^2}\sqrt{D_1}\, dt_1 dt_2 \text{ and } \mu_H(s_1, \delta_0) = \frac{1}{4\pi^2}\sqrt{D_1}\, d\delta_0 ds_1,$$

reflects the fact that $(s_1, \delta_0) = (t_1 + t_2, t_1 - t_2)$ is a transformation of determinant 2, so $d\delta_0 ds_1 = 2dt_1 dt_2$, while the domain of integration $\mathfrak{D}_2$ consists of points (s_1, δ_0) in the upper half of the image of points $(t_1, t_2) \in I_2$, such that $\delta_0 = t_1 - t_2 \geq 0$.

In contrast to the above rank-2 groups, the group Δ is a rank-1 group, for which the support of the Haar measure is restricted to the 1-dimensional subspace $\delta_0 = t_1 - t_2 = 0$.

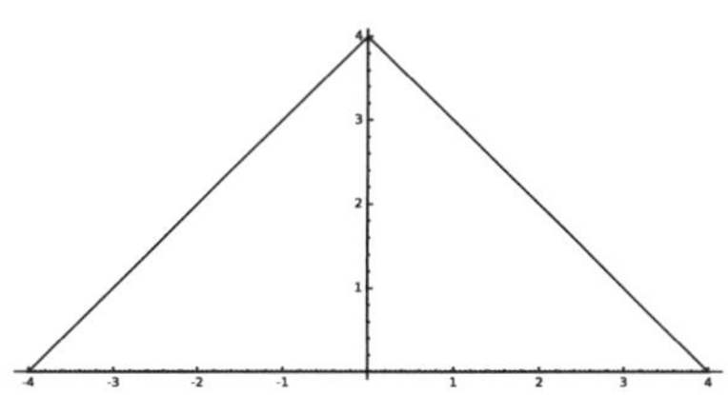

In fact, pulling back the measure for SU(2) by $s_1 = t_1 + t_2 = 2t$ in Corollary 2, we obtain the induced Haar measure in terms of the trace.

COROLLARY 5. *The Haar measure induced by Δ on the trace $s_1 \in [-4, 4]$ is given by*

$$\mu_\Delta(s_1) = \frac{1}{8\pi}\sqrt{16 - s_1^2}\, ds_1.$$

This measure can be viewed as a measure along the subdomain $[-4, 4] \times \{0\} \subset \mathfrak{D}_2$, or as a product distribution on $\mathfrak{D}_2$ such that the measure in δ_0 is a Dirac delta function with density 1 on $\delta_0 = 0$.

In what follows we seek to determine analogous expressions for the induced measure on the trace function s_1 in $[-4, 4]$, coming from the rank-2 groups G and H by integrating in δ_0 along the fibers about s_1. In the next section, we carry this out to develop exact series approximations for the measures in the neighborhood of $s_1 = -4$.

4. Taylor expansions for the trace function

We now introduce a transformation which permits us to develop a series expansion for the Haar measure of the trace s_1 in the neighborhood of the endpoint $s_1 = -4$. We first define the space $\Lambda_2 = [0,4] \times [0,1]$ of points (ε, λ) and a map $\Lambda_2 \to \mathfrak{D}_2$ given by

$$(\varepsilon, \lambda) \longmapsto (\varepsilon - 4, \varepsilon\lambda).$$

This gives a parametrization of the left-half subspace of $\mathfrak{D}_2$.

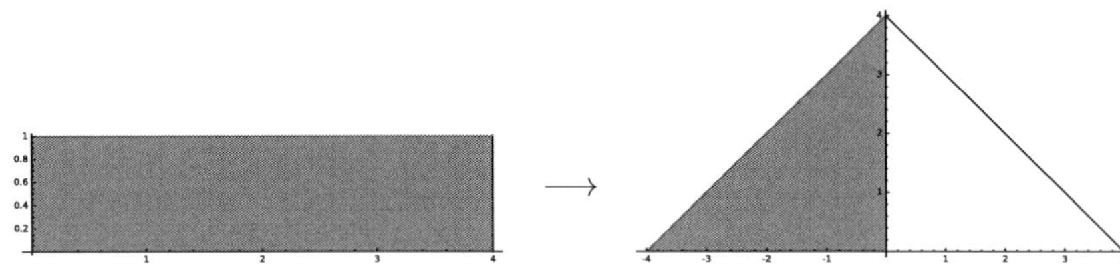

On the space Λ_2, we find the following expression for D_1:

$$D_1 = \frac{(\varepsilon^2 - \delta_0^2)((8-\varepsilon)^2 - \delta_0^2)}{16} = \frac{\varepsilon^2(1-\lambda^2)((8-\varepsilon)^2 - \varepsilon^2\lambda^2)}{16},$$

and setting $\rho = \varepsilon/(8-\varepsilon)$ gives

$$\sqrt{D_1} = \frac{\varepsilon(8-\varepsilon)}{4}\sqrt{(1-\lambda^2)(1-\rho^2\lambda^2)},$$

Using $(s_1, \delta) = (\varepsilon - 4, \varepsilon\lambda)$, it follows that $d\delta_0 ds_1 = (\varepsilon d\lambda + \lambda d\varepsilon)d\varepsilon = \varepsilon d\lambda d\varepsilon$, and we find

$$\mu_G(\varepsilon,\lambda) = \frac{\varepsilon^4(8-\varepsilon)}{32\pi^2}\left(\lambda^2\sqrt{(1-\lambda^2)(1-\rho^2\lambda^2)}\,d\lambda\right)d\varepsilon, \tag{1}$$

and

$$\mu_H(\varepsilon,\lambda) = \frac{\varepsilon^2(8-\varepsilon)}{16\pi^2}\left(\sqrt{(1-\lambda^2)(1-\rho^2\lambda^2)}\,d\lambda\right)d\varepsilon. \tag{2}$$

In a neighborhood of $\varepsilon = 0$, we have $\rho = \varepsilon/8 + O(\varepsilon^2)$, and therefore $1 - \rho^2\lambda^2 = 1 + O(\varepsilon^2)$, giving

(3)

$$\mu_G(\varepsilon,\lambda) \approx \frac{\varepsilon^4(8-\varepsilon)}{32\pi^2}\left(\lambda^2\sqrt{1-\lambda^2}\,d\lambda\right)d\varepsilon \text{ and } \mu_H(\varepsilon,\lambda) \approx \frac{\varepsilon^2(8-\varepsilon)}{16\pi^2}\left(\sqrt{1-\lambda^2}\,d\lambda\right)d\varepsilon.$$

From the values of the inner integrals,

$$\int_0^1 \lambda^2\sqrt{1-\lambda^2}\,d\lambda = \frac{\pi}{16} \quad \text{and} \quad \int_0^1 \sqrt{1-\lambda^2}\,d\lambda = \frac{\pi}{4},$$

we obtain the trace distributions near $\varepsilon = 0$.

THEOREM 6. *The Haar measures associated to the trace function on the groups $G = \mathrm{USp}(4)$ and $H = \mathrm{SU}(2) \times \mathrm{SU}(2)$ have the following approximations at $\varepsilon = s_1 + 4 = 0$.*

$$\mu_G(\varepsilon) = \left(\frac{\varepsilon^4(8-\varepsilon)}{512\pi} + O(\varepsilon^6)\right)d\varepsilon = \left(\frac{\varepsilon^4}{64\pi} - \frac{\varepsilon^5}{512\pi} + O(\varepsilon^6)\right)d\varepsilon \tag{4}$$

and

$$\mu_H(\varepsilon) = \left(\frac{\varepsilon^2(8-\varepsilon)}{64\pi} + O(\varepsilon^4)\right)d\varepsilon = \left(\frac{\varepsilon^2}{8\pi} - \frac{\varepsilon^3}{64\pi} + O(\varepsilon^4)\right)d\varepsilon. \tag{5}$$

In order to simplify compute the full Taylor expansions, we begin by recalling the definition of the Catalan numbers

$$C_n = \frac{1}{n+1}\binom{2n}{n} = \binom{2n}{n} - \binom{2n}{n+1}.$$

noting that in particular $C_0 = C_1 = 1$. With this definition we can state a lemma concerning the form of a class of integrals needed for our series approximation.

LEMMA 7.

$$\int_0^1 \lambda^{2n}\sqrt{1-\lambda^2}\,d\lambda = \frac{\pi}{4}\cdot\frac{C_n}{4^n}.$$

We finally recall the form of a Taylor series for the square root.

LEMMA 8. *The Taylor expansion for* $\sqrt{1-x}$ *is given by the power series*

$$\sqrt{1-x} = 1 - \frac{x}{2}\sum_{n=0}^{\infty}\frac{C_n}{4^n}x^n.$$

Applying Lemma 8 to $\sqrt{1-\rho^2\lambda^2}$, followed by Lemma 7 to the resulting integral summands, we obtain exact Taylor series expansions for the integrals

$$\begin{aligned}(6)\qquad &\int_0^1 \lambda^2\sqrt{(1-\lambda^2)(1-\rho^2\lambda^2)}\,d\lambda \\ &= \int_0^1 \lambda^2\sqrt{(1-\lambda^2)}\,d\lambda \;-\; \frac{\rho^2}{2}\sum_{n=0}^{\infty}\frac{C_n}{4^n}\left(\int_0^1 \lambda^{2(n+2)}\sqrt{1-\lambda^2}\,d\lambda\right)\rho^{2n} \\ &= \frac{\pi}{16}\left(1-\frac{\rho^2}{8}\sum_{n=0}^{\infty}\frac{C_nC_{n+2}}{16^n}\rho^{2n}\right),\end{aligned}$$

and

$$\begin{aligned}(7)\qquad &\int_0^1 \sqrt{(1-\lambda^2)(1-\rho^2\lambda^2)}\,d\lambda \\ &= \int_0^1 \sqrt{(1-\lambda^2)}\,d\lambda \;-\; \frac{\rho^2}{2}\sum_{n=0}^{\infty}\frac{C_n}{4^n}\left(\int_0^1 \lambda^{2(n+1)}\sqrt{1-\lambda^2}\,d\lambda\right)\rho^{2n} \\ &= \frac{\pi}{4}\left(1-\frac{\rho^2}{8}\sum_{n=0}^{\infty}\frac{C_nC_{n+1}}{16^n}\rho^{2n}\right).\end{aligned}$$

Substituting equations (6) and (7) into equations (1) and (2) for the Haar measures arising from $G = \mathrm{USp}(4)$ and $H = \mathrm{SU}(2)\times\mathrm{SU}(2)$, gives the following theorem.

THEOREM 9. *With the notation as above, the Taylor expansions for the trace on* G *and* H, *in the neighborhood of* $\varepsilon = s_1 + 4 = 0$ *are given by*

$$(8)\qquad \mu_G(\varepsilon) = \int_{\lambda=0}^1 \mu_G(\varepsilon,\lambda) = \frac{\varepsilon^4(8-\varepsilon)}{512\pi}\left(1-\frac{\rho^2}{8}\sum_{n=0}^{\infty}\frac{C_nC_{n+2}}{16^n}\rho^{2n}\right)d\varepsilon,$$

and

$$(9)\qquad \mu_H(\varepsilon) = \int_0^1 \mu_H(\varepsilon,\lambda) = \frac{\varepsilon^2(8-\varepsilon)}{64\pi}\left(1-\frac{\rho^2}{8}\sum_{n=0}^{\infty}\frac{C_nC_{n+1}}{16^n}\rho^{2n}\right)d\varepsilon.$$

REMARK. As a consequence of Stirling's formula, the Catalan numbers are known to satisfy the asymptotic growth

$$\frac{C_n}{4^n} \sim \frac{1}{\sqrt{\pi}\, n^{3/2}},$$

from which we can conclude the convergence of the above formulas for ε in the interval $[0, 4]$, since $0 \le \rho = \varepsilon/(8-\varepsilon) \le 1$. This implies convergence for the trace s_1 in the interval $[-4, 0]$, and by symmetry around $s_1 = 0$, extending to the interval $[-4, 4]$.

We conclude this section with the analogous series expansion for the degree-4 diagonal subgroup Δ.

THEOREM 10. The Haar measure induced by Δ in the neighborhood of $\varepsilon = s_1 + 4 = 0$ is given by

$$\mu_\Delta(\varepsilon) = \frac{1}{\sqrt{8}\,\pi}\sqrt{\varepsilon(1-\varepsilon/8)} = \frac{\varepsilon^{1/2}}{\sqrt{8}\,\pi}\left(1 - \frac{\varepsilon}{16}\sum_{n=0}^{\infty}\frac{C_n}{4^n}\left(\frac{\varepsilon}{8}\right)^n\right).$$

PROOF. Substituting $\varepsilon = s_1 + 4$ in Corollary 5 gives

$$\mu_\Delta(\varepsilon) = \frac{1}{8\pi}\sqrt{\varepsilon(8-\varepsilon)}\, ds_1 = \frac{\varepsilon^{1/2}}{\sqrt{8}\pi}\sqrt{1-\varepsilon/8}\, ds_1.$$

Applying Lemma 8, we obtain the desired expansion. □

By the previous remark, this gives convergence for ε in $[0, 8]$ and thus s_1 in $[-4, 4]$.

We recall that the density function of the trace function on Δ is the function $f_\Delta(x)$ such that $\mu_\Delta(s_1) = f_\Delta(s_1)\, ds_1$. Either from the exact form of Corollary 5,

$$f_\Delta(x) = \frac{1}{8\pi}\sqrt{16 - x^2},$$

or the above series expansion, we can graph the density function $f_\Delta(x)$:

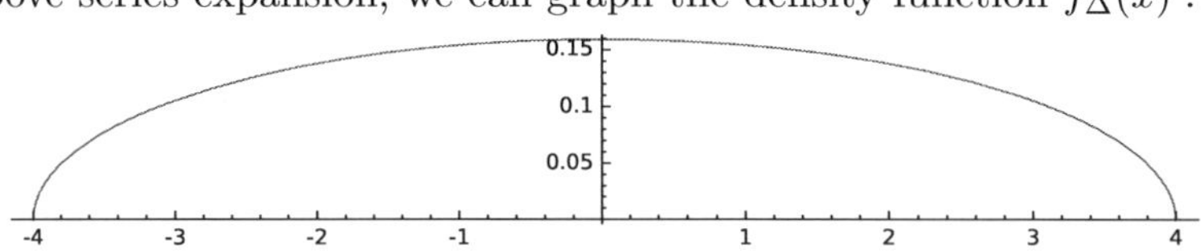

and its cumulative density function $F_\Delta(x) = \int_0^x f_\Delta(t)\, dt$:

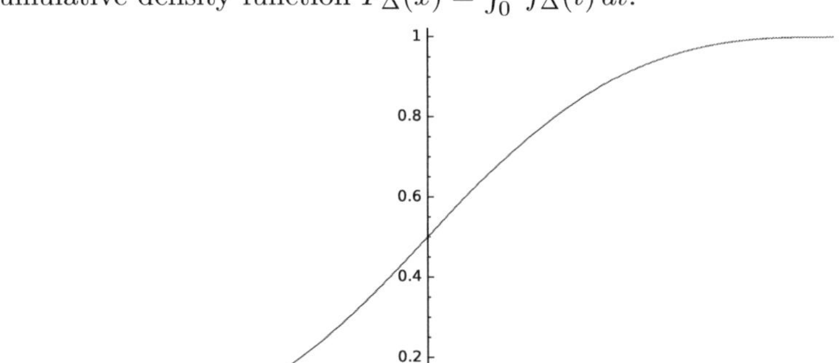

In the next section we recall results of Gilles Lachaud giving exact functional expressions for the density functions of the trace on the groups G and H.

5. On results of Lachaud

In this section we detail results of Gilles Lachaud concerning the global form of the trace function for the groups G and H, growing out of his interest in the theory of compact Lie groups and discussions with the authors during the joint supervision of the Ph.D. project of the second author. His work gives an alternative approach to develop the local expansions for the trace in the vicinity of $s_1 = \pm 4$.

We highlight certain main results in the presentation [**3**] and preprint [**4**] of Lachaud. Following the notation of this work, we denote by $f_G(x)$ and $f_H(x)$ the distribution functions for the trace on G and H such that

$$\mu_G(s_1) = f_G(s_1)ds_1 \text{ and } \mu_H(s_1) = f_H(s_1)ds_1.$$

The first main result is an exact form for the distribution function for $G = \mathrm{USp}(4)$.

THEOREM 11 (Lachaud [**3**]). *For $G = \mathrm{USp}(4)$, if $|x| < 4$, the distribution function f_G of the trace function is given by*

$$f_G(x) = -\frac{64}{15\pi}\sqrt{|x|}\left(1 - \frac{x^2}{16}\right)^2 \mathcal{P}^2_{\frac{1}{2}}\left(\frac{x^2+16}{4x}\right)$$

where

$$\mathcal{P}^a_b(z) = \frac{\Gamma(a+b+1)}{2\pi\Gamma(b+1)}\int_0^{2\pi}\left(z + \sqrt{z^2-1}\cos(\varphi)\right)^b \cos(a\varphi)d\varphi.$$

REMARK. In particular for $z = (x^2+16)/4x$ and $(a,b) = (2,1/2)$, we have

$$\mathcal{P}^2_{\frac{1}{2}}(z) = \frac{15}{8\pi}\int_0^{2\pi}\left(\frac{x^2+16+\sqrt{(x^2+16)^2-16x^2}\cos(\varphi)}{4x}\right)^{1/2}\cos(2\varphi)d\varphi$$

As a corollary, Lachaud finds an alternative expression for the trace distribution f_G in terms of the elliptic integrals of the first kind

$$K(m) = \int_0^{\pi/2}\frac{d\varphi}{\sqrt{1-m\sin^2(\varphi)}} = \int_0^1 \frac{du}{\sqrt{(1-u^2)(1-mu^2)}} = \frac{\pi}{2}\,{}_2F_1\left(\frac{1}{2},\frac{1}{2};1;m\right),$$

and of the second kind:

$$E(m) = \int_0^{\pi/2}\sqrt{1-m\sin^2(\varphi)}\,d\varphi = \int_0^1\sqrt{\frac{1-mu^2}{1-u^2}}\,du = \frac{\pi}{2}\,{}_2F_1\left(\frac{1}{2},-\frac{1}{2};1;m\right).$$

COROLLARY 12 (Lachaud [**3**]). *For $G = \mathrm{USp}(4)$, the distribution function $f_G : [-4,4] \to \mathbb{R}$ of the trace function is given by*

$$f_G(x) = \frac{64}{15\pi}\big((m^2-16m+16)E(m) - 8(m^2-3m+2)K(m)\big),$$

where $m = 1 - x^2/16$.

It should be noted that the functions $K(m)$ and $E(m)$ admit well-known power series representations in m:

$$K(m) = \frac{\pi}{2}\sum_{n=0}^{\infty}\left(\frac{(2n)!}{4^n(n!)^2}\right)^2 m^n \text{ and } E(m) = \frac{\pi}{2}\sum_{n=0}^{\infty}\left(\frac{(2n)!}{4^n(n!)^2}\right)^2 \frac{m^n}{1-2n}$$

which permits one to compute a power series representation in $m = 1-x^2/16$, from which one can derive series expansions around $x = \pm 4$. From the global form for $f_G(x)$, one can compute a graphic representation for the density function:

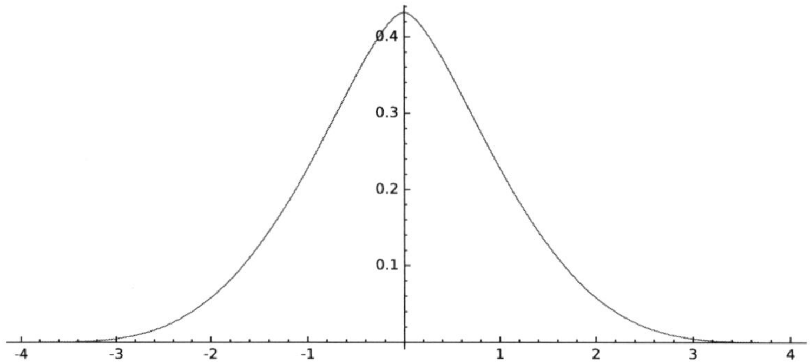

and its cumulative density function $F_G(x) = \int_0^x f_G(x)dx$:

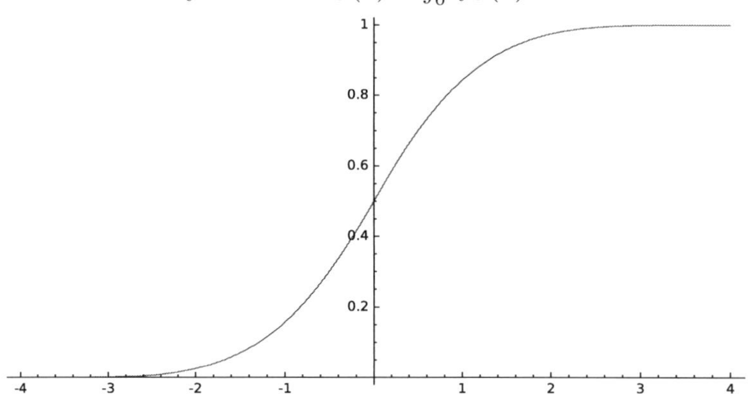

THEOREM 13 (Lachaud [**4**]). *For $H = \mathrm{SU}(2) \times \mathrm{SU}(2)$, the density function $f_H : [-4, 4] \to \mathbb{R}$ of the trace function is given by*

$$f_H(x) = \frac{m^2}{2\pi}\,{}_2F_1\left(\frac{1}{2}, \frac{3}{2}; 3; m\right),$$

where ${}_2F_1(a, b; c; z)$ is the hypergeometric function and $m = 1 - x^2/16$.

From the explicit formula for the $f_H(x)$, we can compute its graphical representation:

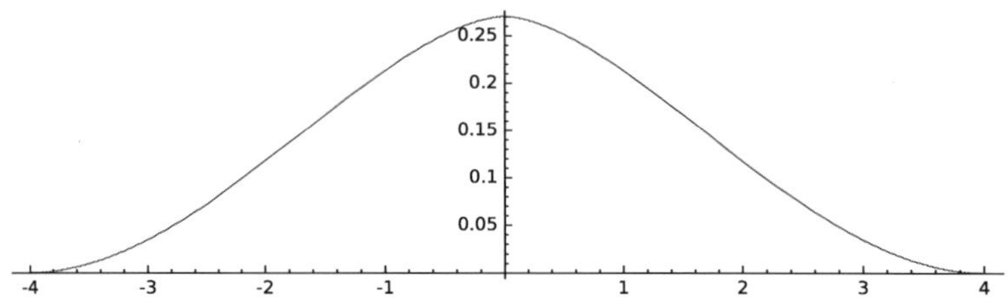

and its cumulative density function $F_G(x) = \int_0^x f_H(x)dx$:

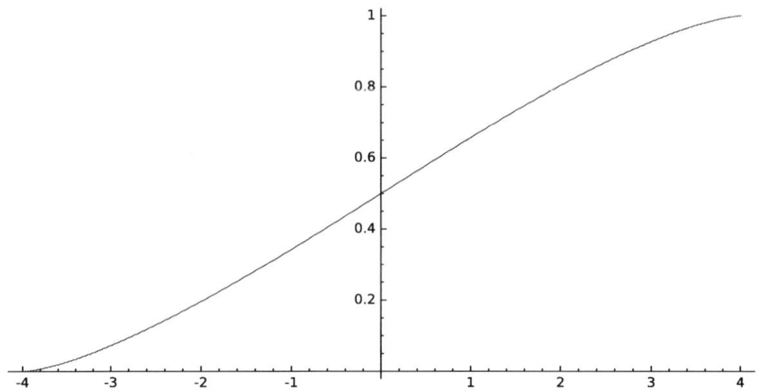

These results of Gilles Lachaud, in particular Theorems 11 and 13, give global expressions for the local power series expressions for $\mu_G = f_G ds_1$ and $\mu_H = f_H ds_1$ in Theorem 9, in terms of classical special functions. Both approaches give exact convergent expressions for the density functions f_G and f_H and cumulative density functions F_G and F_H for s_1 in the interval $[-4,4]$, or equivalently ε in $[0,8]$. In the next section we describe the application to a problem of extremal trace distributions in families which in part motived each of Gilles and the authors to carry out these respective computations of the trace distributions.

6. Distribution of extremal traces in families

Let $\mathcal{M}_2$ be the moduli space of genus-2 curves, $\mathcal{H}_D$ the Humbert surface of genus-2 curves with RM by a real quadratic order of discriminant D and $\mathcal{X}^N$ a Shimura curve of genus-2 curves with QM by an order of discriminant N in which D is inert or ramified. Achter and Howe [**1**] determine bounds on the density of split abelian surfaces among all principally polarized abelian surfaces over a finite field of q elements. Such split surfaces are accounted for by points on Humbert surfaces with D a square. We are interested in the asymptotic contribution of RM and QM points among abelian surfaces, such that that the trace of Frobenius is close to the ends of the Weil interval $[-2\sqrt{q}, 2\sqrt{q}]$. More generally, associated to the inclusions of moduli spaces $\mathcal{X}^N \hookrightarrow \mathcal{H}_D \hookrightarrow \mathcal{M}_2$ we pose the question of the asymptotic contribution to extremal traces of each moduli space and the relative densities.

The respective vertical Sato–Tate groups associated with generic endomorphism ring is $G = \mathrm{USp}(4)$, for real multiplication is $H = \mathrm{SU}(2) \times \mathrm{SU}(2)$ and for quaternionic multiplication is $\Delta \cong \mathrm{SU}(2) \subset H$. A family of genus-2 curves with parametrized RM or QM endomorphism ring structure, whose base les over a Humbert surface or Shimura curve, will have trace distribution characteristic of H or Δ. As the respective trace distribution show (see Corollary 5, Theorems 9, 11 and 13), such families tend to have a higher density of extremal traces.

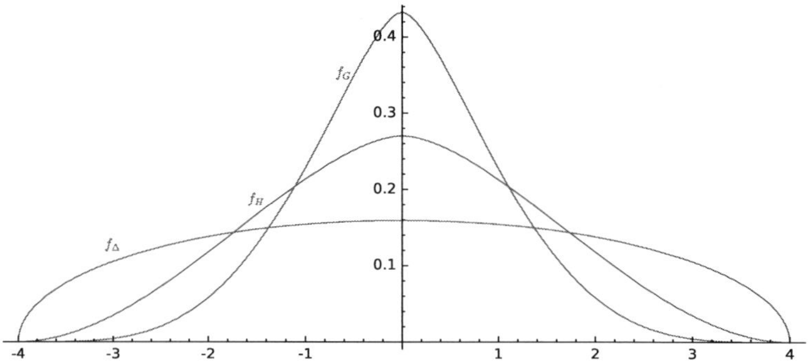

As a consequence, the cumulative density function for the trace of Frobenius in the neighborhood of $-4\sqrt{q}$, grows faster for Δ than for H and faster for H than for the generic group G.

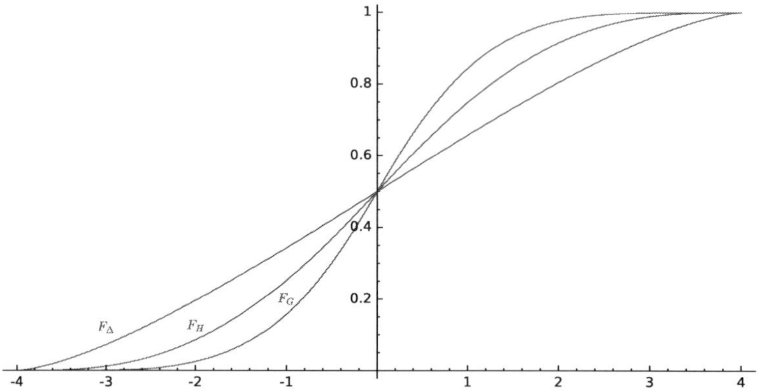

Specifically, in the following table, we summarize the first order approximations to the growth of $\varepsilon = s_1 + 4$ in neighborhood of $\varepsilon = 0$ from Theorems 6 and 10, together with the associated orders of magnitude of points on the associated moduli spaces.

$f_G(\varepsilon) \approx \dfrac{\varepsilon^4}{64\pi}$	$F_G(\varepsilon) \approx \dfrac{\varepsilon^5}{320\pi}$	$\lvert\mathcal{M}_2(\mathbb{F}_q)\rvert \approx q^3$
$f_H(\varepsilon) \approx \dfrac{\varepsilon^2}{8\pi}$	$F_H(\varepsilon) \approx \dfrac{\varepsilon^3}{24\pi}$	$\lvert\mathcal{H}_D(\mathbb{F}_q)\rvert \approx q^2$
$f_\Delta(\varepsilon) \approx \dfrac{\varepsilon^{1/2}}{\sqrt{8}\pi}$	$F_\Delta(\varepsilon) \approx \dfrac{\varepsilon^{3/2}}{3\sqrt{2}\pi}$	$\lvert\mathcal{X}^N(\mathbb{F}_q)\rvert \approx q$

Setting $\varepsilon = d/\sqrt{q}$, where d is the defect from the border $t = -4\sqrt{q}$ of the Weil interval, we observe the relative growth of the cumulative densities:

$$\frac{F_\Delta(\varepsilon)}{F_H(\varepsilon)} = \frac{4\sqrt{2}}{d^{3/2}}\,q^{3/4} \text{ and } \frac{F_H(\varepsilon)}{F_G(\varepsilon)} = \frac{40}{3d^2}\,q.$$

Consequently a search of random points on a Shimura curve is more likely to exhibit curves of extremal trace than on a Humbert surface or for a generic curve. Every genus-2 curve with QM, however, has split Jacobian over a finite field. Similarly, a search of random points on a Humbert surface is more likely to exhibit curves of extremal trace than for a generic curve, and a general curve with moduli on a Humbert surface has absolutely commutative endomorphism ring and nonsplit Jacobian.

Considering all curves over such spaces, we find that in the neighborhood of $\varepsilon = 0$, the rational points $\mathcal{H}_D(\mathbb{F}_q)$ of any single Humbert surface dominates the extremal traces within the full set of points $\mathcal{M}_2(\mathbb{F}_q)$ on the full moduli space. Specifically, with $\varepsilon = d/\sqrt{q}$, as above, we have

$F_G(\varepsilon)\lvert\mathcal{M}_2(\mathbb{F}_q)\rvert \approx \dfrac{\varepsilon^5}{320\pi}q^3 = \dfrac{d^5}{320\pi}\sqrt{q}$
$F_H(\varepsilon)\lvert\mathcal{H}_D(\mathbb{F}_q)\rvert \approx \dfrac{\varepsilon^3}{24\pi}q^2 = \dfrac{d^3}{24\pi}\sqrt{q}$
$F_\Delta(\varepsilon)\lvert\mathcal{X}^N(\mathbb{F}_q)\rvert \approx \dfrac{\varepsilon^{3/2}}{3\sqrt{2}\pi}q = \dfrac{d^{3/2}}{3\sqrt{2}\pi}\sqrt[4]{q}$

Since $1/320\pi < 1/24\pi$, this gives an apparent paradox for any defect $d < \sqrt{40/3} < 3.6515$ from the end of the Weil interval. This shows clearly that the vertical Sato–Tate distribution for the trace of Frobenius on $[-4, 4]$, as a limit in q, does not allow one to conclude existence of points in short intervals (of length $O(1/\sqrt{q})$). Moreover, for particular real discriminant D, we prove a proposition concerning the discrete structure of moduli points in $\mathcal{H}_D(\mathbb{F}_q)$, which shows the dependence of the arithmetic of the Galois representations of Frobenius with the geometry of $\mathcal{H}_D$.

PROPOSITION 14. Suppose $A/\mathbb{F}_q$ is a Jacobian surface such that $\mathrm{End}(A)$ is commutative. Then there exists a unique D such that the moduli point of A lies in $\mathcal{H}_D(\mathbb{F}_q) \subset \mathcal{M}_2(\mathbb{F}_q)$, and the associated point (s_1, δ_0) in $\mathfrak{D}_2$ satisfies

$$\delta_0 = m_0\sqrt{\frac{D}{q}} \leq \varepsilon,$$

where $\varepsilon = 4 - |s_1|$ and $m_0 > 0$ is an integer. In particular, the defect $d = \varepsilon\sqrt{q}$ with respect to the Weil bound satisfies $\sqrt{D} \leq d$.

PROOF. Since $\mathrm{End}(A)$ is commutative, either A is simple and $\mathrm{End}(A)$ is an order in a CM field, or A is isogenous to a product of nonisogenous elliptic curves and $\mathrm{End}(A)$ is a suborder of a product of imaginary quadratic orders. Let $\mathrm{End}^s(A)$ be the subring fixed by the Rosatti involution, containing $\mathbb{Z}[\pi + \bar{\pi}]$. Then $\mathrm{End}^s(A)$ is a quadratic order of discriminant D. If A is simple, then $\mathrm{End}^s(A)$ is a real quadratic order, and otherwise $\mathrm{End}(A)^s$ is an order in the étale algebra $\mathbb{Q} \times \mathbb{Q}$ of discriminant $D = m^2$. The subring $\mathbb{Z}[\pi + \bar{\pi}]$ is then a suborder of index m_0 and discriminant $m_0^2 D$. In view of the optimal embedding of $\mathrm{End}^s(A)$ in $\mathrm{End}(A)$, the moduli point of A lies in $\mathcal{H}_D(\mathbb{F}_q)$. By definition $\delta_0^2 q = m_0^2 D = \mathrm{disc}(\mathbb{Z}[\pi + \bar{\pi}])$, and the bounds follow since (s_1, δ_0) is a point in $\mathfrak{D}_2$. □

REMARK. The condition that $\mathrm{End}(A)$ is commutative includes all absolutely simple abelian surfaces, as well as those isogenous to a product $E_1 \times E_2$ such that E_1 and E_2 are not isogenous. The latter products arise from abelian surfaces whose moduli are on Humbert surfaces of square discriminants. In contrast, any abelian surface A with quaternionic multiplication by a maximal order $\mathcal{O}_B$ in a quaternion algebra B of discriminant N, whose moduli point lies on a Shimura curve X^N, will have larger endomorphism ring over the algebraic closure. Over an extension such an abelian surface is isogenous to $E \times E$, where $\mathcal{O}_K = \mathrm{End}(E)$ is the maximal order in an imaginary quadratic field. In terms of the domain $\mathfrak{D}_2$, for any field $k/\mathbb{F}_q$ such that $\mathrm{End}_k(A) = \mathrm{End}(A)$, the normalized Frobenius characteristic polynomial is of the form $(x^2 - tx + 1)^2$, and the associated point (s_1, δ_0) satisfies $\delta_0 = 0$. Since infinitely many real quadratic orders embed in $\mathcal{O} \subset \mathrm{End}(A) \subseteq \mathbb{M}_2(\mathcal{O}_K)$, the Shimura curve lies in the intersection of infinitely many Humbert surfaces $\mathcal{H}_D$.

Corollary 15. *A Jacobian surface $A/\mathbb{F}_q$ with absolutely commutative endomorphism ring and defect $d = 4\sqrt{q} - |a_1|$ with respect to the Weil bound has moduli in a unique Humbert surface $\mathcal{H}_D(\mathbb{F}_q)$ with $\sqrt{D} \le d$.*

Proposition 14 asserts that the points (s_1, δ_0) in family over $\mathcal{H}_D$ lie in discrete bands, and consequently, there is a zone of exclusion : for $\delta_0 \neq 0$ then

$$\delta_0^2 = s_1^2 - 4s_2 = m_0^2 \frac{D}{q} \ge \frac{D}{q}.$$

Moreover, the dense set of moduli points of Jacobian surfaces with absolutely commutative endomorphism ring lie in disjoint sheets $\mathcal{H}_D$ (which intersect pairwise in unions of the Shimura curves of codimension 1 whose associated Frobenius representations contribute to the domain $\delta_0 = 0$). This gives a stratification of $\mathcal{M}_2$ into Humbert surfaces. The Sato–Tate distribution of USp(4) (over the generic space $\mathcal{M}_2$) can be visualized as layered in the axis D by its contributions of Sato–Tate distributions for $\mathrm{SU}(2) \times \mathrm{SU}(2)$ with zone of exclusion $\delta_0^2 \ge D/q$.

This stratification of the moduli space $\mathcal{M}_2$ by Humbert surfaces $\mathcal{H}_D$ is reflected in the precise algebraic relation between the Haar measures μ_G and μ_H,

$$\frac{\mu_G(t_1,t_2)}{\mu_H(t_1,t_2)} = \frac{\mu_G(s_1,s_2)}{\mu_H(s_1,s_2)} = \frac{\mu_G(s_1,\delta_0)}{\mu_H(s_1,\delta_0)} = \frac{D_0}{2},$$

a scalar multiple of the measure μ_H, which measures the relative number of strata $\mathcal{H}_D$ at a given value of $D_0 = \delta_0^2$, contributing to the Haar measure μ_G over $\mathcal{M}_2$.

Example. The fibration of $\mathcal{M}_2/\mathbb{F}_q$ into strata $\mathcal{H}_D/\mathbb{F}_q$, and the discrete bands $\delta_0 = m_0\sqrt{D/q}$ in $\mathcal{H}_D(\mathbb{F}_q)$ corresponding to each index $m_0 = [\mathcal{O}_D : \mathbb{Z}[\pi + \bar{\pi}]]$ can be visualised by plotting the moduli points in $\mathcal{H}_D(\mathbb{F}_q)$. For the point $q = 47$, and discriminants $D = 5$, 12, 37 and 97, we have:

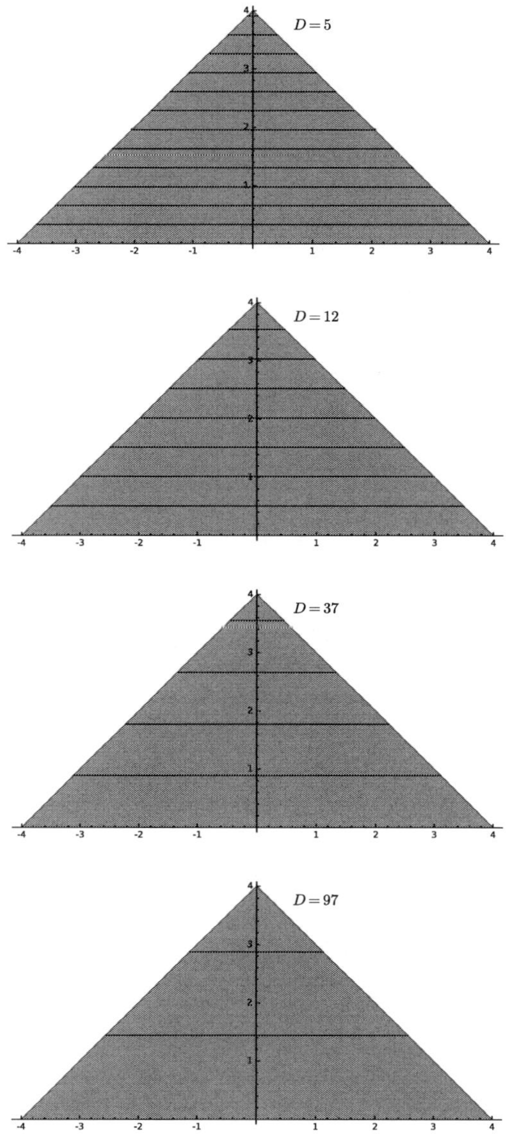

7. Conclusion

Based on the local expansions around the boundary of extremal traces ($s_1 = \pm 4$) for the degree-4 representations of USp(4), SU(2)×SU(2) and SU(2), we derive explicit Taylor series expansions in the neighborhood of the extremal trace $s_1 = -4$ (and by symmetry around $s_1 = 4$). In the application to curves and Jacobians over finite fields, this corresponds to genus-2 curves with many points (and with few points, respectively). Noting that USp(4) is the generic Sato–Tate group for abelian surfaces, and SU(2) × SU(2) is the generic Sato–Tate group for abelian surfaces with RM, we combine this quantitative result with a qualitative analysis of the stratification of the moduli space $\mathcal{M}_2$ (or $\mathcal{A}_2$ of principally polarized abelian surfaces) by Humbert surfaces parametrizing RM. We use this to give a heuristic interpretation of the dominance of small RM (including split abelian surfaces) among the Jacobians of curves with many points.

An interesting question remains the role of RM and more generally exceptional endomorphisms in higher dimensions. From the Weyl integration formulas, one derives similar expressions for the induced Haar measure on Σ_g. As above, setting $t_j = 2\cos(\theta_j)$ and then

$$D_0 = \prod_{j<k}(t_j - t_k)^2 \text{ and } D_1 = \prod_{j=1}^{g}(4 - t_j^2),$$

one finds for $G = \mathrm{USp}(2g)$,

$$\mu_G(s_1,\ldots,s_g) = \frac{1}{(2\pi)^g}\sqrt{D_0 D_1}\, ds_1\cdots ds_g,$$

and for $H = \mathrm{SU}(2)^g$,

$$\mu_H(s_1,\ldots,s_g) = \frac{g!}{(2\pi)^g}\frac{\sqrt{D_1}}{\sqrt{D_0}}\, ds_1\cdots ds_g.$$

This gives an expression similar to that for $g = 2$:

$$\frac{\mu_G(s_1,\ldots,s_g)}{\mu_H(s_1,\ldots,s_g)} = \frac{D_0}{g!}. \tag{10}$$

However, unlike in the case for $g = 2$, the single parameter $D = \mathrm{disc}(\mathcal{O})$ is not sufficient to uniquely characterize an order $\mathcal{O}$ in a totally real field or etale algebra. Nevertheless, the Hilbert moduli space of a totally real order $\mathcal{O}$ maps to a g–dimensional subspace $\mathcal{H}_g(\mathcal{O})$ of the $g(g+1)/2$–dimensional moduli space $\mathcal{A}_g$ of principally polarized abelian varieties of dimension g. In general dimension we expect the ordinary points over $\mathbb{F}_q$ to provide the dominant contribution to the Haar measure, and such points partition themselves into unique strata $\mathcal{H}_g(\mathcal{O})$. For abelian 3–folds, this gives a stratification of the 6–dimensional space $\mathcal{A}_3$ by 3–folds $\mathcal{H}_3(\mathcal{O})$.

While the qualitative description is analogous, the interpretation of the ratio (10) of Haar measures is not as obvious since the discriminant D doesn't characterize totally real orders of rank $g > 2$. Moreover the computation of an explicit Taylor series expansion for the trace to quantify this qualitative contribution would be a much more involved calculation, whose completion would nevertheless be interesting.

Acknowledgments

This work, presented in a previous instance of AGCT in 2015, grew out of discussions in the course of the doctoral work of the second author, jointly supervised by Gilles Lachaud and the first author. Certain results of Gilles in the direction of this work are presented in Section 5. Gilles' departure was both a deep personal and mathematical loss, and his impact continues to be felt in work such the present article and through the international AGCT workshops.

References

[1] Jeffrey D. Achter and Everett W. Howe, *Split abelian surfaces over finite fields and reductions of genus-2 curves*, Algebra Number Theory **11** (2017), no. 1, 39–76, DOI 10.2140/ant.2017.11.39. MR3602766

[2] Nicholas M. Katz and Peter Sarnak, *Random matrices, Frobenius eigenvalues, and monodromy*, American Mathematical Society Colloquium Publications, vol. 45, American Mathematical Society, Providence, RI, 1999, DOI 10.1090/s0273-0979-99-00766-1. MR1659828

[3] G. Lachaud. Distribution asympototique du nombre de points des courbes sur un corps fini, Séminaire ATI, 28 mars 2013.

[4] G. Lachaud. Le groupe $SU(2) \times SU(2)$, preprint, 2014.

[5] Frans Oort, *A stratification of a moduli space of abelian varieties*, Moduli of abelian varieties (Texel Island, 1999), Progr. Math., vol. 195, Birkhäuser, Basel, 2001, pp. 345–416, DOI 10.1007/978-3-0348-8303-0_13. MR1827027

[6] Hermann Weyl, *The Classical Groups. Their Invariants and Representations*, Princeton University Press, Princeton, N.J., 1939. MR0000255

David Kohel, Aix Marseille Univ, CNRS, Centrale Marseille, I2M, Marseille, France, Case 907, 13288 MARSEILLE Cedex 9, France
Email address: david.kohel@univ-amu.fr

Yih-Dar Shieh
Email address: chiapas@gmail.com

Contemporary Mathematics
Volume **770**, 2021
https://doi.org/10.1090/conm/770/15436

La trace et le deltoïde de SU(3)

Gilles Lachaud

ABSTRACT. This article is dedicated to describing random variables on the group $\mathbf{SU}(3)$ associated to the character τ of the standard representation, that is:

(i) The complex-valued character itself τ, and the description of its image, the Deltoid $\Delta \subset \mathbb{C}$.
(ii) The length $\lambda = |\tau|$.
(iii) The norm $\nu = |\tau|^2$.
(iv) The adjoint representation $\alpha(g) = \nu(g) - 1$.
(v) The real part $\xi = 2\,\mathrm{Re}\,\tau$ and the imaginary part $\eta = 2\,\mathrm{Im}\,\tau$.

RÉSUMÉ. Cette note est consacrée à la description des variables aléatoires sur le groupe $\mathbf{SU}(3)$ liées au caractère τ de la représentation standard, à savoir :

(i) Le caractère τ lui-même, à valeurs complexes, et la description du deltoïde $\Delta \subset \mathbb{C}$ image de τ.
(ii) La longueur $\lambda = |\tau|$.
(iii) La norme $\nu = |\tau|^2$.
(iv) La représentation adjointe $\alpha(g) = \nu(g) - 1$.
(v) La partie réelle $\xi = 2\,\mathrm{Re}\,\tau$ et la partie imaginaire $\eta = 2\,\mathrm{Im}\,\tau$.

1. Introduction

Les hypocycloïdes sont liées aux groupes unitaires $\mathbf{SU}(n)$ car les valeurs du caractère de la représentation standard sont exactement les points situés à l'intérieur d'une hypocycloïde avec n points de rebroussement [**4**]. Pour $n = 3$, on obtient un deltoïde, i.e. une hypocycloïde à trois points de rebroussement. C'est une courbe unicursale. (N.B. hypocycloïde est féminin, mais deltoïde (*deltoid* en anglais) est masculin).

NOTATIONS. *Si ν est une variable aléatoire, on note $F_\nu(x)$ la fonction de répartition de ν (*cumulative distribution function, CDF*), $f_\nu(x) = F'_\nu(x)$ la densité de probabilité de ν (*probability density function, PDF*), lorsqu'elle existe,*

$$g_\nu(t) = \int_{\mathbb{R}} e^{itx} f_\nu(x)dx$$

2020 *Mathematics Subject Classification.* 20G05, 22C05, 22E15, 22E45, 33C05.

la fonction caractéristique de ν, et

$$s_\nu(z) = g_\nu(-iz) = \sum_{n=0}^{\infty} M_n(\nu)\frac{z^n}{n!}$$

la série génératrice exponentielle des moments $M_n(\nu)$ de ν.

2. Le groupe unitaire

Le groupe de Lie réel $G = \mathbf{SU}(3)$ est le groupe spécial unitaire

$$\{g \in \mathbf{SL}_3(\mathbb{C}) \mid g^\star g = \mathbf{I}_3\}$$

où $g^\star = {}^t\overline{g}$. Il est de dimension 8. Les matrices

$$\mathbf{t}(\theta) = \begin{bmatrix} e^{i\theta_1} & 0 & 0 \\ 0 & e^{i\theta_2} & 0 \\ 0 & 0 & e^{-i(\theta_1+\theta_2)} \end{bmatrix}, \qquad \theta = (\theta_1, \theta_2) \in \mathbb{R}^2,$$

forment un tore maximal T de G et on peut supposer que $\theta \in X_2 = [0, 2\pi]^2$ pour obtenir un isomorphisme à un ensemble de mesure nulle près. Le caractère $\boldsymbol{\tau}$ de la représentation standard, de degré 3, est à valeurs dans $\mathbb{C}$ et donné sur le tore T par

$$\begin{aligned} \boldsymbol{\tau} \circ \mathbf{t}(\theta) &= e^{i\theta_1} + e^{i\theta_2} + e^{-i(\theta_1+\theta_2)} \\ &= x(\theta) + iy(\theta), \\ x(\theta) &= \cos\theta_1 + \cos\theta_2 + \cos(\theta_1 + \theta_2), \\ y(\theta) &= \sin\theta_1 + \sin\theta_2 - \sin(\theta_1 + \theta_2). \end{aligned}$$

Ce caractère est invariant par le groupe de Weyl W d'ordre 6. On a

$$|\boldsymbol{\tau} \circ \mathbf{t}(\theta)|^2 = 3 + 2\cos(\theta_1 - \theta_2) + 2\cos(2\theta_1 + \theta_2) + 2\cos(\theta_1 + 2\theta_2). \tag{1}$$

et $0 \leq |\boldsymbol{\tau} \circ \mathbf{t}(\theta)|^2 \leq 9$. Voici des exemples de points critiques :

$$\begin{aligned} \boldsymbol{\tau} \circ \mathbf{t}(0,0) &= 3 \quad \text{(maximum)}, \\ \boldsymbol{\tau} \circ \mathbf{t}(2\pi/3, 0) &= 0 \quad \text{(minimum)}, \\ \boldsymbol{\tau} \circ \mathbf{t}(\pi/3, \pi/3) &= \frac{1 + i\sqrt{3}}{2} \quad \text{(point col)}. \end{aligned}$$

3. Le deltoïde

On pose

$$D(x,y) = 4(2x+3)^3 - (x^2 + y^2 + 12x + 9)^2$$

de telle sorte que l'on a également

$$D(x,y) = 8x(x^2 - 3y^2) + 27 - (x^2 + y^2)^2 - 18(x^2 + y^2)$$

et

$$\Delta = \left\{(x,y) \in \mathbb{R}^2 \mid D(x,y) \geq 0\right\}.$$

L'ensemble Δ est compact, non vide, et il est égal à l'adhérence de son intérieur. Le bord du compact Δ est le deltoïde $H = \partial\Delta$. C'est la courbe unicursale

$$\gamma : [0, 2\pi] \longrightarrow \mathbb{C}$$

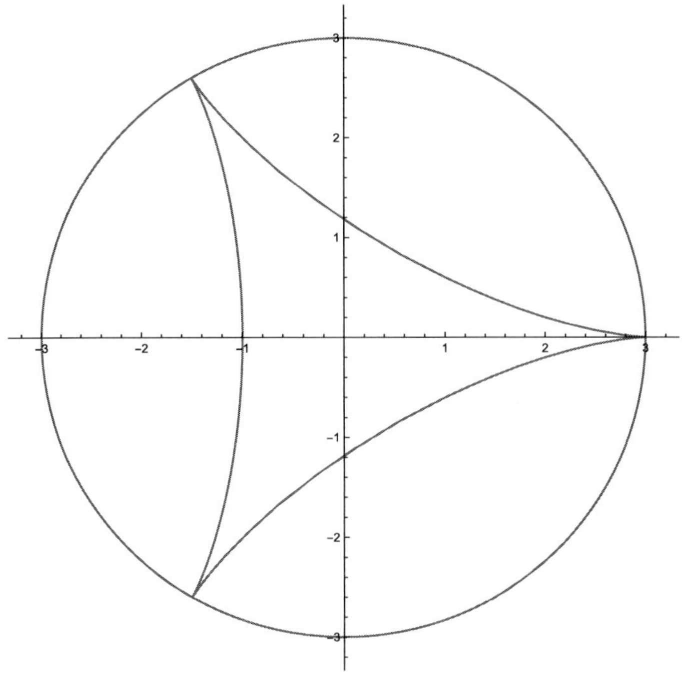

FIGURE 1. Le compact Δ et le deltoïde H.

donnée par

$$\gamma(t) = \alpha(t) + i\beta(t) = 2e^{i}t + e^{-2it},$$
$$\alpha(t) = 2\cos t + \cos 2t, \quad \beta(t) = 2\sin t - \sin 2t.$$

Pour ce qui précède, voir [**1**].

L'ensemble Δ est inscrit dans un cercle de rayon 3 centré à l'origine, et

$$\Delta \subset [-\frac{3}{2}, 3] \times [-3\frac{\sqrt{3}}{2}, 3\frac{\sqrt{3}}{3}].$$

On pose

$$u(x) = \sqrt{-(x^2 + 12x + 9) + 2(2x + 3)^{3/2}},$$
$$v(x) = \sqrt{-(x^2 + 12x + 9) - 2(2x + 3)^{3/2}},$$

de telle sorte que $u(x)v(x) = -D(x, 0)$. Le deltoïde H est la réunion des 4 courbes (voir la figure 2) :

$$S_1 : y = -v(x), \qquad S_2 : y = v(x), \qquad -\frac{3}{2} \leq x \leq 3,$$
$$T_1 : y = -u(x), \qquad T_2 : y = u(x), \qquad -\frac{3}{2} \leq x \leq -1.$$

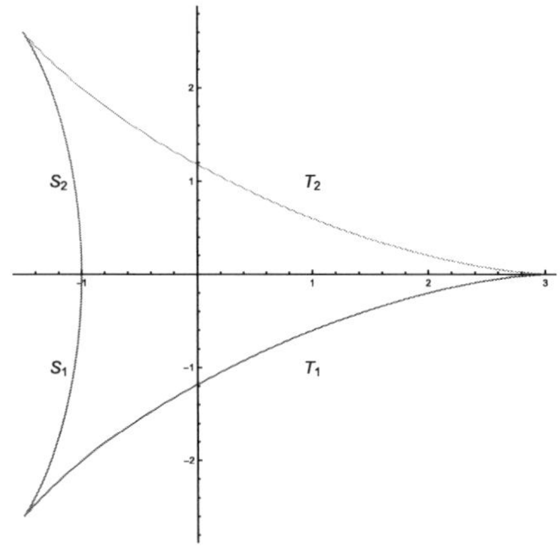

FIGURE 2. Partition de H.

Ces segments sont aussi l'image de γ :

$$[0, \frac{2\pi}{3}] \longrightarrow T_2, \qquad [\frac{2\pi}{3}, \pi] \longrightarrow S_2,$$
$$[\pi, \frac{4\pi}{3}] \longrightarrow S_1, \qquad [\frac{4\pi}{3}, 2\pi] \longrightarrow T_1.$$

THÉORÈME 1. *Posons* $\mathbf{s} = \boldsymbol{\tau} \circ \mathbf{t}$. *La trace définit une surjection*

$$\mathbf{s} : X_2 \longrightarrow \Delta$$

qui est un homéomorphisme de l'alcôve fondamentale $A \subset X_2$ *sur* Δ.

3.1. Densité de Weyl. La matrice Jacobienne de $\mathbf{s}$ est

$$\operatorname{Jac} \mathbf{s}(\theta) = \begin{pmatrix} -\sin\theta_1 - \sin(\theta_1+\theta_2) & -\sin\theta_2 - \sin(\theta_1+\theta_2) \\ \cos\theta_1 - \cos(\theta_1+\theta_2) & \cos\theta_2 - \cos(\theta_1+\theta_2) \end{pmatrix}$$

et

$$J_{\mathbf{s}}(\theta) = \det \operatorname{Jac} \mathbf{s}(\theta) = \sin(\theta_2 - \theta_1) + \sin(2\theta_1 + \theta_2) - \sin(\theta_1 + 2\theta_2).$$

On a

$$D(x(\theta), y(\theta)) = 4J_{\mathbf{s}}(\theta)^2. \tag{2}$$

La densité de Weyl de $\mathbf{SU}(3)$ est donnée par On a

$$\begin{aligned} \delta(\theta) &= \frac{2}{\pi^2}\big(1-\cos(\theta_1-\theta_2)\big)\big(1-\cos(2\theta_1+\theta_2)\big)\big(1-\cos(\theta_1+2\theta_2)\big) \\ &= \frac{16}{\pi^2}\sin^2\frac{\theta_1\theta_2}{2}\sin^2\frac{2\theta_1+\theta_2}{2}\sin^2\frac{\theta_1+2\theta_2}{2} \end{aligned}$$

$$\delta(\theta) = \frac{1}{\pi^2}J_{\mathbf{s}}(\theta)^2. \tag{3}$$

THÉORÈME 2. *Si* φ *est une fonction continue par morceaux sur* Δ, *on a*

$$\int_G \varphi \circ \mathbf{t}(g) dg) = \frac{1}{2\pi^2}\int_\Delta \varphi(x,y)\sqrt{D(x,y)}dxdy.$$

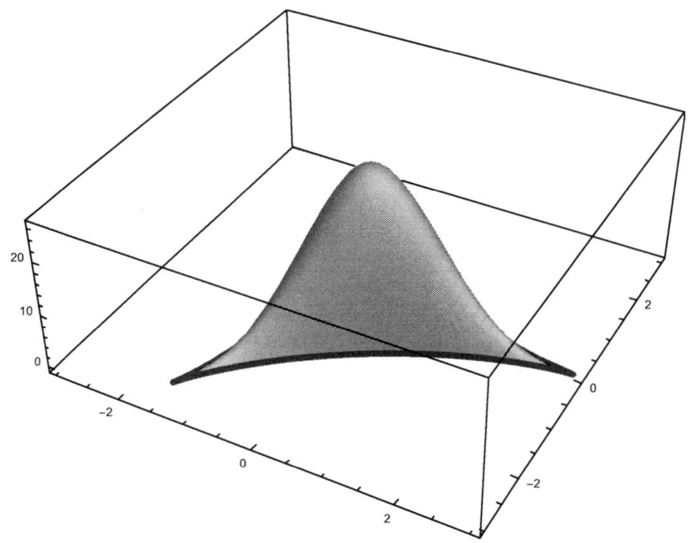

FIGURE 3. Densité dans le deltoïde.

DÉMONSTRATION. Puisque $|W| = 6$, on a, pour toute fonction continue centrale F, la formule de Weyl :

$$\int_G F(g)dg = \frac{1}{6}\int_{X_2} F \circ \mathbf{t}(\theta)\delta(\theta)d\theta,$$

qui entraîne

$$\int_G \varphi \circ \boldsymbol{\tau}(g)dg = \frac{1}{6}\int_{X_2} \varphi \circ \mathbf{s}(\theta)\delta(\theta)d\theta.$$

D'autre part, en utilisant le théorème 1,

$$\int_\Delta \varphi(x,y)dxdy = \frac{1}{6}\int_{X_2} \varphi \circ \mathbf{s}(\theta)|J_{\mathbf{s}}(\theta)|d\theta,$$

ce qui implique, par (2) :

$$\int_\Delta \varphi(x,y)\sqrt{D(x,y)}dxdy = \frac{1}{3}\int_{X_2} \varphi \circ \mathbf{s}(\theta)J_{\mathbf{s}}(\theta)^2 d\theta.$$

On déduit de (3) :

$$\int_{X_2} \varphi \circ \mathbf{s}(\theta)\delta(\theta)d\theta = \frac{1}{\pi^2}\int_{X_2} \varphi \circ \mathbf{s}(\theta)J_{\mathbf{s}}(\theta)^2 d\theta,$$

donc

$$\frac{1}{6}\int_{X_2} \varphi \circ \mathbf{s}(\theta)\delta(\theta)d\theta = \frac{1}{2\pi^2}\int_\Delta \varphi(x,y)\sqrt{D(x,y)}dxdy$$

et le résultat s'ensuit. □

La répartition de la densité se voit sur la figure 3. Le maximum $\Delta(x,y) = 27$ est atteint en $(0,0)$.

4. Loi du caractère standard

Pour le caractère complexe τ, la suite des moments est

$$M_n(\tau) = \int_G \tau(g)^n dg = \frac{1}{6}\int_{X_2} [\tau \circ \mathbf{t}(\theta)]^n \delta(\theta) d\theta.$$

La suite $M_{3n}(\tau)$ est la suite [**10**, A005789] des nombres de Catalan de dimension 3 :

$$1, 0, 0, 1, 0, 0, 5, 0, 0, 42, 0, 0, 462, 0, 0, 6006 \ldots$$

Les termes non multiples de 3 sont nuls, et

$$M_{3n}(\tau) = \frac{2(3n)!}{n!(n+1)!(n+2)!}$$

Une série génératrice pour la suite $(M_n(\tau))$ est celle de Mathar (2015), simplifiée :

$$\sigma_\tau(t) = \frac{1}{3t}\left({}_2F_1\left(-\frac{2}{3}, -\frac{1}{3}; 2; 27t^3\right) - 1\right),$$

on a

$$\sigma_\tau(t) = \sum_{n=0}^{\infty} M_{3n} t^{3n}.$$

Une autre série génératrice de type exponentiel pour la même suite est donnée par

$$r_\tau(t) = {}_2F_2\left(\frac{1}{3}, \frac{2}{3}; 2, 3; 27t^3\right);$$

on a

$$r_\tau(t) = \sum_{n=0}^{\infty} \frac{M_{3n}}{n!} t^{3n}.$$

La fonction caractéristique est

$$g_\tau(t) = {}_0F_2(2, 3; -it^3).$$

La fonction $g_\tau(t)$ est à valeurs complexes et non bornée sur l'axe réel, et la série génératrice exponentielle est

$$s_\tau(z) = g_\tau(-iz) = {}_0F_2(2, 3; z^3).$$

RÉSUMÉ.

— *La densité de τ est à valeurs complexes. Elle se confond avec la densité de Weyl sur le deltoïde.*
— *La fonction caractéristique est à valeurs complexes, et non bornée.*
— *On dispose d'une formule pour les moments $M_n(\tau)$, qui sont des entiers positifs.*
— *On connait la série génératrice usuelle de ces moments, la série génératrice exponentielle, et une autre série du même type.*

5. Loi de la longueur de la trace

La variable aléatoire

$$\nu(g) = |\tau(g)|^2 = \tau(g)\overline{\tau(g)}$$

est un caractère à valeurs dans $[0, 9]$ dont on a donné l'expression en (1). C'est le caractère de la représentation de G dans $V \otimes V^\star$. La variable aléatoire

$$\lambda(g) = |\tau(g)| = (\tau(g)\overline{\tau(g)})^{1/2}$$

est à valeurs dans $[0, 3]$. On a

$$M_n(\lambda) = \int_G |\tau(g)|^n dg = \frac{1}{6}\int_{X_2} |\tau \circ \mathbf{t}(\theta)|^n \delta(\theta) d\theta$$

et $M_n(\nu) = M_{2n}(\lambda)$; en effet

$$M_n(\nu) = \int_0^9 z^n f_\nu(z)dz = \int_0^9 z^n f_\lambda(\sqrt{z})\frac{dz}{2\sqrt{z}} = \int_0^3 t^{2n} f_\lambda(t)dt = M_{2n}(\lambda),$$

où on a posé $z = t^2$. La suite $(M_n(\lambda))$ est donnée par

n	0	1	2	3	4	5	6	7	8	9	10	11	12
$M_n(\lambda)$	1	0,891	1	1,326	2	3,332	6	11,47	23	47,91	103	227,3	513

On note

$$C(t) = \{(r, \theta) \in \Delta \mid x^2 + y^2 \leq t^2\}$$

le disque de rayon t. En appliquant le théorème 2 et en prenant pour φ la fonction caractéristique du disque $C(t)$ avec $0 \leq t \leq 3$, on obtient la CDF de λ :

$$F_\lambda(t) = \mathrm{vol}\{g \in G \mid |\tau(g)| \leq t\} = \frac{1}{2\pi^2}\int_{\Delta(t)} \sqrt{D(x, y)}\, dxdy, \tag{4}$$

où

$$\Delta(t) = \Delta \cap C(t).$$

En coordonnées polaires on a

$$\Delta = \{(r, \theta) \in \mathbb{R}_+ \times [0, 2\pi] \mid D_0(r, \theta) \geq 0\},$$

avec

$$D_0(r, \theta) = 27 - r^2(r^2 + 18) + 8r^3 \cos 3\theta,$$

et la formule du théorème 2 devient

$$\int_G \varphi \circ \mathbf{t}(g)dg = \frac{1}{2\pi^2}\int_\Delta \varphi(r\cos\theta, r\sin\theta)\sqrt{D_0(r, \theta)}\, rdrd\theta. \tag{5}$$

En exprimant (4) à l'aide de (5), on obtient

$$F_\lambda(t) = \frac{1}{2\pi^2}\int_{\Delta(t)} \sqrt{D_0(r, \theta)}\, rdrd\theta.$$

Notons $\Gamma(t) = \partial C(t)$ le cercle de rayon t. La PDF de la distribution λ est donnée par

$$f_\lambda(t) = \frac{d}{dt}F_\lambda(t) = \frac{1}{2\pi^2}\int_{\Delta\cap\Gamma(t)} \sqrt{D_0(t, \theta)}\, d\theta,$$

où l'intégration se fait sur la fibre $\Delta \cap \Gamma(t)$. La fonction f_λ est nulle en dehors de $[0, 3]$.

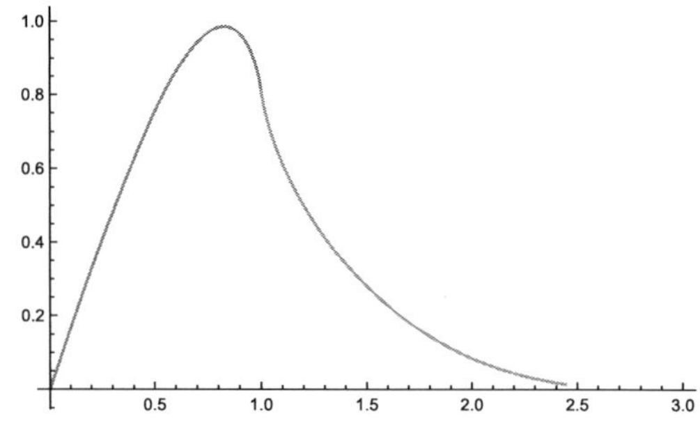

FIGURE 4. Densité de la longueur de la trace.

5.1. Premier cas. Supposons $0 \leq t \leq 1$. On a $C(t) \subset \Delta$ et

$$f_\lambda(t) = \frac{t}{2\pi^2} \int_{\Gamma(t)} \sqrt{D_0(t,\theta)}\, d\theta.$$

Cette intégrale s'exprime à l'aide de l'intégrale de Legendre complète de deuxième espèce $E(k)$, voir [**8**, §19]. Posons

$$u(t) = (3-t)^3(1+t),$$
$$k(t) = 1 - \frac{u(-t)}{u(t)} = \frac{16t^3}{(3-t)^3(1+t)},$$
$$f_1(t) = \frac{2t}{\pi^2}\sqrt{u(t)}E(k(t)).$$

Alors

$$f_\lambda(t) = f_1(t) \quad \text{si} \quad 0 \leq t \leq 1.$$

5.2. Deuxième cas. Supposons $1 \leq t \leq 3$. Ici l'intégrale s'exprime à l'aide de l'intégrale de Legendre incomplète de deuxième espèce $E(\varphi, k)$ [**8**, §19] Posons

$$\psi(t) = \arctan \sqrt{-\frac{u(t)}{u(-t)}}\,.$$

Cette fonction décroit de $\pi/2$ à 0 lorsque t va de 1 à 3, et

$$f_2(t) = \frac{2t}{\pi^2}\sqrt{u(t)}E(\psi(t), k(t)).$$

Alors

$$f_\lambda(t) = f_2(t) \quad \text{si} \quad 1 \leq t \leq 3.$$

On notera que $E(k) = E(\pi/2, k)$.

THÉORÈME 3. *La densité de probabilité de la longueur λ est*

$$f_\lambda(t) = \frac{2t}{\pi^2}\sqrt{u(t)}E(\varphi(t), k(t))$$
$$\textit{avec} \quad \varphi(t) = \begin{cases} \pi/2 & \textit{si } 0 \leq t \leq 1, \\ \psi(t) & \textit{si } 1 \leq t \leq 3. \end{cases}$$

La densité $f_\lambda(t)$ est représentée dans la figure 4.

RÉSUMÉ.

— *On connait la densité et les moments pairs de* λ.
— *On ne connait ni la série génératrice, ni la série génératrice exponentielle de ces moments (ni par conséquent la fonction caractéristique), et on ne connait pas de formule pour les moments impairs.*

6. Loi de la norme de la trace

Le moment

$$M_n(\nu) = \int_G |\tau(g)|^{2n} dg = \frac{1}{6}\int_{X_2} |\tau \circ \mathbf{t}(\theta)|^{2n} \delta(\theta) d\theta$$

est égal à la dimension des invariants de $\mathbf{SL}(3)$ dans $V^n \otimes (V^\star)^n$, où V est la représentation standard de dimension 3 de $\mathbf{SL}(3)$, et où $V^\star$ est son dual. La suite $a_n = M_n(\nu)$ des moments est assez connue [**10**, A005802] :

n	0	1	2	3	4	5	6	7	8	9	10
$M_n(\nu)$	1	1	2	6	23	103	513	2761	15767	94359	586590

C'est aussi :
— Le nombre $u_3(n)$ des permutations de $\mathfrak{S}_n$ dont la plus grande sous-suite croissante est de longueur ≤ 3 (i.e. les permutations évitant 1234) [**2**, p. 280].
— Le nombre des permutations doublement alternées de $\mathfrak{S}_{2n}$ évitant 1234.
La suite (a_n) est définie par les formules suivantes :
— On a $a_0 = a_1 = 1$ et la formule récursive

$$a_n = \frac{(10n^2 + 2n - 3)a_{n-1} + (-9n^2 + 18n - 9)a_{n-2}}{(n+2)^2}. \tag{6}$$

— On a

$$a_n = \frac{1}{(n+1)^2(n+2)} \sum_{k=0}^{n} \binom{2k}{k}\binom{n+1}{k+1}\binom{n+2}{k+1}. \tag{7}$$

— On a

$$a_n = 2\sum_{k=0}^{n} \binom{2k}{k}\binom{n}{k}^2 \frac{2k^2 - 2kn + 2k - n + 1}{(k+1)^2(k+2)(n-k+1)}. \tag{8}$$

La série génératrice de la suite (a_n) est

$$\frac{1}{6t^2}\left(5t + 1 - (1-9t)^{3/4}(1-t)^{1/4}{}_2F_1\left(-\frac{1}{4}, \frac{3}{4}; 1; \frac{64t}{(1-9t)^3(t-1)}\right)\right).$$

La densité de probabilité sera calculée dans la section suivante.

6.1. Formule de Gessel. La série génératrice exponentielle est inconnue, mais on a la spectaculaire formule de Gessel (1990) [**2**, p. 280], [**6**, p. 237], [**12**, p. 1002], qui s'écrit, pour $u_3(n)$:

$$\sum_{n=0}^{\infty} u_3(n) \frac{t^{2n}}{(n!)^2} = U_3(t),$$

où

$$U_3(t) = \det(I_{|i-j|}(2t))_{1 \leq i,j \leq 3}$$

et où

$$I_\nu(2t) = \sum_{n=0}^{\infty} \frac{t^{\nu+2n}}{n!(\nu+n)!} = \frac{t^\nu}{\nu!}{}_0F_1(\nu+1, t^2)$$

est la fonction de Bessel modifiée [**7**, p. 66], [**8**, §10.25]. On a

$$U_3(t) = \frac{2I_0(2t)^2 I_1(2t)}{t} - \frac{I_0(2t)I_1(2t)^2}{t^2} - \frac{2I_1(2t)^3}{t}.$$

6.2. Transformation de Hankel. Pour commencer, observons que l'on a

$$I_0(2t) = \sum_{n=0}^{\infty} \frac{t^{2n}}{(n!)^2} = {}_0F_1(1, t^2).$$

Soit f une fonction intégrable à support compact dans $\mathbb{R}_+^\times$ et posons

$$U_f(t) = \sum_{n=0}^{\infty} M_{2n}(f)\frac{t^{2n}}{(n!)^2} \quad \text{où} \quad M_n(f) = \int_0^\infty x^n f(x)dx.$$

Alors

$$U_f(t) = \int_0^\infty f(x)I_0(2t\sqrt{x})dx = 2\int_0^\infty f(x^2)I_0(2tx)xdx.$$

Par exemple,

$$U_3(t) = 2\int_0^\infty f_\nu(x^2)I_0(2tx)xdx.$$

La *transformée de Hankel* (d'ordre 0) d'une fonction $f(x)$ est

$$\widehat{f}(y) = \int_0^\infty f(x)J_0(xy)xdx \quad \text{et} \quad f(x) = \int_0^\infty \widehat{f}(y)J_0(xy)ydy.$$

Voir [**11**, §8, Th. 8, p. 135], [**7**, p. 397] et [**12**]. Puisque

$$J_0(z) = I_0(-iz) = I_0(iz),$$

on obtient les relations

$$\widehat{f}(y) = \int_0^\infty f(x)I_0(ixy)xdx, \quad f(x) = \int_0^\infty \widehat{f}(y)I_0(ixy)ydy.$$

Par exemple, en posant $f^{(2)}(x) = f(x^2)$, on a

$$\frac{1}{2}U_f\left(\frac{iy}{2}\right) = \int_0^\infty f^{(2)}(x)I_0(ixy)xdx = \widehat{f^{(2)}}(y),$$

$$f(x^2) = \frac{1}{2}\int_0^\infty U_f\left(\frac{iy}{2}\right)I_0(ixy)ydy,$$

$$f(x) = 2\int_0^\infty U_f(iy)I_0(2iy\sqrt{x})ydy$$

sous réserve de convergence. Prenons $f = f_\nu$; alors $U_f = U_3$ et

$$f_\nu(x) = 2\int_0^\infty U_3(iy)I_0(2iy\sqrt{x})ydy.$$

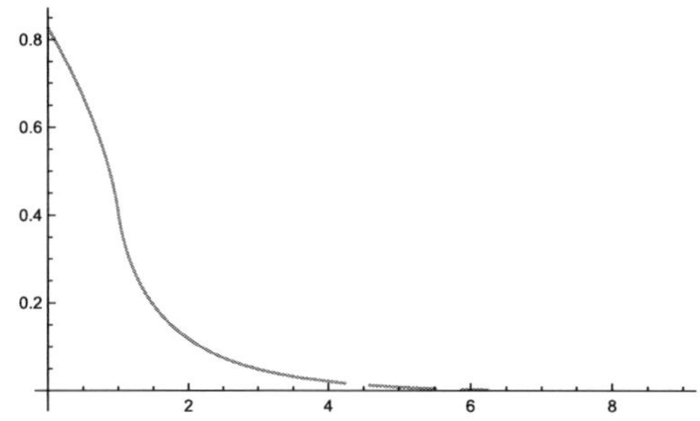

FIGURE 5. Densité de la norme de la trace.

6.3. Densité de probabilité de la norme. Reprenons la variable aléatoire de la norme et appliquons le théorème 2. En prenant pour φ la fonction caractéristique du disque $C(\sqrt{z})$ avec $0 \leq z \leq 9$ on obtient la CDF de ν :

$$F_\nu(z) = \operatorname{vol}\{g \in G \mid |\tau(g)|^2 \leq z\} = \frac{1}{2\pi^2}\int_{\Delta(\sqrt{z})} \sqrt{D(x,y)}dxdy. \tag{9}$$

En exprimant (9) à l'aide de (5), on obtient

$$F_\nu(z) = \frac{1}{2\pi^2}\int_{\Delta(\sqrt{z})} \sqrt{D_0(r,\theta)}rdrd\theta.$$

Pour $0 \leq t \leq 3$, on a

$$F_\lambda(t) = F_\nu(t^2)$$

de telle sorte que, pour $0 \leq z \leq 9$, on a $F_\nu(z) = F_\lambda(\sqrt{z})$. La PDF de la distribution ν est

$$f_\nu(z) = \frac{d}{dz}F_\nu(z) = \frac{1}{2\sqrt{z}}\frac{d}{dz}F_\lambda(\sqrt{z}) = \frac{1}{2\sqrt{z}}f_\lambda(\sqrt{z}). \tag{10}$$

En reportant (10) dans le théorème 3, on obtient :

THÉORÈME 4. *La densité de probabilité de ν est*

$$f_\nu(z) = \frac{1}{\pi^2}\sqrt{u(\sqrt{z})}E\left(\varphi(\sqrt{z}), k(\sqrt{z})\right),$$

$$\textit{avec}\quad \varphi(\sqrt{z}) = \begin{cases} \pi/2 & \textit{si } 0 \leq z \leq 1, \\ \psi(\sqrt{z}) & \textit{si } 1 \leq z \leq 9. \end{cases}$$

La densité de $f_\nu(z)$ est représentée dans la figure 5.

On rencontre pour f_λ et f_ν le phénomène relatif à la différence d'allure entre la fonction de répartition d'une variable aléatoire et son carré, voir par exemple [**3**, Ex. 2.13, p. 62]. La densité $f_\nu(z)$ est continue décroissante sur $[0, 9]$ et on a

$$f_\nu(0) = \frac{3\sqrt{3}}{2\pi} = 0,826993\ldots \qquad f_\nu(1) = \frac{4}{\pi^2} = 0,405285\ldots$$

Lorsque $z \to 9$ on a $f_\nu(z) \sim C(9-z)^3$ avec $C > 0$. On retrouve comme moments les éléments de la suite [**10**, A005802] :

$$M_0(\nu) = 1, \quad M_1(\nu) = 1, \quad M_2(\nu) = 2, \quad M_3(\nu) = 6, \quad M_4(\nu) = 23.$$

Puisque la moyenne ou espérance $\mu = M_1(\nu) \neq 0$, la variable aléatoire n'est pas standardisée. Mais elle est réduite : la variance est

$$V = \sigma^2 = M_2(\nu) - \mu^2 = 1$$

Le *coefficient d'asymétrie* de ν est

$$\gamma_1 = \int_0^9 \left(\frac{z-\mu}{\sigma}\right)^3 f_\nu(z)dz = M_3(\nu) - 3M_2(\nu) + 3M_1(\nu) - 1 = 2.$$

Puisqu'il est strictement positif, la branche droite est plus longue et la masse de la distribution est concentrée à gauche. Le *kurtosis non normalisé* de ν vaut

$$\begin{aligned}\beta_2 &= \int_0^9 \left(\frac{z-\mu}{\sigma}\right)^4 f_\nu(z)dz \\ &= M_4(\nu) - 4M_3(\nu) + 6M_2(\nu) - 4M_1(\nu) + 1 = 8\end{aligned}$$

et *l'excès de kurtosis* de ν est

$$\gamma_2 = \beta_2 - 3 = 5.$$

Puisque $\gamma_2 > 0$, la densité de probabilité de ν est *leptokurtique* (un pic).

RÉSUMÉ.
— *On connait la densité, et on dispose de plusieurs formules pour les moments de* ν*. On connait aussi la série génératrice usuelle de ces moments, et la formule de Gessel.*
— *On ne connait pas la série génératrice exponentielle de ces moments (ni par conséquent la fonction caractéristique).*

7. Loi du caractère de la représentation adjointe

Notons

$$\alpha(g) = \operatorname{Tr}\operatorname{Ad}(g)$$

le caractère de la représentation adjointe, de dimension 8. On a

$$\alpha(g) = \nu(g) - 1.$$

Le caractère α est à valeurs réelles, dans l'intervalle $[-1, 8]$. On sait d'ailleurs, d'après les résultats de Serre [**9**], que

$$\min \operatorname{Tr}\operatorname{Ad}(g) = -1.$$

Voici des exemples de points critiques :

$$\begin{aligned}\boldsymbol{\alpha} \circ \mathbf{t}(0,0) &= 8 \quad \text{(maximum)},\\ \boldsymbol{\alpha} \circ \mathbf{t}(2\pi/3,0) &= -1 \quad \text{(minimum)},\\ \boldsymbol{\alpha} \circ \mathbf{t}(\pi/3,\pi/3) &= 0 \quad \text{(point col)}.\end{aligned}$$

Les moments de α se déduisent de ceux de ν :

$$M_n(\alpha) = \sum_{k=0}^{n} \binom{n}{k} (-1)^{n-k} M_k(\nu).$$

On trouve

n	0	1	2	3	4	5	6	7	8	9
$M_n(\alpha)$	1	0	1	2	8	32	145	702	3598	19280

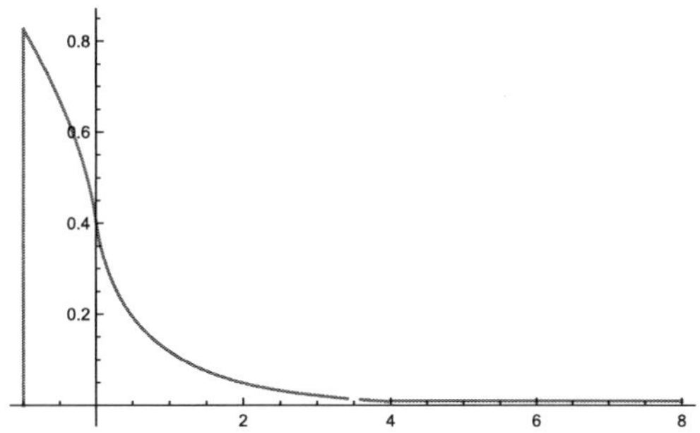

FIGURE 6. Densité du caractère de la représentation adjointe.

Cette suite semble absente de [**10**]. La densité de la loi est donnée par

$$f_\alpha(z) = f_\nu(z+1), \quad -1 \leq z \leq 8.$$

Puisque la moyenne ou espérance $\mu = M_1(\alpha) = 0$, la variable aléatoire est standardisée, et elle est réduite : la variance est

$$V = \sigma^2 = M_2(\alpha) = 1.$$

Le coefficient d'asymétrie de α est

$$\gamma_1 = M_3(\alpha) = 2.$$

Puisqu'il est strictement positif, la branche droite est plus longue, et la masse de la distribution est concentrée à gauche. Le kurtosis non normalisé de ν vaut

$$\beta_2 = M_4(\nu) = 8$$

et l'excès de kurtosis de ν est

$$\gamma_2 = \beta_2 - 3 = 5.$$

Puisque $\gamma_2 > 0$, la densité de probabilité de ν est leptokurtique. La densité $f_\gamma(x)$ est représentée dans la figure 6.

RÉSUMÉ.
— *On connait la densité, et des formules pour les moments de α.*
— *On ne connait pas la série génératrice usuelle, ni la série génératrice exponentielle des moments.*

8. Loi de la partie réelle du caractère standard

La fonction $\xi(g) = 2\operatorname{Re}\tau(g)$ est le caractère de la représentation $V \oplus V^\star$ où V est la représentation standard de dimension 3 de $\mathbf{SL}(3)$ et où $V^\star$ est son dual. Le moment

$$M_n(\xi) = \int_G |2\operatorname{Re}\tau(g)|^n dg = \frac{1}{6}\int_{X_2} |2\operatorname{Re}\tau \circ \mathbf{t}(\theta)|^n \delta(\theta) d\theta$$

est égal à la dimension des invariants de $\mathbf{SL}(3)$ dans $(V \oplus V^\star)^n$. La suite des $M_n(\xi)$ vaut [**10**, A151366] :

(11) $$1, 0, 2, 2, 12, 30, 130, 462, 1946, 7980, 34776 \ldots$$

De plus, $M_n(\xi)$ est égal au nombre de marches aléatoires (*random walks*) de longueur n dans une chambre de Weyl, de l'origine à elle-même, voir [**14**]. Il y a une formule récursive de calcul des $M_n(\xi)$, voir [**10**, A151366]. En prenant pour φ la fonction caractéristique du demi-plan $\operatorname{Re} x \leq z/2$ avec $-3 \leq z \leq 6$, on obtient la CDF de la distribution ξ :

$$F_\xi(z) = \operatorname{vol}\{g \in G \mid 2\operatorname{Re}\tau(g) \leq z\} = \frac{1}{2\pi^2}\int_{\{x\in\Delta \mid x \leq z/2\}} \sqrt{D(x,y)}dxdy.$$

La série génératrice ordinaire est :

$$\begin{aligned} g_\xi(t) = \frac{1}{3t^2}\Bigg(&- 5t^2 - 3t - 1 \\ &+ \frac{2t+1}{3t+1}\left({}_2F_1\left(-\frac{2}{3}, -\frac{1}{3}; 1; k(t)\right) + 4t\,{}_2F_1\left(-\frac{1}{3}, \frac{1}{3}; 2; k(t)\right)\right)\Bigg) \end{aligned}$$

avec $k(t) = 27t^2(2t+1)$. La PDF est donnée par

$$f_\xi(z) = \frac{1}{4\pi^2}\int_{\Delta_{z/2}} \sqrt{D(x,y)}dy$$

où l'intégration se fait sur la fibre verticale

$$\Delta_{z/2} = \Delta \cap D_{z/2}$$

où $D_{z/2}$ est la droite verticale d'abscisse $z/2$. Il est commode de faire certains calculs pour la fonction $\gamma(g) = \operatorname{Re}\tau(g)$. On a

$$\begin{aligned} F_\gamma(z) &= F_\xi(2z) = \frac{1}{2\pi^2}\int_{\{x\in\Delta \mid x\leq z\}} \sqrt{D(x,y)}dxdy, \\ f_\gamma(z) &= 2f_\xi(2z) = \frac{1}{2\pi^2}\int_{\Delta_z} \sqrt{D(x,y)}dy, \\ M_n(\gamma) &= \frac{1}{2^n}M_n(\xi). \end{aligned}$$

D'après la figure 2, il y a deux possibilités :

(i) Si $-3/2 \leq x \leq -1$, on a

$$\Delta_x = \{(x,y) \in X_x \mid -u(x) \leq y \leq -v(x) \text{ et } v(x) \leq y \leq u(x)\}.$$

(ii) Si $-1 \leq x \leq 3$, on a

$$\Delta_x = \{(x,y) \in X_x \mid -u(x) \leq y \leq u(x)\}.$$

Posons

$$p = -(x^2 + 12x + 9), \quad q = 2(2x+3)^{3/2}$$

de telle sorte que

$$u = \sqrt{p+q}, \quad v = \sqrt{p-q}, \quad D = q^2 - (p-y^2)^2.$$

8.1. Premier cas. Si $-3/2 \leq x \leq -1$, on a

$$2\pi^2 f_\gamma(x) = \int_{-u(x)}^{-v(x)} \sqrt{D(x,y)}dy + \int_{v(x)}^{u(x)} \sqrt{D(x,y)}dy.$$

On passe d'une intégrale à l'autre en changeant y de signe, et $f_\gamma(x) = f_1(x)$, avec

$$f_1(x) = \frac{1}{\pi^2}\int_{v(x)}^{u(x)} \sqrt{D(x,y)}dy$$

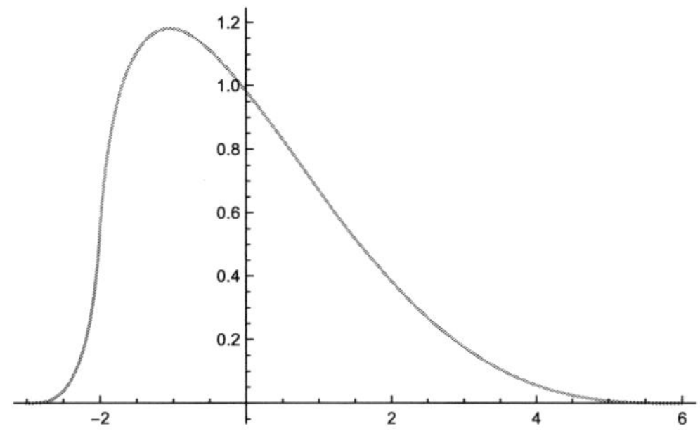

FIGURE 7. Densité de la partie réelle de la trace.

et

$$f_1(x) = \frac{1}{\pi^2} \int_{\sqrt{p-q}}^{\sqrt{p+q}} \sqrt{q^2 - (p - y^2)^2} dy.$$

On trouve, voir [**4**] :

$$f_1(x) = \frac{2}{3\pi^2} \sqrt{p+q} \left(pE\left(\frac{2q}{p+q}\right) - (p-q)K\left(\frac{2q}{p+q}\right)\right),$$

autrement dit

$$f_1(x) = \frac{2}{3\pi^2} u(x) \left(-(x^2 + 12x + 9)E\left(1 - \frac{u(x)^2}{v(x)^2}\right) - v(x)^2 K\left(1 - \frac{u(x)^2}{v(x)^2}\right)\right).$$

8.2. Deuxième cas. Si $-1 \leq x \leq 3$, on a

$$2\pi^2 f_\gamma(x) = \int_{-v(x)}^{u(x)} \sqrt{D(x,y)} dy$$

et $f_\gamma(x) = f_2(x)$ avec

$$f_2(x) = \frac{1}{2\pi^2} \int_{-\sqrt{p+q}}^{\sqrt{p+q}} \sqrt{q^2 - (p - y^2)^2} dy.$$

Si p est réel et si $|\operatorname{Re} p| < q$, cette intégrale s'exprime à l'aide des intégrales de Legendre de première et deuxième espèce $E(k)$ et $K(k)$. On a

$$f_2(x) = \frac{3}{3\pi^2} \sqrt{q-p} \left(pE\left(\frac{p+q}{p-q}\right) + qK\left(\frac{p+q}{p-q}\right)\right),$$

autrement dit

$$f_2(x) = \frac{2}{3i\pi^2} v(x) \left(-(x^2 + 12x + 9)E\left(\frac{u(x)^2}{v(x)^2}\right) + 2(3+2x)^{3/2} K\left(\frac{u(x)^2}{v(x)^2}\right)\right).$$

La densité $f_\gamma(x)$ est représentée dans la figure 7. Enfin

$$f_\xi(x) = \frac{1}{2} f_\gamma\left(\frac{x}{2}\right)$$

et en calculant les moments $M_n(\xi)$ on retrouve la suite (11).

RÉSUMÉ.

— *On connait la densité de ξ, et la série génératrice usuelle des moments. Il y a une formule pour les moments de ξ.*
— *On ne dispose pas de la série génératrice exponentielle de ces moments.*

9. Loi de la partie imaginaire du caractère standard

Les moments de la variable $\eta = 2\,\mathrm{Im}\,\tau(g)$ sont (au signe près) les dimensions des sous-espaces invariants des produits tensoriels de la représentation virtuelle $V \ominus V^\star$. Les moments impairs sont nuls, et la suite des moments $M_{2n}(\eta)$ figure dans [**10**, A241958] :

$$1, 2, 12, 110, 1274, 17136\ldots$$

(il y a une erreur dans [**14**]). On a

$$M_{2n}(\eta) = \frac{(3n+3)!}{(4(n+1)^2-1)((n+1)!)^2(n+2)!}.$$

La série génératrice est

$$s_\eta(z) = {}_4F_3\left(\frac{1}{2}, 1, \frac{4}{3}, \frac{5}{3}; 2, \frac{5}{2}, 3, 27z\right).$$

Résumé.
— *On ne connait pas la densité de η.*
— *On connait la série génératrice usuelle des moments $M_{2n}(\eta)$. Il y a une formule pour les moments de η.*
— *On ne dispose pas de la série génératrice exponentielle de ces moments.*

Références

[1] Robert Bryant, *Boundary of the image of a compact manifold in the complex plane*, MathOverflow, 2017, `https://mathoverflow.net/q/268972`.

[2] Ira M. Gessel, *Symmetric functions and P-recursiveness*, J. Combin. Theory Ser. A **53** (1990), no. 2, 257–285, DOI 10.1016/0097-3165(90)90060-A. MR1041448

[3] Charles M. Grinstead and J. Laurie Snell, *Introduction to probability*, American Mathematical Society, 1997, `https://math.dartmouth.edu/~prob/prob/prob.pdf`.

[4] Gerald Guralnik, Tony Warnock, and Charles Zemach, *Random selection from trace-biased invariant distributions of* SU(3) *matrices*, J. Comput. Phys. **61** (1985), no. 1, 105–118, DOI 10.1016/0021-9991(85)90063-4. MR811563

[5] N. Kaiser, *Mean eigenvalues for simple, simply connected, compact Lie groups*, J. Phys. A **39** (2006), no. 49, 15287–15298, DOI 10.1088/0305-4470/39/49/013. MR2277101

[6] Sergey Kitaev, *Patterns in permutations and words*, Monographs in Theoretical Computer Science. An EATCS Series, Springer, Heidelberg, 2011. With a foreword by Jeffrey B. Remmel, DOI 10.1007/978-3-642-17333-2. MR3012380

[7] Wilhelm Magnus, Fritz Oberhettinger, and Raj Pal Soni, *Formulas and theorems for the special functions of mathematical physics*, Third enlarged edition. Die Grundlehren der mathematischen Wissenschaften, Band 52, Springer-Verlag New York, Inc., New York, 1966. MR0232968

[8] Frank W. J. Olver et al., *NIST digital library of mathematical functions*, `https://dlmf.nist.gov/`.

[9] Jean-Pierre Serre, *On the values of the characters of compact lie groups*, Oberwolfach Reports **1** (2004), 666–667.

[10] N. J. A. Sloane, *An on-line version of the encyclopedia of integer sequences*, Electron. J. Combin. **1** (1994), Feature 1, approx. 5. MR1269167

[11] Edward C. Titchmarsh, *Introduction to the theory of fourier integrals*, Oxford University Press, 1948.

[12] Wikipedia, *Transformation de hankel*, `https://fr.wikipedia.org/wiki/Transformation_de_Hankel`.

[13] Herbert S. Wilf, *Mathematics : An experimental science*, Princeton Companion to Mathematics (Timothy Gowers, ed.), Princeton University Press, 2008, pp. 997–1006.

[14] Qiaochu Yuan, *Sato-tate measure for GL(3) automorphic forms*, MathOverflow, 2012, `https://mathoverflow.net/q/101594`.

Gilles Lachaud, c/o David Kohel, Aix Marseille Univ, CNRS, Centrale Marseille, I2M, Marseille, France, Case 907, 13288 MARSEILLE Cedex 9, France
Email address: `david.kohel@univ-amu.fr`

Contemporary Mathematics
Volume **770**, 2021
https://doi.org/10.1090/conm/770/15437

Stable models of plane quartics with hyperelliptic reduction

Reynald Lercier, Elisa Lorenzo García, and Christophe Ritzenthaler

ABSTRACT. Let $C/K : F = 0$ be a smooth plane quartic over a complete discrete valuation field K. In (Lercier–Liu–Lorenzo-Garcìa–Ritzenthaler, 2019), the authors give various characterizations of the reduction (i.e. non-hyperelliptic genus 3 curve, hyperelliptic genus 3 curve or bad) of the stable model of C: in terms of the existence of a special plane quartic model and in terms of the valuations of the Dixmier-Ohno invariants of C. The last one gives in particular an easy computable criterion for the reduction type. However, it does not produce a stable model, even in the case of good reduction. In this paper we give an algorithm to obtain (an approximation of) the stable model when the reduction of the latter is hyperelliptic and the characteristic of the residue field is not 2. This is based on a new criterion giving the reduction type in terms of the valuations of the theta constants of C. Some examples of the computation of these models are given.

1. Introduction and main result

Let K be a complete discrete valuation field with valuation v and valuation ring $\mathcal{O}$ containing a maximal ideal generated by π. Let $k = \mathcal{O}/\langle \pi \rangle$ be the residual field. In the following, the expression "after a possible extension of K" means that we are allowed to take a finite extension of K and still call the corresponding notions $K, \mathcal{O}, v, \pi$. When F is an integral polynomial, i.e. with coefficients in $\mathcal{O}$, we denote $\bar{F}$ its reduction modulo π.

Given a genus 3 non-hyperelliptic curve C/K, in [**15**], the authors answer the following question: after a possible extension of K, what is the reduction type of the stable model $\mathcal{C}/\mathcal{O}$ of C? By this, we mean to distinguish between

- $\mathcal{C}_k$ is still a non-hyperelliptic curve of genus 3. We say that C has *potentially good quartic reduction*;
- $\mathcal{C}_k$ is a hyperelliptic curve of genus 3. We say that C has *potentially good hyperelliptic reduction*;
- $\mathcal{C}_k$ is not a curve of genus 3. We say that C has *geometrically bad reduction.*

In the first case, the special fiber is again a smooth plane quartic over k, whereas in the second case it is isomorphic over $\bar{k}$ to $y^2 = f(x, z)$ where $f$is a binary octic with

2020 *Mathematics Subject Classification.* Primary 11G20, 14Q05, 14D10, 14D20, 14H25.

Key words and phrases. smooth plane quartic, reduction, hyperelliptic reduction, invariants, valuation.

no multiple roots. If [**15**] gives an algorithm to generically reconstruct the special fiber in the first two cases when $\mathrm{char}(k) > 7$ or $\mathrm{char}(k) = 0$, it does so without computing equations for the stable model over K. This motivates the present work.

We recall that given a curve C/K, a model $\mathcal{C}$ is stable if over $\bar{K}$, it is reduced, connected, has only nodal singularities and all of its irreducible components isomorphic to $\mathbb{P}^1$ meet the other components in at least 3 points. This notion is compatible with base change, and in particular the models obtained in this article will be stable, see remark 4.7. The stable model of a curve of genus $g \geq 2$ is unique up to $\bar{K}$-isomorphism.

EXAMPLE 1.1. The discriminant of the Klein quartic given by $C : x_1^3x_2 + x_2^3x_3 + x_3^3x_1 = 0$ is equal to 7^7, hence the model $x_1^3x_2 + x_2^3x_3 + x_3^3x_1 = 0$ over $\mathbb{Z}$ has good reduction everywhere except at $\pi = 7$. To study the reduction type of the stable model at 7, notice that C is $\bar{\mathbb{Q}}$-isomorphic to the curve [**8**, pp.56]

$$(x_1^2 + x_2^2 + x_3^2)^2 + \sqrt{-7}\alpha^2 \cdot (x_1^2x_2^2 + x_2^2x_3^2 + x_3^2x_1^2) = 0$$

with $\alpha = \frac{-1+\sqrt{-7}}{2}$. Consider now the scheme

$$\mathcal{C} : \begin{cases} y^2 = -(x_1^2x_2^2 + x_2^2x_3^2 + x_3^2x_1^2), \\ \sqrt[4]{-7}\alpha \cdot y = x_1^2 + x_2^2 + x_3^2 \end{cases}$$

in the weighted projective space $\mathbb{P}^{(1,1,1,2)}$ over the ring of integers $\mathcal{O}$ of $K = \mathbb{Q}_7(\sqrt[4]{-7})$. Its generic fiber is isomorphic over K to C whereas $\mathcal{C}_k$ is isomorphic over $\bar{\mathbb{F}}_7$ to

$$\begin{cases} y^2 = -(x_1^2x_2^2 + x_2^2x_3^2 + x_3^2x_1^2), \\ 0 = x_1^2 + x_2^2 + x_3^2 \end{cases}$$

which turns out to be the hyperelliptic curve $y^2 = x^8 + 14x^4z^4 + z^8$.

The shape of the stable model in the previous example is symptomatic of the situation and motivates the following definition.

DEFINITION 1.2. (Def. 1.3, [**15**]) Let C/K be a smooth plane quartic. We say that C admits a *toggle model* if there exist an integer $s > 0$, a primitive (i.e. the gcd of its coefficients is 1) quartic form $G \in \mathcal{O}[x_1, x_2, x_3]$ and a primitive quadric $Q \in \mathcal{O}[x_1, x_2, x_3]$ with $\bar{Q}$ irreducible such that $Q^2 + \pi^s G = 0$ is K-isomorphic to C. If moreover $\bar{Q} = 0$ intersects $\bar{G} = 0$ transversely in 8 distinct $\bar{k}$-points, we say that C admits a *good toggle model.*

PROPOSITION 1.3. *(Prop. 1.2, [**15**]) Suppose $p \neq 2$. Let C/K be a smooth plane quartic having a good toggle model $Q^2 + \pi^{2s}G = 0$. Let us denote $\mathcal{C}/\mathcal{O}$ the subscheme of the weighted projective space $\mathbb{P}^{(1,1,1,2)}$ defined by*

$$\begin{cases} y^2 + G = 0, \\ \pi^s y - Q = 0. \end{cases}$$

Then the generic fiber is isomorphic to the plane smooth quartic C/K which has good hyperelliptic reduction. The special fiber of $\mathcal{C}$ is isomorphic to the double cover of $\bar{Q} = 0$ ramified over the 8 distinct intersection $\bar{k}$-points of $\bar{Q} = 0$ with $\bar{G} = 0$.

REMARK 1.4. In [**15**, Thm. 2.10] there is an equivalent result for the characteristic 2 case.

THEOREM 1.5. *([**15**, Thm. 1.4]) Let C/K be a plane smooth quartic. Then C has good hyperelliptic reduction if and only if C has a good toggle model over K.*

Over $\mathbb{C}$, it is well-known that one can associate to the Jacobian of a genus g curve $N = 2^{g-1}(2^g+1)$ values, which are called (even) theta constants. We will need a fancy version of these values over K (see Section 2) but the intuition remains the same. Now, over a DVR, we can multiply the theta constants by a common factor so that they become integral and that the minimum of their valuations is 0. These new values $(\theta_i)_{i=0,\dots,N-1}$ are uniquely defined up to a unit in $\mathcal{O}$ and we call them the *integral theta constants* of the curve. In this paper we prove the following result:

THEOREM 1.6. *Let C be a smooth plane quartic over K with residue field of characteristic different from 2. The curve C has potentially good hyperelliptic reduction if and only if there is a unique integral theta constant of C with positive valuation.*

The direct implication of Theorem 1.6 is relatively straightforward, for the converse we construct an explicit toggle model given by a Riemann model of the quartic and we show that it is good using the relations between the theta constants given in [**21**]. The construction of the toggle model is completely explicit in the proof and allows us to compute a stable model in this case. We implemented this construction in MAGMA [**3**] and we performed numerical experiments on some example curves.

REMARK 1.7. A similar procedure could actually be realized over a number field (and would be easier to implement) when the 2-structure of the curve is not defined over a too large extension. Unfortunately, in general, this structure is defined over an extension of degree $\#\operatorname{Sp}_6(\mathbb{F}_2) = 1451520$ ([**11**]).

REMARK 1.8. In [**15**], the distinction between potentially good hyperelliptic reduction and geometrically bad reduction in terms of the Dixmier-Ohno is possible only when $p > 7$. Our present algorithm has the advantage to allow the characteristics $p = 3, 5$ and 7.

2. Link with theta constants

Let R be a ring and $S = \operatorname{Spec} R$. Let X/R be an abelian scheme and $\mathcal{L}$ be a relatively ample line bundle on X such that $(-1_X)^*\mathcal{L} \simeq \mathcal{L}$. Fix an isomorphism $\epsilon : 0^*\mathcal{L} \simeq \mathcal{O}_S$ where $0 : S \to X$ is the zero section. To any $s \in \Gamma(X, \mathcal{L})$, Mumford associates (see [**18**, Appendix I]) a morphism $\vartheta_s : X[2] \to \mathbb{A}^1_S$. Following [**18**, Prop. 5.11] (see also loc. cit. Definition. 5.8), in the special case where $S = \operatorname{Spec}\mathbb{C}$ and $\mathcal{L}$ is the basic line bundle on $X_\tau = \mathbb{C}^g/(\mathbb{Z}^g + \tau\mathbb{Z}^g)$ (see loc. cit. p. 36), and after fixing an isomorphism $X[n] \simeq (\frac{1}{n}\mathbb{Z}/\mathbb{Z})^{2g} : x \mapsto (x_1, x_2)$, s is uniquely defined up to a multiplicative constant and there is a unique choice of ϵ such that

$$\vartheta_s(x) = \vartheta^\alpha \left[\begin{smallmatrix} x_1 \\ x_2 \end{smallmatrix}\right](\tau) = e^{-i\pi x_1 . x_2} \cdot \vartheta \left[\begin{smallmatrix} x_1 \\ x_2 \end{smallmatrix}\right](\tau) \tag{2.1}$$

where $\vartheta\left[\begin{smallmatrix} x_1 \\ x_2 \end{smallmatrix}\right](\tau)$ is the value at 0 of the classical theta function with characteristic $\left[\begin{smallmatrix} x_1 \\ x_2 \end{smallmatrix}\right]$ [**17**, p.192].

Let $\mathcal{C}/R$ be a proper and flat scheme whose fibers are smooth curves of genus $g > 0$. Then $\operatorname{Pic}^0(\mathcal{C})/R$ is an abelian scheme and the previous theory can be applied.

Let D_0 be a theta characteristic divisor on $\mathcal{C}$. Recall that a theta characteristic divisor D is a divisor such that $2D \sim \kappa$, where κ is the canonical divisor (relative to S). The divisor $\mathrm{Sym}^{g-1}\mathcal{C} - D_0 \subset \mathrm{Pic}^0(\mathcal{C})$ is symmetric and defines a symmetric line bundle $\mathcal{L}$ on $\mathrm{Pic}^0(\mathcal{C})$ which induces the canonical principal polarization on $\mathrm{Pic}^0(\mathcal{C})$. In particular there exists a unique (up to a multiplicative constant) $s \in \Gamma(\mathrm{Pic}^0(\mathcal{C}), \mathcal{L})$.

Let $R = \Bbbk$ where $\Bbbk$ is a field of characteristic different from 2 and for a divisor D on $\mathcal{C}/\Bbbk$, let us denote $L(D)$ its Riemann-Roch space. Mumford showed in [**16**] that the function $e_* : D \mapsto \dim L(D) \pmod 2$ is a quadratic function on the set of theta characteristic divisors D. We say that D is even (resp. odd) if $e_*(D) = 0$ (resp. 1). Since the map $D \mapsto D - D_0$ is a bijection between the set of theta characteristic divisors and the set of 2-torsion points, one can say that the corresponding $\vartheta_D = \vartheta_s(D - D_0)$ is even (resp. odd). The study of Arf invariants of quadratic forms shows that there are $2^{g-1}(2^g+1)$ even (resp. $2^{g-1}(2^g-1)$ odd) theta characteristic divisors and the corresponding ϑ_D are called *theta constants* –or Thetanullwerte–. As stated in loc. cit. p.182 and refined in [**13**], the classical Riemann singularity theorem over $\mathbb{C}$ extends to the present setting and in particular we get that $\vartheta_D = 0$ if and only if $\dim L(D) > 0$ (the odd ϑ_D are therefore always equal to zero).

In the particular case where $\mathcal{C}/\Bbbk$ is a curve of genus 3, Clifford's theorem [**12**, IV.Th.5.4] shows that $\dim L(D) \leq 2$, the equality being possible only when C is hyperelliptic and for a unique even theta characteristic divisor. Hence one recovers the classical result that C has 36 (resp. 35) non-zero theta constants when C is non-hyperelliptic (resp. hyperelliptic).

Coming back to the case where $R = \mathcal{O}$ is a DVR and C/K a non-hyperelliptic genus 3 we denote (after a possible extension of K) its integral theta constants by $(\theta_0, \ldots, \theta_{35}) \in \mathcal{O}^{36}$.

PROOF. (of Theorem 1.6) Let us assume that C has potentially good hyperelliptic reduction and let $\mathcal{C}/\mathcal{O}$ be a a smooth model of C such that $\mathcal{C}_k$ is hyperelliptic. Since $\mathrm{Pic}^0(\mathcal{C})/R$ is an abelian scheme, we can use the algebraic thetas $\vartheta_s(x) \in \mathcal{O}$ defined above. Mumford [**18**, Appendix I] has proved that these values are equal up to a constant $\lambda \in K^*$ to the ϑ_D on C, therefore there are exactly 36 of them which are non-zero. Moreover they shall also reduce to the theta constants on $\mathcal{C}_k$ which is hyperelliptic of genus 3 and therefore exactly one of the 36 non-zero ones has positive valuation. Therefore after ordering them, the non-zero $\vartheta_s(x)$ coincides up to a unit in $\mathcal{O}$ with $(\theta_0, \ldots, \theta_{35})$. We therefore get that exactly one of the latter has positive valuation.

In the opposite direction, we assume now that among the $(\theta_i)_{i=0,\ldots,35}$, there is a unique one with positive valuation. We are going to use the beautiful work of Weber [**21**] to obtain a good toggle model. Although this work is of course written over $\mathbb{C}$ in the language of classical theta functions, the link (2.1) between classical theta constants and algebraic theta constants shows that all the algebraic homogeneous relations can be used with the latter ones.

Weber ([**21**, p.108], see also [**9**]) introduces 9 values $(a_{ij})_{i,j=1,2,3}$ [1] given as homogeneous quotient of products of theta constants. In Section 4, we will see the relations of these constants with the geometry of the curve (through its bitangents) but we will not need them right now. We will assume that the theta constant

[1] In Weber's notation $a_{1i} = a_i$, $a_{2i} = a'_i$ and $a_{3i} = a''_i$.

denoted $\vartheta\left[\begin{smallmatrix}100\\001\end{smallmatrix}\right]$ in Weber's book is the one which corresponds to the one with positive valuation (this choice can be made without loss of generality after a choice of a right symplectic basis of the 2-torsion for the Weil pairing). The expressions of the a_{ij} in terms of the theta constants imply that only a_{11} and a_{31} have both positive valuation, equal to the one of $\vartheta\left[\begin{smallmatrix}100\\001\end{smallmatrix}\right]$. Weber then gives an explicit construction of a quartic form $F \in K[x_1, x_2, x_3]$ such that $F = 0$ is isomorphic to C from the a_{ij}. This form is

$$F = (x_1u_1 + x_2u_2 - x_3u_3)^2 - 4x_1u_1x_2u_2 \tag{2.2}$$

where the u_i are linear forms, solutions of the linear system

$$\begin{bmatrix} 1 & 1 & 1 \\ \frac{1}{a_{11}} & \frac{1}{a_{12}} & \frac{1}{a_{13}} \\ \frac{1}{a_{21}} & \frac{1}{a_{22}} & \frac{1}{a_{23}} \end{bmatrix} \cdot \begin{bmatrix} u_1 \\ u_2 \\ u_3 \end{bmatrix} = \begin{bmatrix} 1 & 1 & 1 \\ a_{11} & a_{12} & a_{13} \\ a_{21} & a_{22} & a_{23} \end{bmatrix} \cdot \begin{bmatrix} x_1 \\ x_2 \\ x_3 \end{bmatrix}. \tag{2.3}$$

Let us denote $Q = x_1u_1 + x_2u_2 - x_3u_3$ and $G = -4x_1u_1x_2u_2$. A series of computations (see Section 3) using relations between the theta constants shows that

(1) the u_i are defined over $\mathcal{O}$ (Lemma 3.1);
(2) $G = \pi^s G_0$ for $s > 0$ an integer, G_0 integral and primitive and $\bar{G}_0 = 0$ intersects $\bar{Q} = 0$ in 8 distinct points (Lemma 3.2);
(3) $\bar{Q}$ is a non-degenerate quadric (end of Lemma 3.2).

We therefore get that F is a good toggle model of C and so from Proposition 1.3 we get that C has potentially good hyperelliptic reduction. □

3. A Riemann model providing a good toggle model

We resume with the notation of previous section. Recall that a_{11} and a_{31} had both positive valuation, equal to the one of $\vartheta\left[\begin{smallmatrix}100\\001\end{smallmatrix}\right]$. In the sequel, we denote $a_{11} = \pi^{v_0}a'_{11}$ and $a_{31} = \pi^{v_0}a'_{31}$ with $v_0 > 0$ an integer and a'_{11} and a'_{31} units. If $\ell = ax_1 + bx_2 + cx_3$ is a linear form over K, we denote $v(\ell) = \min(v(a), v(b), v(c))$. In particular for the linear form u_i defined in (2.3), we let $v_i = v(u_i)$ for $i = 1, 2, 3$.

LEMMA 3.1. *The valuations v_i are non-negative.*

PROOF. We already know that $v(a_{ij}) \geq 0$ and $v(x_i) = 0$, so it is enough to check that the valuations of all the entries of the inverse of the matrix

$$M = \begin{bmatrix} 1 & 1 & 1 \\ \frac{1}{a_{11}} & \frac{1}{a_{12}} & \frac{1}{a_{13}} \\ \frac{1}{a_{21}} & \frac{1}{a_{22}} & \frac{1}{a_{23}} \end{bmatrix}$$

are also non-negative. We compute the inverse of M by computing the adjoint matrix and dividing by the determinant of M.

The determinant of M is equal to

$$\begin{aligned}&(-a_{11}a_{12}a_{21}a_{23} + a_{11}a_{12}a_{22}a_{23} + a_{11}a_{13}a_{21}a_{22} \\ &\qquad - a_{11}a_{13}a_{22}a_{23} - a_{12}a_{13}a_{21}a_{22} + a_{12}a_{13}a_{21}a_{23}) \,/\, (a_{11}a_{12}a_{13}a_{21}a_{22}a_{23}) = \\ &= \frac{a_{11}(-a_{12}a_{21}a_{23} + a_{12}a_{22}a_{23} + a_{13}a_{21}a_{22} - a_{13}a_{22}a_{23}) - a_{12}a_{13}a_{21}(a_{22} - a_{23})}{a_{11}a_{12}a_{13}a_{21}a_{22}a_{23}}\end{aligned}$$

where the factor $a_{22} - a_{23}$ equals

$$i\frac{\vartheta\left[\begin{smallmatrix}000\\100\end{smallmatrix}\right]\vartheta\left[\begin{smallmatrix}110\\111\end{smallmatrix}\right]\vartheta\left[\begin{smallmatrix}001\\000\end{smallmatrix}\right]\vartheta\left[\begin{smallmatrix}111\\011\end{smallmatrix}\right]}{\vartheta\left[\begin{smallmatrix}101\\111\end{smallmatrix}\right]\vartheta\left[\begin{smallmatrix}011\\100\end{smallmatrix}\right]\vartheta\left[\begin{smallmatrix}111\\101\end{smallmatrix}\right]\vartheta\left[\begin{smallmatrix}001\\110\end{smallmatrix}\right]}$$

by [**21**, 16.14, pp.110]. Hence, $a_{22}-a_{23}$ has valuation 0. Therefore, $v(\det M) = -v_0$. On the other hand, the valuation of the entries of the adjoint matrix are greater or equal to $-v_0$. So, the result follows. □

If we multiply both sides of eq. (2.3) by a_{11} and we look at the valuation of the entries in the equation, we get

$$\begin{bmatrix} v_0 & v_0 & v_0 \\ 0 & v_0 & v_0 \\ v_0 & v_0 & v_0 \end{bmatrix}, \begin{bmatrix} v_1 \\ v_2 \\ v_3 \end{bmatrix} \text{ for the LHS, and } \begin{bmatrix} v_0 & v_0 & v_0 \\ 2v_0 & v_0 & v_0 \\ v_0 & v_0 & v_0 \end{bmatrix}, \begin{bmatrix} 0 \\ 0 \\ 0 \end{bmatrix} \text{ for the RHS,}$$

from which we easily read off that $v_1 \geq v_0$ since v_2, v_3 are non-negative by Lemma 3.1. We can write $u_1 = \pi^{v_1} u_1'$. This proves that the Riemann model

$$F = \underbrace{(\pi^{v_1} x_1 u_1' + x_2 u_2 - x_3 u_3)^2}_{Q} - 4\pi^{v_1} \underbrace{x_1 u_1' x_2 u_2}_{G} = 0,$$

is defined over $\mathcal{O}$. We are going to show that it is a good toggle model.

LEMMA 3.2. *The intersection of $\bar{Q} = 0$ with $\bar{G} = 0$ is transverse and the quadratic form $\bar{Q} = x_2 \bar{u}_2 - x_3 \bar{u}_3$ is non-degenerate.*

PROOF. Let us write down the reduction of eq. (2.3) before studying the intersection points (here $\epsilon = 0$ if $v_1 > v_0$ and 1 otherwise),

$$\bar{u}_2 + \bar{u}_3 = x_1 + x_2 + x_3\,, \tag{3.1}$$

$$\epsilon \frac{\bar{u}_1'}{\bar{a}_{11}'} + \frac{\bar{u}_2}{\bar{a}_{12}} + \frac{\bar{u}_3}{\bar{a}_{13}} = \bar{a}_{12} x_2 + \bar{a}_{13} x_3\,, \tag{3.2}$$

$$\frac{\bar{u}_2}{\bar{a}_{22}} + \frac{\bar{u}_3}{\bar{a}_{23}} = \bar{a}_{21} x_1 + \bar{a}_{22} x_2 + \bar{a}_{23} x_3\,. \tag{3.3}$$

One can also add the following equation (see [**9**, Prop. 2]):

$$\frac{u_1}{a_{31}} + \frac{u_2}{a_{32}} + \frac{u_3}{a_{33}} = a_{31} x_1 + a_{32} x_2 + a_{33} x_3,$$

which reduces to

$$\epsilon \frac{\bar{u}_1'}{\bar{a}_{31}'} + \frac{\bar{u}_2}{\bar{a}_{32}} + \frac{\bar{u}_3}{\bar{a}_{33}} = \bar{a}_{32} x_2 + \bar{a}_{33} x_3. \tag{3.4}$$

We compute the intersection of $\bar{Q} = 0$ successively with $x_2 = 0, x_1 = 0, \bar{u}_2 = 0$ and then $\bar{u}_1' = 0$.

- Intersection with $x_2 = 0$.

$$\{\bar{Q} = 0\} \cap \{x_2 = 0\} = \{x_3 \bar{u}_3 = 0\} \cap \{x_2 = 0\} = \{P_1 = (1:0:0)\} \cup (\{\bar{u}_3 = 0\} \cap \{x_2 = 0\}).$$

To compute the second intersection point above, denoted P_2, we see from equations (3.1) and (3.3) that

$$\left(\frac{1}{\bar{a}_{23}} - \frac{1}{\bar{a}_{22}}\right) \bar{u}_3 = \left(\bar{a}_{21} - \frac{1}{\bar{a}_{22}}\right) x_1 + \left(\bar{a}_{22} - \frac{1}{\bar{a}_{22}}\right) x_2 + \left(\bar{a}_{23} - \frac{1}{\bar{a}_{22}}\right) x_3. \tag{3.5}$$

Letting $x_2 = 0$ and $\bar{u}_3 = 0$ we need to look at the valuation of $a_{21} - 1/a_{22}$. As in Lemma 3.1, Weber [**21**, 16.11, 16.14] gives an expression for this difference in terms of the theta constants

$$(a_{21}a_{22}-1)(a_{22})^{-1} = -\frac{\vartheta\left[\begin{smallmatrix}110\\000\end{smallmatrix}\right]\vartheta\left[\begin{smallmatrix}000\\011\end{smallmatrix}\right]\vartheta\left[\begin{smallmatrix}000\\100\end{smallmatrix}\right]\vartheta\left[\begin{smallmatrix}110\\111\end{smallmatrix}\right]}{\vartheta\left[\begin{smallmatrix}011\\011\end{smallmatrix}\right]\vartheta\left[\begin{smallmatrix}101\\000\end{smallmatrix}\right]\vartheta\left[\begin{smallmatrix}101\\111\end{smallmatrix}\right]\vartheta\left[\begin{smallmatrix}011\\100\end{smallmatrix}\right]}(-i)\frac{\vartheta\left[\begin{smallmatrix}101\\111\end{smallmatrix}\right]\vartheta\left[\begin{smallmatrix}011\\100\end{smallmatrix}\right]}{\vartheta\left[\begin{smallmatrix}110\\001\end{smallmatrix}\right]\vartheta\left[\begin{smallmatrix}000\\010\end{smallmatrix}\right]} = i\frac{\vartheta\left[\begin{smallmatrix}110\\000\end{smallmatrix}\right]\vartheta\left[\begin{smallmatrix}000\\011\end{smallmatrix}\right]\vartheta\left[\begin{smallmatrix}000\\100\end{smallmatrix}\right]\vartheta\left[\begin{smallmatrix}110\\111\end{smallmatrix}\right]}{\vartheta\left[\begin{smallmatrix}011\\011\end{smallmatrix}\right]\vartheta\left[\begin{smallmatrix}101\\000\end{smallmatrix}\right]\vartheta\left[\begin{smallmatrix}110\\001\end{smallmatrix}\right]\vartheta\left[\begin{smallmatrix}000\\010\end{smallmatrix}\right]}$$

and one can check that the valuation is 0. Hence we get $P_2 = \left(\frac{\bar{a}_{23}-1/\bar{a}_{22}}{1/\bar{a}_{22}-\bar{a}_{21}} : 0 : 1\right)$.

• Intersection with $x_1 = 0$. We deal with $\{\bar{Q}=0\}\cap\{x_1=0\}$ in the same way. We use eq. (3.5) in conjunction with

$$\left(\frac{1}{\bar{a}_{22}}-\frac{1}{\bar{a}_{23}}\right)\bar{u}_2 = \left(\bar{a}_{21}-\frac{1}{\bar{a}_{23}}\right)x_1 + \left(\bar{a}_{22}-\frac{1}{\bar{a}_{23}}\right)x_2 + \left(\bar{a}_{23}-\frac{1}{\bar{a}_{23}}\right)x_3, \tag{3.6}$$

and $x_1 = 0$ to get that the coordinates of the intersection points must satisfy

$$a_{22}x_2^2 + (a_{22}+a_{23})x_2x_3 + a_{23}x_3^2 = 0.$$

The discriminant of this quadratic form has valuation 0 since $v(a_{22}-a_{23}) = 0$, still using [**21**, 16.14]. We have that -1 and $-\frac{a_{23}}{a_{22}}$ are the two different roots of this polynomial with $x_3 = 1$, and we get that

$$\{\bar{Q}=0\}\cap\{x_1=0\} = \{P_3 = (0:-1:1),\ P_4 = (0:-\frac{\bar{a}_{23}}{\bar{a}_{22}}:1)\}.$$

• Intersection with $\bar{u}_2 = 0$. The case $\{\bar{Q}=0\}\cap\{\bar{u}_2=0\}$ is also similar. Thanks to [**21**, 16.14], $v(a_{23}a_{22}-1) = v(a_{23}a_{21}-1) = 0$ and so if $x_3 = 0$, eq. (3.6) gives the intersection point

$$P_5 = \left(1 : -\frac{\bar{a}_{23}\bar{a}_{22}-1}{\bar{a}_{23}\bar{a}_{21}-1} : 0\right).$$

If $\bar{u}_3 = 0$, we use equations (3.1) and (3.3) to compute a second intersection point

$$P_6 = (a_{22}-a_{23} : a_{23}-a_{21} : a_{21}-a_{22}).$$

• Intersection with $\bar{u}_1' = 0$. The case $\{\bar{Q}=0\}\cap\{\bar{u}_1'=0\}$ is a bit more delicate as we miss some relations from Weber which we have to work out. Letting $x_3 = 1$ we can write $\bar{u}_3 = x_2\bar{u}_2$. We use this to write $\bar{u}_2$ in terms of x_2 in eq. (3.4) and in eq. (3.2). Equating both expressions for $\bar{u}_2$, we get a degree two equation for x_2:

$$(\bar{a}_{12}\bar{a}_{13}-\bar{a}_{32}\bar{a}_{33})\left(\frac{1}{\bar{a}_{13}\bar{a}_{33}}x_2^2 + \left(\frac{1}{\bar{a}_{12}\bar{a}_{33}}+\frac{1}{\bar{a}_{13}\bar{a}_{32}}\right)x_2 + \frac{1}{\bar{a}_{12}\bar{a}_{32}}\right) = 0.$$

We first need to prove that it indeed defines a degree 2 equation, that is, that the leading coefficient $\bar{a}_{12}\bar{a}_{13}-\bar{a}_{32}\bar{a}_{33}$ is not zero. Secondly, in order to prove that we obtain two different solutions for x_2, we need to prove that the discriminant $\left(\frac{\bar{a}_{13}\bar{a}_{32}-\bar{a}_{12}\bar{a}_{33}}{\bar{a}_{12}\bar{a}_{13}\bar{a}_{32}\bar{a}_{33}}\right)^2$ is also different from zero.

We first prove that the leading coefficient is not 0. By [**21**, 16.14],

$$a_{12}a_{13}-a_{32}a_{33} = \frac{\vartheta\left[\begin{smallmatrix}000\\110\end{smallmatrix}\right]\vartheta\left[\begin{smallmatrix}010\\100\end{smallmatrix}\right]}{\vartheta\left[\begin{smallmatrix}111\\000\end{smallmatrix}\right]\vartheta\left[\begin{smallmatrix}101\\010\end{smallmatrix}\right]}\left(\frac{\vartheta\left[\begin{smallmatrix}010\\001\end{smallmatrix}\right]\vartheta\left[\begin{smallmatrix}000\\011\end{smallmatrix}\right]}{\vartheta\left[\begin{smallmatrix}101\\111\end{smallmatrix}\right]\vartheta\left[\begin{smallmatrix}111\\101\end{smallmatrix}\right]} - \frac{\vartheta\left[\begin{smallmatrix}100\\010\end{smallmatrix}\right]\vartheta\left[\begin{smallmatrix}110\\000\end{smallmatrix}\right]}{\vartheta\left[\begin{smallmatrix}011\\100\end{smallmatrix}\right]\vartheta\left[\begin{smallmatrix}001\\110\end{smallmatrix}\right]}\right).$$

All the theta constants involved in the formula are units, so we only need to check that

$$D = \vartheta\left[\begin{smallmatrix}010\\001\end{smallmatrix}\right]\vartheta\left[\begin{smallmatrix}000\\011\end{smallmatrix}\right]\vartheta\left[\begin{smallmatrix}011\\100\end{smallmatrix}\right]\vartheta\left[\begin{smallmatrix}001\\110\end{smallmatrix}\right] - \vartheta\left[\begin{smallmatrix}100\\010\end{smallmatrix}\right]\vartheta\left[\begin{smallmatrix}110\\000\end{smallmatrix}\right]\vartheta\left[\begin{smallmatrix}101\\111\end{smallmatrix}\right]\vartheta\left[\begin{smallmatrix}111\\101\end{smallmatrix}\right]$$

is also a unit. We claim that

$$D = \pm\vartheta \left[\begin{smallmatrix}011\\000\end{smallmatrix}\right] \vartheta \left[\begin{smallmatrix}001\\010\end{smallmatrix}\right] \vartheta \left[\begin{smallmatrix}010\\101\end{smallmatrix}\right] \vartheta \left[\begin{smallmatrix}000\\111\end{smallmatrix}\right]$$

and so we get the result. The proof of the claim is a well-known game with (classical) theta constants that we will now play. By [**19**, Chap. II, Theorem 18] with the notation from there

$$\begin{bmatrix}\epsilon(1)\\ \epsilon'(1)\end{bmatrix} = \left[\begin{smallmatrix}010\\001\end{smallmatrix}\right], \begin{bmatrix}\epsilon(2)\\ \epsilon'(2)\end{bmatrix} = \left[\begin{smallmatrix}100\\010\end{smallmatrix}\right], \begin{bmatrix}\epsilon(3)\\ \epsilon'(3)\end{bmatrix} = \left[\begin{smallmatrix}011\\000\end{smallmatrix}\right], \left[\begin{smallmatrix}\lambda\\ \lambda'\end{smallmatrix}\right] = \left[\begin{smallmatrix}010\\010\end{smallmatrix}\right] \text{ and } \left[\begin{smallmatrix}\mu\\ \mu'\end{smallmatrix}\right] = \left[\begin{smallmatrix}001\\101\end{smallmatrix}\right]$$

one gets

$$\begin{aligned}\vartheta \left[\begin{smallmatrix}010\\001\end{smallmatrix}\right] \vartheta \left[\begin{smallmatrix}000\\011\end{smallmatrix}\right] \vartheta \left[\begin{smallmatrix}011\\100\end{smallmatrix}\right] \vartheta \left[\begin{smallmatrix}001\\110\end{smallmatrix}\right] \pm \vartheta \left[\begin{smallmatrix}100\\010\end{smallmatrix}\right] \vartheta \left[\begin{smallmatrix}110\\000\end{smallmatrix}\right] \vartheta \left[\begin{smallmatrix}101\\111\end{smallmatrix}\right] \vartheta \left[\begin{smallmatrix}111\\101\end{smallmatrix}\right] \\ \pm\vartheta \left[\begin{smallmatrix}011\\000\end{smallmatrix}\right] \vartheta \left[\begin{smallmatrix}001\\010\end{smallmatrix}\right] \vartheta \left[\begin{smallmatrix}010\\101\end{smallmatrix}\right] \vartheta \left[\begin{smallmatrix}000\\111\end{smallmatrix}\right] = 0.\end{aligned} \tag{3.7}$$

We need to prove that the first sign is negative to get our expression of D. We consider this relation for Riemann matrices[2] $\tau = \begin{pmatrix} * & 0 & * \\ 0 & * & 0 \\ * & 0 & * \end{pmatrix}$, that defines abelian varieties that are the product of an elliptic curve and a principally polarized abelian surface. By [**19**, Chap. I, Theorem 11] the equation (3.7) simplifies to

$$\vartheta^2 \left[\begin{smallmatrix}1\\0\end{smallmatrix}\right] \vartheta^2 \left[\begin{smallmatrix}0\\1\end{smallmatrix}\right] \left(\vartheta^2 \left[\begin{smallmatrix}00\\01\end{smallmatrix}\right] \vartheta^2 \left[\begin{smallmatrix}01\\10\end{smallmatrix}\right] \pm \vartheta^2 \left[\begin{smallmatrix}10\\00\end{smallmatrix}\right] \vartheta^2 \left[\begin{smallmatrix}11\\11\end{smallmatrix}\right] \pm \vartheta^2 \left[\begin{smallmatrix}01\\00\end{smallmatrix}\right] \vartheta^2 \left[\begin{smallmatrix}00\\11\end{smallmatrix}\right]\right) = 0,$$

where the theta constants here are the ones corresponding to the elliptic curve factor and the dimension 2 one. Now we use [**19**, Chap. I, Theorem 5] with the notation there

$$\begin{bmatrix}\epsilon(1)\\ \epsilon'(1)\end{bmatrix} = \left[\begin{smallmatrix}01\\10\end{smallmatrix}\right], \begin{bmatrix}\epsilon(2)\\ \epsilon'(2)\end{bmatrix} = \left[\begin{smallmatrix}10\\00\end{smallmatrix}\right], \begin{bmatrix}\epsilon(3)\\ \epsilon'(3)\end{bmatrix} = \left[\begin{smallmatrix}01\\00\end{smallmatrix}\right], \text{ and } \left[\begin{smallmatrix}\mu\\ \mu'\end{smallmatrix}\right] = \left[\begin{smallmatrix}10\\01\end{smallmatrix}\right]$$

to rewrite the expression inside the parentheses above as

$$\vartheta^2 \left[\begin{smallmatrix}00\\01\end{smallmatrix}\right] \vartheta^2 \left[\begin{smallmatrix}11\\11\end{smallmatrix}\right] \pm \vartheta^2 \left[\begin{smallmatrix}20\\01\end{smallmatrix}\right] \vartheta^2 \left[\begin{smallmatrix}11\\11\end{smallmatrix}\right] \pm \vartheta^2 \left[\begin{smallmatrix}11\\01\end{smallmatrix}\right] \vartheta^2 \left[\begin{smallmatrix}00\\11\end{smallmatrix}\right] = 0.$$

The third term is zero since the characteristic $\left[\begin{smallmatrix}11\\01\end{smallmatrix}\right]$ is odd. Finally, since by [**19**, Chap. I, Theorem 3] $\vartheta \left[\begin{smallmatrix}20\\01\end{smallmatrix}\right] = -\vartheta \left[\begin{smallmatrix}00\\01\end{smallmatrix}\right]$, the previous equation can be realized only if the first sign is a minus sign. This proves our claim on D and we get a degree 2 equation for x_2.

The discriminant of $\bar{Q}$ is equal to

$$\frac{\bar{a}_{21}\bar{a}_{22}\bar{a}_{23}(\bar{a}_{22}\bar{a}_{23} - 1)}{\bar{a}_{22} - \bar{a}_{23}}$$

which is not 0. Hence, the intersection $\{\bar{Q} = 0\} \cap \{\bar{u}'_1 = 0\}$ consists of two distinct points $\{P_7, P_8\}$.

• The intersection is transverse. We need to prove that $P_i \neq P_j$ for all distinct $i, j \in \{1, ..., 8\}$. Recall that:

$$P_1 = (1:0:0),\ P_2 = (\bar{a}_{23} - 1/\bar{a}_{22} : 0 : 1),\ P_3 = (0 : -1 : 1),\ P_4 = (0 : -\bar{a}_{23} : \bar{a}_{22}),$$

$$P_5 = (\bar{a}_{23}\bar{a}_{21} - 1 : 1 - \bar{a}_{23}\bar{a}_{22} : 0),\ P_6 = (\bar{a}_{22} - \bar{a}_{23} : \bar{a}_{23} - \bar{a}_{21} : \bar{a}_{21} - \bar{a}_{22}),\ P_7,\ P_8.$$

The six first points are clearly distinct: we have already seen that $v(a_{22} - a_{23}) = v(a_{22}a_{23} - 1) = v(a_{22}) = 0$ and again [**21**, 16.14] implies $v(a_{21} - a_{22}) = v(a_{21} - a_{23}) = 0$; and P_7 and P_8 are distinct. We will check now that they are also distinct from the first six points. For that we study the following intersections:

[2] using a diagonal matrix τ we would have only got that the second sign is negative, which is not what we want.

$\{\bar{Q}=0\}\cap\{\bar{u}_1'=0\}\cap\{x_2=0\}$: we have here $\bar{u}_1'=x_2=0$ and $x_3=1$, hence $\bar{u}_3=0$ from $\bar{Q}=0$ and from equations (3.2) and (3.4) we get $\bar{a}_{12}\bar{a}_{13}-\bar{a}_{32}\bar{a}_{33}=0$, which we have just seen is not. Hence, the intersection in empty and the points P_7 and P_8 are distinct from the points P_1 and P_2.

$\{\bar{Q}=0\}\cap\{\bar{u}_1'=0\}\cap\{x_2=0\}$: we get $x_1=\bar{u}_1'=0$, $x_3=1$ and $u_3=x_2u_2$. From eq. (3.3) we get $\bar{u}_3=x_2$ and hence equations (3.1) and (3.3) implies $\bar{a}_{13}\bar{a}_{32}-\bar{a}_{12}\bar{a}_{33}=0$. We have also already seen that this is not possible. Therefore, the points P_7, P_8 are distinct from the points P_3, P_4.

$\{\bar{Q}=0\}\cap\{\bar{u}_1'=0\}\cap\{\bar{u}_2=0\}$: in this situation we get $\bar{u}_1'=\bar{u}_2=\bar{u}_3=0$, which is not possible since those forms are linearly independent. We then conclude that P_7, P_8 are also distinct from P_5 and P_6.

• The conic $\bar{Q}=0$ is non-singular. If it were, $\bar{Q}$ would be the product of two linear forms L_1 and L_2. None of these lines $L_i=0$ is equal to $x_1=0$, $x_2=0$, $\bar{u}_1'=0$ or $\bar{u}_2=0$, since the intersection of the conic $\bar{Q}=0$ with each of them is two distinct points. Moreover, L_1 and L_2 define distinct lines. Hence, we have that 4 of the points $P_1,...,P_8$ are on $L_1=0$ and the other 4 on $L_2=0$, where points in the pairs $\{P_1,P_2\}$, $\{P_3,P_4\}$, $\{P_5,P_6\}$, $\{P_7,P_8\}$ are on different lines. Assume that P_1 is on L_1 and P_2 on L_2. Then P_5 is on L_2 since otherwise $L_1=0$ would be $\{x_3=0\}$ but $\{x_3=0\}$ is not contained in $\bar{Q}=0$ by eq. (3.6). Hence, we get

$$\{L_1=0\}=\overline{P_1P_6}:(\bar{a}_{21}-\bar{a}_{22})x_2+(\bar{a}_{21}-\bar{a}_{23})x_3=0.$$

Now, it is easy to check that neither P_3 or P_4 are on $L_1=0$: for P_3 it is enough to check that $a_{22}-a_{23}\neq 0$ and for P_4 that $\bar{a}_{21}(\bar{a}_{22}-\bar{a}_{23})\neq 0$, which is true by [**21**, 16.14].

This gives a contradiction with the assumption $\bar{Q}=L_1\cdot L_2$. Hence, $\bar{Q}=0$ defines a non-singular conic.

□

4. Bitangents of a smooth plane quartic

Let C/K be a smooth plane quartic curve given by $F=0$. As C is canonically embedded in the projective plane, the canonical divisors on C are the intersection of the lines with C. In particular we can describe the 28 odd theta characteristic divisors using bitangents.

DEFINITION 4.1. A line β is called a *bitangent* of C if the intersection divisor $(\beta\cdot C)$ is of the form $2P+2Q$ for some not necessarily distinct points P,Q of C. The divisor $D_\beta:=P+Q$ is the odd theta characteristic divisor associated to β.

REMARK 4.2. The bitangents of a curve can be computed by looking at the singular points of the dual curve, but this is rather expensive in terms of computations since the singularities also contains the tangents at inflexion points. A better approach is to work out the two algebraic conditions in a and b under which the form $F(x_1,ax_1+b,1)$ is a perfect square and to look for the solutions of the corresponding system. Of course, one has to take care of the bitangents which are not of the form $x_2=ax_1+b$.

DEFINITION 4.3. Let $S=\{\beta_i\}$ be a set of bitangents of C. The set S is called *azygetic* if there is no conic Q such that the intersection divisor $Q\cdot C\geq D_{\beta_1}+D_{\beta_2}+D_{\beta_3}$ for $\beta_1,\beta_2,\beta_3\in S$. An azygetic set of 7 elements is called an *Aronhold set*.

Among the $\binom{28}{7}$ subsets of 7 bitangents of C, 288 of them form an Aronhold set, see [**10**, after Cor. 2.5]. Note that if one knows equations for the 28 bitangents and the curve, a tedious but straightforward computation using the definition of azygeticness allow to exhibit an Aronhold set.

We resume with the notation from the previous sections. Let us recall the following result [**20**].

THEOREM 4.4. *Let C be a smooth plane quartic curve and $S = \{\beta_i\}_{i=1,\dots,7}$ be an Aronhold set. After a linear change of variables, we may assume that the β_i are given by the equations:*

$$\beta_i : x_i = 0 \quad \beta_4 : x_1 + x_2 + x_3 = 0 \quad \beta_{4+i} : a'_{i1}x_1 + a'_{i2}x_2 + a'_{i3}x_3 = 0,$$

where $i \in \{1,2,3\}$. The coefficients a'_{ij} are multiples of the a_{ij} (defined in Section 2), i.e. $a'_{ij} = \eta_i a_{ij}$ for some $\eta_i \neq 0$, that are determined, up to sign, by the linear system:

$$\begin{bmatrix} \lambda_1 a'_{11} & \lambda_2 a'_{21} & \lambda_3 a'_{31} \\ \lambda_1 a'_{12} & \lambda_2 a'_{22} & \lambda_3 a'_{32} \\ \lambda_1 a'_{13} & \lambda_2 a'_{23} & \lambda_3 a'_{33} \end{bmatrix} \begin{bmatrix} 1/\eta_1^2 \\ 1/\eta_2^2 \\ 1/\eta_3^2 \end{bmatrix} = \begin{bmatrix} -1 \\ -1 \\ -1 \end{bmatrix},$$

where λ_1, λ_2, λ_3 are given by

$$\begin{bmatrix} \frac{1}{a'_{11}} & \frac{1}{a'_{21}} & \frac{1}{a'_{31}} \\ \frac{1}{a'_{12}} & \frac{1}{a'_{22}} & \frac{1}{a'_{32}} \\ \frac{1}{a'_{13}} & \frac{1}{a'_{23}} & \frac{1}{a'_{33}} \end{bmatrix} \begin{bmatrix} \lambda_1 \\ \lambda_2 \\ \lambda_3 \end{bmatrix} = \begin{bmatrix} -1 \\ -1 \\ -1 \end{bmatrix}.$$

After the previous change of variables, the plane quartic C is given by a Riemann model

$$(x_1u_1 + x_2u_2 - x_3u_3)^2 - 4x_1u_1x_2u_2 = 0,$$

where u_1, u_2, u_3 are given as in Section 2 by

$$\begin{cases} u_1 + u_2 + u_3 + x_1 + x_2 + x_3 = 0, \\ \frac{u_1}{a_{i1}} + \frac{u_2}{a_{i2}} + \frac{u_3}{a_{i3}} + a_{i1}x_1 + a_{i2}x_2 + a_{i3}x_3 = 0. \end{cases}$$

Moreover, we can express all the bitangents for this model as:

$$\begin{array}{llll} \beta_i & : x_i = 0\,, & \beta_{ij} & : u_k = 0\,, \\ \beta_4 & : x_1 + x_2 + x_3 = 0\,, & \beta_{4+i} & : a_{i1}x_1 + a_{i2}x_2 + a_{i3}x_3 = 0\,, \\ \beta_{i4} & : u_i + x_j + x_k = 0\,, & \beta_{i(4+l)} & : u_i/a_{li} + a_{lj}x_j + a_{lk}x_k = 0\,, \end{array}$$

$$\begin{array}{ll} \beta_{4(4+i)} & : \frac{u_1}{a_{i1}(1-a_{i2}a_{i3})} + \frac{u_2}{a_{i2}(1-a_{i3}a_{i1})} + \frac{u_3}{a_{i3}(1-a_{i1}a_{i2})} = 0\,, \\ \beta_{56} & : \frac{u_1}{1-a_{32}a_{33}} + \frac{u_2}{1-a_{33}a_{31}} + \frac{u_3}{1-a_{31}a_{32}} = 0\,, \\ \beta_{57} & : \frac{u_1}{1-a_{22}a_{23}} + \frac{u_2}{1-a_{23}a_{21}} + \frac{u_3}{1-a_{21}a_{22}} = 0\,, \\ \beta_{67} & : \frac{u_1}{1-a_{12}a_{13}} + \frac{u_2}{1-a_{13}a_{11}} + \frac{u_3}{1-a_{11}a_{12}} = 0\,, \end{array}$$

where $i, l \in \{1,2,3\}$ and $\{i,j,k\} = \{1,2,3\}$.

PROPOSITION 4.5. *Assume that C has potentially good hyperelliptic reduction. Given an Aronhold set S as in Theorem 4.4, there exists a constant $v_0 > 0$ such that either*

(1) *two of the a_{ij} (with the same value of j) have positive valuation v_0 and the rest have valuation equal to zero;*
(2) *or two of the a_{ij} (with the same value of j) have negative valuation $-v_0$ and the rest have valuation equal to zero;*

(3) *or all the a_{ij} have valuation equal to zero.*

PROOF. Look at the expressions for the coefficients a_{ij} in terms of the theta constants in page 108 formula (11′) in [**21**] and consider the different cases. For example, if the theta constant that has positive valuation v_0 is $\vartheta\left[\begin{smallmatrix}100\\001\end{smallmatrix}\right]$ as in the proof of Theorem 1.6, we can see that all a_{ij} have zero valuation except

$$a_{11}=\frac{\vartheta\left[\begin{smallmatrix}100\\001\end{smallmatrix}\right]\vartheta\left[\begin{smallmatrix}000\\101\end{smallmatrix}\right]}{\vartheta\left[\begin{smallmatrix}101\\000\end{smallmatrix}\right]\vartheta\left[\begin{smallmatrix}001\\100\end{smallmatrix}\right]}\text{ and }a_{31}=\frac{\vartheta\left[\begin{smallmatrix}110\\110\end{smallmatrix}\right]\vartheta\left[\begin{smallmatrix}100\\001\end{smallmatrix}\right]}{\vartheta\left[\begin{smallmatrix}001\\100\end{smallmatrix}\right]\vartheta\left[\begin{smallmatrix}011\\110\end{smallmatrix}\right]}$$

that have also positive valuation v_0, so we are in case (1) of the Proposition. □

THEOREM 4.6. *Let assume that C has potentially good hyperelliptic reduction. Given an Aronhold set $S=\{\beta_i\}_{i=1,\dots,7}$ we can construct another Aronhold set $\{\gamma_i\}_{i=1,\dots,7}$ for which $v(a'_{11})=v(a'_{31})>0$.*

PROOF. Notice that if we are already in the first case of Proposition 4.5, then we may assume that $v(a_{1j})=v(a_{3j})=v_0$ for $v_0>0$ and some j after permuting the set $\{\beta_1,\beta_2,\beta_3\}$ of bitangents, and indeed, that $j=1$ after permuting the set $\{\beta_4,\beta_5,\beta_6\}$.

If we are in the second case, and after a permutation of the sets $\{\beta_1,\beta_2,\beta_3\}$ and $\{\beta_4,\beta_5,\beta_6\}$, we may assume that a_{11} and a_{31} are the ones with negative valuations, that is, $\vartheta\left[\begin{smallmatrix}001\\100\end{smallmatrix}\right]$ is the theta constant with positive valuation equal to v_0. Then we can take the Aronhold system:

$$\gamma_1=\beta_{23},\ \gamma_2=\beta_2,\ \gamma_3=\beta_3,\ \gamma_4=\beta_{14},$$

$$\gamma_5=\beta_{15},\ \gamma_6=\beta_{16},\ \gamma_6=\beta_{16}.$$

So, $a'_{i1}=1/a_{i1}$ and $a'_{ij}=a_{ij}$ for $i\in\{1,2,3\}$ and $j\in\{2,3\}$, and this falls under the first case with $v(a'_{11})=v(a'_{31})=v_0$.

In the third case we have to distinguish several cases. If any of the differences $1-a_{ij}a_{i(j+1)}$ has positive valuation, then after permutations of the sets $\{\beta_1,\beta_2,\beta_3\}$ and $\{\beta_4,\beta_5,\beta_6\}$, we can assume that $\vartheta\left[\begin{smallmatrix}100\\000\end{smallmatrix}\right]$ is the theta constant with positive valuation, then again by using Weber's formulas [**21**, pp. 109], we have that $\{\gamma_1,\dots,\gamma_7\}=\{\beta_{23},\,\beta_{13}\,\beta_{12},\,\beta_4,\,\beta_{45},\,\beta_{46},\,\beta_{47}\}$ is an Aronhold system with $v(a'_{11})=v(a'_{31})=v_0$. If none of the differences $1-a_{ij}a_{i(j+1)}$ have positive valuation, then the differences $a_{ij}a_{i(j+1)}-a_{(i+1)j}a_{(i+1)(j+1)}$ have positive valuations. Hence, the Aronhold system $\{\gamma_1,\dots,\gamma_7\}=\{\beta_5,\,\beta_6\,\beta_7,\,\beta_4,\,\beta_1,\,\beta_2,\,\beta_3\}$ falls in the first case. □

REMARK 4.7. Let C/K be a plane quartic curve with potentially good hyperellitic reduction. Let $S=\{\beta_i\}_{i=1,\dots,7}$ be an Aronhold set as in Theorem 4.6, then the corresponding model as in (2.2) is a good toggle model with $Q=x_1u_1+x_2u_2-x_3u_3$, $G=-4x_1u_1x_2u_2/\pi^{v(u_1)}$ and $s=v(u_1)/2$. Moreover,

$$\begin{cases}y^2+G=0,\\ \pi^s y-Q=0.\end{cases}$$

is the stable model of C: it is reduced, connected and it does not have singularities or components of genus 0.

5. The algorithm and an example

In this section we start with a smooth plane quartic $C/K : F = 0$, where K is a finite extension of $\mathbb{Q}_p$ with $p \neq 2$. The previous sections lead to the following algorithm to compute a stable model when the curve has potentially good hyperelliptic reduction.

	Computation of the stable model of C
`Input:`	an equation F and a large enough precision e.
`Output:`	a stable model of C when C has potentially good hyperelliptic reduction
1	Compute the set B of bitangents of C over a small extension of K (using remark 4.2)
2	Take any two bitangents b_1, b_2 and let $S = \{b_1, b_2\}$
3	`While` $\#S \neq 7$:
4	`For` $b \in B \setminus S$:
5	`if` $S \cup \{b\}$ is azygetic:
6	$S = S \cup \{b\}$
7	`else`
8	$B = B \setminus \{b\}$
9	Send 4 elements of S by a change of variables to $x_1 = 0$, $x_2 = 0$, $x_3 = 0$ and $x_1 + x_2 + x_3 = 0$ and denote the resulting set by S_0
10	Normalize the coefficients of the elements in S_0 with Theorem 4.4
11	Use Theorem 4.6 and re-label the bitangents to get positive valuations for a_{11} and a_{31}
12	Compute the corresponding Riemann model

REMARK 5.1. The azygetic test on line (5) is perform by checking if the 6 points of intersection of 3 of the bitangents in $S \cup \{b\}$ with the curve lie on a conic. Notice that as we are over a non-exact field, this step is harder than over a number field.

EXAMPLE 5.2. Let us consider the plane smooth quartic $C/\mathbb{Q} : F = 0$ where

$$F = (x_2 + x_3)x_1^3 - (2x_2^2 + x_3x_2)x_1^2 + (x_2^3 - x_3x_2^2 + 2x_3^2x_2 - x_3^3)x_1 - (2x_3^2x_2^2 - 3x_3^3x_2).$$

This curve is an equation of $X_{ns}(13) \simeq X_s(13) \simeq X_0^+(169)$ which is studied in [**1**, Cor.6.8]. Using a general result from [**7**], they prove that this curve has good reduction everywhere and potentially good reduction at 13 after a ramified extension of degree 84. Using the characterizations in [**15**] in terms of the valuations of the Dixmier-Ohno invariants, it is straightforward to see that this is indeed the case and that the stable reduction at 13 is the hyperelliptic curve $y^2 = z(x^7 - z^7)$. We apply our algorithm to find an equation for the stable model at 13.

Starting from F, we get the following Aronhold system modulo 13^2:

$$\begin{gathered} x_1 = 0, \quad x_2 = 0, \quad x_3 = 0, \quad x_1 + x_2 + x_3 = 0, \\ a_{11}\, x_1 + a_{12}\, x_2 + a_{13}\, x_3 = 0, \\ a_{21}\, x_1 + a_{22}\, x_2 + a_{23}\, x_3 = 0, \\ a_{31}\, x_1 + a_{32}\, x_2 + a_{33}\, x_3 = 0, \end{gathered}$$

with

$$\begin{aligned}
a_{11} &= 73\,\pi^6 + (156\,\tau + 27)\,\pi^5 + (4\,\tau + 96)\,\pi^4 + (28\,\tau + 57)\,\pi^3 + \\ &\qquad (8\,\tau + 10)\,\pi^2 + (78\,\tau + 101)\,\pi + 38\,\tau + 26, \\
a_{12} &= 24\,\tau + 94)\,\pi^6 + (13\,\tau + 137)\,\pi^5 + (32\,\tau + 8)\,\pi^4 + (54\,\tau + 76)\,\pi^3 + \\ &\qquad (47\,\tau + 166)\,\pi^2 + (143\,\tau + 75)\,\pi + 77\,\tau + 13, \\
a_{13} &= (153\,\tau + 57)\,\pi^6 + (167\,\tau + 67)\,\pi^5 + (29\,\tau + 35)\,\pi^4 + (83\,\tau + 10)\,\pi^3 + \\ &\qquad (74\,\tau + 94)\,\pi^2 + (161\,\tau + 154)\,\pi + 56\,\tau + 74, \\
a_{21} &= (128\,\tau + 140)\,\pi^6 + (45\,\tau + 69)\,\pi^5 + (110\,\tau + 12)\,\pi^4 + (112\,\tau + 13)\,\pi^3 + \\ &\qquad (60\,\tau + 90)\,\pi^2 + (10\,\tau + 110)\,\pi + 150\,\tau + 159, \\
a_{22} &= (120\,\tau + 40)\,\pi^6 + (77\,\tau + 106)\,\pi^5 + (36\,\tau + 50)\,\pi^4 + (89\,\tau + 100)\,\pi^3 + \\ &\qquad (130\,\tau + 120)\,\pi^2 + (64\,\tau + 64)\,\pi + 17\,\tau + 73, \\
a_{23} &= (142\,\tau + 92)\,\pi^6 + (44\,\tau + 26)\,\pi^5 + (156\,\tau + 145)\,\pi^4 + (88\,\tau + 147)\,\pi^3 + \\ &\qquad (72\,\tau + 109)\,\pi^2 + (104\,\tau + 93)\,\pi + 100\,\tau + 69, \\
a_{31} &= (141\,\tau + 27)\,\pi^6 + (121\,\tau + 39)\,\pi^5 + (6\,\tau + 103)\,\pi^4 + (\tau + 141)\,\pi^3 + \\ &\qquad (51\,\tau + 74)\,\pi^2 + (79\,\tau + 32)\,\pi + 129\,\tau + 39, \\
a_{32} &= (39\,\tau + 11)\,\pi^6 + (8\,\tau + 97)\,\pi^5 + (155\,\tau + 4)\,\pi^4 + (92\,\tau + 36)\,\pi^3 + \\ &\qquad (103\,\tau + 100)\,\pi^2 + (66\,\tau + 45)\,\pi + 77\,\tau + 13, \\
a_{33} &= (154\,\tau + 28)\,\pi^6 + (77\,\tau + 126)\,\pi^5 + (124\,\tau + 5)\,\pi^4 + (151\,\tau + 86)\,\pi^3 + \\ &\qquad (2\,\tau + 149)\,\pi^2 + (141\,\tau + 8)\,\pi + 33\,\tau + 105\,.
\end{aligned}$$

where $\tau^2 + 12\,\tau + 2 = 0 + O(13^2)$ and

$$\pi^7 + 13 \cdot 6\,(\pi^2 + 5\,\pi + 8)(\pi^4 + 2\,\pi^3 + 2\,\pi^2 + 11\,\pi + 12) = 0 + O(13^2)\,.$$

A toggle model model modulo 13^2 is then $Q^2 + \pi G = 0$ where

$$Q = (9\,\tau + 3)\,x_1^2 + (4\,\tau + 2)\,x_1\,x_2 + 2\,x_1\,x_3 + (8\,\tau + 12)\,x_2^2 + x_2\,x_3 + O(13^2)$$

and

$$\begin{aligned}
\mathrm{G} = \quad & O(13^2) + \\
& ((12\,\tau + 11)\,\pi^6 + (18\,\tau + 74)\,\pi^5 + (124\,\tau + 120)\,\pi^4 + (40\,\tau + 94)\,\pi^3 + \\ &\qquad (45\,\tau + 105)\,\pi^2 + (8\,\tau + 61)\,\pi + 145\,\tau + 117)\,x_1^4 + \\
& ((6\,\tau + 122)\,\pi^6 + (68\,\tau + 30)\,\pi^5 + (78\,\tau + 152)\,\pi^4 + (10\,\tau + 142)\,\pi^3 + \\ &\qquad (106\,\tau + 38)\,\pi^2 + (168\,\tau + 47)\,\pi + 162\,\tau + 86)\,x_1^3\,x_2 + \\
& ((23\,\tau + 135)\,\pi^6 + (103\,\tau + 9)\,\pi^5 + (6\,\tau + 160)\,\pi^4 + (20\,\tau + 27)\,\pi^3 + \\ &\qquad (2\,\tau + 155)\,\pi^2 + (146\,\tau + 106)\,\pi + 69\,\tau + 37)\,x_1^3\,x_3 + \\
& ((106\,\tau + 65)\,\pi^6 + (81\,\tau + 26)\,\pi^5 + (92\,\tau + 8)\,\pi^4 + (22\,\tau + 99)\,\pi^3 + \\ &\qquad (6\,\tau + 133)\,\pi^2 + (82\,\tau + 76)\,\pi + 46\,\tau + 104)\,x_1^2\,x_2^2 + \\
& ((116\,\tau + 102)\,\pi^6 + (83\,\tau + 78)\,\pi^5 + (60\,\tau + 152)\,\pi^4 + (97\,\tau + 129)\,\pi^3 + \\ &\qquad (106\,\tau + 44)\,\pi^2 + (92\,\tau + 165)\,\pi + 89\,\tau + 66)\,x_1^2\,x_2\,x_3 +
\end{aligned}$$

$$
\begin{aligned}
&((135\,\tau+10)\,\pi^6+(125\,\tau+97)\,\pi^5+(18\,\tau+93)\,\pi^4+(104\,\tau+92)\,\pi^3+\\
&\qquad\qquad (130\,\tau+151)\,\pi^2+(156\,\tau+101)\,\pi+104+78\,\tau)\,x_1^2\,x_3^2+\\
&((88\,\tau+77)\,\pi^6+(73\,\tau+109)\,\pi^5+(113\,\tau+68)\,\pi^4+(13\,\tau+12)\,\pi^3+\\
&\qquad\qquad (96\,\tau+96)\,\pi^2+(56\,\tau+125)\,\pi+82\,\tau+15)\,x_1\,x_2^3+\\
&((53\,\tau+124)\,\pi^6+(112\,\tau+64)\,\pi^5+(123\,\tau+109)\,\pi^4+(145\,\tau+123)\,\pi^3+\\
&\qquad\qquad (77\,\tau+66)\,\pi^2+(18\,\tau+85)\,\pi+29\,\tau+96)\,x_1\,x_2^2\,x_3+\\
&((8\,\tau+112)\,\pi^6+(20\,\tau+45)\,\pi^5+(73\,\tau+73)\,\pi^4+(96\,\tau+58)\,\pi^3+\\
&\qquad\qquad (138\,\tau+79)\,\pi^2+(130\,\tau+158)\,\pi+156)\,x_1\,x_2\,x_3^2+\\
&((61\,\tau+84)\,\pi^6+(102\,\tau+138)\,\pi^5+(53\,\tau+52)\,\pi^4+(141\,\tau+40)\,\pi^3+\\
&\qquad\qquad (106\,\tau+77)\,\pi^2+143\,\pi+143\,\tau+130)\,x_1\,x_3^3+\\
&((42\,\tau+92)\,\pi^6+(30\,\tau+144)\,\pi^5+(66\,\tau+40)\,\pi^4+(39\,\tau+156)\,\pi^3+\\
&\qquad\qquad (60\,\tau+104)\,\pi^2+(56\,\tau+48)\,\pi+53\,\tau+75)\,x_2^4+\\
&((161\,\tau+125)\,\pi^6+(154\,\tau+139)\,\pi^5+(39\,\tau+40)\,\pi^4+(94\,\tau+40)\,\pi^3+\\
&\qquad\qquad (82\,\tau+106)\,\pi^2+(111\,\tau+156)\,\pi+35\,\tau+119)\,x_2^3\,x_3+\\
&((164\,\tau+105)\,\pi^6+(22\,\tau+24)\,\pi^5+(82\,\tau+58)\,\pi^4+(105\,\tau+147)\,\pi^3+\\
&\qquad\qquad (116\,\tau+126)\,\pi^2+(143\,\tau+135)\,\pi+117\,\tau+13)\,x_2^2\,x_3^2+\\
&((134\,\tau+47)\,\pi^6+(114\,\tau+145)\,\pi^5+(164\,\tau+64)\,\pi^4+(\tau+6)\,\pi^3+\\
&\qquad\qquad (38\,\tau+20)\,\pi^2+(78\,\tau+13)\,\pi+78\,\tau+52)\,x_2\,x_3^3+\\
&((117\,\tau+165)\,\pi^6+(65\,\tau+28)\,\pi^5+(65\,\tau+52)\,\pi^4+(104\,\tau+104)\,\pi^3+\\
&\qquad\qquad 13\,\tau\,\pi^2+(52\,\tau+13)\,\pi+78\,\tau+13)\,x_3^4\,.
\end{aligned}
$$

The Dixmier-Ohno invariants of this model modulo 13^2 coincide with the Dixmier-Ohno invariants of the input curve. This proves the correctness of the output with the given precision.

REMARK 5.3. In order to automatize completely this procedure, one would need packages to work with algebraic systems over the p-adic in a transparent way with control of the errors. Although there are results and fast progress on this topic [**2, 4–6, 14**], such functionality is still not fully implemented.

Acknowledgment

We thank Tristan Vaccon for helpful discussions.

References

[1] Jennifer Balakrishnan, Netan Dogra, J. Steffen Müller, Jan Tuitman, and Jan Vonk, *Explicit Chabauty-Kim for the split Cartan modular curve of level 13*, Ann. of Math. (2) **189** (2019), no. 3, 885–944, DOI 10.4007/annals.2019.189.3.6. MR3961086

[2] Jérémy Berthomieu and Romain Lebreton, *Relaxed p-adic Hensel lifting for algebraic systems*, ISSAC 2012—Proceedings of the 37th International Symposium on Symbolic and Algebraic Computation, ACM, New York, 2012, pp. 59–66, DOI 10.1145/2442829.2442842. MR3206287

[3] Wieb Bosma, John Cannon, and Catherine Playoust, *The Magma algebra system. I. The user language*, J. Symbolic Comput. **24** (1997), no. 3-4, 235–265, DOI 10.1006/jsco.1996.0125. Computational algebra and number theory (London, 1993). MR1484478

[4] Xavier Caruso, David Roe, and Tristan Vaccon, *Tracking p-adic precision*, LMS J. Comput. Math. **17** (2014), no. suppl. A, 274–294, DOI 10.1112/S1461157014000357. MR3240809

[5] Xavier Caruso, David Roe, and Tristan Vaccon, *p-adic stability in linear algebra*, ISSAC'15—Proceedings of the 2015 ACM International Symposium on Symbolic and Algebraic Computation, ACM, New York, 2015, pp. 101–108. MR3388288

[6] Xavier Caruso, David Roe, and Tristan Vaccon, ZpL*: a p-adic precision package*, ISSAC'18—Proceedings of the 2018 ACM International Symposium on Symbolic and Algebraic Computation, ACM, New York, 2018, pp. 119–126, DOI 10.1145/3208976.3208995. MR3840372

[7] Bas Edixhoven, *Minimal resolution and stable reduction of $X_0(N)$* (English, with French summary), Ann. Inst. Fourier (Grenoble) **40** (1990), no. 1, 31–67. MR1056773

[8] Noam D. Elkies, *The Klein quartic in number theory*, The eightfold way, Math. Sci. Res. Inst. Publ., vol. 35, Cambridge Univ. Press, Cambridge, 1999, pp. 51–101. MR1722413

[9] A. Fiorentino, *Weber's formula for the bitangents of a smooth plane quartic*, 2016.

[10] Benedict H. Gross and Joe Harris, *On some geometric constructions related to theta characteristics*, Contributions to automorphic forms, geometry, and number theory, Johns Hopkins Univ. Press, Baltimore, MD, 2004, pp. 279–311. MR2058611

[11] Joe Harris, *Galois groups of enumerative problems*, Duke Math. J. **46** (1979), no. 4, 685–724. MR552521

[12] Robin Hartshorne, *Algebraic geometry*, Springer-Verlag, New York-Heidelberg, 1977. Graduate Texts in Mathematics, No. 52. MR0463157

[13] George Kempf, *On the geometry of a theorem of Riemann*, Ann. of Math. (2) **98** (1973), 178–185, DOI 10.2307/1970910. MR349687

[14] Romain Lebreton, *Relaxed Hensel lifting of triangular sets.* part 2, J. Symbolic Comput. **68** (2015), no. part 2, 230–258, DOI 10.1016/j.jsc.2014.09.012. MR3283864

[15] Reynald Lercier, Qing Liu, Elisa Lorenzo García, and Christophe Ritzenthaler, *Reduction type of non-hyperelliptic genus 3 curves*, Preprint, 2019.

[16] David Mumford, *Theta characteristics of an algebraic curve*, Ann. Sci. École Norm. Sup. (4) **4** (1971), 181–192. MR292836

[17] David Mumford, *Tata lectures on theta. I*, Progress in Mathematics, vol. 28, Birkhäuser Boston, Inc., Boston, MA, 1983. With the assistance of C. Musili, M. Nori, E. Previato and M. Stillman, DOI 10.1007/978-1-4899-2843-6. MR688651

[18] David Mumford, *Tata lectures on theta. III*, Progress in Mathematics, vol. 97, Birkhäuser Boston, Inc., Boston, MA, 1991. With the collaboration of Madhav Nori and Peter Norman, DOI 10.1007/978-0-8176-4579-3. MR1116553

[19] Harry E. Rauch and Hershel M. Farkas, *Theta functions with applications to Riemann surfaces*, The Williams & Wilkins Co., Baltimore, Md., 1974. MR0352108

[20] B. Riemann, *Zur Theorie der Abelschen Funktionen für den Fall $p = 3$*, 1876.

[21] H. Weber, *Theory of Abelian Functions of Genus 3. (Theorie der Abel'schen Functionen vom Geschlecht 3.)*, 1876.

Reynald Lercier, DGA & Univ Rennes, CNRS, IRMAR - UMR 6625, F-35000 Rennes, France

Email address: reynald.lercier@m4x.org

Elisa Lorenzo García, Univ Rennes, CNRS, IRMAR - UMR 6625, F-35000 Rennes, France

Email address: elisa.lorenzogarcia@univ-rennes1.fr

Christophe Ritzenthaler, Univ Rennes, CNRS, IRMAR - UMR 6625, F-35000 Rennes, France

Email address: christophe.ritzenthaler@univ-rennes1.fr

Contemporary Mathematics
Volume **770**, 2021
https://doi.org/10.1090/conm/770/15438

Courbes de genre 3 avec S_3 comme groupe d'automorphismes

Jean-François Mestre

ABSTRACT. We obtain here new results concerning optimal curves of genus 3, in particular for characteristic 3 and 7. We obtain also many optimal curves of genus 3.

RÉSUMÉ. Nous obtenons de nouveaux résultats concernant les courbes optimales de genre 3, notamment en caractéristique 3 et 7. La méthode employée permet aussi d'obtenir de nombreuses courbes optimales de genre 3.

1. Introduction

Nous montrons ici plusieurs résultats nouveaux concernant les courbes optimales de genre 3. En particulier, nous montrons que, pour toute courbe elliptique E définie sur une extension k de $\mathbf{F}_3$ de degré impair et d'invariant non dans $\mathbf{F}_3$, il existe une courbe de genre 3 définie sur k isogène sur k à E^3. Ceci permet de résoudre dans la plupart des cas le problème des courbes optimales sur de tels corps k. De tels résultats sont inconnus pour les autres caractéristiques.

Nous prouvons aussi que, en caractéristique 7, le défaut d'optimalité est ≤ 9. Une majoration de ce défaut d'optimalité en caractéristique p n'était connue jusqu'à présent que pour $p = 2$ et $p = 3$.

Par ailleurs, les courbes que nous construisons permettent expérimentalement d'obtenir des courbes optimales de genre 3 dans la grande majorité des cas (97% des cas pour les corps $\mathbf{F}_p$ avec $p \leq 117109$).

Pour prouver ces résultats, nous construisons des courbes de genre 3 dont le groupe d'automorphismes contient le groupe symétrique S_3. Plus précisément :

Soit k un corps, $a_1, a_2, a_3, a_4, X, Y, Z$ des indéterminées, $T_1 = X + Y + Z$, $T_2 = XY + YZ + ZX$, $T_3 = XYZ$; la courbe projective plane C_{a_1,a_2,a_3,a_4} à coefficients dans $k(a_1, a_2, a_3, a_4)$ d'équation

$$a_1 T_1^4 + a_2 T_1^2 T_2 + a_3 T_1 T_3 + a_4 T_2^2 = 0$$

a comme groupe d'automorphismes le groupe symétrique S_3 agissant de la façon naturelle sur $\{X, Y, Z\}$.

2020 *Mathematics Subject Classification.* 14H37.

Nous étudions ici cette famille de courbes en vue d'applications au problème de la recherche de courbes de genre g, définies sur un corps fini $\mathbf{F}_q$, $q = p^n$, p premier, ayant beaucoup de points.

Plus précisément, notons $N_q(g)$ le maximum du nombre de points de $C(\mathbf{F}_q)$, lorsque C parcourt l'ensemble des courbes de genre g définies sur $\mathbf{F}_q$, et soit m_q la partie entière de $2\sqrt{q}$. La borne de Weil, raffinée par Serre [**12**], donne la majoration $N_q(g) \leq q + 1 + gm_q$. Une courbe C définie sur $\mathbf{F}_q$ dont le nombre de points sur $\mathbf{F}_q$ est égal à $q + 1 + gm_q$ est dite *optimale*, et le *défaut d'optimalité* est la quantité $D_q(g) = q + 1 + gm_q - N_q(g)$.

Notons que, pour q et g donnés, il peut ne pas exister de courbe optimale, y compris dans le cas du genre 1. À ce propos, Serre a posé la question suivante :

> *Pour g fixé, existe-t-il une constante $c(g)$ telle que, pour tout q, $D_q(g) \leq c(g)$?*

Pour $g = 1$ et 2, Serre a prouvé l'existence de $c(g)$; plus précisément, on a $c(1) = 1$ et $c(2) = 2$. Dans le cas du genre 3, rappelons les résultats suivants, en relation avec ceux obtenus ici :

— Supposons q non carré. Si $m_q^2 - 4q \in \{-3, -4, -8, -11\}$, ou si m_q est divisible par p, il n'existe pas de courbe optimale de genre 3 sur $\mathbf{F}_q$ (*cf.* par exemple [**12**] ou [**6**, Appendice]). Sinon, il existe une courbe de genre 3 qui est soit optimale, soit au contraire dont le nombre de points est $q + 1 - 3m_q$.[1] Je ne crois pas que l'on connaisse un nombre premier p tel que $m_p^2 - 4p \notin \{-3, -4, -8, -11\}$ et $D_p(3) \neq 0$. Par contre, pour par exemple $q = 2^5$, on a $m_q = 11$, donc $m_q^2 - 4q = -7$, mais il n'existe pas de courbe optimale : on a $D_{32}(3) = 2$.

— Top [**13**] a trouvé de nombreux cas de courbes de genre 3 optimales en faisant une recherche systématique, en particulier pour $q \leq 100$.

— Lauter [**6**] a prouvé que, pour tout q, il existe une courbe C de genre 3 telle que $|q + 1 - \operatorname{Card} C(\mathbf{F}_q)| \geq 3m_q - 3$.

— Auer et Top [**1**] ont prouvé que, pour n impair, $D_{3^n}(3) \leq 21$.

— Lorsque n est pair, et que $p \equiv 3 \mod 4$, il existe [**4**, p. 2] une courbe optimale de genre 3, qui plus est hyperelliptique. Pour n pair ≥ 4, et $p = 2$, il existe une courbe de genre 3 optimale [**8**]. Pour $p \equiv 1 \mod 4$ et $n \equiv 2 \mod 4$, on sait aussi qu'il existe une courbe optimale.

— Pour $p = 2$, Nart et Ritzenthaler [**9**] ont prouvé qu'il existe une courbe optimale pour une infinité de n impairs.

Plusieurs de ces résultats ont été obtenus en considérant la famille de courbes $ax^4 + by^4 + cz^4 + ey^2z^2 + fz^2x^2 + gx^2y^2 = 0$, dites *quartiques de Ciani* [**2**, **5**], qui possèdent un groupe d'automorphismes isomorphe à $(\mathbf{Z}/2\mathbf{Z})^2$.

Dans ce qui suit, nous démontrons les résultats suivants, obtenus en utilisant plutôt les courbes C_{a_1,a_2,a_3,a_4} ci-dessus :

Théorème 1.1.

— *Soit $q = 3^n$, n impair. Si E est une courbe elliptique définie sur $\mathbf{F}_q$ dont l'invariant modulaire n'est pas dans $\mathbf{F}_3$, il existe une courbe de genre 3 dont la jacobienne est isogène sur $\mathbf{F}_q$ à E^3.*

1. Le fait de ne pas avoir de critère simple permettant de trancher cette alternative est sans doute le problème essentiel de la recherche des courbes optimales en genre 3.

— *Soit $q = 3^n$, avec n impair ≥ 7 ; si m_q n'est pas divisible par 3 (donc pour une infinité de n), il existe une courbe de genre 3 optimale (et aussi une courbe dont le nombre de points est $q + 1 - 3m_q$).*[2] *Dans le cas où n est pair, il existe des courbes optimales non hyperelliptiques.*

— *Soit $q = 7^n$, n impair, et a un entier premier à 7, $|a| \leq m_q$; si a est divisible par 3, et est un carré* $\bmod 7$, *il existe une courbe de genre 3 dont le nombre de points est $q+1-3a$. Si $a \equiv 1, 4, 5, 7, 8$ ou $11 \bmod 12$, il existe une courbe de genre 3 dont le nombre de points est $q + 1 - 3a$ et une courbe de genre 3 dont le nombre de points est $q + 1 + 3a$.*

Corollaire 1.2.

— *Pour tout n, on a $D_{3^n}(3) \leq 3$.*

— *Pour tout n, on a $D_{7^n}(3) \leq 9$*

Par ailleurs, les courbes C_{a_1,a_2,a_3,a_4} permettent fréquemment d'obtenir des courbes optimales : par exemple (cf. 2.2), pour p premier ≤ 117109, on obtient, par spécialisation convenable de courbes de cette famille, une courbe optimale sur $\mathbf{F}_p$ dans plus de 97% des cas.

La première section est consacrée au cas où la caractéristique de k est différente de 2 et 3 ; on y donne d'abord un formulaire concernant les courbes C_{a_1,a_2,a_3,a_4}. Leur jacobienne est isogène à $E_1^2 \times E_2$, où E_1 (resp. E_2) est la courbe quotient de C par l'un quelconque des sous-groupes d'ordre 2 (resp. par le sous-groupe d'ordre 3) de S_3, et il existe un isomorphisme galoisien commutant à l'accouplement de Weil entre les points d'exposant 3 de E_1 et de E_2.

Réciproquement, si E_1 et E_2 sont deux courbes elliptiques sur k, d'invariants distincts et distincts de 0 et 1728, il existe douze courbes de genre 3 sur $\overline{k}$ de la forme C_{a_1,a_2,a_3,a_4} dont la jacobienne est isogène à $E_1^2 \times E_2$.

Dans la sous-section suivante, on applique ces formules au cas où E_1 et E_2 ont le même invariant modulaire, ce qui donne une famille de courbes de genre 3 à un paramètre t. Cela permet en particulier de construire une courbe de genre 3 dont la jacobienne est isogène sur $\mathbf{Q}$ au cube d'une courbe à multiplications complexes, dans le cas où l'anneau d'endomorphismes de cette courbe a comme discriminant $-3, -7, -19, -43, -67, -163, -16$ ou -28.

Nous traitons ensuite le cas où $p = 7$, avant de finir dans la dernière section par le cas de la caractéristique 3.

2. Caractéristique différente de 2 et 3

Notations.

— Soit $j \neq 0, 1728$. Dans ce qui suit, on note $E(j)$ la courbe elliptique d'équation

$$y^2 = x^3 - 3\frac{j}{j - 1728}x - \frac{3j}{4(j - 1728)}.$$

Cette courbe est d'invariant j.

— On note $\rho \in \overline{k}$ une racine de $x^2 + x + 1$. Remarquons que $(\rho : \rho^2 : 1)$ est un point de la courbe C_{a_1,a_2,a_3,a_4}.

— Si E est une courbe elliptique, on note $E[3]$ le groupe de ses points d'exposant 3.

2. Donc, en caractéristique 3, les seules obstructions à l'existence d'une courbe optimale sont $m_q \equiv 0 \bmod 3$ ou $m_q^2 - 4q \in \{-3, -4, -8, -11\}$.

2.1. Isogénies d'ordre 3 des courbes elliptiques. Soit E une courbe elliptique d'équation $y^2 = p(x)$, avec $p(x) = x^3 + ax + b$; l'équation aux abscisses de ses points d'ordre 3 est alors $f_3(x) = 2pp" - p'^2 = 3x^4 + 6ax^2 + 12xb - a^2$. Par une homographie convenable, on peut ramener les quatre racines de f_3 à $\infty, 1, \rho, \rho^2$.

L'ensemble des birapports des 24 permutations de ces racines est égal à $\{-\rho, -\rho^2\}$, donc le groupe des homographies conservant les quatre racines de f_3 est isomorphe au groupe alterné A_4. Si $f : E[3] \to E[3]$ est un isomorphisme préservant l'accouplement de Weil, l'équation aux abscisses de f restreinte à $E[3] - \{0\}$ est donnée par une telle homographie.

Par suite, si E' est une seconde courbe elliptique, d'équation $y^2 = x^3 + a'x + b'$, et si $g_3 = 3x^4 + 6a'x^2 + 12xb' - a'^2$, il existe douze homographies envoyant les racines de f_3 sur celles de g_3, et elles correspondent aux isomorphismes de $E[3]$ sur $E'[3]$ commutant à l'accouplement de Weil.

La courbe elliptique universelle possédant un point d'ordre 3 est la courbe d'équation

$$E_t : y^2 + xy + ty = x^3,$$

le point $(0,0)$ étant d'ordre 3 et t 'etant un paramètre. Son invariant modulaire est égal à

$$j_3(t) = \frac{(24t-1)^3}{t^3(27t-1)}.$$

La courbe quotient de E_t par le groupe d'ordre 3 engendré par $(0,0)$ est la tordue quadratique par -3 de la courbe $E_{-t-1/27}$.

2.2. Formulaire. Si a_3 ou a_4 est nul, la courbe C_{a_1,a_2,a_3,a_4} n'est pas irréductible ; on peut donc supposer $a_4 = 1$ et $a_3 \neq 0$; de plus, comme la caractéristique de k est ici supposée différente de 2 et 3, on peut se ramener à $a_3 = 2$ par une transformation de la forme $X' = X + aT_1, Y' = Y + aT_1, Z' = Z + aT_1$, qui commute aux matrices de permutation et qui est inversible pour $3a + 1 \neq 0$, en prenant $a = \dfrac{2 - a_3}{3a_3}$.

On désigne désormais par C_{a_1,a_2} la courbe d'équation

$$a_1T_1^4 + a_2T_1^2T_2 + 2T_1T_3 + T_2^2 = 0.$$

Son discriminant vaut $256\,(27\,a_1 + 5 + 9\,a_2)\,d^3$, où

$$d = -432\,a_1 + 72\,{a_2}^2 + 76\,{a_2}^3 - 216\,{a_2}^2a_1 + 432\,{a_1}^2 - 432\,a_1\,ta_2 + 27\,{a_2}^4.$$

On note également $q_{a_1,a_3}(x,y)$ la forme non homogène de l'équation de C_{a_1,a_2} obtenue en posant $x = X/Z, y = Y/Z$.

PROPOSITION 2.1.

(1) La courbe E_1 quotient de C_{a_1,a_2} par le groupe engendré par l'involution $(X,Y,Z) \mapsto (Y,X,Z)$ est une courbe elliptique d'équation affine

$$y^2 = x^3 - (3 + 2a_2)x - 4a_1 + 2 + 2a_2 + a_2^2$$

et d'invariant modulaire

$$J_1 = 6912\frac{(2a_2+3)^3}{d}.$$

Le morphisme $\phi_1 : C_{a_1,a_2} \to E_1$ *est donné par*

$$\phi_1 : \begin{cases} x = & \frac{X+Y-Z}{T_1}, \\ y = & \frac{2(Z^2-XY)+a_2T_1^2+4T_2}{T_1^2} \end{cases}$$

Le point à l'infini de E_1 *est l'image par* ϕ_1 *de* $(\rho : \rho^2 : 1)$ *et l'on a*

$$\phi_1^*(\frac{dx}{y}) = \frac{(y-x)dx}{q_y} \qquad \text{où} \qquad q_y = \frac{\partial}{\partial y} q_{a_1,a_2}.$$

Si l'on quotiente C_{a_1,a_2} *par le groupe engendré par l'involution* τ : $(X,Y,Z) \mapsto (X,Z,Y)$ *(resp.* $(X,Y,Z) \mapsto (X,Z,Y)$*), on trouve la même courbe* E_1 *; l'image réciproque de* $\frac{dx}{y}$ *est* $\frac{(x-1)dx}{q_y}$ *(resp.* $\frac{(1-y)dx}{q_y}$*) ;*

La somme de ces trois formes différentielles vaut 0*, et elles sont deux-à-deux linéairement indépendantes.*

La jacobienne de C_{a_1,a_2} *est donc isogène à* $E_1^2 \times E_2$*, où* E_2 *est une courbe elliptique.*

(2) La courbe E_2 *est la jacobienne de de la courbe de genre* 1 *quotient de* C_{a_1,a_2} *par le groupe d'ordre* 3 *engendré par* σ : $(X,Y,Z) \mapsto (Y,Z,X)$*; une équation de* E_2 *est*

$$y^2 = x^3 + Ax + B,$$

avec

$$\begin{aligned} A = \; & -48 - 128\,a_2 - 648\,a_2{}^2 + 3456\,a_1a_2 - 648\,a_2{}^3 + 1944\,a_1a_2{}^2 \\ & -243\,a_2{}^4 + 3168\,a_1 - 3888\,a_1{}^2 \end{aligned}$$

$$\begin{aligned} B = \; & 512\,a_2 - 2400\,a_2{}^2 + 46080\,a_1a_2 - 7200\,a_2{}^3 + 77760\,a_1a_2{}^2 \\ & -9720\,a_2{}^4 - 124416\,a_1{}^2a_2 + 54432\,a_1a_2{}^3 - 5832\,a_2{}^5 - 69984\,a_1{}^2a_2{}^2 \\ & +17496\,a_1a_2{}^4 - 1458\,a_2{}^6 + 128 + 19328\,a_1 - 120960\,a_1{}^2 + 93312\,a_1{}^3 \end{aligned}$$

(3) Le quotient J_1/J_2 *des invariants modulaires de* E_1 *et* E_2 *est un cube* M_3^3*, où*

$$M_3 = \frac{A}{16\,(27\,a_1 + 5 + 9\,a_2)\,(3 + 2\,a_2)}.$$

(4) Il existe un isomorphisme galoisien, commutant à l'accouplement de Weil, entre $E_1[3]$ *et* $E_2[3]$.

Plus précisément, il existe une homographie définie sur k *envoyant les abscisses des points d'ordre* 3 *de* E_1 *sur les abscisses des points d'ordre* 3 *de* E_2.

(5) Soit J_1 *un élément de* $k - \{0, 1728\}$*. L'ensemble des courbes* C_{a_1,a_2} *telles que* E_1 *a comme invariant* J_1 *est une famille à un paramètre* v *; on a*

$$\begin{cases} a_1 = & \frac{N}{16(J_1-1728)^2} \\ a_2 = & \frac{3}{2}\,\frac{v^2J_1-J_1+1728}{J_1-1728} \end{cases}$$

$$\begin{aligned} \text{avec } N = \; & 14929920 - 17280\,J_1 + 5\,J_1{}^2 + 10368\,v^2J_1 - 6\,v^2J_1{}^2 \\ & -13824\,v^3J_1 + 8\,v^3J_1{}^2 + 9\,v^4J_1{}^2. \end{aligned}$$

La courbe E_1 *est tordue de* $E(J_1)$ *par* v.

L'invariant de E_2 *est* $J_2 = J_1\, U^3/V^3$, *avec*

$$\begin{aligned} U = & -2985984 + 35831808\, v + {J_1}^2 - 3456\, J_1 - 93312\, v^2 J_1 + 54\, v^2 {J_1}^2 \\ & +81\, v^4 {J_1}^2 + 108\, v^3 {J_1}^2 + 12\, v {J_1}^2 - 41472\, v J_1 - 186624\, v^3 J_1 - 559872\, v^4 J_1 \\ & \text{et} \\ V = & \; 243\, v^4 {J_1}^2 + 216\, v^3 {J_1}^2 + 54\, v^2 {J_1}^2 - {J_1}^2 + 3456\, J_1 \\ & -2985984 - 93312\, v^2 J_1 - 373248\, v^3 J_1. \end{aligned}$$

La courbe E_2 *a comme équation* $y^2 = x^3 + Cx + D$, *avec*

$$\begin{cases} C = & -3v^2 J_1(J_1 - 1728) J_2 V^3 \\ D = & 2v^3 J_1 (J_1 - 1728)(J_2 - 1728) V^3 \end{cases}$$

(6) *Réciproquement, à tout couple d'invariants distincts* J_1 *et* J_2 *dans* $k - \{0, 1728\}$ *correspondent douze courbes* C_{a_1,a_2} *définies sur* $\overline{k}$, *dont les jacobiennes sont isogènes à* $E(J_1)^2 \times E(J_2)$. *Plus précisément, soit* h *l'une des* 12 *homographies envoyant les abscisses des points d'ordre* 3 *de* $E(J_1)$ *sur les abscisses des points d'ordre* 3 *de* $E(J_2)$; *le paramètre* v *du point* 5) *est donné par*

$$\frac{1}{v} = 3h(\infty).$$

Remarque 2.2. Soit $\Phi_2 = \Phi_1 \circ \sigma : C_{a_1,a_2} \to E_1$, et $\Psi : C_{a_1,a_2} \to E_2 = C_{a_1,a_2}/ < \sigma >$ un revêtement de degré 3 tel que $\Psi(\rho : \rho^2 : 1) = 0$. Soit $f : E_1^2 \times E_2 \to \mathrm{Pic}^0(C_{a_1,a_2})$ définie par

$$(P, Q, R) \mapsto \Phi_1^*((P) - (0)) + \Phi_2^*((Q) - (0)) + \Psi^*((R) - (0))$$

et $g : \mathrm{Pic}^0(C_{a_1,a_2}) \to E_1^2 \times E_2$ définie par

$$D \mapsto (\Phi_1(D), \Phi_2(D), \Psi(D));$$

l'application composée $g \circ f : E_1^2 \times E_2 \to E_1^2 \times E_2$ est donnée par la matrice $\begin{pmatrix} 2 & -1 & 0 \\ -1 & 2 & 0 \\ 0 & 0 & 3 \end{pmatrix}$. Par suite, le degré de f (resp. de g) est égal à 9.

La jacobienne de C_{a_1,a_2} est obtenue par recollement [**12**, p. 37 et seq.] [**6**, p. 95] de E_1^2 muni de la polarisation $\begin{pmatrix} 2 & -1 \\ -1 & 2 \end{pmatrix}$ et de E_2 munie de la polarisation $3\, Id$, via l'isomorphisme galoisien de $E_1[3]$ sur $E_2[3]$ du point 4) ci-dessus.

2.3. Exemples. On peut faire une recherche systématique des courbes optimales de genre 3 de la forme C_{a_1,a_2}, par l'algorithme suivant : soit k un corps fini, de cardinal q, et J l'ensemble, éventuellement vide, des invariants des courbes elliptiques dont la valeur absolue de la trace du Frobenius sur $\mathbf{F}_q$ vaut ${m_q}$[3]. Pour tout couple j, j' d'éléments de J, on calcule les racines de l'équation $J_2(j, v) = j'$, puis le nombre de points de la courbe de genre 3 associée à chacune de ces racines par les formules ci-dessus.

Pour les nombres premiers $p \leq 117109$, i.e. les 11054 premiers nombres premiers, dont on exclut les p tels que $m_p^2 - 4p \in \{-3, -4, -8, -11\}$, cette méthode échoue à trouver une courbe de genre 3 dont le nombre de points est $p + 1 + 3m_p$ dans 330 cas.

3. On exclut les cas $m_q^2 - 4q = -3, -4, -8, -11$, puisqu'il n'existe pas de courbe optimale dans ces cas.

Pour ces cas, le nombre de classes de $m_p^2 - 4p$ est en général égal à 1, sauf dans 40 cas où on trouve $-20, -24, -28, -35, -36, -40, -88, -91, -100, -123, -187, -427$, de nombre de classes 2, et deux cas où l'on trouve -59, de nombre de classes 3.

Les 19 nombres premiers < 1000 pour lequel la méthode précédente ne permet pas de trouver une courbe optimale sont 53, [4]

$$167, 173, 193, 293, 311, 347, 353, 359, 479, 523, 557, 569, 661, 709, 773, 787, 823, 997.$$

On peut faire de même lorsque $q = p^n$ n'est pas premier, et jouer alors avec le fait que J_1 et $J_2 = J_1^p$ peuvent être distincts : soit par exemple $q = 19^3$; en prenant $J_1 = J_2$, où J_1 est l'invariant d'une courbe elliptique optimale sur $\mathbf{F}_q$, on trouve une courbe de genre 3 dont le nombre de points est $q + 1 - 3m_q$; par contre, en prenant $J_2 = J_1^{19}$, on trouve une courbe optimale.

2.4. Cas où $J_1 = J_2$. On cherche ici les courbes C_{a_1,a_2} telles que $J_1 = J_2$; avec les notations précédentes, $J_1/J_2 = M_3^3$, donc M_3 est alors égal à $1, \rho$ ou ρ^2.

2.4.1. $M_3 = 1$. L'équation $M_3 = 1$ s'écrit

$$\begin{aligned}&-288 - 720\, a_2 - 936\, {a_2}^2 + 2592\, a_1 a_2 - 648\, {a_2}^3 + 1944\, a_1 {a_2}^2\\ &-243\, {a_2}^4 + 1872\, a_1 - 3888\, {a_1}^2 = 0\end{aligned}$$

courbe rationnelle que l'on peut paramétrer par

$$\begin{cases} a_1 = \frac{1}{432}\,\frac{112\,t^4+272\,t^3+408\,t^2+296\,t+127}{(t^2+t+1)^2}, \\ a_2 = -\frac{8\,t^2+10\,t+9}{6(t^2+t+1)} \end{cases}$$

Dans ce qui suit, on note C_t la courbe d'équation

$$(t^2 + t + 1)^2(a_1 T_1^4 + a_2 T_1^2 T_2 + 2T_1 T_3 + T_2^2) = 0,$$

a_1 et a_2 étant les fractions rationnelles en t ci-dessus.

On a alors $J_1 = 1728t^3$; la courbe E_2 a comme équation $y^2 = x^3 - 3t(t^3 - 1)x - 2(t^3 - 1)^2$, et la courbe E_1 est tordue de E_2 par $-3(t^2 + t + 1)$.

Le discriminant de C_t est $2^{118}3^{74}(t-1)^9(t^2+t+1)^{43}$.

Remarques 2.3.

(1) D'après [**11**, p. 305], l'invariant modulaire d'une courbe elliptique E est un cube si et seulement si le degré de l'extension obtenue par adjonction des points d'ordre 3 de la courbe est premier à 3.

(2) Supposons que $k = \mathbf{F}_q$ (q impair).
   (a) Si $q \equiv 2 \mod 3$, j est toujours un cube (comme tout élément de k).
   (b) Si $q \equiv 1 \mod 3$, une condition suffisante pour que j soit un cube est que $a_q = q + 1 - N_q(E)$ soit divisible par 3 ; en effet, dans ce cas, ni E ni sa tordue quadratique n'ont de point d'ordre 3 rationnel sur $\mathbf{F}_q$, et le polynôme f_3 des abscisses des points d'ordre 3 de E et de sa tordue n'a donc pas de racine dans $\mathbf{F}_q$. Par suite, il n'a pas de facteur irréductible de degré 3.

   Par ailleurs, la relation

   $$(t^3 - 1)^2 = (t^2 + t + 1)((t)^2 + t + 1)((t\rho^2)^2 + t\rho^2 + 1)$$

   montre que, dès que j est un cube, il existe $t \in k$ tel que $j = 1728t^3$ et $t^2 + t + 1$ est un carré dans k.

4. Comme me l'a indiqué C. Ritzenthaler, M. Nitzberg a montré que $x^4 + y^4 + z^4 = 0$ est optimale pour $p = 53$ [**12**, p. 72].

(3) Lorsque $t^2+t+1=0$, i.e. $t=\rho$ ou ρ^2, on a $J_1=J_2=1728$; la quartique C_t a un discriminant nul, et sa réduction modulo t^2+t+1 est, sur une extension convenable de $k(t)$, la courbe hyperelliptique $y^2=x(x^6-1)$. Pour le voir, il suffit d'écrire l'équation de la quartique sous la forme $A^2+\varepsilon B+O(\varepsilon^2)=0$ [**3**, p. 82], $A=0$ étant l'équation d'une conique Co non singulière ; lorsque ε tend vers 0, la quartique "tend" vers la courbe hyperelliptique revêtement double de la conique Co ramifié aux points d'intersection de Co et de la courbe d'équation $B=0$.

Ici, on paramètre la conique $t^2+t+1=u^2$ par $\begin{cases} t=\rho^2\frac{\varepsilon^2-\rho^2}{\varepsilon^2-1} \\ u=-\sqrt{-3}\frac{\varepsilon}{\varepsilon^2-1}\end{cases}$.

Pour $\varepsilon=0$, on a $(t,u)=(\rho,0)$.

Après le changement de variables $(X,Y,Z)\mapsto(X,Y,S)$, avec $X+Y+Z=2Su$, la quartique C_t s'écrit $A^2-12\varepsilon B+O(\varepsilon^2)=0$, où $A=S^2-\sqrt{-3}(X^2+XY+Y^2)$ et $B=S(X+Y)(A-\sqrt{-3}XY)$.

En paramétrant la conique Co d'équation $A=0$ par $\mu=\sqrt[4]{-3}(X-\rho Y)/S$, les huit paramètres des points d'intersection de Co et de la quartique d'équation $B=0$ sont $0,\infty,\pm\rho,\pm\rho^2,\pm1$; par suite, sur $\overline{k}$, la courbe hyperelliptique "limite" de C_t quand t tend vers ρ a comme équation $y^2=x(x^6-1)$.

2.4.2. *Cas où $k=\mathbf{Q}$ et E_1 de type CM.* Si $k=\mathbf{Q}$, E_1 et E_2 ne sont pas k-isomorphes, puisque tordues l'une de l'autre par $-3(t^2+t+1)$. Néanmoins, si elles sont à multiplications complexes, elles peuvent être isogènes sur $\mathbf{Q}$; dans ce cas, la trace du Frobenius $\bmod p$ de C_t, pour p premier et premier au discriminant de la courbe, est égale au triple de celle de E_1.

Les 13 courbes elliptiques à multiplications complexes définies sur $\mathbf{Q}$ ont comme invariant

$$2^63^3, 2^65^3, 0, -3^35^3, -2^{15}, -2^{15}3^3, -2^{18}3^35^3, -2^{15}3^35^311^3,$$

$$-2^{18}3^35^323^329^3, 2^33^311^3, 2^43^35^3, 3^35^317^3, -3\ 2^{15}5^3,$$

les anneaux d'endomorphismes associés étant ceux de discriminant

$$-4, -8, -3, -7, -11, -19, -43, -67, -163, -16, -12, -28, -27.$$

À part le onzième et le treizième, tous ces invariants sont des cubes, les valeurs de t associées étant

$$1, 5/3, 0, -5/4, -8/3, -8, -80, -440, -53360, 11/2, \frac{85}{4}.$$

La valeur $t=1$ donne une quartique singulière ; pour les dix autres valeurs, on obtient dix courbes de genre 3. À des facteurs carrés près, les quantités $-3(t^2+t+1)$ valent respectivement $-3,-3,-7,-3,-19,-43,-67,-163,-1,-7$, les anneaux d'endomorphismes ayant comme discriminant

$$-8, -3, -7, -11, -19, -43, -67, -163, -16, -28.$$

À part le premier et quatrième cas, les courbes E_1 et E_2 sont donc $\mathbf{Q}$-isogènes, et on obtient ainsi huit quartiques définies sur $\mathbf{Q}$ dont la jacobienne est $\mathbf{Q}$-isogène au cube d'une courbe elliptique à multiplications complexes, les anneaux d'endomorphismes étant $-3,-7,-19,-43,-67,-163,-16,-28$.

EXEMPLES 2.4.

(1) Pour $t = -5/4$, on trouve une quartique $\overline{\mathbf{Q}}$-isomorphe à la courbe de Klein, d'équation $x^3y + y^3 + x = 0$, mais qui ne lui est pas $\mathbf{Q}$-isomorphe (la courbe de Klein n'ayant pas d'involution définie sur $\mathbf{Q}$). Plus précisément, l'isomorphisme entre la courbe de Klein et la quartique ci-dessus est donné par la matrice

$$\begin{pmatrix} u_1 & u_2 & u_3 \\ u_3 & u_1 & u_2 \\ u_2 & u_3 & u_1 \end{pmatrix}$$

avec u_1^4 racine de $X^3 + 26X^2 + 13X + 1$, $u_2 = \frac{1}{29}u_1(5u_1^8 + 126u_1^4 - 30)$ et $u_3 = \frac{1}{29}u_1(17u_1^8 + 440u_1^4 + 159)$.

Si δ est le déterminant de cette matrice, on a $\delta^4 = -14^4$.

La remarque 3 du paragraphe précédent permet de retrouver le fait que, modulo 7, la quartique de Klein se réduit en une courbe hyperelliptique. (De même, les quartiques associées aux courbes à multiplications complexes par l'anneau des entiers de $-19, -43, -67, -163$ se réduisent modulo $19, 43, 67, 163$ en une courbe hyperelliptique).

(2) Pour $t = -5/4$ (resp. $t = 85/4$), qui correspondent aux discriminants -7 et -28, on obtient une quartique dont la jacobienne est $\mathbf{Q}$-isogène au cube de la courbe $y^2 = x^3 - 35x + 98$ (resp. $y^2 = x^3 - 1933155x - 1034488098$). Ces deux courbes sont 2-isogènes sur $\mathbf{Q}(\sqrt{57})$, mais pas sur $\mathbf{Q}$. Par suite, si p est un nombre premier tel que $\left(\frac{p}{57}\right) = -1$, les traces du Frobenius $\bmod p$ sont opposées d'une quartique à l'autre. En particulier, si $p = 4b^2 + 7$, avec $b \geq 4$ (ce qui assure que la partie entière de $2\sqrt{p}$ est égale à $2b$), et si $\left(\frac{p}{57}\right) = -1$, on obtient ainsi une quartique optimale et une quartique minimale sur $\mathbf{F}_p$. Par exemple, les nombres premiers < 10000 vérifiant ces conditions sont

$$\{151, 263, 331, 491, 907, 1031, 1163, 1451, 1607, 2311, 4363, 5483, 5783\}.$$

(3) Dans les cas de discriminant $-19, -43, -67$ et -163, on obtient des modèles sur $\mathbf{Q}$ de courbes trouvées par Ritzenthaler dans [**10**], à savoir celles dont la dernière entrée dans la Table 2 de [*ibid.*] vaut 2×6.

2.4.3. *Le cas où* $k = \mathbf{F}_{7^n}$.

PROPOSITION 2.5. *Soit n un entier impair, $q = 7^n$, et E une courbe elliptique définie sur $k = \mathbf{F}_q$, dont la trace du Frobenius a est congrue à $9, 15$ ou $18 \bmod 21$. Il existe une courbe de genre 3, définie sur k, dont le nombre de points vaut $7^n + 1 - 3a$.*

En effet, on a alors $a \equiv 0 \bmod 3$, donc d'après ce qui précède, l'invariant de E est un cube $1728t^3$, où l'on peut choisir t tel que $t^2 + t + 1$ est un carré dans k.

Il existe donc une courbe de genre 3 définie sur k dont la jacobienne est isogène à E_2^3, où E_2 a comme équation $y^2 = x^3 + Ax + B$, avec

$$A = -3t(t^3 - 1, B = -2(t^3 - 1)^2.$$

Or, si E a comme équation $y^2 = x^3 + \alpha x + \beta$, on sait [**7**, théorème 1] que a est congrue modulo 7 à la norme de son invariant de Hasse, soit $(3\beta)^{1+7+\dots+7^{n-1}}$. Par hypothèse, c'est un carré dans $\mathbf{F}_7$.

Comme la trace du Frobenius de E_2 est aussi un carré, puisque $3B = (t^3 - 1)^2$, E_2 est donc k-isomorphe à E, d'où le résultat.

COROLLAIRE 2.6. *Soit $k = \mathbf{F}_q$ un corps fini de caractéristique 7 ; il existe une courbe de genre 3 définie sur $\mathbf{F}_q$ dont le nombre de points est $\geq q+1+3(m_q-11)$.*

2.4.4. *Le cas $M_3 = \rho$.* On suppose ici que $\rho \in k$; la courbe d'équation $M_3 = \rho$ admet une paramétrisation rationnelle par un paramètre t tel que $J_1 = J_2 = \frac{(-1+24\,t)^3}{t^3(-1+27\,t)}$. On reconnaît l'invariant modulaire de la courbe E_t d'équation $y^2+xy+ty = x^3$, i.e. la courbe elliptique universelle ayant un point d'ordre 3 (cf. 2.1). Par ailleurs, alors que, dans le cas $M_3 = 1$, E_1 et E_2 sont tordues quadratiques l'une de l'autre par $-3(t^2+t+1)$, ici E_1 et E_2 sont $k(t)$-isomorphes, et tordues quadratiques par $(\rho-1)(36t+\rho-1)$ de la courbe E_t.

Supposons que $k = \mathbf{F}_{7^n}$, n impair, et prenons $\rho = 2$. Si E est une courbe elliptique sur k dont le nombre de points est 7^n+1-a, et que $a \equiv -1 \bmod 3$, le courbe E a un point d'ordre 3. Soit F la courbe quotient de E par le groupe engendré par ce point. D'après 2.1, il existe $t \in k$ tel que $E = E_t$ et, -3 étant un carré dans k, $F = F_{-t+1/27}$. La courbe de genre 3 associée à t (resp. $-t+1/27$) est isogène au cube de E ou de sa tordue quadratique, selon que $t+1$ est un carré ou non (resp. au cube de F ou de sa tordue quadratique, selon que $36(1/27-t)+\rho-1 = -t$ est un carré ou non). Comme, à un carré près, le discriminant de E_t est égal à $t(1+t)$, on en déduit que, si a est impair, i.e. si E n'a pas de point d'ordre 2 sur k, ou si $a \equiv 0 \bmod 4$, dans lequel cas, quitte à remplacer E par une courbe 2-isogène, E a trois points d'ordre 2, il existe un unique résidu quadratique dans l'ensemble $\{-t, 1+t\}$.

Donc, si $a \not\equiv -1 \bmod 3$ et $a \not\equiv 2 \bmod 4$, il existe une courbe de genre 3 isogène à E^3 et une autre isogène au cube de la tordue de E. En remplaçant E par sa tordue quadratique, on a le même résultat si $a \equiv 1 \bmod 3$ et $a \not\equiv 2 \bmod 4$, d'où la partie 2 du théorème énoncé en introduction. Le point 2 du corollaire s'ensuit (et peut être précisé : à moins que $m_q \equiv 36, 51, 58$ ou $78 \bmod 84$, le défaut est au plus 6).

3. Le cas de caractéristique 3

Dans ce paragraphe, les corps considérés sont de caractéristique 3.

3.1. Le cas $a_3 \neq 2$. On considère comme précédemment la famille de courbes d'équation

$$a_1T_1^4 + a_2T_1^2T_2 + a_3T_1T_3 + T_2^2.$$

Dans ce paragraphe, on suppose $a_3 \neq 2$. Par une transformation linéaire de la forme

$$X' = X + aT_1, Y' = Y + aT_1, Z' = Z + aT_1,$$

qui est inversible pour tout a, on se ramène à $a_2 = 0$ en prenant $a = -\frac{a_2}{1+a_3}$.

On note désormais, dans ce qui suit, C_{a_1,a_3} la courbe d'équation

$$a_1T_1^4 + a_3T_1T_3 + T_2^2.$$

Son discriminant est $2(a_3+1)^9a_1^3a_3^{12}$.

La courbe E_1 quotient de C par l'une quelconque des involutions naturelles de C a comme équation

$$v^2 = u^3 + a_3(a_3+1)u^2 + 2a_3^2a_1,$$

son invariant modulaire étant

$$J_1 = \frac{a_3(a_3+1)^3}{a_1},$$

un morphisme du modèle affine de C obtenu en posant $x = X/Z, y = Y/Z$ vers E_1 étant donné par

$$u = \frac{-a_3}{T_1}, \quad v = \frac{T_2 - a_3T_1}{T_1^2}.$$

La courbe E_2 quotient de C par le groupe d'ordre 3 engendré par $(x, y, z) \mapsto (y, z, x)$ a comme équation

$$v^2 = u^3 + a_3(a_3+1)u^2 + a_1a_3^5,$$

le morphisme de C vers E_2 étant donné par

$$\begin{cases} u = 2a_3^2 \frac{a_1UV + a_3}{(U+V)(2+a_1(U+V))} \\ v = (U-V)\frac{a_1a_3^4 + u^2}{a_3^2} \end{cases}$$

avec $U = \frac{x^2y+y^2+x}{xy}$ et $V = \frac{y^2x+x^2+y}{xy}$. L'invariant de E_2 est

$$J_2 = \frac{2(a_3+1)^3}{a_1a_3^2}$$

et donc

$$\frac{J_1}{J_2} = -a_3^3.$$

Par ailleurs, les coefficients de u^2 dans les équations de E_1 et E_2 sont les mêmes, et sont non nuls. Comme la trace du Frobenius de E_1 (resp. E_2) est la norme de l'invariant de Hasse de E_1, qui en caractéristique 3 est le coefficient de u^2 dans une équation de Weierstrass de la courbe, les deux courbes E_1 et E_2 ont des traces de Frobenius égales et non nulles modulo 3 ; elles ne peuvent donc pas être opposées.

Par suite, si $k = \mathbf{F}_{3^n}$, et si J_1 et J_2 sont deux éléments distincts et non nuls de k, il existe une courbe de genre 3 définie sur k, dans la famille ci-dessus. dont la jacobienne est isogène à $E_1^2 \times E_2$, où E_1 (resp. E_2) a comme invariant J_1 (resp. J_2) : le coefficient a_3 est déterminé par $J_1/J_2 = -a_3^3$, d'où $a_1 = \frac{a_3(a_3+1)^3}{J_1}$.

Supposons désormais n impair ≥ 3, et $q = 3^n$. Pour tout $j \in k - \mathbf{F}_3$, on obtient ainsi une courbe C associée au couple $(J_1 = j, J_2 = j^3)$, le coefficient a_3 correspondant vérifiant $a_3^3 = -J_1/J_2 = -1/J_1^2$, et une courbe C' associée au couple (J_2, J_1), le coefficient a_3' correspondant étant égal à $1/a_3$.

Le coefficient en x^2 de la courbe E_1 est $a_3(a_3+1)$, alors que celui de la courbe E_1' est $a_3'(1+a_3')$. Le quotient de ces deux termes vaut $a_3^3 = -1/J_1^2$, et n'est donc pas un carré. Par suite, les courbes E_1 et E_1' sont tordues quadratiques l'une de l'autre, et la jacobienne de C' est la tordue quadratique de celle de C.

D'après [**6**, p. 92 et 99] il existe a tel que $a^2 - 4 \times 3^n = -8$ (resp. -11) si et seulement si $n = 1$ ou $n = 3$ (resp. $n = 1$ ou 5). Par suite, pour $n \geq 7$, pour tout entier a non divisible par 3 et tel que $|a| \leq 2\sqrt{q}$, où $q = 3^n$, il existe une courbe elliptique d'invariant dans $\mathbf{F}_q - \mathbf{F}_3$ et dont la trace du Frobenius vaut a.

On en déduit la proposition suivante :

PROPOSITION 3.1. *Soit n un entier impair ≥ 7, et $q = 3^n$. Si m_q n'est pas divisible par 3, il existe deux courbes de genre 3 définies sur $\mathbf{F}_q$, l'une optimale, i.e. dont le nombre de points vaut $q+1+3m_q$, l'autre dont le nombre de points vaut $q+1-3m_q$.*

REMARQUES 3.2.

(1) Il existe une infinité de n impairs pour lesquels $m_q \not\equiv 0 \bmod 3$. En effet, $m_q \equiv 0 \bmod 3$ si et seulement si le $(n-1)/2$-ième terme de la mantisse du développement 3-adique $1.1101120222201212202001\ldots$ de $2\sqrt{3}$ est nul. Comme $2\sqrt{3}$ n'est pas un nombre rationnel, les termes de ce développement ne peuvent pas être tous nuls à partir d'un certain rang.

Dans le cas où $m_q \equiv 0 \bmod 3$, (ce qui, expérimentalement, pour n impair, semble arriver dans une proportion d'environ 1/3), d'après [**12**, p. 7] ou [**6**, Table 1], le défaut est au moins 2, et s'il est 2 les traces du Frobenius sont $-(m_q + 1 - 4\cos^2 \frac{\pi}{7})$ et ses deux conjuguées.

Ceci exige en particulier que la partie fractionnaire de m_q soit supérieure à $1 - 4\cos^2 \frac{3\pi}{7}$. Par exemple, pour n impair ≤ 400, ceci arrive pour

$$n = 15, 47, 53, 69, 159, 329, 349, 375, 383, 399.$$

(2) Il n'existe qu'un nombre fini de n impairs pour lesquels m_q soit la valeur absolue de la trace du Frobenius sur k d'une courbe d'invariant ± 1. Plus généralement, soit E une courbe elliptique ordinaire définie sur un corps fini $\mathbf{F}_q$, de Frobenius π ; si $a_{q^n} = \pi^n + \overline{\pi}^n$, il existe une constante c telle que, pour n impair ≥ 3, on a $|2q^{n/2} - |a_{q^n}|| \geq \frac{q^{n/2}}{n^c}$, quantité qui tend vers l'infini avec n. Ceci découle par exemple du fait que, si β est un nombre algébrique, ici $\pm\frac{\pi^n}{q^{n/2}}$, il existe une constante $c(\beta)$ telle que, pour tout entier $n \geq 2$ tel que $\beta^n \neq 1$, on a

$$|\beta^n - 1| > n^{-c(\beta)}.$$

En fait, dans le cas $j = 1$, où $\pi = \pm\frac{1+\sqrt{-11}}{2}$, (resp. $j = 2$, pour lequel $\pi = \pm 1 + \sqrt{-2}$) il semble que le seul n pour lequel il existe une courbe d'invariant j optimale sur $\mathbf{F}_{3^n}$ soit $n = 5$ (resp. $n = 3$).

Dans ces deux cas, il n'existe pas de courbe de genre 3 optimale sur $\mathbf{F}_{3^n}$, puisqu'on est dans les cas où il n'existe pas de module indécomposable de rang 3 [**6**, Appendice].

(3) Dans le cas où $q = 3^5$ (donc $m_q = 31$), Lauter montre [**6**, p. 14, Prop. 6] que pour toute courbe C de genre 3 sur $k = \mathbf{F}_q$, on a $|q + 1 - \operatorname{Card} C(k)| \leq 3m_q - 3$, et qu'il existe une courbe C_0 pour laquelle on a égalité, le signe de $q + 1 - \operatorname{Card} C_0(k)$ n'étant pas connu (*ibid.*, Note 4.1.5).

Les courbes de genre 3 ci-dessus permettent en fait de montrer qu'il existe, et de construire effectivement, une courbe C de genre 3 telle que $\operatorname{Card} C(k) = 3^5 + 1 + 3m_q - 3$. En effet, l'invariant de la courbe elliptique optimale sur k, dont la trace du Frobenius vaut 31, est $J_1 = 1$; soit z une racine de $z^5 - z + 1$; z engendre k^*, et les invariants des courbes elliptiques dont la trace du Frobenius vaut $31 - 3 = 28$ sont z^{38}, z^{77} et leurs conjugués.

Prenons $J_2 = z^{38\times 3}$. D'après ce qui précède, il existe une courbe C_{a_1,a_3} définie sur k, avec $a_3 = -1/J_2$, dont la jacobienne est k-isogène à $E_1^2 \times E_2$, E_1 et E_2 d'invariants respectifs J_1 et J_2 ; l'invariant de Hasse de chacune de ces deux courbes est $a_3(a_3 + 1)$, c'est-à-dire, à un carré près, $1 - J_2$; sa norme vaut 2, et est donc congrue à $-28 = -(m_q - 3) \bmod 3$. Par suite, la trace du Frobenius de E_1 est -31 et celle de E_2 est -28 ; donc la courbe de genre 3 associée a un nombre de points égal à $3^5 + 1 + 3m_q - 3$.

3.2. Le cas $a_3 = 2$. Dans le paragraphe précédent, on a supposé $a_3 \neq 2$; si $a_3 = -1$, le discriminant de la courbe $a_1T_1^4 + a_2T_1^2T_2 - T_1T_3 + T_2^2 = 0$ est $-a_2^9$; par chacune des trois involutions naturelles, elle a comme quotient la courbe elliptique d'invariant nul et d'équation $y^2 = x^3 + a_2x + 2a_1 + a_2^2$, et la courbe elliptique quotient par le sous-groupe d'ordre 3 engendré par $(X, Y, Z) \mapsto (Y, Z, X)$ est aussi d'invariant nul, et a comme équation

$$y^2 = x^3 + a_2a_1^2x + 2a_1^4.$$

Supposons que $-a_2$ ne soit pas un carré dans $\mathbf{F}_q$, où $q = 3^n$. L'application $x \mapsto x^3 + a_2x$ étant bijective, le nombre de points de chacune des deux courbes précédentes est égal à $q + 1$. Par suite, dans $\mathbf{F}_{3^{2n}}$, ces courbes sont optimales, d'où des courbes de genre 3 sur $\mathbf{F}_{3^{2n}}$ non hyperelliptiques et optimales.

Références

[1] Roland Auer and Jaap Top, *Some genus 3 curves with many points*, Algorithmic number theory (Sydney, 2002), Lecture Notes in Comput. Sci., vol. 2369, Springer, Berlin, 2002, pp. 163–171, DOI 10.1007/3-540-45455-1_13. MR2041081

[2] Edgardo Ciani, *I varii tipi possibili di quartiche piane più volte omologico-armoniche*, Rendiconti del Circolo Matematico di Palermo **13** (1899), no. 2, 347–373.

[3] Noam D. Elkies, *The Klein quartic in number theory*, The eightfold way, Math. Sci. Res. Inst. Publ., vol. 35, Cambridge Univ. Press, Cambridge, 1999, pp. 51–101. MR1722413

[4] Tomoyoshi Ibukiyama, *On rational points of curves of genus 3 over finite fields*, Tohoku Math. J. (2) **45** (1993), no. 3, 311–329, DOI 10.2748/tmj/1178225887. MR1231559

[5] Gilles Lachaud and Christophe Ritzenthaler, *On some questions of Serre on abelian threefolds*, Algebraic geometry and its applications, Ser. Number Theory Appl., vol. 5, World Sci. Publ., Hackensack, NJ, 2008, pp. 88–115, DOI 10.1142/9789812793430_0005. MR2484050

[6] Kristin Lauter, *The maximum or minimum number of rational points on genus three curves over finite fields*, Compositio Math. **134** (2002), no. 1, 87–111, DOI 10.1023/A:1020246226326. With an appendix by Jean-Pierre Serre. MR1931964

[7] Ju. I. Manin, *The Hasse-Witt matrix of an algebraic curve* (Russian), Izv. Akad. Nauk SSSR Ser. Mat. **25** (1961), 153–172. MR0124324

[8] Enric Nart and Christophe Ritzenthaler, *Jacobians in isogeny classes of supersingular abelian threefolds in characteristic 2*, Finite Fields Appl. **14** (2008), no. 3, 676–702, DOI 10.1016/j.ffa.2007.09.006. MR2435055

[9] Enric Nart and Christophe Ritzenthaler, *Genus 3 curves with many involutions and application to maximal curves in characteristic 2*, 2009.

[10] Christophe Ritzenthaler, *Explicit computations of Serre's obstruction for genus-3 curves and application to optimal curves*, LMS J. Comput. Math. **13** (2010), 192–207, DOI 10.1112/S1461157009000576. MR2653834

[11] Jean-Pierre Serre, *Propriétés galoisiennes des points d'ordre fini des courbes elliptiques* (French), Invent. Math. **15** (1972), no. 4, 259–331, DOI 10.1007/BF01405086. MR387283

[12] Jean-Pierre Serre, *Rational points on curves over finite fields*, Harvard University, 1985.

[13] Jaap Top, *Curves of genus 3 over small finite fields*, Indag. Math. (N.S.) **14** (2003), no. 2, 275–283, DOI 10.1016/S0019-3577(03)90011-5. MR2027781

JEAN-FRANÇOIS MESTRE, UNIVERSITÉ PARIS DIDEROT UFR DE MATHÉMATIQUES, BÂTIMENT SOPHIE GERMAIN, 75205 PARIS CEDEX 13, FRANCE

Email address: mestre@math.univ-paris-diderot.fr

Contemporary Mathematics
Volume **770**, 2021
https://doi.org/10.1090/conm/770/15439

Bornes sur le nombre de points rationnels des courbes : en quête d'uniformité

Fabien Pazuki, with an appendix by Sinnou David and Patrice Philippon

ABSTRACT. This article aims to show in detail how a conjectural height lower-bound of Lang-Silverman type leads to an explicit bound on the number of rational points on a genus $g \geq 2$ curve over a number field. This bound is uniform in the height of the curve.

RÉSUMÉ. L'objectif de ce texte est de montrer en détail comment un résultat conjectural de minoration de hauteur de type Lang-Silverman permet d'obtenir une borne explicite sur le nombre de points rationnels d'une courbe de genre $g \geq 2$ sur un corps de nombres. La borne est uniforme en la hauteur de la courbe.

1. Introduction

Soit C une courbe projective lisse de genre $g \geq 2$ définie sur un corps de nombres K. La conjecture de Mordell, qui est depuis 1983 un théorème célèbre de Faltings [**12**], donne la finitude de l'ensemble des points K-rationnels $C(K)$. Une question naturelle a alors émergé : à quel point le cardinal $\#C(K)$ dépend-il de la courbe C ? Est-il possible de produire un majorant explicite de $\#C(K)$ et à quel point ce majorant peut-il être uniforme ? C'est à cette question d'uniformité du majorant que ce texte est consacré.

On sait produire depuis Rémond [**23**] des bornes totalement explicites et inconditionnelles. Citons une forme plus récente de son résultat, donnée dans [**24**].

THÉORÈME 1.1. *Soit C une courbe lisse, projective, géometriquement connexe et de genre $g \geq 2$ définie sur un corps de nombres K. On fixe un plongement projectif de la variété jacobienne $Jac(C)$ relatif à la puissance seizième du fibré associé à un translaté symétrique du diviseur thêta. Alors on a*

$$\#C(K) \leq (2^{38+2g} \cdot [K:\mathbb{Q}] \cdot g \cdot \max\{1, h_\theta\})^{(m_K+1)\cdot g^{20}},$$

où h_θ est la hauteur thêta de $Jac(C)$ dans ce plongement et m_K est le rang de Mordell-Weil de $Jac(C)(K)$.

2020 *Mathematics Subject Classification.* Primary 11G50, 14G40.

L'auteur remercie Gaël Rémond pour ses commentaires précis et précieux. Merci à Amos Turchet pour ses remarques. L'auteur est soutenu par la Chaire DNRF Niels Bohr de Lars Hesselholt et les Projets ANR-14-CE25-0015 Gardio et ANR-17-CE40-0012 Flair.

On cherche à savoir dans quelle mesure on peut se passer de la dépendance en la jacobienne de la courbe dans l'expression du majorant. L'uniformité la plus forte est conjecturée par Caporaso, Harris, Mazur en page 2 de [**5**] :

Conjecture 1.2. *Soit $g \geq 2$ un entier naturel et soit K un corps de nombres. Il existe une quantité $c_0(g,K) > 0$ telle que pour toute courbe projective lisse C de genre g définie sur K, on a*

$$\#C(K) \leq c_0(g,K).$$

Ce dernier énoncé est impliqué par une conjecture très générale de Lang qui s'énonce comme suit : les points rationnels sur une variété de type général ne sont pas Zariski denses. La déduction se fait via un procédé dit de corrélation qui est bien expliqué dans [**5**]. On consultera aussi avec profit les textes [**1, 19**] sur le sujet. Ce niveau d'uniformité maximal peut sembler difficile à atteindre. On pourra essayer dans un premier temps [1] de répondre à la question classique suivante (trouvée par exemple dans [**8**] en introduction).

Conjecture 1.3. *Soit $g \geq 2$ un entier naturel et soit K un corps de nombres. Il existe une quantité $c_1(g,K) > 0$ telle que pour toute courbe projective lisse C de genre g définie sur K, on a*

$$\#C(K) \leq c_1(g,K)^{m_K+1},$$

où m_K est le rang de Mordell-Weil de la jacobienne de C sur K.

Citons aussi une version plus souple de la conjecture 1.3 et trouvée sous forme de question dans [**18**] page 223 (voir aussi le premier paragraphe de la page 234 de [**17**]) :

Conjecture 1.4. *Soit $g \geq 2$ un entier naturel, soit K un corps de nombres, soit m un entier naturel. Il existe une quantité $c_2(g,K,m) > 0$ telle que pour toute courbe projective lisse C de genre g définie sur K, dont le rang de Mordell-Weil de la jacobienne de C sur K est m, on a*

$$\#C(K) \leq c_2(g,K,m).$$

David, Nakamaye et Philippon donnent dans le théorème 3.6 du texte [**8**] des familles infinies de courbes définies sur $\mathbb{Q}$ et vérifiant la conjecture 1.3 (des familles infinies de courbes sont proposées pour chaque $g \geq 4$ fixé). Notons qu'il est classique de penser que la dépendance en le corps K des quantités c_0, c_1, c_2 ne fait intervenir que le degré de K sur $\mathbb{Q}$.

De nouveaux progrès, suivant une approche de ces questions basée sur l'intégration p-adique, concernent les courbes C telles que la jacobienne $\mathrm{Jac}(C)$ possède un rang de Mordell-Weil sur K petit par rapport au genre de C. Initiée par des idées de Chabauty, puis de Coleman, la stratégie p-adique de majoration du cardinal de $C(K)$ fait voir le jour à un résultat d'uniformité par Stoll [**26**] qui traite des courbes hyperelliptiques. Cette attaque a été généralisée à toutes les courbes (pour $g \geq 3$) dans [**14**] par Katz, Rabinoff, Zureick-Brown, dont voici le résultat :

Théorème 1.5. *Soit $d \geq 1$ et soit $g \geq 3$ des entiers. Il existe une constante $c_3(g,d) > 0$ telle que pour tout corps de nombres K de degré d et pour toute*

1. Ajouté après arbitrage : une preuve de la conjecture 1.3 est annoncée par Dimitrov, Gao, Habegger dans [**11**] !

courbe projective lisse C de genre g, définie sur K et telle que le rang m_K de la jacobienne $\mathrm{Jac}(C)$ *sur K vérifie $m_K \leq g-3$, on a $\#C(K) \leq c_3(g,d)$.*

Notons que la quantité $c_3(g,d)$ peut être explicitée, comme cela est proposé dans les textes [**26**] et [**14**]. Lorsqu'elle est applicable, la méthode de Chabauty-Coleman fournit une dépendance polynomiale en le paramètre g, voire quasi-linéaire.

REMARQUE 1.6. Le théorème 1.5 pourra donc être vu comme un pas vers la conjecture 1.2, ou comme un pas vers les conjectures 1.3 ou 1.4. En effet la conjecture 1.3 (ou la conjecture 1.4) combinée à une inégalité du type $m_K \leq c_4(g)$ pour une constante $c_4(g) > 0$ ne dépendant que de g impliquent la conjecture 1.2 *restreinte aux courbes dont la jacobienne vérifie* $m_K \leq c_4(g)$ bien entendu.

REMARQUE 1.7. La conjecture 1.3 ou la conjecture 1.4 couplée à une borne uniforme sur le rang de la forme $m_K \leq c(g,K)$ implique bien entendu la conjecture 1.2. La question de savoir si le rang des variétés abéliennes de dimension fixée $g \geq 1$ sur un corps de nombre fixé K est borné uniformément est un problème encore largement ouvert. De récents travaux [**3**] tendent à suggérer qu'une borne uniforme est plausible pour les courbes elliptiques sur $\mathbb{Q}$.

Notre modeste contribution ici est de montrer en détail qu'une conjecture de minoration de hauteur dans l'esprit Lang-Silverman implique aussi la conjecture 1.3, sans besoin d'hypothèse sur le rang. C'est une idée naturelle après la lecture de [**10**], qui traite cependant uniquement le cas des jacobiennes isogènes à des produits de courbes elliptiques (voir l'hypothèse $(*)$ page 110 de [**10**]). On traite le cas général ici en suivant le chemin euclidien tracé par Rémond dans ses articles [**23, 24**], et en ajoutant comme ingrédient une minoration à la Lang-Silverman. La déduction de la conjecture 1.3 s'obtient alors moyennant quelques efforts techniques supplémentaires. Ce faisant, nous améliorons donc la proposition 1.9 page 23 de [**20**] qui est basée sur une version obsolète de Lang-Silverman. Voici une version plus moderne de cette conjecture.

CONJECTURE 1.8. *([**25**]) Pour tout couple d'entiers naturels (d,g) il existe un réel $c_5 = c_5(d,g) > 0$ tel que pour tout corps de nombres K de degré d, pour toute variété abélienne A/K de dimension g et pour toute polarisation L sur A, il existe une sous-variété abélienne stricte B de A de degré $\deg_L(B) \leq c_5 \deg_L(A)$ et telle que pour tout point $P \in A(K)$, au moins l'une des deux assertions suivantes est vérifiée :*

(1) il existe un entier $1 \leq N \leq c_5$ tel que $[N]P \in B$,

(2)

$$\widehat{h}_{A,L}(P) \geq c_5^{-1} \max\left\{ h_{\mathrm{F}^+}(A/K),\, 1\right\},$$

où $\widehat{h}_{A,L}(.)$ est la hauteur de Néron-Tate associée à L et $h_{\mathrm{F}^+}(A/K)$ est la hauteur de Faltings de la variété A/K.

La formulation originale de cette conjecture présente uniquement la minoration du second cas, la conjecture 1.8 donne une description plus complète. Notons que la sous-variété B ne dépend pas de P. L'utilisation de cette minoration en famille de la hauteur canonique sur les variétés jacobiennes fournit alors le résultat suivant.

THÉORÈME 1.9. *Supposons vraie la conjecture* 1.8. *Soit* K *un corps de nombres de degré* d *et soit* $g \geq 2$ *un entier. Il existe une constante* $c_6 = c_6(g, K) > 0$ *telle que pour toute courbe* C *de genre* g *définie sur* K,

$$\#C(K) \leq c_6^{m_K+1},$$

où m_K *désigne le rang de Mordell-Weil de la jacobienne de* C *sur* K. *On peut choisir*

$$c_6 = c_5^{2g+2} 2^{120g} g^{14+g},$$

où c_5 *est la quantité introduite dans la conjecture* 1.8.

On trouve des résultats partiels en direction de la conjecture 1.8 dans les travaux [**21**] en dimension 2 et [**6, 16**] en dimension générale. Ceci fournit donc le corollaire immédiat suivant, non couvert par [**8, 10**].

COROLLAIRE 1.10. *Les familles de courbes de genre* 2 *satisfaisant aux hypothèses du corollaire 1.14 de* [**21**] *et de genre* $g \geq 2$ *décrites dans* [**6, 16**] *vérifient la conjecture* 1.8 *et sont donc des exemples où la borne du théorème* 1.9 *est démontrée inconditionnellement.*

La section 2 contient les arguments de comptage euclidien et la preuve du théorème 1.9, la section 3 présente une variante de Lang-Silverman et les modifications que cela entraîne dans le comptage, la section A est un appendice écrit par Sinnou David et Patrice Philippon qui corrigent un lemme de leur article [**9**] et fournissent par la même occasion un calcul utile à la comparaison de hauteurs donnée en (2.6).

2. Comptage euclidien

On donne dans cette section une preuve détaillée du fait qu'une version forte de Lang-Silverman implique une borne uniforme sur le nombre de points rationnels des courbes de genre $g \geq 2$ sur les corps de nombres.

Dans toute la suite, on se place dans le cadre suivant. Soit C une courbe projective lisse et géométriquement connexe, de genre $g \geq 2$ sur un corps de nombres K. On suppose tout d'abord que la courbe possède un point rationnel $P_0 \in C(K)$, dans le cas contraire toute borne supérieure positive donnée sur le cardinal de $C(K)$ sera trivialement valide. La courbe C est alors plongée dans sa jacobienne $A = \mathrm{Jac}(C)$ par le plongement $j : P \to (P) - (P_0)$, et on définit un diviseur thêta par $\Theta = j(C) + \ldots + j(C)$ où on somme $g-1$ termes $j(C)$, diviseur fournissant une polarisation principale sur $\mathrm{Jac}(C)$. On considère la jacobienne plongée dans $\mathbb{P}^{4^{2g}-1}$ via le fibré L correspondant à la puissance 16-ième du fibré associé au translaté symétrique de ce diviseur thêta. Par le théorème 5.8 page 118 de [**15**], on a $\deg_\Theta(C) = g$. On aura donc $\deg_L(A) = 2^{4\dim(A)} \deg_\Theta(A) = 16^g g!$ et $\deg_L(C) = 2^{4\dim(C)} \deg_\Theta(C) = 16g$.

Si on note $\phi : A \to \mathbb{P}^{4^{2g}-1}$ le plongement associé, la hauteur thêta de A est alors définie par $h_\Theta(A) = h(\phi(0_A))$, où $h(.)$ est la hauteur du vecteur coordonnées comme donnée dans [**24**] page 762. On utilise aussi la hauteur de Faltings $h_{\mathrm{F}^+}(A)$ d'une variété abélienne A, normalisée comme dans [**22**]. On utilise enfin la classique hauteur de Néron-Tate d'un point algébrique sur une variété abélienne A polarisée par un fibré ample et symmétrique L, qui est une forme quadratique notée $\widehat{h}_{A,L}$ ou simplement $\widehat{h}$ quand le contexte est clair.

Nous allons faire usage de la proposition 3.7 page 527 de [**23**] dans une version valable uniquement pour les courbes et reprenant les résultats du paragraphe baptisé “Cas des courbes” page 529 de [**23**], lequel explique comment améliorer les valeurs des constantes dans le cas particulier qui nous intéresse ici. Cet énoncé est donc numériquement plus efficace qu'une banale spécialisation de la proposition 3.7 page 527 au cas des courbes.

PROPOSITION 2.1. *(Rémond) Soit C une courbe de genre $g \geq 2$ plongée dans sa jacobienne $A = \mathrm{Jac}(C)$ comme précédemment dans un espace projectif ambiant $\mathbb{P}^n$ avec $n = 4^{2g} - 1$, soit Γ un sous-groupe de $A(\overline{\mathbb{Q}})$ de rang fini r. Soit c_{NT} une constante majorant la différence entre la hauteur de Weil et la hauteur de Néron-Tate sur A, relativement au plongement donné. Soit h_1 la hauteur de la famille de polynômes définissant l'addition sur A, relativement au plongement donné. On définit alors c_7 par la formule*

$$c_7^2 = 2^{187+8g} g^{21} \max\{1, c_{NT}, h_1\},$$

et on a l'alternative suivante, où au moins une des assertions est vérifiée :

— *ou*

$$\#\Big(C \cap \Gamma\Big) \leq (2^{170+8g} g^{12})^r,$$

— *ou il existe un point $y \in C \cap \Gamma$ tel que*

$$\#\Big\{x \in C \cap \Gamma \mid \widehat{h}(x-y) \geq 4c_7^2\Big\} \ \leq \ (2^{170+8g} g^{12})^r.$$

DÉMONSTRATION. Le cœur de la preuve est bien entendu détaillé dans [**23**] pages 527-529. On demandera au lecteur de se référer au texte original pour les arguments clefs de la preuve de cette proposition. Nous donnons ici les valeurs des quantités correspondant à notre contexte. Tout d'abord comme $g \geq 2$, l'union des translatés de sous-variétés abéliennes contenues dans C est vide, on a donc $Z_C = \emptyset$ dans la notation de [**23**]. La quantité c_5 apparaissant dans [**23**] vient de la Proposition 3.6 page 526, mais est améliorée par le paragraphe traitant le cas particulier des courbes page 529, on peut donc choisir comme valeur pour c_7 la quantité donnée par $c_7^2 = 2^{20}\Lambda^2 c_{NT} + 2(n+1)\Lambda^{28/3} \max\{((n+1)D+1)(c_{NT} + 3\log(n+1)), \Lambda^{-1}h_1\}$, où $\Lambda = \max\{(\deg_L(C))^2, 2^{14}\deg_L(C)\}$ et $\deg_L(C) = 16g$. En majorant légèrement plus et en tenant compte de $n = 4^{2g} - 1$ et de $D = \max\{2^{14}, \deg_L(C)\} \leq 2^{14}g$, on voit que $2^{187+8g} g^{21} \max\{1, c_{NT}, h_1\}$ convient.

Le majorant figurant dans l'alternative est donné par $(3(n+1)^2 D^{12})^r$ dans [**23**] page 529, ce qui fournit la valeur proposée ici là encore via $n = 4^{2g} - 1$ et $D = \max\{2^{14}, \deg_L(C)\} \leq 2^{14}g$. □

On donne à présent un lemme utile, une estimation sur la taille du rayon discriminant les points “petits” des points “grands” dans le comptage euclidien qui suivra.

LEMME 2.2. *Soit C une courbe de genre $g \geq 2$. Nous conservons le même cadre que précédemment, un plongement thêta de la jacobienne $\mathrm{Jac}(C)$ dans $\mathbb{P}^n$, avec $n = 4^{2g} - 1$, correspondant à un translaté symétrique du diviseur thêta (multiplié par 16). Soit $h_\Theta(A)$ la hauteur thêta de la variété abélienne $A = \mathrm{Jac}(C)$ dans ce plongement et considérons la quantité c_7^2 pour ce plongement de A dans $\mathbb{P}^n$. Alors en utilisant les notations précédentes*

$$c_7^2 \leq 2^{196+10g} g^{22} \max\{1, h_\Theta(A)\}.$$

Démonstration. On part de la valeur explicite de la proposition précédente

$$c_7^2 = 2^{187+8g} g^{21} \max\{1, c_{NT}, h_1\},$$

et on trouve dans [**9**] les majorations (page 652, page 662 et page 665),

$$h_1 \leq 2(4^g - 1) h_\Theta(A) \; , \qquad c_{NT} \leq (4^{g+2} h_\Theta(A) + 6g \log 2) \deg_L(C).$$

Comme plus haut, C est plongée dans sa jacobienne par un plongement thêta, on a alors dans $\mathbb{P}^{4^{2g}-1}$ la valeur $\deg_L(C) = 16g$, où L correspond à la puissance seizième du fibré associé au diviseur thêta. Ceci implique

$$c_7^2 \leq 2^{187+8g} g^{21} \max\{1, 2^{2g+9} g, 2^{2g+1}\} \max\{1, h_\Theta(A)\},$$

et on obtient donc

$$c_7^2 \leq 2^{196+10g} g^{22} \max\{1, h_\Theta(A)\}.$$

□

Nous sommes à présent en mesure de démontrer le théorème 1.9 de l'introduction.

Démonstration. (du théorème 1.9) Soit A la jacobienne de C et soit m_K le rang de $A(K)$. On considère $C(K)$ comme un sous-ensemble de $A(K)$. Notons comme toujours $\widehat{h}$ la hauteur de Néron-Tate sur A associée au plongement thêta dans $\mathbb{P}^{4^{2g}-1}$. Nous allons en fait borner $\#C(K)$ en majorant $\#(C \cap A(K))$.

D'après la proposition 2.1, il suffit de traiter le cas où il existe un point $y \in C(K)$ tel que

$$\#\Big\{x \in C(K) \mid \widehat{h}(x-y) \geq 4c_7^2\Big\} \;\leq N,$$

où N est en fait un entier explicite. On divise les points rationnels en deux clans :

$$\begin{aligned} S_1 &= \left\{x \in C(\overline{\mathbb{Q}}) \cap A(K) \,\middle|\, \widehat{h}(x-y) \leq 4c_7^2\right\}, \\ S_2 &= \left\{x \in C(\overline{\mathbb{Q}}) \cap A(K) \,\middle|\, \widehat{h}(x-y) \geq 4c_7^2\right\}. \end{aligned}$$

Notre entreprise de majoration de $\#C(K)$ commence par la remarque triviale $C(K) \subset S_1 \cup S_2$. On majore alors le cardinal des deux ensembles auxiliaires S_1 et S_2.

Invoquons tout d'abord la proposition 2.1, qui fournit directement

$$\#S_2 \leq (2^{170+8g} g^{12})^{m_K}. \tag{2.1}$$

Pour borner le cardinal de S_1, il faut travailler un peu plus. On adapte ici deux techniques respectivement exposées dans [**10**] et dans [**23**] à notre situation. L'ensemble $(A(K) \otimes \mathbb{R}, \widehat{h})$ est un espace euclidien, on gardera les mêmes notations pour l'image des points de $A(K)$ dans cet espace. Le fait que $\widehat{h}$ est une forme quadratique définie positive sur cet espace euclidien est classique (mais non trivial de prime abord, comme remarqué par Cassels), une preuve est donnée par exemple par la proposition B.5.3 de [**13**] page 201.

Soit $B_o(P, \rho)$ la boule ouverte euclidienne de centre P et rayon ρ. Une version du lemme géométrique élémentaire suivant est démontrée par Rémond et figure au lemme 6.1 page 541 de [**23**].

LEMME 2.3. *Soit r un entier et $\rho > 0$ un réel. Soit S un sous-ensemble de $\mathbb{R}^r$ contenu dans une boule euclidienne de rayon ρ. Pour tout nombre réel $\gamma \geq 1$, on peut trouver $\lfloor (2\gamma + 1)^r \rfloor$ boules de rayon ρ/γ, centrées en des points de S et telles que leur union recouvre tout S.*

Nous pouvons illustrer la situation sur un dessin (en deux dimensions, certes). Les points noirs désignent les points de S. On recouvre tout S en choisissant des points centrés sur les “paquets” de points éventuellement regroupés et en faisant grandir le rayon des petits cercles jusqu'à ρ/γ.

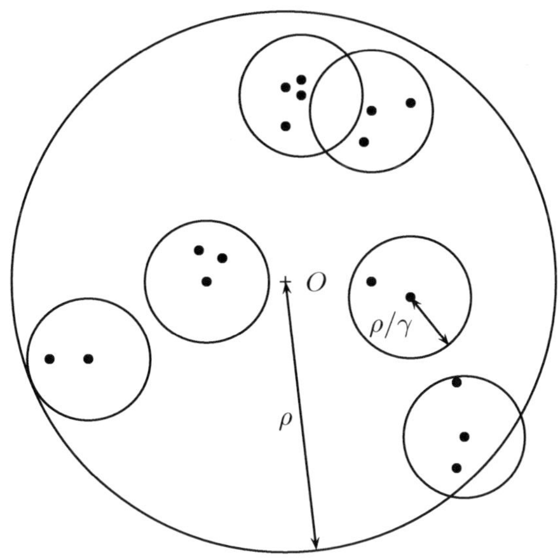

On va appliquer ce lemme à $S = S_1 \subset B_o(y, 2c_7)$. Ici $\rho = 2c_7$ et $r = m_K$, donc en prenant $\gamma = \max\{1, 2c_7/\sqrt{c_5^{-1}\max\{1, h_{\mathrm{F}^+}(A/K)\}}\}$ on obtient

$$\frac{\rho}{\gamma} = \frac{2c_7}{\max\{1, 2c_7/\sqrt{c_5^{-1}\max\{1, h_{\mathrm{F}^+}(A/K)\}}\}} = \min\{2c_7, \sqrt{c_5^{-1}\max\{1, h_{\mathrm{F}^+}(A/K)\}}\}$$

ainsi que l'existence d'un ensemble fini $I \subset S_1$ de cardinal inférieur à

$$\lfloor (2\gamma + 1)^r \rfloor \leq \left(1 + \max\{2, 4c_7/\sqrt{c_5^{-1}\max\{1, h_{\mathrm{F}^+}(A/K)\}}\}\right)^{m_K} \tag{2.2}$$

tel que

$$S_1 \subset \bigcup_{P_0 \in I} B_o(P_0, \min\{2c_7, \sqrt{c_5^{-1}\max\{1, h_{\mathrm{F}^+}(A/K)\}}\}).$$

Comme $S_1 \subset C(K)$, on obtient

$$S_1 \subset \bigcup_{P_0 \in I} B_o(P_0, \min\{2c_7, \sqrt{c_5^{-1}\max\{1, h_{\mathrm{F}^+}(A/K)\}}\}) \cap C(K).$$

Or tout point P de $B_o(P_0, \min\{2c_7, \sqrt{c_5^{-1}\max\{1, h_{\mathrm{F}^+}(A/K)\}}\})$ vérifie en particulier l'inégalité

$$\hat{h}(P - P_0) < c_5^{-1}\max\{1, h_{\mathrm{F}^+}(A/K)\},$$

(la hauteur étant le carré de la norme) donc par la conjecture 1.8 il existe une sous-variété abélienne stricte B avec $\deg_L(B) \leq c_5 \deg_L(A)$ et un entier naturel $1 \leq N \leq c_5$ tel que $[N](P - P_0) \in B$. Tout point $P \in C(K)$ tel que $[N]P$ appartient à $B + [N]P_0$ est en fait un point vivant sur $[N]^{-1}(B + [N]P_0) \cap C$. On a bien sûr $\dim([N]^{-1}(B + [N]P_0) \cap C) \leq 1$. Comme la courbe C est plongée dans sa jacobienne $\mathrm{Jac}(C)$ via j, la somme de g copies $j(C) + \ldots + j(C)$ couvre tout $\mathrm{Jac}(C)$. Cette courbe ne peut donc pas, dès que $g \geq 2$, être contenue dans un translaté de sous-variété abélienne stricte. Donc $\dim([N]^{-1}(B + [N]P_0) \cap C) = 0$ et on majore le degré de l'intersection par le théorème de Bézout :

$$\deg_L([N]^{-1}(B + [N]P_0) \cap C) \leq \deg_L([N]^{-1}(B + [N]P_0)) \deg_L(C).$$

Par la proposition 2.3 page 9 de [**7**] on obtient

$$\deg_L([N]^{-1}(B + [N]P_0)) = N^{2\,\mathrm{codim}(B)} \deg_L(B).$$

On déduit donc

$$\deg_L([N]^{-1}(B + [N]P_0) \cap C) \leq N^{2g} \deg_L(B) \deg_L(C).$$

On obtient alors la majoration suivante, en vertu de $\deg_L(B) \leq c_5 \deg_L(A)$, de $N \leq c_5$ et de $\deg_L(C) = 16g$:

$$\#S_1 \leq \#I \; c_5^{2g+1} \deg_L(A) \; 16g \tag{2.3}$$

et donc avec $\deg_L(A) = 16^g g!$

$$\#S_1 \leq 16^{g+1} g! g \, c_5^{2g+1} \#I. \tag{2.4}$$

On va à présent majorer le terme $\#I$ dans le membre de droite de (2.4). En reprenant (2.2), le majorant de c_7^2 donné dans le lemme 2.2 montre que

$$\#I \leq \left(1 + \max\left\{2, \frac{4 \cdot 2^{98+5g} g^{11} \sqrt{\max\{1, h_\Theta(A)\}}}{\sqrt{c_5^{-1}\max\{1, h_{\mathrm{F}^+}(A/K)\}}}\right\}\right)^{m_K}. \tag{2.5}$$

Pour comparer la hauteur thêta et la hauteur de Faltings explicitement, on invoque [**4, 20**] afin d'obtenir

$$h_{\mathrm{F}^+}(A/K) \geq h_{\mathrm{F}^+}(A/\overline{\mathbb{Q}}) \geq 2h_\Theta(A) - \frac{g}{2}\log\max\{1, h_\Theta(A)\} - 2M(g), \tag{2.6}$$

où $M(g) = g4^{2g}$ convient (voir le Corollaire A.2 dans l'appendice du présent article). On divise en deux cas.

<u>Premier cas :</u> si $h_\Theta(A) > 2M(g) + \frac{g}{2}\log\max\{1, h_\Theta(A)\}$, alors

$$2h_\Theta(A) - 2M(g) - \frac{g}{2}\log\max\{1, h_\Theta(A)\} > h_\Theta(A),$$

donc $h_{\mathrm{F}^+}(A/K) > h_\Theta(A)$ et par (2.5) on obtient (on rappelle que $c_5 \geq 1$ puisque c_5 majore l'entier $N \geq 1$)

$$\#I \leq \left(1 + \frac{2^{100+5g} g^{11}}{\sqrt{c_5^{-1}}}\right)^{m_K}.$$

<u>Second cas :</u> si $h_\Theta(A) \leq 2M(g) + \frac{g}{2}\log\max\{1, h_\Theta(A)\}$, alors $h_\Theta(A) \leq 4M(g)$ et on utilise cela pour majorer le numérateur dans (2.5) (tout en minorant le terme au dénominateur par $\sqrt{c_5^{-1}}$) et ainsi trouver

$$\#I \leq \left(1 + 2^{100+5g} g^{11}\sqrt{4c_5 M(g)}\right)^{m_K}.$$

Dans les deux cas, en utilisant $M(g) = g4^{2g}$ (voir le Corollaire A.2 dans l'appendice du présent article), on a montré la majoration suivante (où on utilise $c_5^{-1} \leq 1$) :

$$\#I \leq \left(\sqrt{c_5} 2^{102+7g} g^{12}\right)^{m_K}. \tag{2.7}$$

On peut donc revenir à (2.4) pour obtenir :

$$\#S_1 \leq 16^{g+1} g! g\, c_5^{2g+1} \left(\sqrt{c_5} 2^{102+7g} g^{12}\right)^{m_K} \tag{2.8}$$

Combinant (2.1) et (2.8), une dernière majoration donne le résultat. □

3. Variante de Lang-Silverman

On regarde dans ce paragraphe une autre variante de Lang-Silverman, issue de [**22**].

CONJECTURE 3.1. *Soit $g \geq 1$ un entier. Pour tout corps de nombres K, il existe trois nombres $c_8 = c_8(g, K) > 0$, $c_9 = c_9(g, K) > 0$ et $c_{10} = c_{10}(g, K) > 0$ tels que pour toute variété abélienne A/K de dimension g et pour tout fibré en droites ample et symétrique L sur A, pour tout point $P \in A(K)$, au moins l'une des deux assertions suivantes est vérifiée :*

- *il existe une sous-variété abélienne $B \subset A$, $B \neq A$, de degré $\deg_L(B) \leq c_9 \deg_L(A)$ et telle que le point P est d'ordre borné par c_{10} modulo B,*
- *on a $\mathrm{End}(A){\cdot}P$ est Zariski dense et*

$$\widehat{h}_{A,L}(P) \geq c_8 \max\left\{h_{\mathrm{F}^+}(A/K),\, 1\right\},$$

où $\widehat{h}_{A,L}(.)$ est la hauteur de Néron-Tate associée à L et $h_{\mathrm{F}^+}(A/K)$ est la hauteur de Faltings de la variété A/K.

C'est un énoncé qui dit sensiblement la même chose que la conjecture 1.8 : les points rationnels dont la hauteur est trop petite proviennent en fait d'un translaté d'une sous-variété abélienne stricte. Cette version est là encore motivée par le théorème 1.4 page 511 de [**6**] et les théorèmes 1.8 et 1.13 de [**21**]. Des exemples de familles de jacobiennes de courbes vérifiant cette conjecture sont donnés dans [**16**] et [**21**] : ces exemples sont aussi valables pour cette forme forte, puisque les jacobiennes associées sont simples.

Si on parcourt le raisonnement exposé plus haut, la même preuve vaut (en faisant de petites retouches aisées, que nous faisons plus bas), le décompte est un peu moins pratique car il est à présent nécessaire de savoir estimer le nombre de sous-variétés abéliennes de degré borné pour tenir compte de l'alternative, alors que dans la conjecture 1.8 n'y a qu'une seule variété obstructrice B.

REMARQUE 3.2. Notons une différence entre la conjecture 1.8 et la conjecture 3.1. Prenons le cas d'un point $P = (P_1, 0) \in A = A_1 \times A_2$ avec la polarisation L produit de L_1 sur A_1 et L_2 sur A_2, avec A_2 indépendante de A_1 (la hauteur de A_2 de dépend pas de la hauteur de A_1, la variété A_2 n'est donc pas dans la classe d'isogénie de A_1, ni isogène à une puissance de A_1, etc). On suppose que P_1 n'est pas un point de torsion. On choisit A_2 telle que $h_{\mathrm{F}^+}(A_2/K) > 1 + \widehat{h}_{A,L}(P)$, ce qui est possible puisque $\widehat{h}_{A,L}(P) = \widehat{h}_{A_1,L_1}(P_1)$ est indépendant de A_2. Donc P doit être dans le premier cas. Regardons à présent l'ensemble des itérés de P, la variété A_2 étant fixée. A partir d'un certain rang, la suite de réels positifs $\widehat{h}_{A,L}([k]P)$ dépasse $\max\{1, h_{\mathrm{F}^+}(A_2/K)\}$, et donc $[k]P$ passe dans le deuxième cas. C'est un phénomène qui ne se produit pas dans la conjecture 3.1 où un point exclu du second cas voit tous ses itérés exclus *de facto*.

REMARQUE 3.3. La conjecture classique (la forme "faible") propose une minoration de la hauteur canonique d'un point rationnel par la hauteur de Faltings de la variété en imposant des hypothèses nécessaires sur le point, et ne dit rien sur les points ne satisfaisant pas ces hypothèses, qui se retrouvent exclus. Un geste d'ouverture envers ces points a été fait et la conjecture a ainsi été fortifiée pour la première fois dans [**20**], l'inspiration venant du théorème 1.4 page 511 de [**6**], puis la forme forte a été corrigée dans [**22**] pour donner la conjecture 3.1 et dans [**25**] pour donner la conjecture 1.8.

Voyons à présent ce qui change dans le décompte si on fait usage de cette conjecture 3.1.

THÉORÈME 3.4. *Supposons vraie la conjecture* 3.1. *Soit K un corps de nombres et soit $g \geq 2$ un entier. Il existe une constante $c_{11} = c_{11}(g, K) > 0$ telle que pour toute courbe C de genre g définie sur K,*

$$\#C(K) \leq c_{11}^{m_K+1},$$

où m_K désigne le rang de Mordell-Weil de la jacobienne de C sur K. On peut choisir

$$c_{11} = (2g)^{42g^4} \max\left\{c_9^{(2g+1)4g^2+1}, c_{10}^{2g}, \frac{1}{\sqrt{c_8}}\right\},$$

avec $c_8 = c_8(g, K) > 0$, $c_9 = c_9(g, K) \geq 1$ et $c_{10} = c_{10}(g, K) > 0$ données dans la conjecture 3.1.

DÉMONSTRATION. Soit $\mathcal{B}$ l'ensemble des sous-variétés abéliennes de A de degré borné par $c_9 \deg_L(A)$. C'est un ensemble fini, une borne sur son cardinal est donnée un peu plus bas. On reprend la majoration à l'étape (2.3) :

$$\#S_1 \leq \deg_L(C) N^{2g} \sum_{P_0 \in I} \sum_{B \in \mathcal{B}} \deg_L(B), \tag{3.1}$$

donc comme pour tout $B \in \mathcal{B}$ on a $\deg_L(B) \leq c_9 \deg_L(A)$, et comme $\deg_L(C) = 16g$ et $N \leq c_{10}$ il vient

$$\#S_1 \leq 16g\, c_{10}^{2g}\, \#I\, \#\mathcal{B}\, c_9 \deg_L(A) \tag{3.2}$$

et donc avec $\deg_L(A) = 16^g g!$

$$\#S_1 \leq 16^{g+1} g! g\, c_{10}^{2g}\, c_9\, \#I\, \#\mathcal{B}. \tag{3.3}$$

L'équivalent de (2.7) est ici

$$\#I \leq \left(\frac{1}{\sqrt{c_8}} 2^{102+6g} g^{12} \right)^{m_K} . \tag{3.4}$$

Majorons les autres termes du membre de droite de (3.3). Par la proposition 4.1 page 529 de [**23**] on obtient la borne $\#\mathcal{B} \leq (4g!(2g)!(2^g c_9 \deg_L(A))^{2g+1})^{4g^2}$, le majorant donné ici étant légèrement moins fin que celui de Rémond, mais plus facile à manipuler. On a encore une fois $\deg_L(A) = 16^g g!$, ce qui donne en majorant brutalement $\#\mathcal{B} \leq 2^{40g^4} g^{24g^4} c_9^{(2g+1)4g^2}$ et par (3.3) et (3.4) on obtient

$$\#S_1 \leq 2^{41g^4} g^{25g^4} c_{10}^{2g} c_9^{(2g+1)4g^2+1} \left(\frac{1}{\sqrt{c_8}} 2^{102+6g} g^{12} \right)^{m_K} \tag{3.5}$$

Combinant (2.1) et (3.5) et $g \geq 2$, un calcul facile donne le résultat. □

On conclut avec une remarque supplémentaire, donnant la mesure de la difficulté de la conjecture 3.1 (et la même remarque vaut pour la conjecture 1.8).

REMARQUE 3.5. Supposons vraie la conjecture 3.1. Alors pour tout $g \geq 1$, pour tout corps de nombres K, il existe une quantité $c_{12}(g, K) > 0$ telle que pour toute variété abélienne A de dimension g et définie sur K, on a la majoration $\#A(K)_{tors} \leq c_{12}(g, K)$, où la quantité $c_{12}(g, K)$ est exprimable explicitement en termes des quantités apparaissant dans la conjecture 3.1.

Appendice A. Corrigendum à “Minorations des hauteurs normalisées des sous-variétés de variétés abéliennes II” par Sinnou David et Patrice Philippon

Une erreur s'est glissée dans la preuve du lemme 6.7 (lemme matriciel) et du corollaire 6.9 de l'appendice de [**9**]. Nous présentons ici une version corrigée de ces énoncés, les constantes numériques y sont changées. Nous indiquons d'emblée que ces changements n'altèrent pas les autres énoncés de [**9**]. En effet, le lemme 6.7 n'est invoqué que pour établir le lemme suivant 6.8 et le corollaire 6.9 et c'est au lemme 6.8 qu'il est fait appel dans le texte. Or, la preuve du lemme 6.8 utilise en réalité une version faible de l'inégalité du lemme 6.7, dont la véracité est confirmée et que nous avons détachée dans l'énoncé A.1 de la version corrigée ci-dessous.

Nous remercions Pascal Autissier et Fabien Pazuki de nous avoir signalé cette erreur. Le lecteur pourra se reporter à [**2**] pour une version presque optimale du lemme matriciel ; en effet le seul terme susceptible d'être amélioré est le terme additif ne dépendant que de g. Pour une version accessible de la comparaison entre la hauteur thêta et la hauteur différentielle, le lecteur pourra consulter [**20**].

Nous reprenons ici les mêmes conventions et notations que dans le texte initial. En particulier, pour tout élément τ de l'espace de Siegel $\mathfrak{S}_g$ nous notons k_τ le corps de définition de l'origine de A_τ, c'est-à-dire du point de coordonnées projectives $(\theta_p(\tau, 0))_{p \in \mathcal{Z}_2^2}$. On note de plus F_g le domaine fondamental de Siegel tel que dans le texte initial. Rappelons enfin que, selon la notation 3.2 de [**9**], $h(A)$ désigne la hauteur projective de l'origine de A dans le plongement induit par la puissance 16-ième de la polarisation principale implicite dans le choix de τ.

LEMME A.1. *Soit τ un élément de $\mathfrak{S}_g$ tel que k_τ soit contenu dans un corps de nombres k, de degré d sur $\mathbb{Q}$. Soit σ un plongement complexe de k et $\tau(\sigma)$ un* *élément de* $\mathfrak{S}_g$ *satisfaisant* $A_\tau^\sigma = A_{\tau(\sigma)}$. *Soit enfin* $\tau'(\sigma)$ *un représentant de* $\tau(\sigma)$ *dans* F_g *et* $y'(\sigma)$ *la partie imaginaire de* $\tau'(\sigma)$. *Alors pour tout plongement* σ *on a :*

$$\frac{\pi}{8}\|y'(\sigma)\| \leq dh(A) + 2g^2\log(4g) \ .$$

De plus,

$$\frac{\pi}{8d}\sum_\sigma \|y'(\sigma)\| \leq 4^g h(A) + 2g^2\log(4g) \ .$$

DÉMONSTRATION. Quitte à faire une extension de k, on peut supposer que $\tau'(\sigma) = \tau(\sigma)$; c'est ce que nous ferons. Soit donc σ un plongement complexe de k, choisissons des indices $p(\sigma), q(\sigma) \in \mathcal{Z}_2^2$ tels que $\left|\theta_{p(\sigma)}(\tau(\sigma),0)\right|$ soit minimal, non nul, tandis que $\left|\theta_{q(\sigma)}(\tau(\sigma),0)\right|$ est maximal. En tenant compte des lemmes 6.5 et 6.6 de [**9**] on a

$$\text{(A.1)} \quad \left|\theta_{q(\sigma)}(\tau(\sigma),0)\right| \geq 1 \quad \text{et} \quad 0 \neq |\theta_{p(\sigma)}(\tau(\sigma),0)| \leq (4g)^{2g^2}\exp(-\frac{\pi}{8}\|y'(\sigma)\|) \ .$$

Maintenant, pour $p \in \mathcal{Z}_2^2$ nous notons $I(p)$ l'ensemble des plongements tels que $p(\sigma) = p$. La définition de la hauteur de Weil nous assure alors que :

$$\text{(A.2)} \quad \begin{aligned} \frac{1}{d}\sum_{\sigma\in I(p)} \log\left|\frac{\theta_{q(\sigma)}(\tau(\sigma),0)}{\theta_p(\tau(\sigma),0)}\right| &= \frac{1}{d}\sum_{\sigma\in I(p)} \log\max_{q\in\mathcal{Z}_2^2}\left\{\left|\frac{\theta_q(\tau(\sigma),0)}{\theta_p(\tau(\sigma),0)}\right|\right\} \\ &\leq \frac{1}{d}\sum_\sigma \log\max_{q\in\mathcal{Z}_2^2}\left\{\left|\frac{\theta_q(\tau(\sigma),0)}{\theta_p(\tau(\sigma),0)}\right|\right\} \\ &\leq h_{W,\mathcal{L}^{\otimes 16}}(A_\tau) = h(A) \ , \end{aligned}$$

car $\frac{\theta_q(\tau(\sigma),0)}{\theta_p(\tau(\sigma),0)} = \sigma\left(\frac{\theta_q(\tau,0)}{\theta_p(\tau,0)}\right)$ pour tous les plongements σ. On déduit en particulier de (A.1) et (A.2) pour tout plongement σ :

$$\frac{\pi}{8}\|y'(\sigma)\| - 2g^2\log(4g) \leq \log\left|\frac{\theta_{q(\sigma)}(\tau(\sigma),0)}{\theta_{p(\sigma)}(\tau(\sigma),0)}\right| \leq dh(A) \ ,$$

ce qui démontre la première inégalité du lemme.

Par ailleurs, il suit de (A.1) :

$$\frac{1}{d}\sum_\sigma \log\left|\frac{\theta_{q(\sigma)}(\tau(\sigma),0)}{\theta_{p(\sigma)}(\tau(\sigma),0)}\right| \geq \frac{\pi}{8d}\sum_\sigma \|y'(\sigma)\| - 2g^2\log(4g) \ .$$

Ainsi, le principe des tiroirs combiné à cette dernière inégalité montre qu'il existe un élément $p \in \mathcal{Z}_2^2$ tel que

$$\text{(A.3)} \quad \begin{aligned} \frac{\mathrm{Card}(\mathcal{Z}_2^2)}{d}.\sum_{\sigma\in I(p)} \log\left|\frac{\theta_{q(\sigma)}(\tau(\sigma),0)}{\theta_{p(\sigma)}(\tau(\sigma),0)}\right| &\geq \frac{1}{d}\sum_\sigma \log\left|\frac{\theta_{q(\sigma)}(\tau(\sigma),0)}{\theta_{p(\sigma)}(\tau(\sigma),0)}\right| \\ &\geq \frac{\pi}{8d}\sum_\sigma \|y'(\sigma)\| - 2g^2\log(4g) \ . \end{aligned}$$

Reportant (A.3) dans (A.2) et observant $\mathrm{Card}(\mathcal{Z}_2^2) = 4^g$, on aboutit à la seconde inégalité du lemme A.1 qui est donc complètement démontré. □

On notera que si l'on tient compte du théorème 1.1 de [**4**], *voir* [**20**], théorème 1.1, on déduit immédiatement du lemme A.1 et du lemme 6.4(i) de [**9**] le corollaire suivant :

COROLLAIRE A.2. *Soit A une variété abélienne de dimension g, principalement polarisée, définie sur un corps de nombres k sur lequel elle admet une réduction semi-stable, et soit $h_F(A)$ sa hauteur de Faltings. Alors :*

$$\left| h(A) - \frac{1}{2} h_F(A) \right| \leq \frac{g}{4} \log(\max\{1; h(A)\}) + g4^{2g} .$$

Références

[1] Dan Abramovich, *Uniformité des points rationnels des courbes algébriques sur les extensions quadratiques et cubiques* (French, with English and French summaries), C. R. Acad. Sci. Paris Sér. I Math. **321** (1995), no. 6, 755–758. MR1354720

[2] Pascal Autissier, *Un lemme matriciel effectif* (French, with English and French summaries), Math. Z. **273** (2013), no. 1-2, 355–361, DOI 10.1007/s00209-012-1008-x. MR3010164

[3] Manjul Bhargava and Arul Shankar, *Binary quartic forms having bounded invariants, and the boundedness of the average rank of elliptic curves*, Ann. of Math. (2) **181** (2015), no. 1, 191–242, DOI 10.4007/annals.2015.181.1.3. MR3272925

[4] J.-B. Bost et S. David, *Notes on the comparison of heights of abelian varieties.* Manuscrit, 1999.

[5] Lucia Caporaso, Joe Harris, and Barry Mazur, *Uniformity of rational points*, J. Amer. Math. Soc. **10** (1997), no. 1, 1–35, DOI 10.1090/S0894-0347-97-00195-1. MR1325796

[6] Sinnou David, *Minorations de hauteurs sur les variétés abéliennes* (French, with English and French summaries), Bull. Soc. Math. France **121** (1993), no. 4, 509–544. MR1254751

[7] Sinnou David and Marc Hindry, *Minoration de la hauteur de Néron-Tate sur les variétés abéliennes de type C. M* (French, with English summary), J. Reine Angew. Math. **529** (2000), 1–74, DOI 10.1515/crll.2000.096. MR1799933

[8] Sinnou David, Michael Nakamaye, and Patrice Philippon, *Bornes uniformes pour le nombre de points rationnels de certaines courbes* (French, with English summary), Diophantine geometry, CRM Series, vol. 4, Ed. Norm., Pisa, 2007, pp. 143–164. MR2349652

[9] Sinnou David and Patrice Philippon, *Minorations des hauteurs normalisées des sous-variétés de variétés abeliennes. II* (French, with English and French summaries), Comment. Math. Helv. **77** (2002), no. 4, 639–700, DOI 10.1007/PL00012437. MR1949109

[10] Teresa de Diego, *Points rationnels sur les familles de courbes de genre au moins 2* (French, with English and French summaries), J. Number Theory **67** (1997), no. 1, 85–114, DOI 10.1006/jnth.1997.2146. MR1485428

[11] V. Dimitrov, Z. Gao et P. Habegger, *Uniformity in Mordell-Lang for curves*, Annals of Mathematics, to appear.

[12] G. Faltings, *Endlichkeitssätze für abelsche Varietäten über Zahlkörpern* (German), Invent. Math. **73** (1983), no. 3, 349–366, DOI 10.1007/BF01388432. MR718935

[13] Marc Hindry and Joseph H. Silverman, *Diophantine geometry*, Graduate Texts in Mathematics, vol. 201, Springer-Verlag, New York, 2000. An introduction, DOI 10.1007/978-1-4612-1210-2. MR1745599

[14] Eric Katz, Joseph Rabinoff, and David Zureick-Brown, *Uniform bounds for the number of rational points on curves of small Mordell-Weil rank*, Duke Math. J. **165** (2016), no. 16, 3189–3240, DOI 10.1215/00127094-3673558. MR3566201

[15] Serge Lang, *Fundamentals of Diophantine geometry*, Springer-Verlag, New York, 1983, DOI 10.1007/978-1-4757-1810-2. MR715605

[16] D. W. Masser, *Large period matrices and a conjecture of Lang*, Séminaire de Théorie des Nombres, Paris, 1991–92, Progr. Math., vol. 116, Birkhäuser Boston, Boston, MA, 1993, pp. 153–177, DOI 10.1007/978-1-4757-4273-2_10. MR1300888

[17] Barry Mazur, *Arithmetic on curves*, Bull. Amer. Math. Soc. (N.S.) **14** (1986), no. 2, 207–259, DOI 10.1090/S0273-0979-1986-15430-3. MR828821

[18] Barry Mazur, *Abelian varieties and the Mordell-Lang conjecture*, Model theory, algebra, and geometry, Math. Sci. Res. Inst. Publ., vol. 39, Cambridge Univ. Press, Cambridge, 2000, pp. 199–227. MR1773708

[19] Patricia L. Pacelli, *Uniform boundedness for rational points*, Duke Math. J. **88** (1997), no. 1, 77–102, DOI 10.1215/S0012-7094-97-08803-7. MR1448017

[20] Fabien Pazuki, *Theta height and Faltings height* (English, with English and French summaries), Bull. Soc. Math. France **140** (2012), no. 1, 19–49, DOI 10.24033/bsmf.2623. MR2903770

[21] Fabien Pazuki, *Minoration de la hauteur de Néron-Tate sur les surfaces abéliennes* (French, with English and French summaries), Manuscripta Math. **142** (2013), no. 1-2, 61–99, DOI 10.1007/s00229-012-0593-7. MR3081000

[22] F. Pazuki, *Heights, ranks and regulators of abelian varieties.* Ramanujan Math. Society, Lecture Notes 26, to appear.

[23] Gaël Rémond, *Décompte dans une conjecture de Lang* (French, with English summary), Invent. Math. **142** (2000), no. 3, 513–545, DOI 10.1007/s002220000095. MR1804159

[24] Gaël Rémond, *Nombre de points rationnels des courbes* (French, with English summary), Proc. Lond. Math. Soc. (3) **101** (2010), no. 3, 759–794, DOI 10.1112/plms/pdq005. MR2734960

[25] Gaël Rémond, *Conjectures uniformes sur les variétés abéliennes* (French, with English summary), Q. J. Math. **69** (2018), no. 2, 459–486, DOI 10.1093/qmath/hax042. MR3815154

[26] Michael Stoll, *Uniform bounds for the number of rational points on hyperelliptic curves of small Mordell-Weil rank*, J. Eur. Math. Soc. (JEMS) **21** (2019), no. 3, 923–956, DOI 10.4171/JEMS/857. MR3908770

Fabien Pazuki, University of Copenhagen, Institute of Mathematics, Universitetsparken 5, 2100 Copenhagen, Denmark, and Université de Bordeaux, IMB, 351, cours de la Libération, 33400 Talence, France.

Email address: fpazuki@math.ku.dk

Contemporary Mathematics
Volume **770**, 2021
https://doi.org/10.1090/conm/770/15440

The quadratic hull of a code and the geometric view on multiplication algorithms

Hugues Randriambololona

ABSTRACT. We introduce the notion of quadratic hull of a linear code, and give some of its properties. We then show that any symmetric bilinear multiplication algorithm for a finite-dimensional algebra over a field can be obtained by evaluation-interpolation at simple points (i.e. of degree and multiplicity 1) on a naturally associated space, namely the quadratic hull of the corresponding code. This also provides a geometric answer to some questions such as: which linear maps actually are multiplication algorithms, or which codes come from supercodes (as asked by Shparlinski, Tsfasman, and Vladut in [**38**]. We illustrate this with examples, in particular we describe the quadratic hull of all the optimal algorithms computed by Barbulescu, Detrey, Estibals, and Zimmermann in [**6**] for small algebras.

In our presentation we actually work with multiplication reductions. This is a generalization of multiplication algorithms, that allows for instance evaluation-interpolation at points of higher degree and/or with multiplicities, and also includes the notion of "reverse multiplication-friendly embedding" recently introduced by Cascudo, Cramer, Xing, and Yuan in [**11**]. All our results hold in this more general context.

1. Introduction

1.1. — Early remarks of Goppa [**23**] and Lachaud [**26**, §5.10], made more precise by Pellikaan, Shen, and van Wee in [**33**], show that any linear code is an evaluation code on an algebraic curve. In some sense this work presents an analogue for (symmetric bilinear) multiplication algorithms: any such multiplication algorithm is an evaluation-interpolation algorithm on some algebraic space.

1.2. — The theory of bilinear complexity started with the celebrated algorithms of Karatsuba [**25**], that allows to multiply two 2-digit numbers with 3 elementary multiplications instead of 4, and of Strassen [**39**], that allows to multiply two 2×2 matrices with 7 field multiplications instead of 8. Used recursively, these algorithms then allow to multiply numbers with a large number of digits, or matrices of large size, with a dramatic improvement on complexity over the naive methods.

General considerations on bilinear algorithms, in particular in relation with tensor decompositions, can be found in [**40**][**8**]. In [**21**][**42**] the theory is developed

2020 *Mathematics Subject Classification.* Primary 94B27, 68W30, 12E20, 14A25.

in terms of multiplicative complexity of algebras, with a special focus on quotient algebras of a polynomial ring in one indeterminate, especially over a finite field.

A bilinear algorithm of length n for a bilinear map B over a field $\mathbf{F}$ reduces the computation of B to n multiplications in $\mathbf{F}$, plus some fixed linear operations. Expressed in coordinates, the bilinear map B gives rise to a collection of bilinear forms $B_1, \dots, B_k$. The goal is then to find n products of two linear forms, i.e. n bilinear forms of rank 1, whose linear span contains $B_1, \dots, B_k$. This is how the problem is usually formulated in these early works.

In this text the bilinear map B will always be the multiplication map in a finite dimensional commutative algebra $\mathcal{A}$. Also we will consider only algorithms whose structure reflects this commutativity. Rephrasing the definition in this particular context and in a more coordinate-free way, a *symmetric bilinear multiplication algorithm* of length n for $\mathcal{A}$ over $\mathbf{F}$ is a pair of linear maps $\varphi : \mathcal{A} \to \mathbf{F}^n$ and $\omega : \mathbf{F}^n \to \mathcal{A}$, such that for all $a, a' \in \mathcal{A}$ we have

$$aa' = \omega(\varphi(a) * \varphi(a'))$$

where $*$ denotes componentwise multiplication in $\mathbf{F}^n$.

1.3. — In their important paper [**14**], Chudnovsky and Chudnovsky highlighted several links between multiplication algorithms and codes. Then, using a construction similar to Goppa's algebraic geometry codes, they showed that, at least over a base finite field of (not too small) square order, *evaluation-interpolation* on curves allows to produce multiplication algorithms in an extension field of arbitrary large degree, whose complexity remains linear with the degree.

Several improvements were then proposed by Shparlinski, Tsfasman, and Vladut in [**38**]: they showed how to deduce linearity of the complexity over an arbitrary base finite field; they introduced the notion of supercode, that captures the part of the procedure that is mere linear algebra, and they reduced what remains to a cleanly posed problem in algebraic geometry; and they corrected certain statements or made some arguments in the proofs of [**14**] more precise, in particular concerning the choice of the curves in the construction. Then in [**3**], Ballet replaced certain geometric conditions with more manageable numerical criteria.

In [**5**], Ballet and Rolland extended the Chudnovsky-Chudnovsky method by allowing evaluation-interpolation at points of higher degree, while [**1**] used points with multiplicities, and [**12**] a combination of both.

Further contributions to the method were then proposed in [**34**]: it improved the use of points of higher degree and with multiplicities on curves (as a particular case of evaluation-interpolation at arbitrary closed subschemes on arbitrary schemes); it distinguished symmetric and asymmetric evaluation-interpolation algorithms; and it fixed a geometric construction of [**38**], that allows to improve the asymptotic bound of [**14**].

In another direction, works such as [**32**][**6**][**16**] aimed at exhaustively finding multiplication algorithms of optimal length for algebras of small cardinality.

For recent results on bilinear multiplication algorithms in extensions of finite fields we refer to the extensive survey [**4**]. For a more general introduction to bilinear complexity we refer to [**9**, Ch. 14].

1.4. — Multiplication algorithms are closely related to multiplicative (or "arithmetic") linear secret sharing schemes. We refer to [**17**][**13**] for foundational works on this topic, and to the book [**18**] for more on secure multi-party computation and secret sharing. Roughly speaking, a multiplicative linear secret sharing scheme is a multiplication algorithm that further admits certain threshold properties.

Because of this link, plain multiplication algorithms (without these threshold properties) can be used as a technical tool in the construction of multiplicative secret sharing schemes [**10**]: in this context, a multiplication algorithm that reduces a multiplication in $\mathbb{F}_{q^k}$ to n multiplications in $\mathbb{F}_q$ is also called a "multiplication-friendly embedding" of $\mathbb{F}_{q^k}$ into $\mathbb{F}_q^n$. In [**11**] the notion of "reverse multiplication-friendly embedding" (RMFE) is also introduced, which goes in the opposite direction: a RMFE from $\mathbb{F}_q^k$ to $\mathbb{F}_{q^n}$ reduces k multiplications in $\mathbb{F}_q$ to one multiplication in $\mathbb{F}_{q^n}$. It turns out such RMFE can be constructed by evaluation-interpolation, in a way that duplicates the Chudnovsky-Chudnovsky method.

This situation made it desirable to introduce a more general notion of *multiplication reduction* (Definition 2.1), from an arbitrary finite-dimensional commutative algebra $\mathcal{A}$ to another such arbitrary algebra $\mathcal{B}$, and to which evaluation-interpolation constructions can be applied in a uniform framework (Proposition 2.9). As a by-product, this new notion of multiplication reduction also allows to reinterpret the evaluation-interpolation algorithms at points of higher degree and with multiplicities of [**5**][**1**][**12**][**34**] discussed above.

For simplicity, in the remaining of this introduction we describe our results only for multiplication algorithms. Full statements and proofs for general multiplication reductions will be found in the main text. This will be best expressed in the language of abstract algebraic geometry, although for ease of the reader we will also give alternative descriptions, in coordinates.

1.5. — Assume that X is an algebraic space (e.g. a curve, a surface, etc.) together with a point $Q \in X(\mathcal{A})$ with coordinates in $\mathcal{A}$, and a collection $\mathcal{P} = (P_1, \dots, P_n)$ of rational points. In order to multiply $a, a' \in \mathcal{A}$, we can proceed by evaluation-interpolation, as follows:

(i) Lift a, a' to functions $f_a, f_{a'}$ on X such that $f_a(Q) = a, f_{a'}(Q) = a'$, and then evaluate these functions at $\mathcal{P}$ to get vectors $c_a = (f_a(P_1), \dots, f_a(P_n))$, $c_{a'} = (f_{a'}(P_1), \dots, f_{a'}(P_n))$.

(ii) Multiply c_a and $c_{a'}$ componentwise, i.e. compute $y_1 = f_a(P_1)f_{a'}(P_1)$, ..., $y_n = f_a(P_n)f_{a'}(P_n)$.

(iii) By "Lagrange interpolation", find a function h on X that takes these values $h(P_1) = y_1$, ..., $h(P_n) = y_n$, and then evaluate h at Q. Under proper hypotheses, we will get $h(Q) = aa'$ as wished.

This actually provides a multiplication algorithm for $\mathcal{A}$: step (i) gives the map φ, step (ii) gives the n multiplications in $\mathbf{F}$, and step (iii) gives ω.

Our main result is that any multiplication algorithm is of this sort. Actually we will provide two proofs for this result (and for its generalization to multiplication reductions). Our first version (Corollary 2.14) is somewhat tautological, deriving from a formal play with the notion of supercode (Proposition 2.13).

1.6. — In a second approach, inspired by the so-called geometric view on coding theory, we try to get a description with a more meaningful geometric content. Let's first briefly review on which sort of spaces evaluation-interpolation is usually performed.

Early multiplication algorithms, such as those of Karatsuba or Toom-Cook, are evaluation-interpolation algorithms on the projective line. Chudnovsky and Chudnovsky [**14**] introduced evaluation-interpolation algorithms on curves. In [**34**, Ex. 2.5] one can find an evaluation-interpolation algorithm on the projective plane.

In section 3 we will construct a multiplication algorithm for $\mathbb{F}_{2^5}$ over $\mathbb{F}_2$, of optimal length $n = 13$, by evaluation-interpolation on a del Pezzo surface.

In general we show (second half of Theorem 6.1) that any multiplication algorithm φ can be realized as an evaluation-interpolation algorithm on a naturally associated space, namely the *quadratic hull* of the linear code image of φ, defined below.

A converse result (first half of Theorem 6.1) characterizes whether a given code is the image of a multiplication algorithm, also in terms of its quadratic hull. Informally, it says that any quadratic identity satisfied in the code should also be satisfied in $\mathcal{A}$ (Remark 6.2). It then translates as a geometric characterization of codes that come from a supercode (Corollary 6.3).

Again, all this generalizes to multiplication reductions.

We observe a similarity between these results and those from [**33**], in that they both are *a posteriori*. They are of no help if one's goal is to build better codes, or better multiplication algorithms. What they provide is only a better abstract understanding of these objects.

1.7. — The quadratic hull (French: *enveloppe quadratique*) of a non-degenerate linear code $C \subseteq \mathbf{F}^n$ can be seen either algebraically or geometrically.

Starting from the algebraic point of view, let $\mathbf{G} \in \mathbf{F}^{k \times n}$ be a generator matrix for C, and let $P_1, \dots, P_n$ be the columns of $\mathbf{G}$. In [**28**, sec. 3], Márquez-Corbella, Martínez-Moro, and Pellikaan set

$$I_2(C) = \{q \in \mathbf{F}[x_1, \dots, x_k]_2 : \ q(P_1) = \cdots = q(P_n) = 0\},$$

the space of homogeneous quadratic forms that vanish at $P_1, \dots, P_n$, which up to some linear transformation depends only on C. Put more intrinsically,

$$I_2(C) = \ker(S^2 C \xrightarrow{*} C^{\langle 2 \rangle}),$$

where $S^2 C$ is the second symmetric power of C, and $C^{\langle 2 \rangle}$ is the *square* of C, the linear span of pairwise $*$-products of codewords of C (and similar spaces $I_t(C)$ are defined likewise for higher powers $C^{\langle t \rangle}$). They then show [**30**, Th. 2] that if C is an AG code with suitable numerical parameters, $I_2(C)$ allows to reconstruct the curve from which C is defined. Later on, these $I_t(C)$ were also considered in [**35**, §§1.11-1.36] as part of an intrinsic description of the geometric view on codes and their powers.

Turning to geometry, we define the quadratic hull $Z_2(C)$ of C as the zero locus of $I_2(C)$ in projective space:

$$Z_2(C) = Z(I_2(C)) \subseteq \mathbf{P}^{k-1},$$

i.e. the intersection of all quadrics in $\mathbf{P}^{k-1}$ that pass through $P_1, \dots, P_n$.

These two points of view, algebraic and geometric, are equivalent, i.e. $I_2(C)$ and $Z_2(C)$ carry exactly the same information, provided $Z_2(C)$ is considered with its scheme structure. Indeed, giving a closed subscheme in $\mathbf{P}^{k-1}$ is the same as giving its defining ideal.

When generalizing from multiplication algorithms to multiplication reductions, codes have to be replaced with what we call $\mathcal{B}$-codes. The quadratic hull can also be defined in this context.

1.8. — In general the quadratic hull of a code can be quite pathological, and we have no a priori control on it. It is thus interesting to have plenty of examples.

In [**6**] (see also [**16**]) an algorithm is given that allows to compute the exhaustive list of multiplication algorithms of optimal length for algebras of small cardinality. In section 7, Tables 1-4, we describe the quadratic hull of the corresponding codes. Moreover, Proposition 7.4, Corollary 7.5, and Corollary 7.6 derive informations on the quadratic hull directly from the linear span of the quadratic forms of rank 1 that constitute the multiplication algorithm.

We complement this with several properties of the quadratic hull, of independent interest:

(i) The rational points P of $Z_2(C)$ parameterize *square-preserving* extensions of C, i.e. those $[n+1,k]$-codes C_P with generator matrix $\mathbf{G}_P = (\ \mathbf{G}\ |P)$ such that
$$\dim C_P^{\langle 2\rangle} = \dim C^{\langle 2\rangle}.$$

(ii) They also parameterize hyperplanes $H \subseteq C$ such that
$$H \cdot C \subsetneq C^{\langle 2\rangle}.$$

(iii) The rational points of the secant variety of $Z_2(C)$ parameterize a subset of (and possibly all) the locus of hyperplanes $H \subseteq C$ that satisfy
$$H^{\langle 2\rangle} \subsetneq C^{\langle 2\rangle}.$$

The following is a slightly stronger version of the result from [**28**][**30**] already mentioned:

(iv) If C is the evaluation code of some $L(D)$ at n distinct rational points $P_1,\dots,P_n$ on a smooth projective curve X of genus g, with $\deg(D) > 2g+1$ and $L(2D-\sum_i P_i) = 0$, then
$$Z_2(C) = X.$$

All these properties actually extend to $\mathcal{B}$-codes, as covered in Propositions 8.1, 8.2, 8.3, and 8.4 respectively.

Properties (iii)(iv)(v) are especially meaningful in the analysis of McEliece-type cryptosystems. Beside [**28**][**30**], we refer to [**41**][**29**][**15**][**31**] for a sample of related works.

In (iv), it seems to be a difficult question to characterize those codes for which the secant variety exactly coincides with the locus of hyperplanes whose square does not fill the square of C. This is not true for all codes, but it holds at least for certain MDS codes and for certain evaluation codes on curves. Partial results of this sort will be found in [**36**].

1.9. *Conventions* — In this work, by an algebra we will always mean an algebra that is commutative, associative, with unity, nonzero, and of finite dimension (as a vector space) over a field $\mathbf{F}$.

Also we will say “multiplication algorithm (resp. reduction)” as short for “symmetric bilinear multiplication algorithm (resp. reduction)”. It would not be difficult to devise asymmetric extensions of our results, but we will stick to the symmetric setting for simplicity.

If $\mathcal{B}$ is an algebra and $\mathcal{P} : \operatorname{Spec} \mathcal{B} \to X$ is a morphism of schemes, then for any invertible sheaf $\mathcal{L}$ on X we set $\mathcal{L}|_{\mathcal{P}} = \mathcal{P}^*\mathcal{L}$, the pullback of $\mathcal{L}$ by $\mathcal{P}$. Observe that, since $\mathcal{B}$ is finite, $\mathcal{L}|_{\mathcal{P}} \simeq \mathcal{B}$ admits a trivialization.

2. Multiplication reductions

2.1. DEFINITION. A (symmetric bilinear) multiplication reduction

$$\varphi : \mathcal{A} \rightsquigarrow \mathcal{B}$$

from an algebra $\mathcal{A}$ to an algebra $\mathcal{B}$ over $\mathbf{F}$, is a linear map $\varphi : \mathcal{A} \to \mathcal{B}$ that admits an "adjoint" linear map $\omega : \mathcal{B} \to \mathcal{A}$ that makes the following diagram commute:[1]

$$\begin{array}{ccc} \mathcal{A} \times \mathcal{A} & \xrightarrow{m_{\mathcal{A}}} & \mathcal{A} \\ {\scriptstyle \varphi\times\varphi}\downarrow & & \uparrow{\scriptstyle \omega} \\ \mathcal{B} \times \mathcal{B} & \xrightarrow{m_{\mathcal{B}}} & \mathcal{B} \end{array}$$

where $m_{\mathcal{A}}$ and $m_{\mathcal{B}}$ denote multiplication in $\mathcal{A}$ and $\mathcal{B}$, respectively.

Equivalently, φ and ω should satisfy

$$a \cdot a' = \omega(\varphi(a) \cdot \varphi(a'))$$

for all $a, a' \in \mathcal{A}$. In words, multiplication of a and a' in $\mathcal{A}$ "reduces" (up to some fixed linear operations) to multiplication of $\varphi(a)$ and $\varphi(a')$ in $\mathcal{B}$.

2.2. LEMMA. *A multiplication reduction $\varphi : \mathcal{A} \rightsquigarrow \mathcal{B}$ is always injective.*

PROOF. Apply the formula above with $a \in \ker \varphi$ and $a' = 1$. □

2.3. — Here are three important instances of multiplication reductions:

- A multiplication *algorithm* of length n for an algebra $\mathcal{A}$ over $\mathbf{F}$ is a multiplication reduction

  $$\mathcal{A} \rightsquigarrow \mathbf{F}^n$$

  where $\mathcal{B} = \mathbf{F}^n$ is equipped with componentwise multiplication $*$.

  Equivalently, if $\varphi_1, \dots, \varphi_n \in \mathcal{A}^\vee$ are the components of $\varphi : \mathcal{A} \to \mathbf{F}^n$, and $\omega_1, \dots, \omega_n \in \mathcal{A}$ are those of $\omega : \mathbf{F}^n \to \mathcal{A}$, one asks

  $$aa' = \varphi_1(a)\varphi_1(a')\omega_1 + \cdots + \varphi_n(a)\varphi_n(a')\omega_n$$

  for all $a, a' \in \mathcal{A}$. In words, a multiplication in $\mathcal{A}$ reduces (up to some fixed linear operations) to n multiplications in $\mathbf{F}$.

  This also amounts to a decomposition

  $$T_{\mathcal{A}} = \omega_1 \otimes \varphi_1^{\otimes 2} + \cdots + \omega_n \otimes \varphi_n^{\otimes 2}$$

  of the multiplication tensor of $\mathcal{A}$ into n elementary (symmetric) tensors.

  As already discussed in the Introduction, multiplication algorithms are closely related to multiplicative linear secret sharing schemes. In this context they are sometimes called "multiplication-friendly embeddings" [**10**].

  A large part of the literature on multiplication algorithms (surveyed in [**4**]) is devoted to the case where $\mathcal{A} = \mathbb{F}_{q^m}[t]/(t^l)$ is a monogeneous local algebra over $\mathbf{F} = \mathbb{F}_q$ a finite field. This includes the cases $\mathcal{A} = \mathbb{F}_{q^k}$ (finite field extension) and $\mathcal{A} = \mathbb{F}_q[t]/(t^k)$ (truncated polynomials). However, the

[1] It would make no fundamental difference to define a multiplication reduction as the data of the pair (φ, ω), instead of φ alone. See §2.5 below for a discussion.

case of a more general local algebra $\mathcal{A}$ over an arbitrary field $\mathbf{F}$, for instance $\mathcal{A} = \mathbf{F}[[t_1, \ldots, t_m]]/(t_1^l, \ldots, t_m^l)$, is also of interest.

As for non-local algebras, the issue is related to Strassen's direct sum conjecture [**40**][**8**]. Indeed, such an algebra decomposes as the product of its localizations: $\mathcal{A} = \prod_i \mathcal{A}_i$. The conjecture asserts that a multiplication algorithm for $\mathcal{A}$ of minimal length should decompose as a product (in the sense of 2.7 below) of multiplication algorithms for each $\mathcal{A}_i$.

- More general multiplication reductions often occur as an intermediate step in the construction of multiplication algorithms. This includes generalized Chudnovsky-Chudnovsky type constructions that use evaluation-interpolation on curves at points of higher degree and/or with multiplicities. For instance (the symmetric part of) [**34**, Th. 3.5] can be analyzed in two steps: first, a multiplication *reduction*

$$\mathcal{A} = \mathbb{F}_{q^m}[t]/(t^l) \rightsquigarrow \mathcal{B} = \prod_i \mathbb{F}_{q^{d_i}}[t]/(t^{u_i})$$

is constructed, and then it is composed (in the sense of 2.8 below) with the product of multiplication algorithms $\mathbb{F}_{q^{d_i}}[t]/(t^{u_i}) \rightsquigarrow \mathbb{F}_q^{\mu_q^{\mathrm{sym}}(d_i,u_i)}$, to get a multiplication *algorithm*

$$\mathcal{A} = \mathbb{F}_{q^m}[t]/(t^l) \rightsquigarrow \mathbb{F}_q^N$$

with $N = \sum_i \mu_q^{\mathrm{sym}}(d_i, u_i)$.

- Last, reverse multiplication-friendly embeddings (RMFE), recently introduced in [**11**], also are multiplication reductions, namely of the form

$$\mathcal{A} = \mathbb{F}_q^k \rightsquigarrow \mathcal{B} = \mathbb{F}_{q^n}.$$

As such, the results presented in this work apply to them as well.

2.4. Definition. Let $\mathcal{B}$ be an algebra over $\mathbf{F}$. By a $\mathcal{B}$-code we mean a linear subspace

$$C \subseteq \mathcal{B}.$$

If C has dimension k over $\mathbf{F}$, we also say C is a $[\mathcal{B}, k]$-code.

In the particular case $\mathcal{B} = \mathbf{F}^n$, we say C is a $[n, k]$-code, compatibly with the established literature.

2.5. — Let $\varphi : \mathcal{A} \rightsquigarrow \mathcal{B}$ be a multiplication reduction, and consider its image $C_\varphi = \varphi(\mathcal{A}) \subseteq \mathcal{B}$. It is a $[\mathcal{B}, k]$-code, where $k = \dim \mathcal{A}$.

If $x \in \mathcal{B}$ is of the form $x = cc'$, where $c = \varphi(a), c' = \varphi(a')$ for $a, a' \in \mathcal{A}$, then necessarily one should have $\omega(x) = aa'$ in $\mathcal{A}$. By linearity, this condition *uniquely determines* the values of ω on the linear subspace

$$C_\varphi^{\langle 2\rangle} = \langle\{cc' \,:\, c, c' \in C_\varphi\}\rangle \subseteq \mathcal{B}$$

spanned by pairwise products of codewords (under $m_\mathcal{B}$), called the *square* of C_φ.

By a slight abuse, the resulting uniquely determined map

$$\omega : C_\varphi^{\langle 2\rangle} \longrightarrow \mathcal{A}$$

will also be called the adjoint of φ.

Conversely, values of ω outside of $C_\varphi^{\langle 2\rangle}$ do not matter in Definition 2.1.

This means we can take the commutativity of the diagram

$$\begin{array}{ccc} \mathcal{A}\times\mathcal{A} & \xrightarrow{m_{\mathcal{A}}} & \mathcal{A} \\ {\scriptstyle\varphi\times\varphi}\downarrow & & \uparrow{\scriptstyle\omega} \\ C_\varphi\times C_\varphi & \xrightarrow{m_{\mathcal{B}}} & C_\varphi^{\langle 2\rangle} \end{array}$$

as an alternative, equivalent definition for a multiplication reduction. Indeed, once this holds, we can pick an arbitrary linear extension of ω from $C_\varphi^{\langle 2\rangle}$ to the whole of $\mathcal{B}$, and the condition in Definition 2.1 will be satisfied.

It is customary in bilinear complexity theory to focus solely on the number of bilinear operations in the algorithms, ignoring linear operations as costless. From this point of view, specifying the choice of ω in Definition 2.1 would be somehow redundant. However, if one is interested in practical implementations, it could happen that some clever choice of an extension of ω from $C_\varphi^{\langle 2\rangle}$ to the whole of $\mathcal{B}$ would give a better total complexity.

2.6. — Let $\varphi:\mathcal{A}\rightsquigarrow\mathcal{B}$ be a multiplication reduction. Given invertible elements $u\in\mathcal{A}^\times$ and $v\in\mathcal{B}^\times$, define $\tilde{\varphi}:\mathcal{A}\to\mathcal{B}$ by

$$\tilde{\varphi}(a)=v\varphi(u^{-1}a)$$

for all $a\in\mathcal{A}$. Then $\tilde{\varphi}$ is a multiplication reduction $\mathcal{A}\rightsquigarrow\mathcal{B}$. Indeed, if φ has adjoint $\omega: C_\varphi^{\langle 2\rangle}\to\mathcal{A}$, then $\tilde{\varphi}$ has adjoint $\tilde{\omega}: C_{\tilde{\varphi}}^{\langle 2\rangle}\to\mathcal{A}$ where $\tilde{\omega}(b)=u^2\omega(v^{-2}b)$ for all $b\in C_{\tilde{\varphi}}^{\langle 2\rangle}=v^2C_\varphi^{\langle 2\rangle}\subseteq\mathcal{B}$.

Definition. We say two multiplication reductions $\varphi,\tilde{\varphi}:\mathcal{A}\rightsquigarrow\mathcal{B}$ are *diagonally equivalent*,[2] and we write

$$\varphi\sim\tilde{\varphi},$$

if they can be related as above for some $u\in\mathcal{A}^\times$, $v\in\mathcal{B}^\times$.

This defines an equivalence relation on the set of multiplication reductions from $\mathcal{A}$ to $\mathcal{B}$.

2.7. — There is a notion of product for multiplication reductions: given $\varphi_1:\mathcal{A}_1\rightsquigarrow\mathcal{B}_1$ and $\varphi_2:\mathcal{A}_2\rightsquigarrow\mathcal{B}_2$, we derive $\varphi_1\times\varphi_2:\mathcal{A}_1\times\mathcal{A}_2\rightsquigarrow\mathcal{B}_1\times\mathcal{B}_2$, where each product algebra is equipped with componentwise multiplication. If φ_1 has adjoint ω_1 and φ_2 has adjoint ω_2, then $\varphi_1\times\varphi_2$ has adjoint $\omega_1\times\omega_2$.

This is compatible with diagonal equivalence: if $\varphi_1\sim\tilde{\varphi}_1$ and $\varphi_2\sim\tilde{\varphi}_2$, then $\varphi_1\times\varphi_2\sim\tilde{\varphi}_1\times\tilde{\varphi}_2$.

2.8. — Likewise, multiplication reductions can be composed (or concatenated): given $\varphi:\mathcal{A}\rightsquigarrow\mathcal{B}$ and $\psi:\mathcal{B}\rightsquigarrow\mathcal{C}$, we derive $\psi\circ\varphi:\mathcal{A}\rightsquigarrow\mathcal{C}$. If φ has adjoint ω and ψ has adjoint ϖ, then $\psi\circ\varphi$ has adjoint $\omega\circ\varpi$.

However, in general, composition is *not* compatible with diagonal equivalence, i.e. it could happen that $\varphi\sim\tilde{\varphi}$ and $\psi\sim\tilde{\psi}$ but $\psi\circ\varphi\not\sim\tilde{\psi}\circ\tilde{\varphi}$.

2.9. Proposition. *Let X be a scheme over $\mathbf{F}$, together with points $\mathcal{Q}\in X(\mathcal{A})$ and $\mathcal{P}\in X(\mathcal{B})$ with coordinates in $\mathcal{A}$ and $\mathcal{B}$ respectively, i.e. morphisms $\mathcal{Q}:\operatorname{Spec}\mathcal{A}\to X$ and $\mathcal{P}:\operatorname{Spec}\mathcal{B}\to X$.*

[2] one could devise another natural notion of equivalence, coarser, by further allowing composition on the right with automorphisms of $\mathcal{A}$ and on the left with automorphisms of $\mathcal{B}$.

Let $\mathcal{L}$ be an invertible sheaf on X, and $V \subseteq \Gamma(X, \mathcal{L})$ a finite dimensional linear system. Let also $V^{\langle 2\rangle} \subseteq \Gamma(X, \mathcal{L}^{\otimes 2})$ be its square, i.e. the linear system spanned by all pairwise products of elements of V. Consider the natural restriction (i.e. pullback) maps from V to $\mathcal{L}|_Q$ and from $V^{\langle 2\rangle}$ to $\mathcal{L}^{\otimes 2}|_{\mathcal{P}}$, and assume

(i) $V \longrightarrow \mathcal{L}|_Q$ is surjective

(ii) $V^{\langle 2\rangle} \longrightarrow \mathcal{L}^{\otimes 2}|_{\mathcal{P}}$ is injective.

Then there is a multiplication reduction

$$\varphi : \mathcal{A} \rightsquigarrow \mathcal{B}$$

given by the following construction:

- *choose trivializations $\mathcal{L}|_Q \simeq \mathcal{A}$ and $\mathcal{L}|_{\mathcal{P}} \simeq \mathcal{B}$ (recall $\mathcal{A}, \mathcal{B}$ finite over $\mathbf{F}$)*
- *choose a right-inverse $\sigma : \mathcal{A} \to V$ of the surjective map $V \to \mathcal{L}|_Q \simeq \mathcal{A}$*
- *compose σ with the "evaluation-at-$\mathcal{P}$" map $V \to \mathcal{L}|_{\mathcal{P}} \simeq \mathcal{B}$, to get the linear map $\varphi : \mathcal{A} \to \mathcal{B}$.*

Moreover, if we assume that $V \to \mathcal{L}|_Q$ is bijective, *then the diagonal equivalence class of φ so constructed is* independent *of the choices made.*

(Observe that if we're interested in multiplication algorithms, then $\mathcal{B} = \mathbf{F}^n$, and we can see $\mathcal{P}$ as an ordered collection $\mathcal{P} = (P_1, \dots, P_n)$ of n rational points $P_i \in X(\mathbf{F})$. This explains our notation. But on the other hand, if we're interested in RMFE $\mathcal{A} = \mathbb{F}_q^k \rightsquigarrow \mathcal{B} = \mathbb{F}_{q^n}$, then $Q = (Q_1, \dots, Q_k)$ is a collection of k rational points, while $\mathcal{P}$ is a point defined over an extension. Our formalism encompasses both situations, and even more general ones, in a uniform way.)

PROOF. Given $f \in V$, write $f(Q)$ and $f(\mathcal{P})$ for the images of f under the evaluation maps $V \to \mathcal{L}|_Q \simeq \mathcal{A}$ and $V \to \mathcal{L}|_{\mathcal{P}} \simeq \mathcal{B}$, respectively.[3]

Likewise, given $h \in V^{\langle 2\rangle}$, write $h(Q)$ and $h(\mathcal{P})$ for the images of h under $V^{\langle 2\rangle} \to \mathcal{L}^{\otimes 2}|_Q \simeq \mathcal{A}$ and $V \to \mathcal{L}^{\otimes 2}|_{\mathcal{P}} \simeq \mathcal{B}$, where the trivializations $\mathcal{L}^{\otimes 2}|_Q \simeq \mathcal{A}$ and $\mathcal{L}^{\otimes 2}|_{\mathcal{P}} \simeq \mathcal{B}$ are deduced from the ones chosen previously by passing to the square. This makes evaluation commute with multiplication: for $f, f' \in V$ we have $(ff')(Q) = f(Q)f'(Q)$ in $\mathcal{A}$, and likewise $(ff')(\mathcal{P}) = f(\mathcal{P})f'(\mathcal{P})$ in $\mathcal{B}$.

Now we construct the adjoint linear map $\omega : \mathcal{B} \to \mathcal{A}$, and then we check that φ, ω indeed satisfy the condition in Definition 2.1.

Since $V^{\langle 2\rangle} \to \mathcal{L}^{\otimes 2}|_{\mathcal{P}} \simeq \mathcal{B}$ is injective, we can choose a left-inverse $\rho : \mathcal{B} \to V^{\langle 2\rangle}$. We then compose ρ with the evaluation-at-Q map $V^{\langle 2\rangle} \to \mathcal{L}^{\otimes 2}|_Q \simeq \mathcal{A}$ to get a linear map $\omega : \mathcal{B} \to \mathcal{A}$.

Let $a, a' \in \mathcal{A}$. By construction, $\sigma(a), \sigma(a')$ are elements $f_a, f_{a'} \in V$ with $f_a(Q) = a$, $f_{a'}(Q) = a'$, and then $\varphi(a) = f_a(\mathcal{P})$, $\varphi(a') = f_{a'}(\mathcal{P})$ in $\mathcal{B}$. Since evaluation and multiplication commute, we get $\varphi(a)\varphi(a') = f_a(\mathcal{P})f_{a'}(\mathcal{P}) = (f_a f_{a'})(\mathcal{P})$. By construction, $\rho((f_a f_{a'})(\mathcal{P}))$ is then an element $h \in V^{\langle 2\rangle}$ such that $h(\mathcal{P}) = (f_a f_{a'})(\mathcal{P})$. By injectivity of $V^{\langle 2\rangle} \to \mathcal{L}^{\otimes 2}|_{\mathcal{P}}$, this forces $h = f_a f_{a'}$. And then, commuting evaluation and multiplication again, we get $\omega(\varphi(a)\varphi(a')) = (f_a f_{a'})(Q) = f_a(Q)f_{a'}(Q) = aa'$, as desired.

As for the last statement, observe that in the first step of the construction, another choice of trivializations multiplies the evaluation-at-Q and -at-$\mathcal{P}$ maps by some $u \in \mathcal{A}^\times$ and $v \in \mathcal{B}^\times$, respectively. Moreover if $V \to \mathcal{L}|_Q$ is bijective, then in the second step there is only one choice for the right inverse, which is actually the

[3] e.g. when $\mathcal{B} = \mathbf{F}^n$ and $\mathcal{P} = (P_1, \dots, P_n)$, we can write $f(\mathcal{P}) = (f(P_1), \dots, f(P_n)) \in \mathbf{F}^n$.

inverse of the evaluation-at-Q map. All in all this replaces φ with an equivalent map precisely as in 2.6. $\square$

2.10. DEFINITION. A multiplication reduction $\varphi : \mathcal{A} \rightsquigarrow \mathcal{B}$ is geometric, or is of evaluation-interpolation type, if it can be obtained through Construction 2.9.

Conversely, in this situation, we say that $(X, Q, \mathcal{P}, \mathcal{L}, V)$ is a *geometric realization* of the reduction φ.

This can be compared with the generalized Chudnovsky-Chudnovsky constructions surveyed in [**4**], that use evaluation-interpolation on curves at points of higher degree and/or with multiplicities. This fits in our formalism of multiplication reductions. But if we insist on having a multiplication *algorithm*, so $\mathcal{B} = \mathbf{F}^n$, the key point then is that we ask for evaluation-interpolation at $\mathcal{P} = (P_1, \ldots, P_n)$, a collection of *rational points, without multiplicities*. But on the other hand, we allow X to be a higher dimensional variety, or even an arbitrary scheme.

2.11. REMARK. Some arguments suggest that "good" geometric multiplication reductions should actually be low dimensional.

For instance, suppose given an algebra $\mathcal{A}$, of dimension k, and consider a geometric multiplication algorithm $\varphi : \mathcal{A} \rightsquigarrow \mathbf{F}^n$, of length n as small as possible. Then condition (i) forces $\dim V \geq k$, and condition (ii) forces $\dim V^{\langle 2 \rangle} \leq n$. Thus, if X is a variety (i.e. an *integral* scheme), and if we assume

$$n \leq 3k - 4,$$

then the field analogue [**2**] of Freiman's theorem applies and shows that the subfield generated by $V^{-1}V \subseteq \mathbf{F}(X)$ is the function field $\mathbf{F}(Y)$ of a curve Y. With a little extra work, one could then prove that φ can be realized over this curve Y.

Interestingly, if $\mathbf{F}$ is the finite field $\mathbb{F}_p$ for $p \geq 7$ prime, and $\mathcal{A} = \mathbb{F}_{p^k}$ is its degree k field extension for k large, then the best constructions [**34**, §6] allow to get $n \approx 3\left(1 + \frac{2}{p-2}\right) k$, which is slightly above this $3k - 4$ bound. So the argument does not apply there.

2.12. — The notion of *supercode* introduced in [**38**] for multiplication algorithms is easily extended in the context of multiplication reductions. We define a supercode as a $\mathbf{F}$-linear subspace

$$\widehat{C} \subseteq \mathcal{A} \times \mathcal{B}$$

such that the first projection $\pi_{\mathcal{A}} : \widehat{C} \to \mathcal{A}$ is surjective, and the second projection $\pi_{\mathcal{B}} : \widehat{C}^{\langle 2 \rangle} \to \mathcal{B}$ is injective, where as usual $\widehat{C}^{\langle 2 \rangle}$ is the $\mathbf{F}$-linear span of pairwise products of elements of $\widehat{C}$ in the product algebra $\mathcal{A} \times \mathcal{B}$.

If $\varphi : \mathcal{A} \rightsquigarrow \mathcal{B}$ is a multiplication reduction, then $\widehat{C} = \operatorname{im}(\mathrm{id}_{\mathcal{A}} \times \varphi) \subseteq \mathcal{A} \times \mathcal{B}$ is a supercode.

2.13. — Conversely, if $\widehat{C} \subseteq \mathcal{A} \times \mathcal{B}$ is a supercode, then $\widehat{C}$ gives rise to a multiplication reduction. This can be seen in our geometric formalism.

Set $X = \operatorname{Spec}(\mathcal{A} \times \mathcal{B})$, and consider the points $Q \in X(\mathcal{A})$ given by the first projection $\mathcal{A} \times \mathcal{B} \to \mathcal{A}$, and $\mathcal{P} \in X(\mathcal{B})$ given by the second projection $\mathcal{A} \times \mathcal{B} \to \mathcal{B}$.

Also set $\mathcal{L} = \mathcal{O}_X$, the structure sheaf, and consider then $\widehat{C}$ as an incomplete linear system $\widehat{C} \subseteq \Gamma(X, \mathcal{O}_X) = \mathcal{A} \times \mathcal{B}$.

PROPOSITION. *With these notations,* $(X, Q, \mathcal{P}, \mathcal{O}_X, \widehat{C})$ *is the geometric realization of a multiplication reduction* $\mathcal{A} \rightsquigarrow \mathcal{B}$.

Moreover, if $\widehat{C} = \mathrm{im}(\mathrm{id}_{\mathcal{A}} \times \varphi)$ *comes from a multiplication reduction* $\varphi : \mathcal{A} \rightsquigarrow \mathcal{B}$, *then actually* $(X, Q, \mathcal{P}, \mathcal{O}_X, \widehat{C})$ *is a geometric realization for* φ.

PROOF. Proposition 2.9 applies, since the surjectivity and injectivity conditions in the definition of a supercode reflect conditions (i) and (ii).

Moreover, if $\widehat{C} = \mathrm{im}(\mathrm{id}_{\mathcal{A}} \times \varphi)$, then we have bijectivity in (i), and lifting the first projection from $\mathcal{A}$ back to $\widehat{C}$ and projecting to $\mathcal{B}$ indeed gives φ, by construction. □

It follows at once:

2.14. COROLLARY. *Any multiplication reduction is geometric.*

Unfortunately, this result, based on the tautological, zero-dimensional space $\mathrm{Spec}(\mathcal{A} \times \mathcal{B})$, is quite disappointing.[4] One aim of this work will be to present geometric realizations with a more interesting structure.

This will allow us to also address the following question: how can one tell if a given linear map $\varphi : \mathcal{A} \to \mathcal{B}$ is a multiplication reduction? Or instead of specifying the whole map φ, one can give only its image C. The question then reduces to the following, from [**38**]: how can one tell if a given code $C \subseteq \mathcal{B}$ is the second projection of a supercode $\widehat{C} \subseteq \mathcal{A} \times \mathcal{B}$? To these questions we will provide a geometric answer, albeit perhaps not an algorithmic criterion, as could be even more desirable.

3. An example in dimension 2

3.1. — Here we work over the base field $\mathbf{F} = \mathbb{F}_2$. We will construct a geometric multiplication algorithm

$$\mathbb{F}_{2^5} \rightsquigarrow \mathbb{F}_2^{13}$$

by evaluation-interpolation on a surface.

It turns out this length $n = 13$ is best possible for a multiplication algorithm for $\mathbb{F}_{2^5}$ over $\mathbb{F}_2$, i.e. it reaches the *(symmetric) bilinear complexity*

$$\mu_2^{\mathrm{sym}}(5) = 13.$$

Indeed, no such algorithm exists in length 12 or less, because otherwise, it would imply the existence of a $[12, 5, 5]$ binary linear code [**14**, Cor 5.1][**38**, Prop. 1.4], which is known to be false [**20**][**24**].

3.2. — A smooth conic $\mathcal{Q}$ in $\mathbf{P}^2$ over $\mathbb{F}_2$ contains precisely three rational points P_1, P_2, P_3 and one pair of conjugate quadratic points $Q, \overline{Q}$, no three of which lie on a line.

Blowing-up $\mathbf{P}^2$ at these five points gives a (nonsingular) del Pezzo surface X of degree 4 [**27**, Th. 24.4], with

$$|X(\mathbb{F}_2)| = 13.$$

More precisely, there are 3 rational points on each of the exceptional divisors E_1, E_2, E_3 above P_1, P_2, P_3, and then 4 other rational points above the remaining points P_4, P_5, P_6, P_7 in $\mathbf{P}^2(\mathbb{F}_2)$.

---

[4] A similarly tautological statement is that any linear code C is the evaluation code, at the n points of $\mathrm{Spec}(\mathbf{F}^n)$, of C itself, seen as an incomplete linear system in $\Gamma(\mathrm{Spec}(\mathbf{F}^n), \mathcal{O}_{\mathrm{Spec}(\mathbf{F}^n)})$.

The canonical sheaf of X is

$$\mathcal{K}_X = (\pi^*\mathcal{O}(-3))(E_1 + E_2 + E_3 + F + \overline{F})$$

where $\pi : X \to \mathbf{P}^2$ is the blow-up map, and $F, \overline{F}$ are the exceptional divisors over $Q, \overline{Q}$, respectively.

The anticanonical embedding is projectively normal [**19**, Th. 8.3.4] and realizes X as an intersection of two quadrics in $\mathbf{P}^4$ [**19**, Th. 8.6.2], thus:

- $\dim \Gamma(X, \mathcal{K}_X^{-1}) = 5$
- $\dim \Gamma(X, \mathcal{K}_X^{-2}) = 13$.

Here π_* gives an identification

$$\Gamma(X, \mathcal{K}_X^{-1}) = \Gamma(\mathbf{P}^2, \mathfrak{I}_{P_1,P_2,P_3,Q,\overline{Q}}\mathcal{O}(3))$$

of the space of sections of the anticanonical sheaf with the linear system of cubic forms on $\mathbf{P}^2$ vanishing at $P_1, P_2, P_3, Q, \overline{Q}$ (which, indeed, has dimension 5, e.g. by Cayley-Bacharach).

Likewise we have an identification

$$\Gamma(X, \mathcal{K}_X^{-2}) = \Gamma(\mathbf{P}^2, \mathfrak{I}^2_{P_1,P_2,P_3,Q,\overline{Q}}\mathcal{O}(6))$$

where the right-hand side is the linear system of sextic forms on $\mathbf{P}^2$ vanishing at $P_1, P_2, P_3, Q, \overline{Q}$ with multiplicity at least 2.

3.3. LEMMA. *The natural restriction map, from $\Gamma(X, \mathcal{K}_X^{-2})$ to the product of the fibers of $\mathcal{K}_X^{-2}$ at the 13 rational points of X, is injective.*

PROOF. Let $s \in \Gamma(X, \mathcal{K}_X^{-2})$ vanishing at the 13 rational points of X. We have to show $s = 0$.

First, we use the vanishing of s at the 9 rational points in E_1, E_2, E_3. Since $\mathcal{K}_X^{-2} = (\pi^*\mathcal{O}(6))(-2E_1 - 2E_2 - 2E_3 - 2F - 2\overline{F})$, the restriction $s|_{E_i}$ defines a section of

$$\mathcal{O}(-2E_i)|_{E_i} \simeq \mathcal{O}_{\mathbf{P}^1}(2).$$

The fact that this degree 2 section vanishes at the 3 rational points of $E_i \simeq \mathbf{P}^1$ then forces $s|_{E_i} = 0$, so actually s is a section of

$$\mathcal{K}_X^{-2}(-E_1 - E_2 - E_3) = (\pi^*\mathcal{O}(6))(-3E_1 - 3E_2 - 3E_3 - 2F - 2\overline{F}).$$

This means $f = \pi_* s \in \Gamma(\mathbf{P}^2, \mathcal{O}(6))$ vanishes with multiplicity at least 3 at P_1, P_2, P_3 and 2 at $Q, \overline{Q}$.

In turn, the vanishing of s at the remaining 4 rational points of X means that f vanishes at their images P_4, P_5, P_6, P_7 in $\mathbf{P}^2$.

Recollecting, we have established so far that $f = \pi_* s \in \Gamma(\mathbf{P}^2, \mathcal{O}(6))$ is a sextic form that vanishes:

- at P_1, P_2, P_3 with multiplicity at least 3
- at $Q, \overline{Q}$ with multiplicity at least 2
- at P_4, P_5, P_6, P_7 with multiplicity at least 1.

We have to show $f = 0$.

On the conic $\mathcal{Q} = (P_1 P_2 P_3 Q \overline{Q})$, the sextic f vanishes at $3P_1 + 3P_2 + 3P_3 + 2Q + 2\overline{Q}$, a divisor of total degree 13. By Bézout, this forces $f|_{\mathcal{Q}} = 0$, i.e. $f = qt$ where $q \in \Gamma(\mathbf{P}^2, \mathcal{O}(2))$ is an equation for $\mathcal{Q}$, and $t \in \Gamma(\mathbf{P}^2, \mathcal{O}(4))$ is a quartic that has to vanish further:

- at P_1, P_2, P_3 with multiplicity at least 2

- at $Q, \overline{Q}$ and P_4, P_5, P_6, P_7 with multiplicity at least 1.

Set $L = (Q\overline{Q})$. It is a rational line that does not meet P_1, P_2, P_3 (and actually it is the only such line). Relabelling, we may assume $P_4 = L \cap (P_1P_2)$, $P_5 = L \cap (P_1P_3)$, $P_6 = L \cap (P_2P_3)$.

Then, the quartic t vanishes at a divisor of degree 5 on each of the following lines:

- at $2P_1 + 2P_2 + P_4$ on (P_1P_2)
- at $2P_1 + 2P_3 + P_5$ on (P_1P_3)
- at $2P_2 + 2P_3 + P_6$ on (P_2P_3)
- at $Q + \overline{Q} + P_4 + P_5 + P_6$ on L.

This implies that, up to multiplication by a constant, t is the product of the equations of these four lines. However, t also vanishes at P_7, while none of these lines passes through P_7. Thus the only possibility is $t = 0$, hence $f = 0$, $s = 0$. □

3.4. — Now consider another smooth conic $\mathcal{Q}'$, that passes through P_1, P_2 but not through $P_3, Q, \overline{Q}$.

For instance, setting $L = (Q\overline{Q})$, $P_4 = L \cap (P_1P_2)$, $P_5 = L \cap (P_1P_3)$, and letting $Q', \overline{Q'}$ be the conjugate quadratic points of the line (P_3P_4), we can take $\mathcal{Q}' = (P_1P_2P_5Q'\overline{Q'})$.

Let then $R_1 \in \mathcal{Q}'(\mathbb{F}_{2^5})$ be a point of degree 5 on $\mathcal{Q}'$, non-rational (there are 30 of them), and write R_2, R_3, R_4, R_5 for its conjugates.

LEMMA. *The natural restriction map $\Gamma(X, \mathcal{K}_X^{-1}) \to \mathcal{K}_X^{-1}|_{R_1}$ is surjective.*

PROOF. Since both sides have dimension 5, it suffices to prove that this restriction map is injective. So consider a section $s \in \Gamma(X, \mathcal{K}_X^{-1})$ that vanishes at R_1. We have to show $s = 0$.

Assume the contrary. Then $f = \pi_* s \in \Gamma(\mathbf{P}^2, \mathcal{O}(3))$ defines a cubic $\mathcal{C} \subseteq \mathbf{P}^2$ that passes through $P_1, P_2, P_3, Q, \overline{Q}$, and also through R_1 and its conjugates R_2, R_3, R_4, R_5, because of Galois invariance. Then $\mathcal{C} \cap \mathcal{Q}'$ contains the 7 points $P_1, P_2, R_1, R_2, R_3, R_4, R_5$, so by Bézout we have $\mathcal{C} = \mathcal{Q}' \cup D$ for a certain line D. Then D has to pass through $P_3, Q, \overline{Q}$, a contradiction since these three points do not lie on a line. □

3.5. PROPOSITION. *With these notations, $(X, R_1, X(\mathbb{F}_2), \mathcal{K}_X^{-1}, \Gamma(X, \mathcal{K}_X^{-1}))$ is the geometric realization of a multiplication algorithm $\mathbb{F}_{2^5} \rightsquigarrow \mathbb{F}_2^{13}$.*

PROOF. The surjectivity condition (i) in Proposition 2.9 is satisfied by 3.4, and the injectivity condition (ii), by 3.3; actually, both evaluation maps in (i) and (ii) turn out to be bijective here. □

3.6. — All this can be made explicit. Set $\mathbb{F}_4 = \mathbb{F}_2[\alpha]/(\alpha^2 + \alpha + 1)$ and $\mathbb{F}_{2^5} = \mathbb{F}_2[\gamma]/(\gamma^5 + \gamma^2 + 1)$. Let x, y, z be linear coordinates on $\mathbf{P}^2$.

Take $P_1 = (1:0:0)$, $P_2 = (0:1:0)$, $P_3 = (0:0:1)$, $L = \{x + y + z = 0\}$, $Q = (1 : \alpha : \alpha^2)$, $\overline{Q} = (1 : \alpha^2 : \alpha)$, $\mathcal{Q} = \{xy + xz + yz = 0\}$.

Then $P_4 = (1:1:0)$, $P_5 = (1:0:1)$, $(P_3P_4) = \{x = y\}$, $Q' = (1:1:\alpha)$, $\overline{Q'} = (1:1:\alpha^2)$, $\mathcal{Q}' = \{xy + xz + z^2 = 0\}$, and we can take $R_1 = (\gamma^2 : \gamma^3 : 1)$.

As a basis for $V = \Gamma(X, \mathcal{K}_X^{-1}) = \Gamma(\mathbf{P}^2, \mathfrak{I}_{P_1,P_2,P_3,Q,\overline{Q}}\mathcal{O}(3))$ we take:

$$\begin{aligned} v_0 &= x(xy + xz + yz) \\ v_1 &= y(xy + xz + yz) \\ v_2 &= xy(x + y + z) + x(xy + z^2) \\ v_3 &= (x + z)(xy + xz + yz) \\ v_4 &= x(xy + z^2). \end{aligned}$$

Indeed, these cubics are easily seen to vanish at $P_1, P_2, P_3, Q, \overline{Q}$, and were chosen through linear algebra so as to evaluate at R_1 as

$$v_i(R_1) = \gamma^i$$

for $i = 0, \ldots, 4$.

Now we evaluate these v_i at the 13 points in $X(\mathbb{F}_2)$ to get φ. Evaluation at P_4, P_5, P_6, P_7 is straightforward, but how to evaluate v_i at the points that lie in an exceptional divisor, say, at the three points in E_1 above P_1? These values are given by the derivatives $(\partial_y v_i)(P_1)$, $(\partial_z v_i)(P_1)$, and $((\partial_y + \partial_z)v_i)(P_1)$.

Doing so, we find that $\varphi : \mathbb{F}_{2^5} \rightsquigarrow \mathbb{F}_2^{13}$ is given by the matrix

$$\begin{pmatrix} 1&1&0&0&0&0&0&0&0&1&1&0&1 \\ 0&0&0&1&1&0&0&0&0&1&0&1&1 \\ 1&1&0&1&0&1&0&1&1&0&1&1&1 \\ 1&1&0&0&0&0&1&1&0&1&0&1&0 \\ 1&0&1&0&0&0&1&0&1&1&1&0&0 \end{pmatrix}$$

relative to the basis $1, \gamma, \gamma^2, \gamma^3, \gamma^4$ of $\mathbb{F}_{2^5}$ (with row vector notation).

As observed in 2.5, the adjoint $\omega : \mathbb{F}_2^{13} \to \mathbb{F}_{2^5}$ of φ is then uniquely determined, by mere linear algebra (i.e. we need not follow the proof of Proposition 2.9). We find it is given by (the transpose of) the matrix

$$\begin{pmatrix} 1&0&0&0&1&1&0&1&1&1&0&1&1 \\ 1&1&0&1&0&0&1&0&1&0&1&0&1 \\ 0&1&0&0&0&0&0&1&1&1&1&1&1 \\ 1&0&1&1&1&1&1&1&1&0&1&0&0 \\ 1&1&0&1&0&1&0&1&1&0&0&1&0 \end{pmatrix}$$

so $\omega_1 = 1 + \gamma + \gamma^3 + \gamma^4$, $\omega_2 = \gamma + \gamma^2 + \gamma^4$, $\omega_3 = \gamma^3$, etc.

3.7. Remark. The space of quadratic relations between v_0, v_1, v_2, v_3, v_4 is generated by

$$v_0^2 + v_0v_2 + v_1v_4$$

and

$$v_0(v_0 + v_1 + v_3 + v_4) + v_3(v_2 + v_3 + v_4).$$

Thus, now seeing v_0, v_1, v_2, v_3, v_4 as indeterminates that define linear coordinates in $\mathbf{P}^4$, we find that $X \subseteq \mathbf{P}^4$ is the surface cut out by these two quadrics.

3.8. Remark. The binary linear code C_φ image of φ has parameters $[13, 5, 5]$, which is optimal. Its automorphism group has size 48; it acts transitively on the first 12 coordinates, but fixes the last coordinate (which corresponds to evaluation at $P_7 = (1 : 1 : 1)$).

The square code $C_\varphi^{\langle 2 \rangle}$ is the whole of $\mathbb{F}_2^{13}$.

4. The canonical point

4.1. — Following the Russian school of coding theory, any (non-degenerate) linear code can be obtained by evaluation of linear functions on a system of rational points in a projective space. Here we consider what happens when we apply this point of view to multiplication algorithms, or more generally to multiplication reductions.

If V is a finite dimensional $\mathbf{F}$-vector space, recall that $\mathbf{P}V = \operatorname{Proj} S^{\cdot}V$ is the ("dual") projective space that parameterizes invertible quotients of V. More precisely, for any algebra $\mathcal{R}$, there is a natural bijection between the set $\mathbf{P}V(\mathcal{R})$ of points of $\mathbf{P}V$ with coordinates in $\mathcal{R}$, and the set of equivalence classes of surjective maps

$$q : V \otimes_{\mathbf{F}} \mathcal{R} \twoheadrightarrow L$$

with L a locally free $\mathcal{R}$-module of rank 1.

In coordinates, the choice of a basis $v_1, \dots, v_k$ of V identifies the symmetric algebra $S^{\cdot}V$ with the polynomial algebra $\mathbf{F}[x_1, \dots, x_k]$ (where $v_i \leftrightarrow x_i$), hence it identifies $\mathbf{P}V$ with the standard projective space $\mathbf{P}^{k-1}$. If moreover $\mathcal{R}$ is finite over $\mathbf{F}$, then L is actually free. Choosing a trivialization $L \simeq \mathcal{R}$, we see that the quotient map $q : V \otimes_{\mathbf{F}} \mathcal{R} \twoheadrightarrow L \simeq \mathcal{R}$ then corresponds to the point

$$(q(v_1 \otimes 1) : \cdots : q(v_k \otimes 1))$$

in $\mathbf{P}^{k-1}(\mathcal{R})$.

4.2. — Let now $\mathcal{A}$ be an algebra, finite over $\mathbf{F}$. We apply what precedes with $V = \mathcal{R} = \mathcal{A}$ and the multiplication map

$$m_{\mathcal{A}} : \mathcal{A} \otimes_{\mathbf{F}} \mathcal{A} \twoheadrightarrow \mathcal{A}.$$

Definition. The *canonical point* of $\mathcal{A}$ is the point

$$Q_{\mathcal{A}} \in \mathbf{P}\mathcal{A}(\mathcal{A})$$

corresponding to $m_{\mathcal{A}}$.

In coordinates, if we choose a basis $a_1, \dots, a_k$ of $\mathcal{A}$, identifying $\mathbf{P}\mathcal{A}$ with $\mathbf{P}^{k-1}$ accordingly, then

$$Q_{\mathcal{A}} = (a_1 : \cdots : a_k) \in \mathbf{P}^{k-1}(\mathcal{A}).$$

4.3. — We say that a $\mathcal{B}$-code $C \subseteq \mathcal{B}$ is *non-degenerate* if it generates the unit ideal in $\mathcal{B}$:

$$\mathcal{B} \cdot C = \mathcal{B}.$$

(For instance, if $C \subseteq \mathbf{F}^n$ is a linear code in the usual sense, it means C has full support.)

If C is non-degenerate, then the surjective map

$$m_{\mathcal{B}} : C \otimes \mathcal{B} \twoheadrightarrow C \cdot \mathcal{B} = \mathcal{B}$$

defines a point

$$\mathcal{P}_C \in \mathbf{P}C(\mathcal{B}),$$

which we call the *projective system* of C.

In coordinates, let $c_1, \dots, c_k$ be a basis of C, and identify $\mathbf{P}C$ with $\mathbf{P}^{k-1}$ accordingly. Then

$$\mathcal{P}_C = (c_1 : \cdots : c_k) \in \mathbf{P}^{k-1}(\mathcal{B}).$$

Furthermore, write $\mathcal{B} = \mathcal{B}_1 \times \cdots \mathcal{B}_m$ as a product of local algebras, and for $1 \leq j \leq m$ let $\pi_j : \mathcal{B} \to \mathcal{B}_j$ be the j-th projection. This allows to define the generator matrix of C (with respect to the given basis) as the $k \times m$ matrix $\mathbf{G}$ whose (i,j)-th entry is $\pi_j(c_i) \in \mathcal{B}_j$. Then we can view $\mathcal{P}_C$ as an ordered collection

$$\mathcal{P}_C = (P_1, \ldots, P_m)$$

where

$$P_j = (\pi_j(c_1) : \cdots : \pi_j(c_k)) \in \mathbf{P}^{k-1}(\mathcal{B}_j)$$

is the point defined by the j-th column of $\mathbf{G}$. (For instance, if $\mathcal{B} = \mathbf{F}^n$, then $\mathcal{P}_C$ is the projective system of rational points associated to C in the usual sense.)

Then we retrieve C as the image of the evaluation-at-$\mathcal{P}_C$ map

$$\Gamma(\mathbf{P}^{k-1}, \mathcal{O}(1)) \longrightarrow \mathcal{B}$$

which sends a linear function $l \in \Gamma(\mathbf{P}^{k-1}, \mathcal{O}(1)) = \mathbf{F}[x_1, \ldots, x_k]_1$ to its value $l(c_1, \ldots, c_k) \in \mathcal{B}$, or equivalently to $(l(P_1), \ldots, l(P_m)) \in \mathcal{B}_1 \times \cdots \mathcal{B}_m$.

4.4. — We also say a linear map $\varphi : \mathcal{A} \to \mathcal{B}$ is non-degenerate if it is injective and its image $C_\varphi = \varphi(\mathcal{A}) \subseteq \mathcal{B}$ is a non-degenerate $\mathcal{B}$-code. For instance:

- If $\varphi : \mathcal{A} \rightsquigarrow \mathbf{F}^n$ is a multiplication algorithm, it means its components $\varphi_1, \ldots, \varphi_n \in \mathcal{A}^\vee$ are all nonzero. We can always assume this holds: otherwise, just discard the bad coordinates to get a shorter, non-degenerate algorithm.
- A reverse multiplication-friendly embedding $\varphi : \mathbb{F}_q^k \rightsquigarrow \mathbb{F}_{q^n}$ is always non-degenerate, since it is nonzero and its image lives in $\mathbb{F}_{q^n}$, a field.
- For a general multiplication reduction $\varphi : \mathcal{A} \rightsquigarrow \mathcal{B}$, write $\mathcal{B} = \mathcal{B}_1 \times \cdots \times \mathcal{B}_m$ as a product of local algebras, and let $\varphi_j : \mathcal{A} \to \mathcal{B}_j$ be the components of φ. Then φ non-degenerate means for all j, the image $\varphi_j(\mathcal{A})$ contains an invertible element in $\mathcal{B}_j^\times$.

Since φ is injective, it identifies $\mathcal{A}$ with its image: $\mathcal{A} \simeq C_\varphi$, so it identifies the corresponding projective spaces: $\mathbf{P}\mathcal{A} \simeq \mathbf{P}C_\varphi$. We then let

$$\mathcal{P}_\varphi \in \mathbf{P}\mathcal{A}(\mathcal{B})$$

correspond to $\mathcal{P}_{C_\varphi} \in \mathbf{P}C_\varphi(\mathcal{B})$ under this identification.

Then we retrieve φ as the evaluation-at-$\mathcal{P}_\varphi$ map

$$\mathcal{A} = \Gamma(\mathbf{P}\mathcal{A}, \mathcal{O}(1)) \longrightarrow \mathcal{B}.$$

We also call $\mathcal{P}_\varphi$ the projective system of φ.

In coordinates, let $a_1, \ldots, a_k$ be a basis of $\mathcal{A}$, and identify $\mathbf{P}\mathcal{A}$ with $\mathbf{P}^{k-1}$ accordingly. Then

$$\mathcal{P}_\varphi = (\varphi(a_1) : \cdots : \varphi(a_k)) \in \mathbf{P}^{k-1}(\mathcal{B}).$$

Furthermore, write $\mathcal{B} = \mathcal{B}_1 \times \cdots \mathcal{B}_m$ as a product of local algebras, and for $1 \leq j \leq m$ let $\varphi_j : \mathcal{A} \to \mathcal{B}_j$ be the j-th component of φ. Then we can view $\mathcal{P}_\varphi$ as an ordered collection

$$\mathcal{P}_\varphi = (P_1, \ldots, P_m)$$

where

$$P_j = (\varphi_j(a_1) : \cdots : \varphi_j(a_k)) \in \mathbf{P}^{k-1}(\mathcal{B}_j).$$

4.5. — Let $\varphi : \mathcal{A} \rightsquigarrow \mathcal{B}$ be a non-degenerate multiplication reduction. Let $Q_{\mathcal{A}} \in \mathbf{P}\mathcal{A}(\mathcal{A})$ be the canonical point of $\mathcal{A}$, and let $\mathcal{P}_{\varphi} \in \mathbf{P}\mathcal{A}(\mathcal{B})$ as introduced just above. Also consider the complete linear system $V = \Gamma(\mathbf{P}\mathcal{A}, \mathcal{O}(1)) = \mathcal{A}$.

Then,

$$(\mathbf{P}\mathcal{A}, Q_{\mathcal{A}}, \mathcal{P}_{\varphi}, \mathcal{O}(1), V)$$

is "almost" a geometric realization for φ.

By this, we mean that φ indeed results from the construction given in Proposition 2.9 applied to this geometric data. This is because evaluation of $V = \mathcal{A}$ at $Q_{\mathcal{A}}$ is the identity map of $\mathcal{A}$, and evaluation of $V = \mathcal{A}$ at $\mathcal{P}_{\varphi}$ is φ.

However, to have a geometric realization proper, one should check conditions (i) and (ii). Condition (i) is clearly satisfied, but turning to condition (ii), for the complete linear system $V = \Gamma(\mathbf{P}\mathcal{A}, \mathcal{O}(1))$ we have $V^{\langle 2\rangle} = \Gamma(\mathbf{P}\mathcal{A}, \mathcal{O}(2)) = S^2\mathcal{A}$, which might fail to evaluate injectively at $\mathcal{P}_{\varphi}$.

More precisely, the defect of injectivity is measured by the kernel of the map

$$S^2\mathcal{A} \overset{S^2\varphi}{\simeq} S^2C_{\varphi} \xrightarrow{m_{\mathcal{B}}} C_{\varphi}^{\langle 2\rangle} \subseteq \mathcal{B},$$

which will be studied below.

5. The quadratic hull

5.1. — As explained in the Introduction, the quadratic kernel $I_2(C)$ of a code was first considered in [**28**, sec. 3], and also later in [**35**, §1.11].

In full generality, if $f : Y \to \mathbf{P}$ is a morphism from a scheme to a projective space over $\mathbf{F}$, we define $I_2(f)$ as the degree 2 part of the homogeneous ideal of the (schematic) image of f. Equivalently:

$$I_2(f) = \ker(\Gamma(\mathbf{P}, \mathcal{O}(2)) \xrightarrow{f^*} \Gamma(Y, f^*\mathcal{O}(2))).$$

And then:

5.2. Definition. The quadratic hull of f is the closed subscheme

$$Z_2(f) = Z(I_2(f)) \subseteq \mathbf{P}$$

defined by $I_2(C)$.

Equivalently, $Z_2(f)$ is the (schematic) intersection of the quadratic hypersurfaces through which f factorizes.

It is the smallest closed subscheme of $\mathbf{P}$ that contains the (schematic) image of f and is defined by quadratic equations.

5.3. — If $C \subseteq \mathcal{B}$ is a non-degenerate $\mathcal{B}$-code, we set $I_2(C) = I_2(\mathcal{P}_C)$ and define the quadratic hull of C as

$$Z_2(C) = Z_2(\mathcal{P}_C) \subseteq \mathbf{P}C,$$

where $\mathcal{P}_C$ from 4.3 is seen as a morphism $\mathcal{P}_C : \operatorname{Spec}\mathcal{B} \to \mathbf{P}C$.

In this situation, 5.1 specializes to

$$\begin{aligned} I_2(C) &= \ker(S^2C \xrightarrow{m_{\mathcal{B}}} \mathcal{B}) \\ &= \ker(S^2C \xrightarrow{m_{\mathcal{B}}} C^{\langle 2\rangle}). \end{aligned}$$

And thus, the quadratic hull of C is the closed subscheme

$$Z_2(C) = Z(I_2(C)) \subseteq \mathbf{P}C$$

defined by this $I_2(C)$.

In coordinates, as in 4.3, let $c_1, \dots, c_k$ be a basis of C, write $\mathcal{B} = \mathcal{B}_1 \times \cdots \mathcal{B}_m$ as a product of local algebras, let $\mathbf{G}$ be the corresponding $k \times m$ generator matrix for C, and for $1 \leq j \leq m$ let $P_j \in \mathbf{P}^{k-1}(\mathcal{B}_j)$ be the point defined by the j-th column of $\mathbf{G}$. Then $C^{\langle 2 \rangle}$ is the image of the evaluation-at-$\mathcal{P}_C$ map

$$\Gamma(\mathbf{P}^{k-1}, \mathcal{O}(2)) \longrightarrow \mathcal{B}$$

which sends a homogeneous quadratic form $q \in \Gamma(\mathbf{P}^{k-1}, \mathcal{O}(2)) = \mathbf{F}[x_1, \dots, x_k]_2$ to its value $q(c_1, \dots, c_k) = (q(P_1), \dots, q(P_m))$, and $I_2(C)$ is the kernel of this evaluation map:

$$\begin{aligned} I_2(C) &= \{q \in \mathbf{F}[x_1, \dots, x_k]_2 : \ q(c_1, \dots, c_k) = 0\} \\ &= \{q \in \mathbf{F}[x_1, \dots, x_k]_2 : \ q(P_1) = 0, \dots, q(P_m) = 0\}. \end{aligned}$$

And then $Z_2(C) = Z_2(\{P_1, \dots, P_m\})$ is the intersection of all quadrics in $\mathbf{P}^{k-1}$ that contain $P_1, \dots, P_m$. This intersection should be understood scheme-theoretically, so in general $Z_2(C)$ need not be a nice smooth, irreducible, nor even reduced variety.

5.4. Example. Let $\mathbf{F}$ be a field in which 2 admits two square roots (e.g. $\mathbf{F} = \mathbb{F}_7$, $\pm\sqrt{2} = \pm 3$). Let $C \subseteq \mathbf{F}^7$ be the linear code with generator matrix

$$\mathbf{G} = \begin{pmatrix} 1 & 1 & 1 & 1 & 1 & 1 & 1 \\ 0 & 1 & 1 & 0 & 0 & 1 & 1 \\ 0 & 0 & 0 & 1 & 1 & 1 & 1 \\ 0 & 1 & -1 & 1 & -1 & \sqrt{2} & -\sqrt{2} \end{pmatrix}$$

and let $P_1, \dots, P_7 \in \mathbf{P}^3(\mathbf{F})$ be the points defined by the columns of $\mathbf{G}$.

If w, x, y, z are linear coordinates on $\mathbf{P}^3$, then by linear algebra we find

$$I_2(C) = \langle x^2 - xw,\ y^2 - yw,\ x^2 + y^2 - z^2 \rangle.$$

Set-theoretically, there is no nonzero solution for $w = 0$, while for $w = 1$ the first two relations force $x, y \in \{0, 1\}$. From this one easily derives that $Z_2(C)$ has dimension 0 and $P_1, \dots, P_7$ are its only points.

However, by Bézout, $Z_2(C)$ should have length 8. It follows that $Z_2(C)$ is *not* reduced.

Closer inspection shows that, indeed, P_1 has multiplicity 2 in $Z_2(C)$ along the z-direction, while its other points $P_2, \dots, P_7$ are simple.

We might write this informally as $Z_2(\{P_1, \dots, P_7\}) = \{P_1 + \vec{z}, P_2, \dots, P_7\}$.

5.5. — The quadratic hull $Z_2(C)$ (or equivalently, its ideal $I_2(C)$) contains a lot of information about the square code $C^{\langle 2 \rangle}$. First, by construction, we have

$$\dim C^{\langle 2 \rangle} = \binom{k+1}{2} - \dim I_2(C)$$

where $k = \dim C$.

Further important properties will be given in section 8 at the end of this work. These properties are not needed in the application to geometric realizations of multiplication reductions, but they illustrate the significance of the quadratic hull.

6. Application to geometric realizations

6.1. — We finish the discussion started in 4.5. If $\varphi : \mathcal{A} \to \mathcal{B}$ is a non-degenerate linear map with projective system $\mathcal{P}_\varphi \in \mathbf{P}\mathcal{A}(\mathcal{B})$, we set $I_2(\varphi) = I_2(\mathcal{P}_\varphi)$. Recall that φ induces an identification $\mathcal{A} \simeq C_\varphi$, $S^{\cdot}\mathcal{A} \simeq S^{\cdot}C_\varphi$, hence $\mathbf{P}\mathcal{A} \simeq \mathbf{P}C_\varphi$, under which $\mathcal{P}_\varphi$ corresponds to $\mathcal{P}_{C_\varphi}$. It follows

$$\begin{aligned} I_2(\varphi) &= (S^2\varphi)^{-1} I_2(C_\varphi) \\ &= \ker(S^2\mathcal{A} \overset{S^2\varphi}{\simeq} S^2 C_\varphi \xrightarrow{m_\mathcal{B}} C_\varphi) \\ &= \ker(S^2\mathcal{A} \xrightarrow{S^2\varphi} S^2\mathcal{B} \xrightarrow{m_\mathcal{B}} \mathcal{B}). \end{aligned}$$

We then define the quadratic hull of φ as

$$Z_2(\varphi) = Z_2(\mathcal{P}_\varphi) = Z(I_2(\varphi)) \subseteq \mathbf{P}\mathcal{A},$$

which is the isomorphic image of $Z_2(C_\varphi) \subseteq \mathbf{P}C_\varphi$ under the identification above.

THEOREM. *Let $\varphi : \mathcal{A} \to \mathcal{B}$ be a non-degenerate linear map. Let $Q_\mathcal{A} \in \mathbf{P}\mathcal{A}(\mathcal{A})$ be the canonical point of $\mathcal{A}$, and let $\mathcal{P}_\varphi \in \mathbf{P}\mathcal{A}(\mathcal{B})$ be the projective system of φ, and $Z_2(\varphi) \subseteq \mathbf{P}\mathcal{A}$ its quadratic hull.*

Then φ is a multiplication reduction $\mathcal{A} \rightsquigarrow \mathcal{B}$ if and only if $Z_2(\varphi)$ contains $Q_\mathcal{A}$. Moreover, if this holds, then

$$(Z_2(\varphi), Q_\mathcal{A}, \mathcal{P}_\varphi, \mathcal{O}(1)|_{Z_2(\varphi)}, V)$$

is a geometric realization of φ, where the linear system $V \subseteq \Gamma(Z_2(\varphi), \mathcal{O}(1)|_{Z_2(\varphi)})$ is the isomorphic image of $\mathcal{A} = \Gamma(\mathbf{P}\mathcal{A}, \mathcal{O}(1))$ under the restriction-to-$Z_2(\varphi)$ map.

PROOF. By elementary linear algebra, a necessary and sufficient condition for the existence of an $\omega : \mathcal{B} \to \mathcal{A}$ that completes the diagram

$$\begin{array}{ccc} S^2\mathcal{A} & \xrightarrow{m_\mathcal{A}} & \mathcal{A} \\ {\scriptstyle S^2\varphi}\downarrow & & \\ S^2\mathcal{B} & \xrightarrow{m_\mathcal{B}} & \mathcal{B} \end{array}$$

is that $\ker(S^2\mathcal{A} \xrightarrow{S^2\varphi} S^2\mathcal{B} \xrightarrow{m_\mathcal{B}} \mathcal{B}) = I_2(\varphi)$ is contained in $\ker(S^2\mathcal{A} \xrightarrow{m_\mathcal{A}} \mathcal{A}) = I_2(Q_\mathcal{A})$. This means precisely that $Q_\mathcal{A}$ lies in $Z_2(\varphi)$.

Now observe that the map $\mathcal{A} = \Gamma(\mathbf{P}\mathcal{A}, \mathcal{O}(1)) \xrightarrow{\mathrm{ev}_{Q_\mathcal{A}}} \mathcal{A}$ is the identity of $\mathcal{A}$, and factorizes through V as

$$\mathcal{A} = \Gamma(\mathbf{P}\mathcal{A}, \mathcal{O}(1)) \xrightarrow{\mathrm{res}_{Z_2(\varphi)}} V \subseteq \Gamma(Z_2(\varphi), \mathcal{O}(1)|_{Z_2(\varphi)}) \xrightarrow{\mathrm{ev}_{Q_\mathcal{A}}} \mathcal{A},$$

which shows that condition (i) in Proposition 2.9 is satisfied (moreover, with bijectivity).

Likewise the map $\mathcal{A} = \Gamma(\mathbf{P}\mathcal{A}, \mathcal{O}(1)) \xrightarrow{\mathrm{ev}_{\mathcal{P}_\varphi}} \mathcal{B}$ is φ, and factorizes through V as

$$\mathcal{A} = \Gamma(\mathbf{P}\mathcal{A}, \mathcal{O}(1)) \xrightarrow{\mathrm{res}_{Z_2(\varphi)}} V \subseteq \Gamma(Z_2(\varphi), \mathcal{O}(1)|_{Z_2(\varphi)}) \xrightarrow{\mathrm{ev}_{\mathcal{P}_\varphi}} \mathcal{B},$$

which shows that if condition (ii) is satisfied, then $(Z_2(\varphi), Q_\mathcal{A}, \mathcal{P}_\varphi, \mathcal{O}(1)|_{Z_2(\varphi)}, V)$ is a geometric realization for φ.

In order to check (ii), we observe that the map $S^2\mathcal{A} = \Gamma(\mathbf{P}\mathcal{A}, \mathcal{O}(2)) \overset{\mathrm{ev}_{\mathcal{P}_\varphi}}{\twoheadrightarrow} \mathcal{B}$ has kernel $I_2(\varphi)$, and factorizes through $V^{\langle 2\rangle}$ as

$$S^2\mathcal{A} = \Gamma(\mathbf{P}\mathcal{A}, \mathcal{O}(2)) \xrightarrow{\mathrm{res}_{Z_2(\varphi)}} V^{\langle 2\rangle} \subseteq \Gamma(Z_2(\varphi), \mathcal{O}(2)|_{Z_2(\varphi)}) \xrightarrow{\mathrm{ev}_{\mathcal{P}_\varphi}} \mathcal{B}.$$

But by definition $I_2(\varphi)$ is also the kernel of $\Gamma(\mathbf{P}\mathcal{A}, \mathcal{O}(2)) \longrightarrow V^{\langle 2\rangle}$. Passing to the quotient we find that $V^{\langle 2\rangle} \overset{\mathrm{ev}_{\mathcal{P}_\varphi}}{\twoheadrightarrow} \mathcal{B}$ is injective, as desired. □

6.2. REMARK. The first part of the Theorem (together with its proof) is easily rephrased in elementary terms.

We are given a linear map $\varphi : \mathcal{A} \to \mathcal{B}$, and we ask whether it is a multiplication algorithm, i.e. whether it admits an adjoint $\omega : C_\varphi^{\langle 2\rangle} \to \mathcal{A}$.

Choose a basis $a_1, \dots, a_k$ of $\mathcal{A}$, and let $c_1 = \varphi(a_1)$, ..., $c_k = \varphi(a_k)$ be the corresponding basis of C.

Any $y \in C_\varphi$ is of the form $y = q(c_1, \dots, c_k)$, for $q \in \mathbf{F}[x_1, \dots, x_k]_2$ a quadratic form in k variables. Then necessarily one should have $\omega(y) = q(a_1, \dots, a_k)$. As in 2.5 this shows the *unicity* of ω, if it exists.

But now we're interested in the *existence*. For this we observe that $y \in C_\varphi$ can possibly be represented by several quadratic forms. Then $\omega(y)$ will be well defined precisely when, for any such multiple representation $y = q_1(c_1, \dots, c_k) = q_2(c_1, \dots, c_k)$, the corresponding values $q_1(a_1, \dots, a_k) = q_2(a_1, \dots, a_k)$ coincide. Setting $q = q_1 - q_2$, this is equivalent to:

Any quadratic form $q \in \mathbf{F}[x_1, \dots, x_k]_2$ vanishing at $(c_1, \dots, c_k)$ also vanishes at $(a_1, \dots, a_k)$.

This exactly means that $Z_2(\varphi)$ contains $Q_\mathcal{A}$.

6.3. COROLLARY. *Let C be a $[\mathcal{B}, k]$-code, where $k = \dim \mathcal{A}$. Then C is the image of a supercode $\widehat{C} \subseteq \mathcal{A} \times \mathcal{B}$ under the second projection $\pi_\mathcal{B}$ if and only if its quadratic hull $Z_2(C) \subseteq \mathbf{P}C$ contains a point $Q \in Z_2(C)(\mathcal{A})$ with coordinates in $\mathcal{A}$ that does not lie in a rational hyperplane of $\mathbf{P}C$.*

PROOF. Choose a basis $c_1, \dots, c_k$ of C, and identify $\mathbf{P}C \simeq \mathbf{P}^{k-1}$ accordingly. Assume there exists such a point $Q = (a_1 : \cdots : a_k) \in Z_2(C)(\mathcal{A}) \subseteq \mathbf{P}^{k-1}(\mathcal{A})$ not lying in a rational hyperplane. Then $a_1, \dots, a_k$ are linearly independent, so they form a basis of $\mathcal{A}$. Identifying $\mathbf{P}\mathcal{A} \simeq \mathbf{P}^{k-1}$ accordingly, we then have $Q = Q_\mathcal{A}$. Now if $\varphi : \mathcal{A} \to C \subseteq \mathcal{B}$ is the linear map that sends a_i to c_i, then φ satisfies the conditions in the Theorem, so φ is a multiplication reduction. Then $\widehat{C} = \mathrm{im}(\mathrm{id}_\mathcal{A} \times \varphi)$ is a supercode, and $C = C_\varphi = \pi_\mathcal{B}(\widehat{C})$ as wished.

The converse works similarly. □

6.4. REMARK. The condition in the Corollary can be rephrased as the existence of a rational point on (a Zariski open set of) a certain scheme. Indeed, points in $Z_2(C)(\mathcal{A})$ correspond to rational points of the Weil restriction of scalars of the base change to $\mathcal{A}$ of $Z_2(C)$. And the points not lying in a rational hyperplane of $\mathbf{P}^{k-1}$ are those in the Zariski open defined by the non-vanishing of the determinant, in this Weil restriction.

6.5. REMARK. If $\mathbf{F} = \mathbb{F}_q$ is a finite field, then there are only finitely many rational hyperplanes in $\mathbf{P}^{k-1}$, so their complement U is a Zariski open set. Then the condition in the Corollary also asks

$$(Z_2(C) \cap U)(\mathcal{A}) \neq \emptyset.$$

7. Experimental results

7.1. — In this section we describe the quadratic hull of all the multiplication algorithms of minimal length for some algebras of small cardinality over $\mathbb{F}_2$ and $\mathbb{F}_3$. Let us first recall how these algorithms can be found exhaustively.

Let $\mathcal{A}$ be an algebra over a field $\mathbf{F}$. Given a basis $a_1, \dots, a_k$ of $\mathcal{A}$, the product of two elements $x, y \in \mathcal{A}$ can be expanded as

$$xy = B_1(x,y)a_1 + \cdots + B_k(x,y)a_k,$$

which uniquely determines a family of k symmetric bilinear forms $B_1, \dots, B_k$ on $\mathcal{A}$. In order to fix notations, we state as a formal lemma what is only a rephrasing of the definitions (and, as explained in the Introduction, was already implicit in the very first papers on the topic):

LEMMA. *A linear map $\varphi = (\varphi_1, \dots, \varphi_n) : \mathcal{A} \to \mathbf{F}^n$ is a multiplication algorithm if and only if*

$$B_1, \dots, B_k \in \langle \varphi_1^{\otimes 2}, \dots, \varphi_n^{\otimes 2} \rangle.$$

Here $\langle \varphi_1^{\otimes 2}, \dots, \varphi_n^{\otimes 2} \rangle$ is the linear span of $\varphi_1^{\otimes 2}, \dots, \varphi_n^{\otimes 2}$ inside $(\mathcal{A}^\vee \otimes \mathcal{A}^\vee)^{\mathrm{sym}}$, the space of symmetric tensors in $\mathcal{A}^\vee \otimes \mathcal{A}^\vee$, identified with the space of symmetric bilinear forms on $\mathcal{A}$.

PROOF OF THE LEMMA. Let $T_\mathcal{A} \in \mathcal{A} \otimes (\mathcal{A}^\vee \otimes \mathcal{A}^\vee)^{\mathrm{sym}}$ be the multiplication tensor. Then φ is a multiplication algorithm if and only if it has an adjoint ω, i.e. if and only if one can write

$$T_\mathcal{A} = \omega_1 \otimes \varphi_1^{\otimes 2} + \cdots + \omega_n \otimes \varphi_n^{\otimes 2} \ \in \ \mathcal{A} \otimes \langle \varphi_1^{\otimes 2}, \dots, \varphi_n^{\otimes 2} \rangle.$$

However, by definition of $B_1, \dots, B_k$, we also have

$$T_\mathcal{A} = a_1 \otimes B_1 + \cdots + a_k \otimes B_k.$$

Writing $\omega_1, \dots, \omega_n$ in the basis $a_1, \dots, a_k$ identifies these two expressions and concludes. □

Now in $V = (\mathcal{A}^\vee \otimes \mathcal{A}^\vee)^{\mathrm{sym}}$ set:

- $T = \langle B_1, \dots, B_k \rangle$, or more intrinsically, $T = \{l \circ m_\mathcal{A} : l \in \mathcal{A}^\vee\}$
- $\mathcal{E} = \{l^{\otimes 2} : l \in \mathcal{A}^\vee \setminus \{0\}\}$, the set of elementary symmetric tensors.

By the Lemma, if $\varphi = (\varphi_1, \dots, \varphi_n)$ is a multiplication algorithm, then $W = \langle \varphi_1^{\otimes 2}, \dots, \varphi_n^{\otimes 2} \rangle$ contains T; moreover, if we assume that no shorter algorithm exists, then $\varphi_1^{\otimes 2}, \dots, \varphi_n^{\otimes 2}$ are linearly independent, and $\dim W = n$. Conversely, if $W \subseteq V$ is a subspace of dimension n containing T and admitting a basis $\varphi_1^{\otimes 2}, \dots, \varphi_n^{\otimes 2}$ made of elements of $\mathcal{E}$, then $\varphi = (\varphi_1, \dots, \varphi_n)$ is a multiplication algorithm of length n for $\mathcal{A}$.

Thus, finding all these algorithms reduces to the following:

7.2. PROBLEM. Let V be a vector space over $\mathbf{F}$, and k, n two integers. Given:

- $T \subseteq V$ a "target" subspace of dimension $\dim T = k$
- $\mathcal{E} \subseteq V$ a set of generating elements.

Find all subspaces $W \subseteq V$ such that

(i) $T \subseteq W$
(ii) W is generated by elements in $\mathcal{E}$
(iii) $\dim W = n$.

A key observation of Oseledets [**32**] is that these W are of the form $W = T \oplus \langle e_1, \dots, e_{n-k} \rangle$ for some $e_1, \dots, e_{n-k} \in \mathcal{E}$. Thus they can be found by a search among $(n-k)$-tuples of elements of $\mathcal{E}$, instead of a naive search among n-tuples.

This was implemented systematically by Barbulescu, Detrey, Estibals and Zimmermann in [**6**], which allowed them, among other results, to find all the optimal (symmetric) multiplication algorithms for $\mathbb{F}_{2^k}$ over $\mathbb{F}_2$ for $k \leq 6$, for $\mathbb{F}_{3^k}$ over $\mathbb{F}_3$ for $k \leq 5$, for $\mathbb{F}_2[t]/(t^k)$ over $\mathbb{F}_2$ for $k \leq 6$, for $\mathbb{F}_3[t]/(t^k)$ over $\mathbb{F}_3$ for $k \leq 5$, for $\mathbb{F}_2[t]/(t^k - 1)$ over $\mathbb{F}_2$ for $k \leq 7$, and for $\mathbb{F}_3[t]/(t^k - 1)$ over $\mathbb{F}_3$ for $k \leq 6$. They wrote a computer program [**43**] that solves Problem 7.2 for arbitrary $V, T, \mathcal{E}$ over $\mathbb{F}_2$ or $\mathbb{F}_3$.

Then in [**16**] Covanov gives further improvements, using specifically the geometry and symmetries of spaces of bilinear forms.

7.3. — In the approach above, one space W might admit several bases in $\mathcal{E}$, hence correspond to several multiplication algorithms. However, an interesting fact is that all these algorithms will have the same quadratic hull, i.e. this quadratic hull only depends on W.

Indeed, first recall that there is a natural duality between $(\mathcal{A}^\vee \otimes \mathcal{A}^\vee)^{\mathrm{sym}}$ and $S^2\mathcal{A} = \Gamma(\mathbf{P}\mathcal{A}, \mathcal{O}(2))$.

In coordinates, if $S \in (\mathcal{A}^\vee \otimes \mathcal{A}^\vee)^{\mathrm{sym}}$ is written as a symmetric matrix

$$S = (s_{i,j})_{\substack{1 \leq i \leq k \\ 1 \leq j \leq k}}$$

and if $Q \in \Gamma(\mathbf{P}\mathcal{A}, \mathcal{O}(2))$ is written as a quadratic form

$$Q = \sum_{1 \leq i \leq j \leq k} q_{i,j} x_i x_j,$$

then this duality is given by

$$\langle Q, S \rangle = \sum_{1 \leq i \leq j \leq k} q_{i,j} s_{i,j}.$$

More intrinsically, if $S = l^{\otimes 2}$ for $l \in \mathcal{A}^\vee$, we have

$$\langle Q, l^{\otimes 2} \rangle = Q(l),$$

and then this is extended by linearity for general $S \in (\mathcal{A}^\vee \otimes \mathcal{A}^\vee)^{\mathrm{sym}}$.

7.4. PROPOSITION. *Let $\varphi = (\varphi_1, \dots, \varphi_n) : \mathcal{A} \rightsquigarrow \mathbf{F}^n$ be a multiplication algorithm, let $W = \langle \varphi_1^{\otimes 2}, \dots, \varphi_n^{\otimes 2} \rangle$ be the linear span of $\varphi_1^{\otimes 2}, \dots, \varphi_n^{\otimes 2}$ in $(\mathcal{A}^\vee \otimes \mathcal{A}^\vee)^{sym}$, and let $I_2(\varphi) \subseteq S^2\mathcal{A}$ be the space of defining equations for the quadratic hull $Z_2(\varphi)$. Then, under the duality above, we have*

$$I_2(\varphi) = W^\perp.$$

PROOF. By definition, $I_2(\varphi) = \{Q \in S^2\mathcal{A} : Q(\varphi_1) = \cdots = Q(\varphi_n) = 0\}$. □

7.5. COROLLARY. *The elementary tensors in W are precisely the $l^{\otimes 2}$ for $l \in Z_2(\varphi)$ (up to a multiplicative constant).*

As a consequence, we have $\#Z_2(\varphi)(\mathbb{F}_q) \geq n$, with equality if and only if φ is the only multiplication algorithm of length n corresponding to W.

PROOF. Indeed, $l^{\otimes 2} \in W = I_2(\varphi)^\perp$ if and only if $Q(l) = 0$ for all $Q \in I_2(\varphi)$. □

7.6. COROLLARY. *Set $k = \dim \mathcal{A}$, and let $\varphi : \mathcal{A} \rightsquigarrow \mathbf{F}^n$ be a multiplication algorithm of minimal length, i.e. with n equal to the (symmetric) bilinear complexity $\mu_{\mathbf{F}}^{\mathrm{sym}}(\mathcal{A})$. Then $Z_2(\varphi)$ has codimension at most $\binom{k+1}{2} - n$ in $\mathbf{P}^{k-1}$, with equality if and only if it is a complete intersection.*

PROOF. Minimality of n implies that $\varphi_1^{\otimes 2}, \ldots, \varphi_n^{\otimes 2}$ are linearly independent, so $\dim W = n$, so $\dim I_2(\varphi) = \binom{k+1}{2} - n$. Thus $Z_2(\varphi)$ is defined by $\binom{k+1}{2} - n$ linearly independent quadratic equations, and we conlude. □

There are many examples of algebras $\mathcal{A}$ for which one knows multiplication algorithms φ of (possibly not minimal) length n much smaller than $\binom{k+1}{2}$ — for instance, n linear in k. This makes one expect $Z_2(\varphi)$ to be low dimensional.

7.7. — Zimmermann [**43**] kindly gave to the author a copy of the computer program that solves Problem 7.2. The output of this program is the list of the subspaces W, each given by a basis in row echelon form, thus in general not made from elements of $\mathcal{E}$. By Proposition 7.4, this suffices to find the quadratic hull of the corresponding multiplication algorithms, with no need to compute these multiplication algorithms explicitly.

For $q = 2$, $\mathcal{A}_k = \mathbb{F}_{q^k}$, one gets the minimal length $n = \mu_q^{\mathrm{sym}}(\mathcal{A}_k)$, the number of subspaces W, and the quadratic hull of the corresponding algorithms, given as follows:

Table 1: quadratic hull of minimal multiplication algorithms for $q = 2$, $\mathcal{A}_k = \mathbb{F}_{q^k}$

k	n	nb. of W	$Z_2(\varphi)$	$\#Z_2(\varphi)(\mathbb{F}_q)$
2	3	1	$\mathbf{P}^1$	3
3	6	1	$\mathbf{P}^2$	7
4	9	25	quadric $\simeq \mathbf{P}^1 \times \mathbf{P}^1$	9
5	13	2015	{ 310 : smooth del Pezzo surface 1705 : singular del Pezzo surface	13
6	15	21	(connected) union of a pair of conjugate conics and five $\mathbf{P}^1$	15

For $k \leq 5$, each of these quadratic hulls is a complete intersection. In particular for $k = 5$ (first case, smooth del Pezzo surface) one retrieves the example of section 3.

For $k = 6$ we have $\operatorname{codim} Z_2(\varphi) = 4 < \binom{k+1}{2} - n = 6$, so we do not get a complete intersection. Each of the two conjugate conics intersects each of the $\mathbf{P}^1$ at one point. Otherwise no other pair of components intersect. The canonical point is in one of the two conics.

7.8. — For $q = 3$, $\mathcal{A}_k = \mathbb{F}_{q^k}$, we find likewise:

Table 2: quadratic hull of minimal multiplication algorithms for $q=3$, $\mathcal{A}_k=\mathbb{F}_{q^k}$

k	n	nb. of W	$Z_2(\varphi)$	$\#Z_2(\varphi)(\mathbb{F}_q)$
2	3	1	$\mathbf{P}^1$	4
3	6	1	$\mathbf{P}^2$	13
4	9	234	$\begin{cases} 84: & \text{quadric} \simeq \mathbf{S}^2 \\ 150: & \text{quadric} \simeq \mathbf{P}^1\times\mathbf{P}^1 \end{cases}$	10 16
5	11	121	{11 rational points and one point of degree 5}	11

These are all complete intersections. In particular for $k=5$, $Z_2(\varphi)$ is finite of length 16, intersection of four quadrics; actually it is the trivial geometric representation from 2.13.

7.9. — For $q=2$, $\mathcal{A}_k=\mathbb{F}_q[t]/(t^k)$, we get:

Table 3: quadratic hull of minimal multiplication algorithms for $q=2$, $\mathcal{A}_k=\mathbb{F}_q[t]/(t^k)$

k	n	nb. of W	$Z_2(\varphi)$	$\#Z_2(\varphi)(\mathbb{F}_q)$
2	3	1	$\mathbf{P}^1$	3
3	5	2	$\mathbf{P}^1\cup\mathbf{P}^1$	5
4	8	4	$\mathbf{P}^2\cup\mathbf{P}^1$	9
5	11	112	$\begin{cases} 96: & \text{quadric} \cup \mathbf{P}^1 \\ 16: & \mathbf{P}^2\cup\mathbf{P}^2\cup\mathbf{P}^1 \end{cases}$	11 13
6	14	384	$\begin{cases} 256: & \text{quadric} \cup \mathbf{P}^2\cup\mathbf{P}^1 \\ 128: & \text{quadric} \cup\mathbf{P}^1\cup\mathbf{P}^1\cup\mathbf{P}^1 \end{cases}$	15

These are complete intersections for $k=2$ or 3, but not for $k=4,5,6$.

These are all connected.

For $k=3$, $k=4$, and the first case of $k=5$, the two components intersect precisely at a rational point, which is the support of the canonical point.

For the second case of $k=5$, the two $\mathbf{P}^2$ intersect along a line, and then the $\mathbf{P}^1$ intersects them at a point on this line, which is the support of the canonical point.

Likewise for the first case of $k=6$, the quadric and the $\mathbf{P}^2$ intersect along a line, and then the $\mathbf{P}^1$ intersects them at a point on this line, which is the support of the canonical point.

Last, for the second case of $k=6$, all components concur at one point, which is the support of the canonical point.

7.10. — For $q=3$, $\mathcal{A}_k=\mathbb{F}_q[t]/(t^k)$, we get:

Table 4: quadratic hull of minimal multiplication algorithms for $q=3$, $\mathcal{A}_k=\mathbb{F}_q[t]/(t^k)$

k	n	nb. of W	$Z_2(\varphi)$	$\#Z_2(\varphi)(\mathbb{F}_q)$
2	3	1	$\mathbf{P}^1$	4
3	5	3	$\mathbf{P}^1\cup\mathbf{P}^1$	7
4	8	252	$\begin{cases} 243: & \text{conic} \cup\mathbf{P}^1\cup\mathbf{P}^1 \\ 9: & \mathbf{P}^2\cup\mathbf{P}^1 \end{cases}$	10 16
5	10	243	conic $\cup\,\mathbf{P}^1\cup\mathbf{P}^1\cup\mathbf{P}^1$	13

These are complete intersections for $k = 2$ or 3, but not for $k = 4$ or 5.

These are all connected. For $k = 3, 4, 5$, the components all concur at one point, which is the support of the canonical point.

7.11. — Minimal algorithms were also computed for $\mathcal{A}_k = \mathbb{F}_q[t]/(t^k - 1)$, with $q = 2$ and $k \leq 7$, and with $q = 3$ and $k \leq 6$. In general $\mathbb{F}_q[t]/(t^k - 1)$ is not a local algebra, so it decomposes as the product of its local factors. It turns out all these examples follow Strassen's (strong) direct sum conjecture: one finds that the minimal algorithms for $\mathcal{A}_k$ are obtained by taking products of minimal algorithms for the local factors. And then, the quadratic hull of the product algorithm is the disjoint union of the quadratic hulls of the factors.

For instance, we have $\mathbb{F}_2[t]/(t^7 - 1) \simeq \mathbb{F}_2 \times \mathbb{F}_{2^3} \times \mathbb{F}_{2^3}$, and as predicted by the conjecture, one finds 49 multiplication algorithms for $\mathbb{F}_2[t]/(t^7 - 1)$ of minimal length $n = 13$, all of which decomposing as a product of a minimal algorithm for $\mathbb{F}_2$ (1 choice, length 1), and a minimal algorithm for each copy of $\mathbb{F}_{2^3}$ (7 choices each, length 6). All these 49 multiplication algorithms come from the same subspace W, hence have the same quadratic hull, which is a disjoint union of a point and two copies of $\mathbf{P}^2$.

8. Further properties of the quadratic hull

8.1. — Given a $[\mathcal{B}, k]$-code $C \subseteq \mathcal{B}$, and a point $P \in \mathbf{P}C(\mathbf{F})$ representing a nonzero linear form $\lambda_P : C \twoheadrightarrow \mathbf{F}$ (up to a multiplicative constant), the extended code $C_P \subseteq \mathcal{B} \times \mathbf{F}$ is the $[\mathcal{B} \times \mathbf{F}, k]$-code made of the words of the form

$$(c, \lambda_P(c))$$

for $c \in C$. (In particular, if $\mathcal{B} = \mathbf{F}^n$ and C is a $[n, k]$-code, then C_P is the extended $[n + 1, k]$-code of C by λ_P in the usual sense.)

In coordinates, if $\mathbf{G}$ is a generator matrix for C, and if $P \in \mathbf{P}^{k-1}(\mathbf{F})$ is represented as a column vector in $\mathbf{F}^k$, then C_P is the code with generator matrix $\mathbf{G}_P = (\ \mathbf{G}\ | P)$.

Proposition. *With these notations, we have either* $\dim C_P^{\langle 2 \rangle} = \dim C^{\langle 2 \rangle}$ *or* $\dim C_P^{\langle 2 \rangle} = \dim C^{\langle 2 \rangle} + 1$.

Moreover, the rational points of $Z_2(C)$ parameterize square-preserving *extensions C_P of C, i.e. those such that*

$$\dim C_P^{\langle 2 \rangle} = \dim C^{\langle 2 \rangle}.$$

Proof. Take $P \in \mathbf{P}^{k-1}(\mathbf{F})$, write $C^{\langle 2 \rangle}$ as the image of the evaluation map

$$\begin{array}{ccc} \Gamma(\mathbf{P}^{k-1}, \mathcal{O}(2)) & \longrightarrow & \mathcal{B} \\ q & \mapsto & (q(P_1), \ldots, q(P_m)) \end{array}$$

whose kernel is $I_2(C) = I_2(\{P_1, \ldots, P_m\})$, and write $C_P^{\langle 2 \rangle}$ as the image of the evaluation map

$$\begin{array}{ccc} \Gamma(\mathbf{P}^{k-1}, \mathcal{O}(2)) & \longrightarrow & \mathcal{B} \times \mathbf{F} \\ q & \mapsto & (q(P_1), \ldots, q(P_m), q(P)) \end{array}$$

whose kernel is $I_2(C_P) = I_2(\{P_1, \ldots, P_m, P\}) = I_2(C) \cap I_2(P)$.However, $I_2(P)$ is a hyperplane in $\Gamma(\mathbf{P}^{k-1}, \mathcal{O}(2)$, so

- either $I_2(C) \not\subseteq I_2(P)$, and then $I_2(C_P) \subsetneq I_2(C)$ has codimension 1, and $\dim C_P^{\langle 2\rangle} = \dim C^{\langle 2\rangle} + 1$
- or $I_2(C) \subseteq I_2(P)$, thus $I_2(C_P) = I_2(C)$ and $\dim C_P^{\langle 2\rangle} = \dim C^{\langle 2\rangle}$.

The latter means precisely that P lies in $Z_2(C)$. $\square$

8.2. — Recall that points of $\mathbf{P}^{k-1} \simeq \mathbf{P}C$ also parameterize hyperplanes of C. If $P \in \mathbf{P}^{k-1}(\mathbf{F})$, its orthogonal $P^\perp$ is a hyperplane in $\mathbf{F}^k$. The image of $P^\perp$ under the generator matrix $\mathbf{G}$ is then the hyperplane $H \subseteq C$ corresponding to P.

Equivalently, linear forms vanishing at P form a hyperplane $\Gamma(\mathbf{P}^{k-1}, \mathfrak{I}_P\mathcal{O}(1)) \subseteq \Gamma(\mathbf{P}^{k-1}, \mathcal{O}(1))$. Then we have

$$H = \mathrm{ev}_{\mathcal{P}_C}(\Gamma(\mathbf{P}^{k-1}, \mathfrak{I}_P\mathcal{O}(1))).$$

Proposition. *Rational points of $Z_2(C)$ parameterize hyperplanes $H \subseteq C$ such that*

$$H \cdot C \subsetneq C^{\langle 2\rangle}.$$

Proof. Evaluation at $\mathcal{P}_C = (P_1, \dots, P_m)$ identifies $\Gamma(\mathbf{P}^{k-1}, \mathcal{O}(1))$ with C, and maps $\Gamma(\mathbf{P}^{k-1}, \mathcal{O}(2))$ onto $C^{\langle 2\rangle}$, compatibly with multiplication.

Writing $H = \mathrm{ev}_{\mathcal{P}_C}(\Gamma(\mathbf{P}^{k-1}, \mathfrak{I}_P\mathcal{O}(1)))$ we find

$$\begin{aligned} H \cdot C &= \mathrm{ev}_{\mathcal{P}_C}(\Gamma(\mathbf{P}^{k-1}, \mathfrak{I}_P\mathcal{O}(1))) \cdot \mathrm{ev}_{\mathcal{P}_C}(\Gamma(\mathbf{P}^{k-1}, \mathcal{O}(1))) \\ &= \mathrm{ev}_{\mathcal{P}_C}(\Gamma(\mathbf{P}^{k-1}, \mathfrak{I}_P\mathcal{O}(1)) \cdot \Gamma(\mathbf{P}^{k-1}, \mathcal{O}(1))) \\ &= \mathrm{ev}_{\mathcal{P}_C}(\Gamma(\mathbf{P}^{k-1}, \mathfrak{I}_P\mathcal{O}(2))) \end{aligned}$$

where

$$\Gamma(\mathbf{P}^{k-1}, \mathfrak{I}_P\mathcal{O}(2)) = I_2(P)$$

is the hyperplane in $\Gamma(\mathbf{P}^{k-1}, \mathcal{O}(2))$ of quadratic forms vanishing at P.

We thus have a commutative diagram

$$\begin{array}{ccccccccc} 0 & \longrightarrow & I_2(\{P_1, \dots, P_m\}) & \longrightarrow & \Gamma(\mathbf{P}^{k-1}, \mathcal{O}(2)) & \xrightarrow{\mathrm{ev}_{\mathcal{P}_C}} & C^{\langle 2\rangle} & \longrightarrow & 0 \\ & & \uparrow & & \uparrow & & \uparrow & & \\ 0 & \longrightarrow & I_2(\{P_1, \dots, P_m, P\}) & \longrightarrow & I_2(P) & \xrightarrow{\mathrm{ev}_{\mathcal{P}_C}} & H \cdot C & \longrightarrow & 0 \end{array}$$

with exact rows, and injective vertical maps. Since $I_2(P)$ has codimension 1 in $\Gamma(\mathbf{P}^{k-1}, \mathcal{O}(2))$, then

- either $H \cdot C = C^{\langle 2\rangle}$ and $I_2(\{P_1, \dots, P_m, P\}) \subsetneq I_2(\{P_1, \dots, P_m\})$ has codimension 1
- or $H \cdot C \subsetneq C^{\langle 2\rangle}$ has codimension 1 and $I_2(\{P_1, \dots, P_m, P\}) = I_2(\{P_1, \dots, P_m\})$.

The latter occurs precisely when P lies in $Z_2(\{P_1, \dots, P_m\}) = Z_2(C)$. $\square$

8.3. — Recall that set-theoretically, the secant variety $\mathrm{Sec}(Z)$ of a closed subscheme Z in a projective space, is the union of all lines, possibly defined over an extension field, whose intersection with Z has length at least 2. This intersection could consist of non-rational points, or of points with multiplicities, so $\mathrm{Sec}(Z)$ really depends on the scheme structure of Z.

PROPOSITION. *Rational points of the secant variety* $\mathrm{Sec}(Z_2(C))$ *parameterize a subset of (and possibly all) the locus of hyperplanes* $H \subseteq C$ *that satisfy*

$$H^{\langle 2\rangle} \subsetneq C^{\langle 2\rangle}.$$

PROOF. Let $P \in \mathrm{Sec}(Z_2(C))(\mathbf{F})$ parameterize a hyperplane $H \subsetneq C$, and assume by contradiction $H^{\langle 2\rangle} = C^{\langle 2\rangle}$.

We have $H = \mathrm{ev}_{\mathcal{P}_C}(\Gamma(\mathbf{P}^{k-1}, \mathfrak{I}_P\mathcal{O}(1)))$, so

$$H^{\langle 2\rangle} = \mathrm{ev}_{\mathcal{P}_C}(\Gamma(\mathbf{P}^{k-1}, \mathfrak{I}_P^2\mathcal{O}(2)))$$

is the evaluation of the space of quadratic forms vanishing at P with multiplicity 2.

From the diagram

$$\begin{array}{ccccccccc} 0 & \longrightarrow & I_2(\{P_1,\dots,P_m\}) & \longrightarrow & \Gamma(\mathbf{P}^{k-1},\mathcal{O}(2)) & \xrightarrow{\mathrm{ev}_{\mathcal{P}_C}} & C^{\langle 2\rangle} & \longrightarrow & 0 \\ & & & & \uparrow & & \| & & \\ & & & & \Gamma(\mathbf{P}^{k-1},\mathfrak{I}_P^2\mathcal{O}(2)) & \xrightarrow{\mathrm{ev}_{\mathcal{P}_C}} & H^{\langle 2\rangle} & \longrightarrow & 0 \end{array}$$

we deduce that

- $\Gamma(\mathbf{P}^{k-1}, \mathfrak{I}_P^2\mathcal{O}(2))$ maps onto $\Gamma(\mathbf{P}^{k-1}, \mathcal{O}(2))/I_2(\{P_1,\dots,P_m\})$
- thus, $\Gamma(\mathbf{P}^{k-1}, \mathfrak{I}_P^2\mathcal{O}(2)) + I_2(\{P_1,\dots,P_m\}) = \Gamma(\mathbf{P}^{k-1}, \mathcal{O}(2))$
- thus, $I_2(\{P_1,\dots,P_m\})$ maps onto $\Gamma(\mathbf{P}^{k-1}, \mathcal{O}(2))/\Gamma(\mathbf{P}^{k-1}, \mathfrak{I}_P^2\mathcal{O}(2))$

(actually these three statements are equivalent).

After possibly extension of scalars, we restrict the last assertion to a line L passing through P and having intersection with $Z_2(C)$ of length at least 2. It gives that $I_2(\{P_1,\dots,P_m\})|_L$ maps onto $\Gamma(L, \mathcal{O}(2))/\Gamma(L, \mathfrak{I}_P^2\mathcal{O}(2))$, which has dimension 2.

However, since $I_2(\{P_1,\dots,P_m\})$ is a space of defining equations for $Z_2(C)$, its restriction $I_2(\{P_1,\dots,P_m\})|_L$ lives in $\Gamma(L, \mathfrak{I}_{Z_2(C)\cap L}\mathcal{O}(2))$. But $Z_2(C) \cap L$ has length at least 2, so $\Gamma(L, \mathfrak{I}_{Z_2(C)\cap L}\mathcal{O}(2))$ has dimension at most 1, a contradiction. □

8.4. — In [**30**, Th. 2], Márquez-Corbella, Martínez-Moro, and Pellikaan showed that if C is the evaluation code of some $L(D)$ at n distinct rational points on a smooth projective curve X of genus g, with $2g+1 < \deg(D) < n/2$, then $Z_2(C) = X$. We slightly strengthen this result, and also generalize it to $\mathcal{B}$-codes:

PROPOSITION. *Let X be a smooth projective curve of genus g, let D be a divisor on X, and $\mathcal{P} \in X(\mathcal{B})$. Choose a trivialization $\mathcal{O}(D)|_{\mathcal{P}} \simeq \mathcal{B}$, and let*

$$C = \mathrm{im}(L(D) \xrightarrow{\mathrm{res}_{\mathcal{P}}} \mathcal{O}(D)|_{\mathcal{P}} \simeq \mathcal{B})$$

be the evaluation $\mathcal{B}$-code of $L(D)$ at $\mathcal{P}$. Let $\mathrm{im}(\mathcal{P})$ be the schematic image of $\mathcal{P}$ in X, which is a finite closed subscheme of X, and let $[\mathrm{im}(\mathcal{P})]$ be its fundamental cycle [**22**, §1.5], *seen as a divisor on X. Assume*

$$\deg(D) > 2g+1$$

and

$$L(2D - [\mathrm{im}(\mathcal{P})]) = 0.$$

Then

$$Z_2(C) = X.$$

PROOF. Set $k = \dim(C) = \dim L(D)$, and identify X with its image under the embedding $\iota : X \hookrightarrow \mathbf{P}^{k-1}$ defined by $L(D)$. Observe that we then have $\mathcal{P}_C = \iota \circ \mathcal{P}$.

By [**37**] and the hypothesis $\deg(D) > 2g+1$, $X \subseteq \mathbf{P}^{k-1}$ is defined by quadratic equations, and it contains $\mathcal{P}_C$, thus

$$Z_2(C) \subseteq X.$$

Conversely, let $f \in I_2(C) \subseteq \Gamma(\mathbf{P}^{k-1}, \mathcal{O}(2))$. Then $f|_X$ lives in $L(2D)$ and vanishes at the image of $\mathcal{P}$, so actually $f|_X \in L(2D - [\mathrm{im}(\mathcal{P})])$, which is 0 by hypothesis. This forces $f|_X = 0$, which means $f \in I_2(X)$. This proves $I_2(C) \subseteq I_2(X)$, hence

$$Z_2(C) \supseteq X.$$

□

If $\mathcal{B} = \mathbb{F}^n$ and $\mathcal{P}$ is a collection of *distinct* rational points $P_1, \dots, P_n$, then $[\mathrm{im}(\mathcal{P})] = P_1 + \cdots + P_n$. If moreover $\deg(D) < n/2$ then $L(2D - [\mathrm{im}(\mathcal{P})]) = 0$ for degree reasons, so Proposition 8.4 implies [**30**, Th. 2]. Is this slight improvement useful?

Let $q \geq 49$ be a square. As $k \to \infty$, the best multiplication algorithms for $\mathbb{F}_{q^k}$ over $\mathbb{F}_q$ in the literature [**34**, Th. 6.4] use evaluation-interpolation on a curve X at $n = 2k + g - 1$ points, for a divisor D of degree $\deg(D) = k + g - 1$ (actually this is also the case in older results such as [**3**, Th. 1.1], that require curves with a larger number of points but only use $2k + g - 1$ of them), with $g \approx \frac{2k}{\sqrt{q}-2}$. As a consequence, condition $\deg(D) > 2g + 1$ is satisfied, but condition $\deg(D) < n/2$ is *not*. So [**30**, Th. 2] does not apply to the associated code C. However, the very construction of these algorithms ensures $L(2D - (P_1 + \cdots + P_n)) = 0$, so the conclusion $Z_2(C) = X$ still holds thanks to Proposition 8.4.

Acknowledgments

The author greatly benefited from discussions with R. Blache and E. Hallouin on del Pezzo surfaces, while preparing [**7**]. This influenced the presentation of the example developed in section 3.

The author also thanks P. Zimmermann and S. Covanov for giving details about their results [**6**][**16**]. In particular P. Zimmermann provided the computer program that allowed the author to get the list of explicit optimal formulae on which section 7 is based.

References

[1] N. Arnaud. Évaluation dérivée, multiplication dans les corps finis et codes correcteurs. Thèse de doctorat, Univ. Aix-Marseille 2, 2006.

[2] Christine Bachoc, Alain Couvreur, and Gilles Zémor, *Towards a function field version of Freiman's theorem*, Algebr. Comb. **1** (2018), no. 4, 501–521, DOI 10.5802/alco. MR3875074

[3] Stéphane Ballet, *Curves with many points and multiplication complexity in any extension of* $\mathbf{F}_q$, Finite Fields Appl. **5** (1999), no. 4, 364–377, DOI 10.1006/ffta.1999.0255. MR1711833

[4] S. Ballet, J. Chaumine, J. Pieltant, M. Rambaud, H. Randriambololona, R. Rolland. On the tensor rank of multiplication in finite extensions of finite fields and related issues in algebraic geometry. To appear in Moscow Math. J. — `arXiv:1906.07456`

[5] S. Ballet and R. Rolland, *Multiplication algorithm in a finite field and tensor rank of the multiplication*, J. Algebra **272** (2004), no. 1, 173–185, DOI 10.1016/j.jalgebra.2003.09.031. MR2029030

[6] Razvan Barbulescu, Jérémie Detrey, Nicolas Estibals, and Paul Zimmermann, *Finding optimal formulae for bilinear maps*, Arithmetic of finite fields, Lecture Notes in Comput. Sci., vol. 7369, Springer, Heidelberg, 2012, pp. 168–186, DOI 10.1007/978-3-642-31662-3_12. MR2992378

[7] Régis Blache, Alain Couvreur, Emmanuel Hallouin, David Madore, Jade Nardi, Matthieu Rambaud, and Hugues Randriam, *Anticanonical codes from del Pezzo surfaces with Picard rank one*, Trans. Amer. Math. Soc. **373** (2020), no. 8, 5371–5393, DOI 10.1090/tran/8119. MR4127880

[8] Roger W. Brockett and David Dobkin, *On the optimal evaluation of a set of bilinear forms*, Linear Algebra Appl. **19** (1978), no. 3, 207–235, DOI 10.1016/0024-3795(78)90012-5. MR495183

[9] Peter Bürgisser, Michael Clausen, and M. Amin Shokrollahi, *Algebraic complexity theory*, Grundlehren der Mathematischen Wissenschaften [Fundamental Principles of Mathematical Sciences], vol. 315, Springer-Verlag, Berlin, 1997. With the collaboration of Thomas Lickteig, DOI 10.1007/978-3-662-03338-8. MR1440179

[10] Ignacio Cascudo, Hao Chen, Ronald Cramer, and Chaoping Xing, *Asymptotically good ideal linear secret sharing with strong multiplication over any fixed finite field*, Advances in cryptology—CRYPTO 2009, Lecture Notes in Comput. Sci., vol. 5677, Springer, Berlin, 2009, pp. 466–486, DOI 10.1007/978-3-642-03356-8_28. MR2556974

[11] Ignacio Cascudo, Ronald Cramer, Chaoping Xing, and Chen Yuan, *Amortized complexity of information-theoretically secure MPC revisited*, Advances in cryptology—CRYPTO 2018. Part III, Lecture Notes in Comput. Sci., vol. 10993, Springer, Cham, 2018, pp. 395–426, DOI 10.1002/jgt.22193. MR3847903

[12] Murat Cenk and Ferruh Özbudak, *On multiplication in finite fields*, J. Complexity **26** (2010), no. 2, 172–186, DOI 10.1016/j.jco.2009.11.002. MR2607731

[13] Hao Chen and Ronald Cramer, *Algebraic geometric secret sharing schemes and secure multi-party computations over small fields*, Advances in cryptology—CRYPTO 2006, Lecture Notes in Comput. Sci., vol. 4117, Springer, Berlin, 2006, pp. 521–536, DOI 10.1007/11818175_31. MR2422182

[14] D. V. Chudnovsky and G. V. Chudnovsky, *Algebraic complexities and algebraic curves over finite fields*, J. Complexity **4** (1988), no. 4, 285–316, DOI 10.1016/0885-064X(88)90012-X. MR974928

[15] Alain Couvreur, Philippe Gaborit, Valérie Gauthier-Umaña, Ayoub Otmani, and Jean-Pierre Tillich, *Distinguisher-based attacks on public-key cryptosystems using Reed-Solomon codes*, Des. Codes Cryptogr. **73** (2014), no. 2, 641–666, DOI 10.1007/s10623-014-9967-z. MR3237950

[16] Svyatoslav Covanov, *Improved method for finding optimal formulas for bilinear maps in a finite field*, Theoret. Comput. Sci. **790** (2019), 41–65, DOI 10.1016/j.tcs.2019.04.029. MR4000704

[17] Ronald Cramer, Ivan Damgård, and Ueli Maurer, *General secure multi-party computation from any linear secret-sharing scheme*, Advances in cryptology—EUROCRYPT 2000 (Bruges), Lecture Notes in Comput. Sci., vol. 1807, Springer, Berlin, 2000, pp. 316–334, DOI 10.1007/3-540-45539-6_22. MR1772026

[18] R. Cramer, I. Damgård, J. Nielsen. Secure multiparty computation and secret sharing. Cambridge University Press, Cambridge, 2015.

[19] Igor V. Dolgachev, *Classical algebraic geometry: A modern view*, Cambridge University Press, Cambridge, 2012, DOI 10.1017/CBO9781139084437. MR2964027

[20] A. Fontaine, W. Peterson, *Group code equivalence and optimum codes*. IRE Trans. Inform. Theory 5 (1959) 60–70.

[21] C. M. Fiduccia and Y. Zalcstein, *Algebras having linear multiplicative complexities*, J. Assoc. Comput. Mach. **24** (1977), no. 2, 311–331, DOI 10.1145/322003.322014. MR483695

[22] William Fulton, *Intersection theory*, Ergebnisse der Mathematik und ihrer Grenzgebiete (3) [Results in Mathematics and Related Areas (3)], vol. 2, Springer-Verlag, Berlin, 1984, DOI 10.1007/978-3-662-02421-8. MR732620

[23] V. Goppa. Algebraico-geometric codes (English translation). Math. USSR Izv. 21 (1983) 75–91.

[24] M. Grassl. Code Tables: bounds on the parameters of various types of codes. Online: `http://www.codetables.de`

[25] A. Karatsuba, Yu. Ofman. Multiplication of multi-digit numbers on automata (English translation). Sov. Phys. Dokl. 7 (1963) 595–596.

[26] Gilles Lachaud, *Les codes géométriques de Goppa* (French), Astérisque **133-134** (1986), 189–207. Seminar Bourbaki, Vol. 1984/85. MR837220

[27] Yu. I. Manin, *Cubic forms: Algebra, geometry, arithmetic*, 2nd ed., North-Holland Mathematical Library, vol. 4, North-Holland Publishing Co., Amsterdam, 1986. Translated from the Russian by M. Hazewinkel. MR833513

[28] I. Márquez-Corbella, E. Martínez-Moro, R. Pellikaan. Evaluation of public-key cryptosystems based on algebraic geometry codes. Proc. Third International Castle Meeting on Coding Theory and Applications (3ICMTA, Cardona Castle, Barcelona, Spain, September 11-15, 2011), pp. 199–204.

[29] Irene Márquez-Corbella, Edgar Martínez-Moro, and Ruud Pellikaan, *The non-gap sequence of a subcode of a generalized Reed-Solomon code*, Des. Codes Cryptogr. **66** (2013), no. 1-3, 317–333, DOI 10.1007/s10623-012-9694-2. MR3016572

[30] Irene Márquez-Corbella, Edgar Martínez-Moro, and Ruud Pellikaan, *On the unique representation of very strong algebraic geometry codes*, Des. Codes Cryptogr. **70** (2014), no. 1-2, 215–230, DOI 10.1007/s10623-012-9758-3. MR3147595

[31] Irene Márquez-Corbella, Edgar Martínez-Moro, Ruud Pellikaan, and Diego Ruano, *Computational aspects of retrieving a representation of an algebraic geometry code*, J. Symbolic Comput. **64** (2014), 67–87, DOI 10.1016/j.jsc.2013.12.007. MR3170323

[32] Ivan Oseledets, *Optimal Karatsuba-like formulae for certain bilinear forms in* GF(2), Linear Algebra Appl. **429** (2008), no. 8-9, 2052–2066, DOI 10.1016/j.laa.2008.06.004. MR2446640

[33] R. Pellikaan, B.-Z. Shen, and G. J. M. van Wee, *Which linear codes are algebraic-geometric?*, IEEE Trans. Inform. Theory **37** (1991), no. 3, 583–602, DOI 10.1109/18.79915. MR1145820

[34] Hugues Randriambololona, *Bilinear complexity of algebras and the Chudnovsky-Chudnovsky interpolation method*, J. Complexity **28** (2012), no. 4, 489–517, DOI 10.1016/j.jco.2012.02.005. MR2925903

[35] Hugues Randriambololona, *On products and powers of linear codes under componentwise multiplication*, Algorithmic arithmetic, geometry, and coding theory, Contemp. Math., vol. 637, Amer. Math. Soc., Providence, RI, 2015, pp. 3–78, DOI 10.1090/conm/637/12749. MR3364442

[36] H. Randriambololona. Squares of hyperplanes and the secant variety. In preparation.

[37] Bernard Saint-Donat, *Sur les équations définissant une courbe algébrique* (French), C. R. Acad. Sci. Paris Sér. A-B **274** (1972), A324–A327. MR289516

[38] Igor E. Shparlinski, Michael A. Tsfasman, and Serge G. Vladut, *Curves with many points and multiplication in finite fields*, Coding theory and algebraic geometry (Luminy, 1991), Lecture Notes in Math., vol. 1518, Springer, Berlin, 1992, pp. 145–169, DOI 10.1007/BFb0087999. MR1186422

[39] Volker Strassen, *Gaussian elimination is not optimal*, Numer. Math. **13** (1969), 354–356, DOI 10.1007/BF02165411. MR248973

[40] Volker Strassen, *Vermeidung von Divisionen* (German, with English summary), J. Reine Angew. Math. **264** (1973), 184–202. MR521168

[41] Christian Wieschebrink, *Cryptanalysis of the Niederreiter public key scheme based on GRS subcodes*, Post-quantum cryptography, Lecture Notes in Comput. Sci., vol. 6061, Springer, Berlin, 2010, pp. 61–72, DOI 10.1007/978-3-642-12929-2_5. MR2776311

[42] S. Winograd, *Some bilinear forms whose multiplicative complexity depends on the field of constants*, Math. Systems Theory **10** (1976/77), no. 2, 169–180, DOI 10.1007/BF01683270. MR468322

[43] P. Zimmermann. Personal communication.

Hugues Randriambololona, Laboratoire de cryptographie, Agence nationale de sécurité des systèmes d'information, 51 boulevard de La Tour-Maubourg, 75700 PARIS 07 SP, France

Email address: `randriam@telecom-paris.fr`

Contemporary Mathematics
Volume **770**, 2021
https://doi.org/10.1090/conm/770/15441

Serre's genus fifty example

Jaap Top

ABSTRACT. This note presents explicit equations (up to birational equivalence over $\mathbb{F}_2$) for a complete, smooth, absolutely irreducible curve X over $\mathbb{F}_2$ of genus 50 satisfying $\#X(\mathbb{F}_2) = 40$. In his 1985 Harvard lecture notes on curves over finite fields, J-P. Serre already showed the existence of such a curve: he used class field theory to describe the function field $\mathbb{F}_2(X)$ as a certain abelian extension of the function field $\mathbb{F}_2(E)$ of some elliptic curve $E/\mathbb{F}_2$. Although various more recent texts recall Serre's construction, explicit equations as well as a description of intermediate curves $X \to Y \to E$ over $\mathbb{F}_2$ seem to be new. We also describe explicit equations for a curve over $\mathbb{F}_2$ of genus 8 with 11 rational points, and for a curve over $\mathbb{F}_2$ of genus 22 with 21 rational points.

1. Introduction

Given a smooth and complete, absolutely irreducible curve X of genus g defined over a finite field $\mathbb{F}_q$ of cardinality q, one can bound its number of rational points $\#X(\mathbb{F}_q)$ in terms of the two integers g, q. As a consequence, the integer $N_q(g)$, defined as the maximal $\#X(\mathbb{F}_q)$ that occurs when X runs over all such curves of genus g, exists. Much of J-P. Serre's 1985 Harvard lectures [**8**] discuss these integers $N_q(g)$. In particular, the second part of his lectures considers $g \mapsto N_q(g)$ for fixed prime power q, finishing with a treatment of $N_2(g)$. One of the results is the estimate

$$N_2(g) \leq 0.6272 \cdot g + 9.562$$

which, in the case $g = 50$, shows that any complete, smooth, absolutely irreducible curve $X/\mathbb{F}_2$ of genus 50 satisfies $\#X(\mathbb{F}_2) \leq 40$.

The last pages of [**8**] contain a proof of the equality $N_2(50) = 40$. Serre starts from an elliptic curve $E/\mathbb{F}_2$ with $\#E(\mathbb{F}_2) = 5$. He shows using class field theory that a covering $\pi\colon X \to E$ over $\mathbb{F}_2$ of degree 8 exists, ramifying only over the Galois orbit of some point in $E(\mathbb{F}_{128}) \setminus E(\mathbb{F}_2)$, and all 40 points in $\pi^{-1}(E(\mathbb{F}_2))$ are defined over $\mathbb{F}_2$, and X has genus 50.

The aim of this note is to make this explicit. Section 2 briefly describes some properties of the elliptic curve $E/\mathbb{F}_2$. Section 3 constructs the desired covering $X \to E$ as a tower $X = X_{123} \to X_{12} \to X_1 \to E$ of covers of degree 2. The final section discusses a plane model of X, and gives remarks on intermediate curves $X \to Y \to E$. It ends with an explicit description of curves with the same genus as X_{12} and X_1 respectively, but with more $\mathbb{F}_2$-rational points.

2020 *Mathematics Subject Classification.* 14H45, 14G15, 14H25.

2. The elliptic curve

An elliptic curve over $\mathbb{F}_2$ with $5 = 2 + 1 + 2$ rational points is supersingular, hence has j-invariant 0. Up to isomorphism over $\mathbb{F}_2$ there exist precisely 3 elliptic curves over $\mathbb{F}_2$ with j-invariant 0 (see, e.g., [**3**, Prop. 3.1]), with 1, 3, 5 rational points, respectively. So there is a unique curve $E/\mathbb{F}_2$ with $\#E(\mathbb{F}_2) = 5$. Another way to see this is by applying [**4**, Thm. 4.6]. In the remainder of this paper we give this elliptic curve by the equation

$$E: \quad y^2 + y = x^3 + x.$$

Let $\iota \in \mathrm{End}(E)$ be given by

$$\iota: \quad (x, y) \mapsto (x + 1, y + x + 1).$$

Note that $\iota^2 = [-1]$. Denoting by $F \in \mathrm{End}(E)$ the Frobenius endomorphism $F(x, y) = (x^2, y^2)$, an easy calculation shows $F = [-1] + \iota$ and hence

$$\#E(\mathbb{F}_{2^n}) = \deg(1 - F^n) = \begin{cases} 2^n + 1 - 2^{m+1} & \text{if } n = 2m \equiv 0 \bmod 8; \\ 2^n + 1 + 2^{m+1} & \text{if } n = 2m + 1 \equiv \pm 1 \bmod 8; \\ 2^n + 1 & \text{if } n \equiv \pm 2 \bmod 8; \\ 2^n + 1 - 2^{m+1} & \text{if } n = 2m + 1 \equiv \pm 3 \bmod 8; \\ 2^n + 1 + 2^{m+1} & \text{if } n = 2m \equiv 4 \bmod 8. \end{cases}$$

The case $n = 7$ gives $E(\mathbb{F}_{128}) = 145$, so $E(\mathbb{F}_{128}) \setminus E(\mathbb{F}_2)$ consists of 20 orbits of points under the action of $\mathrm{Gal}(\mathbb{F}_{128}/\mathbb{F}_2)$. One of these orbits is described in the PhD thesis [**1**, § 9.3] of R. Auer, and we will use it in the next section. It is identified as follows.

On E, consider the function $y + x^6 + x^5 + x^2 + x$. Its only pole is $O = (0 : 1 : 0)$, of order 12. The zeroes of the function are the points $(a, a^6 + a^5 + a^2 + a) \in E$ where a satisfies $(a^6 + a^5 + a^2 + a)^2 + (a^6 + a^5 + a^2 + a) + a^3 + a = 0$, i.e.,

$$a^3 \cdot (a + 1)^2 (a^7 + a + 1) = 0.$$

So the zeroes are $(0, 0)$ (with multiplicity 3), and $(1, 0)$ (with multiplicity 2), and one Galois orbit consisting of the 7 points $(a, a^6 + a^5 + a^2 + a) \in E(\mathbb{F}_{128})$ where a runs over the zeroes of $T^7 + T + 1 \in \mathbb{F}_2[T]$.

Write D for the divisor on E consisting of these 7 points:

$$D := \sum_{a \text{ such that } a^7 + a + 1 = 0} (a, a^6 + a^5 + a^2 + a) \in \mathrm{Div}(E).$$

3. Constructing the curve via Artin-Schreier extensions

Given are a prime number p and a curve (absolutely irreducible) $C/\mathbb{F}_p$ with function field $\mathbb{F}_p(C)$. Any $f \in \mathbb{F}_p(C)$ such that the polynomial $T^p - T + f$ is irreducible in $\overline{\mathbb{F}_p}(C)[T]$ yields a cyclic extension $C_f \to C$ of degree p, namely the curve $C_f/\mathbb{F}_p$ with function field the (separable, cyclic) extension of $\mathbb{F}_p(C)$ obtained by adjoining the zeros of $T^p - T + f$. The covering $C_f \to C$ is called an Artin-Schreier covering; its branch points are the points $P \in C$ that are poles of f and that have the additional property that no $g \in \overline{\mathbb{F}_p}(C)$ exists such that $f + g^p - g$ is regular in P. In particular, all poles of f with pole order prime to p will be branch points of $C_f \to C$. If $P \in C(\mathbb{F}_p)$ is one of the branch points, then the unique point in C_f mapping to it is $\mathbb{F}_p$-rational. If $Q \in C_f(\mathbb{F}_p)$ does not map to a branch point, then either it lies over a (rational) zero of f, or it lies over a pole $P \in C(\mathbb{F}_p)$ of f

and g exists such that $f + g^p - g$ has a zero in P. In particular, over each of the $\mathbb{F}_p$-rational zeros of f one finds p points in $C_f(\mathbb{F}_p)$.

The above well known facts will now be used, continuing the discussion in Section 2, to construct suitable $C_f \to E$ over $\mathbb{F}_2$ of degree 2. Since we want two rational points over each of the points in $E(\mathbb{F}_2)$ we look for $f \in \mathbb{F}_2(E)$ with zeros in all points of $E(\mathbb{F}_2)$. To this end, put $R := \sum P \in \mathrm{Div}(E)$, the divisor consisting of the 5 rational points $P \in E(\mathbb{F}_2)$, and consider the Riemann-Roch space $L(D-R) \subset \mathbb{F}_2(E)$ consisting of all functions regular away from the points in D, with at most simple poles in the points of D, and with zeros in the rational points of E. The Riemann-Roch theorem implies that $L(D-R)$ is a vector space over $\mathbb{F}_2$ of dimension $\deg(D-R) = 2$. In the present case, $L(D-R) = \{0, f_1, f_2, f_1 + f_2\}$ with

$$f_1 := \frac{(x^5 + x) \cdot y + (x^2 + x)}{x^7 + x + 1}$$

and

$$f_2 := \frac{(x^5 + x^4 + x^3 + x) \cdot y + (x^6 + x^4)}{x^7 + x + 1}.$$

Since the nonzero functions in $L(D-R)$ have a pole of order 1 in the points of D, they are not of the form $g^2 + g$ for any $g \in \overline{\mathbb{F}_2}(E)$. Hence adjoining a zero w_j of $T^2 + T + f_j$ to $\mathbb{F}_2(E)$, one obtains a quadratic extension $\mathbb{F}_2(E)(w_j)$ which is the function field $\mathbb{F}_2(X_j)$ of a smooth and absolutely irreducible curve X_j over $\mathbb{F}_2$.

It turns out that the genus $g(X_j)$ of X_j equals 8: indeed, consider the differential $dw_j = df_j$ on X_j. Since every pole of f_j in E has order 1, the differential df_j has divisor of poles $2D$. The relation $w_j(w_j + 1) = f_j$ shows that $1/w_j$ is a uniformizer at each of the points of X_j lying over a point of D. The equality $dw_j = w_j^2 d(1/w_j)$ then shows that the divisor of poles of dw_j has degree 14. Since df_j (as differential on E has degree $2g(E) - 2 = 0$, it follows that the divisor of zeros of df_j (on E) has degree 14. As the degree two map $X_j \to E$ is unramified away from the poles of f_j, the divisor of zeros of df_j on X_j has degree 28. As a consequence $2g(X_j) - 2 = \deg(\mathrm{div}(dw_j)) = 28 - 14 = 14$, showing that $g(X_j) = 8$. Alternatively, this genus (as well as the other genera discussed here) can be computed using the '*Führerdiskriminantenproduktformel*' of Artin and Hasse; see [**6**, VI §3 Cor. 2] and [**5**, Prop. 8.2]. In fact this is how Serre in [**8**] finds the relevant genera.

The polynomial $T^2 + T + f_2$ is even irreducible in $\overline{\mathbb{F}_2}(X_1)[T]$: it has no zero in $\overline{\mathbb{F}_2}(E)$, so a zero in $\overline{\mathbb{F}_2}(X_1)$ would be of the form $a + bw_1$ for certain $a, b \in \overline{\mathbb{F}_2}(E)$ with $b \neq 0$. The form of the zero then shows $b = 1$ and $a^2 + a = f_1 + f_2$. Since $f_1 + f_2 \in \mathbb{F}_2(E)$ has a pole of exact order 1 in the points of the divisor D, no such $a \in \overline{\mathbb{F}_2}(E)$ exists. As a consequence the extension $\mathbb{F}_2(E)(w_1, w_2) \supset \mathbb{F}_2(E)$ is a Galois extension of degree 4 (with group isomorphic to $\mathbb{Z}/2\mathbb{Z} \times \mathbb{Z}/2\mathbb{Z}$) and it is the function field of a smooth and absolutely irreducible curve X_{12} over $\mathbb{F}_2$.

Analogous to the argument above one obtains $g(X_{12}) = 22$. Namely, the differential $dw_2 = df_2$ on X_{12} has its only poles (each of order 2) at the points over D, so the divisor of poles of dw_2 has degree 14. The zeros of df_2, seen as differential on E, form an effective divisor on E of degree 14. Since $X_{12} \to E$ is unramified over the points in this divisor, one concludes that the divisor of zeros of df_2 on X_{12} has degree $4 \cdot 14 = 56$. So $2g(X_{12}) - 2 = \deg(\mathrm{div}(dw_j)) = 56 - 14 = 42$, hence $g(X_{12}) = 22$.

The next and last step in explicitly constructing a curve as desired (genus 50, covering of E with group isomorphic to $\mathbb{Z}/2\mathbb{Z} \times \mathbb{Z}/2\mathbb{Z} \times \mathbb{Z}/2\mathbb{Z}$, 40 points over $\mathbb{F}_2$)

is done by finding a suitable $f_3 \in L(2D-R)$. We demand that T^2+T+f_3 is irreducible, not only as polynomial over $\overline{\mathbb{F}_2}(E)$ but even in $\overline{\mathbb{F}_2}(E)(w_1,w_2)[T]$. To show that f_3 exists, substitute $a+bw_1+cw_2+dw_1w_2$ with $a,b,c,d\in\overline{\mathbb{F}_2}(E)$ into T^2+T+f_3. This leads to the condition

$$f_3 \notin \wp(\mathbb{F}_2(E)) + L(D-R)$$

where $\wp\colon \mathbb{F}_2(E)\to\mathbb{F}_2(E)$ denotes the Artin-Schreier map $\xi\mapsto\xi^2+\xi$. Since we demand $f_3\in L(2D-R)$, a function $a\in\mathbb{F}_2(E)$ can only satisfy $\wp(a)\in f_3+L(D-R)\subset L(2D-R)$ if $a\in L(D)$. So for $f_3\in L(2D-R)$ to have T^2+T+f_3 irreducible over $\overline{\mathbb{F}_2}(E)(w_1,w_2)$, a necessary and sufficient condition is that

$$f_3\notin\wp(L(D))+L(D-R).$$

Note that any $a\in L(D)\subset\mathbb{F}_2(E)$ is regular in the points $P\in E(\mathbb{F}_2)$ and $a(P)\in\mathbb{F}_2$. Hence $\wp(L(D))\subset L(2D-R)$. The map $\wp\colon L(D)\to L(2D-R)$ is linear over $\mathbb{F}_2$ with as kernel the constant functions $\mathbb{F}_2\subset L(D)$. Moreover $\wp$ maps non-constant functions in $L(D)$ to functions in $L(2D-R)$ with a pole of order 2 in the points of D. As a consequence $\wp(L(D))\cap L(D-R)=\{0\}$ and

$$\dim_{\mathbb{F}_2}\left(\wp(L(D))+L(D-R)\right)=(7-1)+2=8.$$

As $\dim_{\mathbb{F}_2} L(2D-R)=\deg(2D-R)=9$, an f_3 as requested exists, and in fact all $f_3\in L(2D-R)\setminus(\wp(L(D))+L(D-R))$ yield the same quadratic extension of $\mathbb{F}_2(E)(w_1,w_2)$.

One of the possible choices, which we take from now on, is

$$f_3:=\frac{(x^6+x^5)y+(x^{10}+x^6+x^2+x)}{x^{14}+x^2+1}.$$

Adjoining a zero w_3 of T^2+T+f_3 to $\mathbb{F}_2(E)(w_1,w_2)$, one obtains the extension $\mathbb{F}_2(E)(w_1,w_2,w_3)\supset\mathbb{F}_2(E)$ which by construction is Galois and of degree 8, with Galois group isomorphic to $\mathbb{Z}/2\mathbb{Z}\times\mathbb{Z}/2\mathbb{Z}\times\mathbb{Z}/2\mathbb{Z}$, and which is the function field of a smooth and absolutely irreducible curve X_{123} over $\mathbb{F}_2$.

The genus of X_{123} can be determined in the same way as it was done for the other curves mentioned above: the function $f_3\in\mathbb{F}_2(E)$ turns out to have divisor $8\cdot O+3\cdot(1,1)+(0,0)+(0,1)+(1,0)-2D$, hence the divisor df_3 on E has a zero and therefore also a pole. The only points where it can have a pole is in the support of D, and as f_3 has poles of order 2 here, the pole order of df_3 must necessarily be 2. So the degree of the divisor of poles of df_3 on E equals 14. By construction $X_{123}\to E$ is totally ramified at the points of D, so the degree divisor of poles of df_3 on X_{123} is also 14. The divisor of zeros of df_3 on X_{123} has degree $14\cdot 8=112$, hence on X_{123} one concludes $\deg(\mathrm{div}(df_3))=112-14=98$. This implies $g(X_{123})=50$.

In fact the differential df_3 on E has divisor $12\cdot O+2\cdot(1,1)-2D$. Since there are 8 $\mathbb{F}_2$-rational points on X_{123} over each of the points in $E(\mathbb{F}_2)$, one obtains $\#X_{123}(\mathbb{F}_2)=5\cdot 8=40$. This completes the explicit construction of a genus 50 curve X over $\mathbb{F}_2$ with $\#X(\mathbb{F}_2)=40$.

4. More equations and intermediate curves

For explicit calculations, it is convenient to present the curve X as a plane curve. This is done below, starting from the tower of fields

$$\mathbb{F}_2(x)\subset\mathbb{F}_2(x,y)\subset\mathbb{F}_2(x,y,w_1)\subset\mathbb{F}_2(x,y,w_1,w_2)\subset\mathbb{F}_2(x,y,w_1,w_2,w_3)$$

in which all successive extensions have degree 2. The equations are

$$\begin{cases} y^2 + y = x^3 + x, \\ (x^7 + x + 1)(w_1^2 + w_1) = (x^5 + x) \cdot y + x^2 + x, \\ (x^7 + x + 1)(w_2^2 + w_2) = (x^5 + x^4 + x^3 + x) \cdot y + x^6 + x^4, \\ (x^{14} + x^2 + 1)(w_3^2 + w_3) = (x^6 + x^5) \cdot y + x^{10} + x^6 + x^2 + x. \end{cases}$$

To obtain a plain curve equation, one may start by observing that the conjugates $(w_1 + a_1)(w_2 + a_2)(w_3 + a_3)$ (for all $a_1, a_2, a_3 \in \mathbb{F}_2$) of $w_1w_2w_3$ over $\mathbb{F}_2(x, y)$ are pairwise distinct, hence $\mathbb{F}_2(x, y, w_1, w_2, w_3) = \mathbb{F}_2(x, y, w_1w_2w_3)$. A straightforward calculation shows that the minimal polynomial of $w_1w_2w_3$ over $\mathbb{F}_2(x, y)$ equals

$$T^8 + T^7 + a_6T^6 + a_5T^5 + a_4T^4 + a_5f_1f_2f_3T^3 + a_6f_1^2f_2^2f_3^2T^2 + f_1^3f_2^3f_3^3T + f_1^4f_2^4f_3^4$$

where the f_j are as in Section 3 and

$$\begin{aligned} a_6 &:= f_1 + f_2 + f_3, \\ a_5 &:= f_1f_2f_3 + f_1f_2 + f_1f_3 + f_2f_3, \\ a_4 &:= f_1^2f_2^2 + f_1^2f_3^2 + f_2^2f_3^2 + f_1f_2f_3. \end{aligned}$$

This polynomial explicitly defines an extension of $\mathbb{F}_2(E) = \mathbb{F}_2(x, y)$ as originally discussed by Serre in [**8**].

The polynomial obtained here is not fixed under the action of $\mathrm{Gal}(\mathbb{F}_2(x, y)/\mathbb{F}_2(x))$ on its coefficients. This is because several of those coefficients have poles at the points of the divisor D and are regular in all other points of E; the generator γ of $\mathrm{Gal}(\mathbb{F}_2(x, y)/\mathbb{F}_2(x))$ corresponds to $[-1] \in \mathrm{End}(E)$ and the support of $[-1]^*D$ is disjoint from that of D. Multiplying the above polynomial by its conjugate under γ results in an irreducible degree 16 element of $\mathbb{F}_2(x)[T]$, defining $\mathbb{F}_2(x, y, w_1, w_2, w_3)$ as an extension of $\mathbb{F}_2(x)$ by a single equation. We do not present it here; it is a rather long expression in $\mathbb{F}_2[x, \frac{1}{x^7+x+1}, T]$, of the form

$$T^{16} + \frac{x^{14} + x^{11} + x^{10} + x^6 + x^3 + x^2 + 1}{(x^7 + x + 1)^2}T^{14} + \frac{x^{22} + x^{20} + x^{14} + x^{13} + x^{11} + x^6}{(x^7 + x + 1)^4}T^{13} + \dots$$

$$\dots + \frac{(x^2 + x)^{31}(x^{13} + x^{12} + x^{11} + x^{10} + x^8 + x^5 + x^4 + x + 1)}{(x^7 + x + 1)^{16}}T + \frac{(x^2 + x)^{36}}{(x^7 + x + 1)^{16}}.$$

The MAGMA code given below calculates, among other things, this polynomial F.

```
K0<x>:=FunctionField(GF(2)); P0<Y>:=PolynomialRing(K0);
KE<y>:=ext<K0 | Y^2+Y+x^3+x>; PE<W>:=PolynomialRing(KE);
D:=Divisor(Zeros(KE!x^7+x+1)[1]); R:=&+Places(KE,1);
V,a:=RiemannRochSpace(D-R); RR1:={a(v) : v in V};
f1:=((x^5+x)*y+x^2+x)/(x^7+x+1); f1 in RR1;
f2:=((x^5+x^4+x^3+x)*y+x^6+x^4)/(x^7+x+1); f2 in RR1;
K1<w1>:=ext<KE | W^2+W+f1>; PK1<Z>:=PolynomialRing(K1);
K12<w2>:=ext<K1 | Z^2+Z+K1!f2>; PK12<T>:=PolynomialRing(K12);
V2,b:=RiemannRochSpace(2*D-R); RR2:={b(v) : v in V2};
V3,c:=RiemannRochSpace(D);
RRred:={a(g)+c(h)+c(h)^2 : g in V , h in V3};
f3:=((x^6+x^5)*y+x^10+x^6+x^2+x)/(x^14+x^2+1); f3 in RR2;
f3 in RRred; K123<w3>:=ext<K12 | T^2+T+K12!f3>;
MinimalPolynomial(w1*w2*w3, KE);
F:=MinimalPolynomial(w1*w2*w3, K0);
L<z>:=ext<K0 | F>; Genus(L); #Places(L,1);
```

Put $S := \mathbb{F}_2 f_1 + \mathbb{F}_2 f_2 + \mathbb{F}_2 f_3$. The linear (over $\mathbb{F}_2$) subspaces $R \subset S$ correspond to the fields F with $\mathbb{F}_2(E) \subset F \subset \mathbb{F}_2(X_{123})$: given R, adjoining the zeros in $\mathbb{F}_2(X_{123})$ of T^2+T+r (for all $r \in R$) to $\mathbb{F}_2(E)$ yields the intermediate field $\mathbb{F}_2(X_R)$, and all intermediate fields are obtained in this way. The smooth and absolutely irreducible, complete curves X_R over $\mathbb{F}_2$ satisfy $g(X_R) = 1 + 7 \cdot (-1 + \#R)$ and $\#(X_R) = 5 \cdot \#R$. Clearly $X_{\{0\}} = E$ and $X_S = X_{123}$. Moreover inclusion of subspaces $R \subset R'$ translates into inclusion $\mathbb{F}_2(X_R) \subset \mathbb{F}_2(X_{R'})$ of intermediate fields. Is $R \subset R'$, then $[\mathbb{F}_2(X_{R'}) : \mathbb{F}_2(X_R)] = \#(R'/R)$.

Although both E and X_{123} reach the upper bound $N_2(g)$ for the number of $\mathbb{F}_2$-rational points of curves of genus g over $\mathbb{F}_2$, the 'intermediate' curves do not have this property. If $\dim_{\mathbb{F}_2} R = 1$ then $g(X_R) = 8$ and $\#X_R(\mathbb{F}_2) = 8 < 11 = N_2(8)$ (for the latter equality, see [**7**] as well as [**9**, §2.2]). And in case $\dim_{\mathbb{F}_2} R = 2$ one has $g(X_R) = 22$ and by [**5**, Exc. 8.D] as well as [**9**, §2.2] $N_2(22) \in \{21, 22\}$, whereas $\#X_R(\mathbb{F}_2) = 20$.

Quite similar to the construction of the functions $f_1, f_2, f_3 \in \mathbb{F}_2(E)$ and the resulting explicit curves X_R, one finds equations for curves over $\mathbb{F}_2$ having $N_2(8) = 11$ and $21 \le N_2(22)$ rational points. We briefly sketch how this is done, and we present the resulting curves. The function $\frac{x^2+x}{x^3+x+1}$ on $\mathbb{P}_1$ over $\mathbb{F}_2$ has zeros in the three points of $\mathbb{P}^1(\mathbb{F}_2)$. Hence the (hyperelliptic, genus 2) curve H corresponding to

$$H\colon y^2 + y = \frac{x^2 + x}{x^3 + x + 1}$$

satisfies $\#H(\mathbb{F}_2) = 6$. We will construct abelian coverings of H with groups $\mathbb{Z}/2\mathbb{Z}$ and $\mathbb{Z}/2\mathbb{Z} \times \mathbb{Z}/2\mathbb{Z}$, respectively, in which 5 points of $H(\mathbb{F}_2)$ split completely, whereas the covering is totally ramified over the sixth rational point and has no other branch points. As a consequence, the covering has $1 + 5 \cdot d$ rational points, where d is the degree of the covering map.

As the branch point we take the common zero P of the two functions $1/x$ and $y(x^3 + x + 1)/x^3$. Let $R \in \mathrm{Div}(H)$ be the divisor (of degree 5) consisting of the remaining five points in $H(\mathbb{F}_2)$. The space $L(9P - R) \subset \mathbb{F}_2(H)$ contains the function

$$a_1 := (x^9 + x^7 + x^4 + x^3)y + x^9 + x^8 + x^7 + x^2$$

satisfying $\mathrm{ord}_P(a_1) = 9$, which implies that $T^2 + T + a_1 \in \overline{\mathbb{F}_2}(H)[T]$ is irreducible. Adjoining a zero of this polynomial to $\mathbb{F}_2(H)$ therefore yields the function field of a smooth, complete and absolutely irreducible curve C_1 over $\mathbb{F}_2$. It has $g(C_1) = 8$ and $\#C_1(\mathbb{F}_2) = 11$. Note that an explicit curve over $\mathbb{F}_2$ reaching $N_2(8) = 11$ was also found by Ísabel Piršić [**2**].

Next, $L(11P - R) \subset \mathbb{F}_2(H)$ contains

$$a_2 := (x^{11} + x^9 + x^7 + x^6 + x^5 + x^4 + x^3 + x^2)y + x^{11} + x^{10} + x^9 + x^7 + x^6 + x^4$$

and $\mathrm{ord}_P(a_2) = 11$. The polynomial $T^2 + T + a_2$ is not only irreducible over $\overline{\mathbb{F}_2}(H)$ but even over the quadratic extension $\overline{\mathbb{F}_2}(C_1) \supset \overline{\mathbb{F}_2}(H)$. Hence adjoining a zero of $T^2 + T + a_2$ to $\mathbb{F}_2(C_1)$, an extension is obtained which is Galois over $\mathbb{F}_2(H)$ with group $\mathbb{Z}/2\mathbb{Z} \times \mathbb{Z}/2\mathbb{Z}$, and which is the function field of a smooth, complete and absolutely irreducible curve C_2 over $\mathbb{F}_2$. It turns out that $g(C_2) = 22$ and by construction $\#C_2(\mathbb{F}_2) = 21$. So we obtain the following models, up to birational equivalence over $\mathbb{F}_2$.

$$C_1\colon \quad \begin{cases} (x^3+x+1)(y^2+y) = x^2+x, \\ z^2+z = a_1 \end{cases}$$

and

$$C_2\colon \quad \begin{cases} (x^3+x+1)(y^2+y) = x^2+x, \\ z^2+z = a_1, \\ w^2+w = a_2. \end{cases}$$

Acknowledgment

It is a pleasure to thank Gerard van der Geer, Enis Kaya, Jean-Pierre Serre, and the referee of an earlier version of this text for various useful remarks. I thank Alp Bassa and René Schoof: a conversation with them during AGC^2T-17 in Luminy encouraged me to write the present note.

References

[1] Roland Auer, *Ray class fields of global function fields with many rational places*, Acta Arith. **95** (2000), no. 2, 97–122, DOI 10.4064/aa-95-2-97-122. MR1785410

[2] Ísabel Piršić, *Entry on the website* manypoints.org, 2012.

[3] Max Kronberg, Muhammad Afzal Soomro, and Jaap Top, *Twists of elliptic curves*, SIGMA Symmetry Integrability Geom. Methods Appl. **13** (2017), Paper No. 083, 13, DOI 10.3842/SIGMA.2017.083. MR3715838

[4] René Schoof, *Nonsingular plane cubic curves over finite fields*, J. Combin. Theory Ser. A **46** (1987), no. 2, 183–211, DOI 10.1016/0097-3165(87)90003-3. MR914657

[5] René Schoof, *Algebraic curves and coding theory*, 2000. (Lectures given at the National Mathematical Centre in Abuja, Nigeria during the summer of 1990).

[6] Jean-Pierre Serre, *Local fields*, Graduate Texts in Mathematics, vol. 67, Springer-Verlag, New York-Berlin, 1979. Translated from the French by Marvin Jay Greenberg. MR554237

[7] Jean-Pierre Serre, *Sur le nombre des points rationnels d'une courbe algébrique sur un corps fini* (French, with English summary), C. R. Acad. Sci. Paris Sér. I Math. **296** (1983), no. 9, 397–402. MR703906

[8] J-P. Serre, *Rational points on curves over finite fields*, 1985. (Lectures given at Harvard University, September–December 1985; Notes by Fernando Q. Gouvêa). Revised and completed as J-P. Serre, *Rational Points on Curves over Finite Fields* (with contributions by Everett Howe, Joseph Oesterlé and Christophe Ritzenthaler). Documents Math., Vol. 18. Paris: Soc. Math. de France, 2020.

[9] Gerard van der Geer, *Hunting for curves with many points*, Coding and cryptology, Lecture Notes in Comput. Sci., vol. 5557, Springer, Berlin, 2009, pp. 82–96, DOI 10.1007/978-3-642-01877-0_9. MR2836234

Jaap Top, Bernoulli Institute, University of Groningen, Nijenborgh 9, 9747 AG Groningen, The Netherlands

Email address: j.top@rug.nl

Selected Published Titles in This Series

770 **Stéphane Ballet, Gaetan Bisson, and Irene Bouw, Editors,** Arithmetic, Geometry, Cryptography and Coding Theory, 2021

769 **Kiyoshi Igusa, Alex Martsinkovsky, and Gordana Todorov, Editors,** Representations of Algebras, Geometry and Physics, 2021

768 **Dražen Adamović, Andrej Dujella, Antun Milas, and Pavle Pandžić, Editors,** Lie Groups, Number Theory, and Vertex Algebras, 2021

767 **Moshe Jarden and Tony Shaska, Editors,** Abelian Varieties and Number Theory, 2021

766 **Paola Comparin, Eduardo Esteves, Herbert Lange, Sebastián Reyes-Carocca, and Rubí E. Rodríguez, Editors,** Geometry at the Frontier, 2021

765 **Michael Aschbacher,** Quaternion Fusion Packets, 2021

764 **Gabriel Cunningham, Mark Mixer, and Egon Schulte, Editors,** Polytopes and Discrete Geometry, 2021

763 **Tyler J. Jarvis and Nathan Priddis, Editors,** Singularities, Mirror Symmetry, and the Gauged Linear Sigma Model, 2021

762 **Atsushi Ichino and Kartik Prasanna,** Periods of Quaternionic Shimura Varieties. I., 2021

761 **Ibrahim Assem, Christof Geiß, and Sonia Trepode, Editors,** Advances in Representation Theory of Algebras, 2021

760 **Olivier Collin, Stefan Friedl, Cameron Gordon, Stephan Tillmann, and Liam Watson, Editors,** Characters in Low-Dimensional Topology, 2020

759 **Omayra Ortega, Emille Davie Lawrence, and Edray Herber Goins, Editors,** The Golden Anniversary Celebration of the National Association of Mathematicians, 2020

758 **Jan Šťovíček and Jan Trlifaj, Editors,** Representation Theory and Beyond, 2020

757 **Kaïs Ammari and Stéphane Gerbi, Editors,** Identification and Control: Some New Challenges, 2020

756 **Joeri Van der Veken, Alfonso Carriazo, Ivko Dimitrić, Yun Myung Oh, Bogdan D. Suceavă, and Luc Vrancken, Editors,** Geometry of Submanifolds, 2020

755 **Marion Scheepers and Ondřej Zindulka, Editors,** Centenary of the Borel Conjecture, 2020

754 **Susanne C. Brenner, Igor Shparlinski, Chi-Wang Shu, and Daniel B. Szyld, Editors,** 75 Years of Mathematics of Computation, 2020

753 **Matthew Krauel, Michael Tuite, and Gaywalee Yamskulna, Editors,** Vertex Operator Algebras, Number Theory and Related Topics, 2020

752 **Samuel Coskey and Grigor Sargsyan, Editors,** Trends in Set Theory, 2020

751 **Ashish K. Srivastava, André Leroy, Ivo Herzog, and Pedro A. Guil Asensio, Editors,** Categorical, Homological and Combinatorial Methods in Algebra, 2020

750 **A. Bourhim, J. Mashreghi, L. Oubbi, and Z. Abdelali, Editors,** Linear and Multilinear Algebra and Function Spaces, 2020

749 **Guillermo Cortiñas and Charles A. Weibel, Editors,** K-theory in Algebra, Analysis and Topology, 2020

748 **Donatella Danielli and Irina Mitrea, Editors,** Advances in Harmonic Analysis and Partial Differential Equations, 2020

747 **Paul Bruillard, Carlos Ortiz Marrero, and Julia Plavnik, Editors,** Topological Phases of Matter and Quantum Computation, 2020

746 **Erica Flapan and Helen Wong, Editors,** Topology and Geometry of Biopolymers, 2020